SpringerWienNewYork

Andrey L. Rogach (Ed.)

Semiconductor Nanocrystal Quantum Dots
Synthesis, Assembly, Spectroscopy and Applications

SpringerWienNewYork

Dr. Andrey Rogach
Department of Physics and CeNS
University of Munich
Germany

This work is subject to copyright.
All rights are reserved, whether the whole or part of the material is concerned, specifically those of translation, reprinting, re-use of illustrations, broadcasting, reproduction by photocopying machines or similar means, and storage in data banks.

Product Liability: The publisher can give no guarantee for all the information contained in this book. This does also refer to information about drug dosage and application thereof. In every individual case the respective user must check its accuracy by consulting other pharmaceutical literature. The use of registered names, trademarks, etc. in this publication does not imply, even in the absence of a specific statement, that such names are exempt from the relevant protective laws and regulations and therefore free for general use.

© 2008 Springer-Verlag/Wien
Printed in Austria

SpringerWienNewYork is part of
Springer Science + Business Media
springer.at

Typesetting: Thomson Press (India) Ltd., Chennai
Printing: Holzhausen Druck & Medien, 1140 Wien

Printed on acid-free and chlorine-free bleached paper

SPIN: 12100002

With 194 partly coloured Figures

Library of Congress Control Number: 2008925753

ISBN 978-3-211-75235-7 SpringerWienNewYork

To my parents and grandparents

Foreword

When investigations on semiconductor nanocrystal quantum dots started more than a quarter of a century ago, no one ever believed that nanoparticle research would develop into one of the major fields in modern science. The basis was laid by studies of photocatalysis and artificial water splitting driven by the former oil crisis. These euphorically started activities ebbed away more and more when on one side oil brimmed over again and the scientists on the other did not succeed in the concomitant formation of hydrogen and oxygen.

At the same time size quantisation was discovered in nanocrystals initiating a fruitful research field on scaling laws of physical and chemical properties of quantum dots. Especially optical investigations of semiconductor nanocrystals led to fascinating scientific results and to applications in optoelectronics and biolabeling. Advances in spectroscopic measurements were always correlated with advances in synthesis. The better the size, shape and surface control of the particles was developed, the more detailed and precise was the spectroscopic information obtained. Applications of nanocrystal quantum dots often require asssembly processes for the formation of polymer hybrids or thin films. For this as well as for the use in biomedical applications new ligand chemistry needed to be developed during the recent past.

This book gives a very competent view on all these facets of nanocrystal quantum dot research. Twelve chapters are written by experts in the fields in a way introducing the respective concepts and providing comprehensive overview on the current state of the art.

I hope the reader will appreciate this book as much as I do when I read the chapters.

Hamburg, May 7, 2008 *Horst Weller*

Preface

The enormous potential of science and technology on nanoscale to impact on industrial output has been recognised all over the world. One emerging area of nanoscience being at the interface of chemistry, physics, biology and materials science is the field of semiconductor nanocrystals, also known as colloidal quantum dots, whose unique properties have attracted great attention by researchers during the last two decades. Different strategies for the synthesis of nanocrystals have been developed, so that their composition, size, shape, and surface protection can be controlled nowadays with an exceptionally high degree. Control over nanocrystal's size at the synthetic stage is a straightforward way to obtain semiconductor materials with specifically designed optical properties. The surface chemistry of nanocrystals is another key parameter, in many respects determining their properties related to their assembly. A surface layer of organic capping molecules (typically thiols, amines, phosphines, or encapsulating polymer shells) provides the fluorescent core with solubility, stability, and processability from solution. Advanced spectroscopy techniques applied to colloidal quantum dots address exciting physical phenomena and provide deep insights into electronic structure and recombination dynamics of these low-dimensional objects. Semiconductor nanocrystals, being highly efficient fluorophores with a strong band-gap luminescence are particularly interesting for the applications in biology as luminescent labels, and this is in fact the area of research where quantum dots products are now offered by several companies.

This book specifically focuses on semiconductor nanocrystal quantum dots, and addresses their synthesis, assembly, optical properties and spectroscopy, and the applications in biology and medicine. It starts with four chapters on the synthesis and morphology of quantum dots. The first chapter by Kudera, Carbone, Manna and Parak covers general issues of growth mechanism, shape and composition control of semiconductor nanocrystals. The next two chapters consider in details the synthesis of nanocrystals in organic solvents (Reiss) and in water (Gaponik and Rogach). These three chapters also provide a review on the basic optical properties of colloidal quantum dots tuneable by size as a result of the quantum confinement effect. The fourth chapter (by Dorfs and Eychmüller) introduces in details a particular case of multishell semiconductor nanocrystals. The following three chapters are devoted to self-assembly of colloidal quantum dots (Shevchenko and Talapin), to their hybrid structures with polymers (Wang), and to the layer-by-layer assembly technique in application to nanocrystals and nanowires (Srivastava and Kotov). The chapter by Vasilevskiy provides the theoretical description of electronic structure and exciton-phonon interaction in semiconductor nanocrystals. The next three chapters deal with the optical spectroscopy of colloidal quantum dots, and address Anti-Stokes photoluminescence (Rakovich and Donegan), exciton dynamics and energy transfer processes in semiconductor nanocrystals (Meijerink), and single particle fluorescence spectroscopy on an example of CdSe-based nanocrystals (Lupton and Müller). The last chapter by Choi and Maysinger critically reviews applications of quantum dots in biomedicine.

The collection of chapters in this book will be of interest to a multidisciplinary audience of chemists, physicists, engineers, biologists and material scientists engaged in rapidly expanding research on semiconductor nanocrystal quantum dots, and those scientists from neighbouring disciplines and graduate students who look for a comprehensive account of the current state of quantum dots related research. I would like to thank all contributors to this book for taking valuable time from their busy schedules to put together stimulating and informative chapters, and Horst Weller, one of the distinguished pioneers of the semiconductor nanocrystals research, for writing the foreword. I would like to express my gratitude to Stephen Soehnlen at Springer for inviting me to bring this exciting field to a wider audience.

Munich, May 2008   *Andrey Rogach*

Contents

Stefan Kudera, Luigi Carbone, Liberato Manna, Wolfgang J. Parak
Growth mechanism, shape and composition control of semiconductor
nanocrystals 1

Peter Reiss
Synthesis of semiconductor nanocrystals in organic solvents 35

Nikolai Gaponik, Andrey L. Rogach
Aqueous synthesis of semiconductor nanocrystals 73

Dirk Dorfs, Alexander Eychmüller
Multishell semiconductor nanocrystals 101

Elena V. Shevchenko, Dmitri V. Talapin
Self-assembly of semiconductor nanocrystals into ordered superstructures .. 119

Dayang Wang
Semiconductor nanocrystal–polymer composites: using polymers
for nanocrystal processing 171

Sudhanshu Srivastava, Nicholas A. Kotov
Layer-by-layer (LBL) assembly with semiconductor nanoparticles and
nanowires 197

M. I. Vasilevskiy
Exciton–phonon interaction in semiconductor nanocrystals 217

Yury P. Rakovich, John F. Donegan
Anti-Stokes photoluminescence in semiconductor nanocrystal
quantum dots 257

Andries Meijerink
Exciton dynamics and energy transfer processes in semiconductor
nanocrystals 277

John M. Lupton, Josef Müller
Fluorescence spectroscopy of single CdSe nanocrystals 311

Angela O. Choi, Dusica Maysinger
Applications of quantum dots in biomedicine 349

Subject index 367

Growth mechanism, shape and composition control of semiconductor nanocrystals

By

Stefan Kudera[1], Luigi Carbone[2], Liberato Manna[3], Wolfgang J. Parak[4]

[1] Max Planck Institute for Metals Research, Department of New Materials and Biosystems & University of Heidelberg, Department of Biophysical Chemistry, Stuttgart, Germany
[2] Nano-Bio-Technology Group, Institut für Physikalische Chemie, Mainz, Germany
[3] National Nanotechnology Laboratories of CNR-INFM, Lecce, Italy
[4] Fachbereich Physik, Philipps Universität Marburg, Marburg, Germany

1. Introduction

In the last few years the development of colloidal nanocrystals has been extended from the pure adjustment of the size of the particles to the control of more sophisticated properties as their shape and composition. The hope is that through these features different fields of applications might be opened for the use of nanocrystals. In this chapter we will introduce a concept for the preparation of colloidal nanocrystals with a narrow size distribution and then discuss the shape control of colloidal nanocrystals as well as a technique for composition control, i.e. the formation of hybrid materials.

In Sect. 1, some very general issues related to semiconductor nanocrystals are shortly discussed, followed by Sect. 2 where the general growth model of nanocrystals is introduced. It involves two steps, which ideally should occur separately: the nucleation of the nanocrystals and the actual process of growth. Sections 3–5 describe deviations from these two standard processes. Under certain conditions the nucleation event can be slowed down as to occur only stepwise (Sect. 3). This process is an intermediate between the growth and the nucleation event: single monolayers are nucleated in a sequential manner, leading to the formation of magic size clusters. In Sect. 4 shape control of nanocrystals is introduced. The possibility to produce anisotropic nanocrystals at first presents a deviation from the general growth model which predicts almost spherical particles. Additionally, the control of the shape can be achieved by an alternation of the nucleation event. This enables the growth of branched structures such as tetrapods. In Sect. 5 we will describe the composition control as an example for a modified nucleation event.

1.1 Size dependence. Especially in the case of semiconductor nanocrystals the effect of the size on the optical properties of the particles is very striking. The smallest CdSe nanocrystals with a diameter of less than 2 nm show for example a blue fluorescence. Larger nanocrystals of the same material (diameter ca. 6 nm) emit red light. In a simple model this can be understood through the confinement of the exciton, i.e. the light generated electron-hole pair whose recombination is responsible for the fluorescence emission, into the volume of the nanocrystal. Like in the classical textbook example of a particle in a box, a stronger confinement leads to a larger separation of the energetic levels [1].

In a more detailed analysis, colloidal quantum dots can be considered as an intermediate species between atoms or molecules on the one hand and bulk material on the other hand. The energetic levels of a simple molecule can be calculated by the tight binding or LCAO (linear combination of atomic orbitals) approach [1, 2]. In this model the outer orbitals of the participating atoms are combined to new molecular orbitals. Generally one observes orbitals with energy lower than that of the atomic orbitals, the so-called binding orbitals, and orbitals of higher energy, termed antibinding. Extension of this model to a larger number of atoms leads to the formation of bands. The binding orbitals form the valence band, whereas the antibinding orbitals are combined to the conduction band. The spacing between the bands, i.e. the bandgap, decreases with the number of atoms added to the molecule. The lower limit for the bandgap is the value of bulk material. This value is reached when the radius of the crystals is of the order of the Bohr radius of the exciton.

This trend can be observed when comparing semiconductor nanocrystals of the same material, but of different sizes. The larger the nanocrystals are, the more the fluorescence colour is shifted towards the red, i.e. towards lower energies (Fig. 1).

1.2 Shape dependence. Spherical nanocrystals can be considered as zero-dimensional objects. Confinement is exerted in all three dimensions. For this the density of states is discontinuous. If one axis of the nanocrystal is extended, the density of states changes slightly, as confinement is now exerted only in two dimensions [1, 4]. In these dimensions the levels are still quantised. Levels attributed

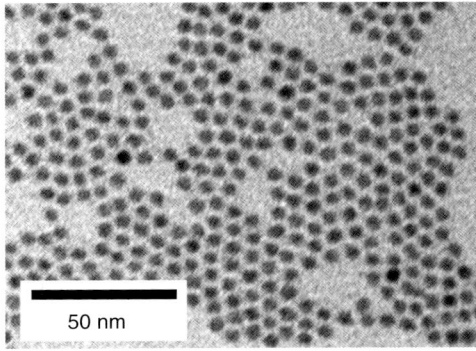
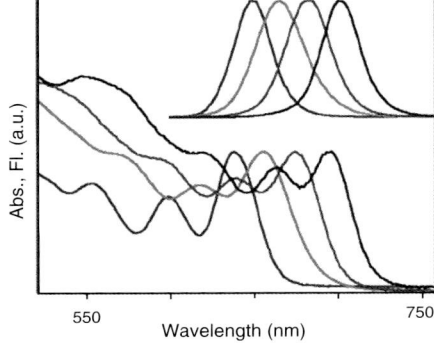

Fig. 1. Spherical semiconductor nanocrystals. The transmission electron micrograph (TEM) shows nanocrystals of CdTe. In the right panel absorption ("Abs") and fluorescence ("Fl", shifted upwards for clarity) spectra of samples with different particle sizes are shown. Reprinted from [3], with permission from Elsevier

to wave vectors parallel to the long axis of the structure are quasi continuous. Therefore each of the quantised levels is effectively broadened by the absence of confinement in this long axis of the structure. The positions of the electronic levels are still dominated by the size of the smallest axes.

It can be shown that rod-shaped semiconductor nanocrystals, so-called nanorods, emit polarized light and that the degree of polarization scales with their aspect ratio [5]. With increasing aspect ratio of the nanorods, only a slight shift of the bandgap is observed, which saturates at aspect ratios of the order of ten [6, 7]. This makes nanorods an appealing system for electronic devices. The long axis facilitates to contact the structure, while the short axes preserve the quantised nature of the electronic levels. For instance nanowires, i.e. nanorods with very high aspect ratios, of different materials can be embedded into electric circuits to act as transistors or other active elements [8, 9]. Electroluminescence is observed from CdSe nanorods [10]. Their symmetry facilitates the alignment of nanorods. In some cases it is sufficient to slowly evaporate the solvent to obtain large areas of aligned nanorods [11–14]. Better results are achieved when the nanorods are oriented by an electric field [15–20]. Furthermore nanorods were proven to be of advantage in solar cells [21, 22].

Tetrapods, i.e. nanoparticles with four rod-shaped arms that are combined to the shape of a tetraeder, offer the possibility to expand this spectrum of applications. When three of the arms are contacted electrically, one of the arms can be used as a gate to control the current through the entire structure [23]. Also, tetrapod-shaped semiconductor nanocrystals can be of advantage in solar cells. They exhibit a large surface on which charges can be separated, and still provide a pathway to transport charges to an electrode due to their complex and extended geometry [24]. Furthermore, tetrapods exhibit an interesting structure of their excited states. Electrons and holes can be localised in the core region where the arms are linked and the overlap between the two wave functions differs considerably from one excited state to another [25].

1.3 Synthesis techniques. Nanocrystals can be synthesised from a variety of different materials and in several different surroundings. For instance the synthesis of gold colloids and CdTe nanocrystals is frequently carried out in aqueous solutions [26–28]. However, in this chapter we will focus on the synthesis of semiconductor nanocrystals in organic solutions. The principle for this technique was demonstrated ca. 15 years ago by Murray et al. [29]. It involves the decomposition of molecular precursors, i.e. molecules that deliver the monomers of the nanoparticles, at relatively high temperatures. Precursors are injected swiftly into a hot solvent. By this, the monomers, i.e. the atomic species that constitute the nanocrystal, are freed rapidly leading to a high oversaturation of monomers. In this surrounding the growth of nanocrystals is highly favoured as will be discussed in Sect. 2. The use of organic solvents has the advantage that one can tune the reaction temperature over a wide range, and also in this environment the specific reactants are exhaustively explored. As we will explain later, temperature and composition of the solvent exert strong influence on the growth kinetics and on the shape of the nanocrystals [30, 31]. Also, through the reaction environment the crystalline phase of the material can be influenced. For instance, CdTe nanocrystals being synthesised at high temperature in organic solvents generally grow in the hexagonal wurtzite phase [29]. However, by a careful control of the reaction conditions they can also be grown in the cubic zinc blende phase [32].

In general the solvent serves two purposes. Evidently its main purpose is to solubilise and disperse the nanocrystals and the reactants involved in the growth. The second task is to control the speed of the reaction. To do so, the solvent molecules need to bind and unbind dynamically on the surface of the growing crystals. Once a molecule detaches from the surface of the nanocrystal, new atomic species (monomers) can be incorporated into the nanocrystal, and thus it can grow. When referring to these characteristics of the solvent molecules, they are termed "surface ligands", or "surfactants". In a synthesis the solvent can be a mixture of different species, including pure solvent and pure surfactants.

Especially the role as surfactant is of great interest. In general, the surfactant molecules exhibit two domains, one non-polar, generally a long alkyl chain, and a polar head group. As it will be described in Sect. 4 of this chapter their functionality depends on both domains. The shape of the non-polar group as well as the binding strength of the polar group influences the growth dynamics. Briefly, the non-polar tail biases the diffusion properties, whereas the polar head group mainly affects the binding efficiency.

In their seminal work Murray et al. [29] initially proposed tri-n-octylphosphine oxide (TOPO), and tri-n-octylphosphine (TOP) as surfactants, which are still frequently used. Other types of ligands are different amines and carboxylic acids [33–35]. Most of these compounds are solid at room temperature and melt only at ca. 50°C.

The reaction is typically performed in a three-necked flask connected to a Schlenk line with one of the necks. The remaining two necks are sealed with rubber septa and serve for the measurement of the temperature inside the flask and for the injection of reactants. The reaction is usually carried out under an inert atmosphere, as some of the reactants may be pyrophoric and also some types of nanocrystals may be sensitive to air. After loading the flask with the organic solvent and surfactant molecules, these organic molecules are molten from powder state to a liquid and the flask is evacuated and kept under vacuum at ca. 130–180°C for 10–20 min to remove volatile impurities. This step is crucial for most syntheses. Before starting the actual reaction the flask is flushed with an inert gas (generally nitrogen or argon).

So far we have been describing the reaction surrounding, but not the actual constituents of the nanocrystals, the monomers. They can be introduced in many different ways. At least one species should be in liquid form. Through this one can start the formation of nanocrystals by the quick injection of this compound and thus determine a sharp nucleation event. For the growth of II/VI semiconductor nanocrystals the elemental chalcogens are introduced into the reaction as a complex with either TOP or tri-n-butylphosphine (TBP). The complex is formed by dissolving the chalcogen in liquid TOP or TBP. Generally, for the formation of the complex it is sufficient to simply vortex this solution for some minutes. Only for the preparation of a Te-solution heat has to be applied for at least 1 h.

In the early syntheses of CdE (E = S, Se or Te) nanocrystals dimethyl cadmium is used as precursor [29]. This liquid compound along with Se:TOP is injected into the reaction solution to initiate the growth of CdSe nanocrystals, but unfortunately it is unstable, highly toxic and pyrophoric. In the modern syntheses this compound is usually replaced by CdO [36]. At a temperature of ca. 300°C the Cd ions bind to the surfactants, which are either phosphonic acids like dodecyl-, tetradecyl- or octadecyl-phosphonic acids [33, 36, 37] or oleic acid [33, 38]. This reaction is accompanied by formation of steam and a colour change of the solution, which turns from dark red to translucent. In this reaction scheme only the chalcogen complex is injected and the nucleation sets in shortly after the injection.

2. General growth mechanism of nanocrystals

In this section we will present a simple description of the growth process of colloidal nanocrystals. The argumentation is divided into two parts: the nucleation event in which nanocrystals spontaneously form through an assembly of freely dispersed atoms and the actual growth process. We will regard spherical nanocrystals as model system.

2.1 Nucleation. The first step in the growth of any sort of (nano)crystal is evidently the nucleation. Through a density fluctuation of the medium several atoms assemble to a small crystal that is thermodynamically stable, and thus does not decay to free atoms or ions. In that sense the nucleation can be understood as the overcoming of a barrier. This section will explain briefly the origin of this barrier. In the following we will distinguish between the crystalline phase, in which the atoms are bound to a crystal, and the solution phase, in which the atoms are dispersed freely in the solution. The nucleation in a solution at constant temperature and constant pressure is driven by the difference in the free energy between the two phases. At the simplest the driving forces in the nucleation event can be reduced to two, the gain in the chemical potential and the increase of the total surface energy. The gain in chemical potential can be understood as the energy freed by the formation of the bonds in the growing crystal. The surface term takes into account the correction for the incomplete saturation of the surface bonds. Upon formation of a spherical nucleus consisting of n atoms the total free energy of the system changes by the value

$$\Delta G = n(\mu_c - \mu_s) + 4\pi r^2 \sigma \tag{1}$$

(μ_c and μ_s are the chemical potentials of the crystalline phase and the solution phase, respectively, r is the radius of the nucleus and σ the surface tension[1].) In this equation the surface term constitutes the main difference between nanomaterials and bulk crystals. Bulk material is dominated by volume effects and thus the surface energy term in Eq. (1) can be neglected, whereas in nanocrystals a non-negligible portion of atoms might be situated on the surface and thus this term is of importance.

---

[1] Sometimes in literature in this context σ is termed *surface tension*, sometimes *surface energy*, where the term "surface energy" should be understood as an energy density (energy per unit area). The physical quantity referred to is the same in both cases, however the term "surface tension" is more likely to be used in the description of liquids, whereas "surface energy" refers more to solids. The reasoning presented here is in fact inspired from the formation of liquid droplets, and therefore also the terminology is usually borrowed from this field. Therefore, we will use the terms "surface tension" and "total surface energy", where the latter refers to the surface tension multiplied by the surface area.

In Eq. (1) the surface tension σ is assumed to be constant for any size and morphology of the crystal, which is a very rough approximation. This uniformity of the surface tension σ refers to the drop model of the nucleation. In a more detailed discussion one would have to take into account an effect related to the small size of the nanocrystal and an effect of the faceting of the nanocrystal, i.e. the existence of the defined crystalline surfaces. One can gain a qualitative understanding of the size effect, when considering the surface tension as a result of the interaction between the surface atoms and the bulk of the crystal. It is evident that this interaction is actually weaker for smaller crystallites as there are no long-range interactions. It has been calculated that under the assumption of a Lennard-Jones interaction between the atoms the total surface energy of a cluster of 13 atoms (as sketched in Fig. 6) is reduced by 15% with respect to the total surface energy of a flat surface [39]. In nanocrystals, the effect of the presence of facets on the total surface energy is related to the precise arrangement of the atoms on the individual facets. Between different facets the densities of atoms vary and also the arrangements of the dangling bonds. One can assume that in general those facets which exhibit a closer packing of the atoms and a smaller number of unsaturated bonds are more stable, and thus have a lower tension. An example of this effect is discussed qualitatively in Sect. 3 of this chapter.

In the following discussion in this section we will assume a spherical shape for the nanocrystals and neglect any variation of the surface tension σ (i.e. we will assume an isotropic crystal without facets). In this case the number of atoms n in the first term of Eq. (1) can be expressed by the radius r of the nanocrystals, taking into account the density d_m of atoms in the nanocrystals. Then the equation reads:

$$\Delta G = \frac{4\pi d_m}{3} r^3 (\mu_c - \mu_s) + 4\pi r^2 \sigma \qquad (2)$$

In the case that the chemical potential of an atom in the solution is smaller compared to that of an atom within the crystal, the minimum of the free energy is given when all

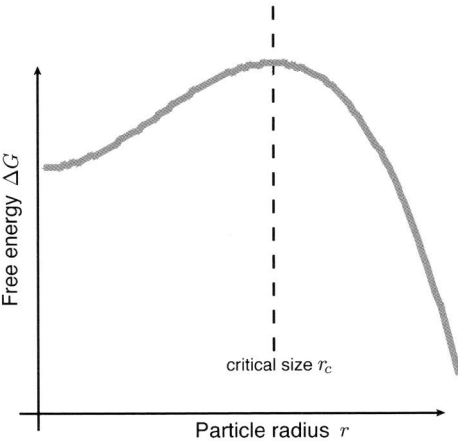

Fig. 2. Sketch of the potential landscape for the nucleation. Only at small values of the radius r in Eq. (2) the r^2 of the surface energy term outcompetes the r^3 contribution of the chemical potential, so that a barrier is imposed at the critical size r_c

atoms are unbound, and so no stable crystals are formed. Here we are interested in the opposite case with the chemical potential of atoms in solution being larger than that of bound atoms. In that case the first term becomes negative, and thus the free energy reaches a maximum for a certain radius r_c, termed *critical size*, at which a nucleation barrier is imposed, see Fig. 2. For small nuclei the surface energy term dominates the free energy, whereas only for crystals much larger than r_c the growth is driven by the gain in chemical potential and in principle the crystals can grow to an infinite size. The amplitude of the nucleation barrier controls the rate at which crystals nucleate [40].

2.2 Growth. The actual process of the deposition of monomers onto the growing nanocrystal can be split into two steps. First the monomers have to be transported towards the surface of the nanocrystal and in a second process they have to react with the nanocrystal. Generally the first process is accomplished through diffusion and thus at a rate dominated by the diffusion constant D, whereas the speed of the second process is given by the rate of reaction between free monomers and the crystal surface.

In the following the dynamics of the growth process will be outlined briefly. The discussion will start from a mechanistic view of the growth in which the growth rate $\dot{r} = \mathrm{d}r/\mathrm{d}t$ of a crystal of radius r depends only on the rate at which monomers are incorporated into the crystal. This latter rate is simply the time derivative of the number of monomers n in the crystal and it describes the number of monomers that go through the two processes mentioned above (diffusion and reaction) during a unit of time. We can write the growth rate $\dot{r}$ as

$$\frac{\mathrm{d}r}{\mathrm{d}t} = \frac{\dot{n}}{4\pi r^2 d_\mathrm{m}} \tag{3}$$

In this equation d_m denotes the density of monomers in the crystal, thus the inverse of the volume occupied by one monomer. The last equation is the time derivative $\dot{n} = \mathrm{d}n/\mathrm{d}t$ of the number n of monomers in the crystal solved for $\dot{r}$. In the following the behaviour of the deposition rate of monomers $\dot{n}$ will be discussed for two extreme examples.

In a typical synthesis, an excess of free monomers is injected to initiate the growth process. Under these conditions the effect of the diffusion process can be virtually neglected due to the high concentration of monomers. Monomers are available whenever there is a free site for their incorporation into a growing crystal. In this case the incorporation rate $\dot{n}$ depends only on the rate of the reaction monomer – crystal. This rate is proportional to the surface area of the crystal. Therefore the growth rate $\dot{r}$ in Eq. (3) is independent of the radius of the crystal. This growth regime is called *reaction controlled growth* and it is important only at very high concentrations of monomers. In this regime the width of the size distribution Δr does not vary with time. Only the relative width $\Delta r / \bar{r}$ decreases with time ($\bar{r}$ denotes the mean radius of the crystals) [41].

After a while the reservoir of monomers is partially depleted and the growth rate is dictated by the rate at which monomers reach the surface of the crystal. The following can be understood as the limit of a infinite diffusion layer in the argumentation presented by Sugimoto [41]. The flux J of monomers towards a growing crystal is driven by a gradient of the concentration C of the monomers. The surface of a nanocrystal represents a sink for free monomers and thus a minimum for the monomer concentration. In the model the monomer concentration is assumed to be constant on any sphere of radius x (greater than r) around the crystal. On the surface of the crystal the flux J is equal to the incorporation rate $\dot{n}$ of monomers:

$$J(x=r) = \dot{n} \qquad (4)$$

For any radius greater than r the flux through a spherical surface is determined by Fick's law of diffusion:

$$J(x>r) = 4\pi x^2 D \frac{dC}{dx} \qquad (5)$$

Here D is the diffusion constant. In a steady state, this flux J is independent of the distance x:

$$\frac{dJ}{dx}(x>r) = 4\pi D \left(2x \frac{dC}{dx} + x^2 \frac{d^2 C}{dx^2} \right) = 0$$
$$\Rightarrow \frac{d^2 C}{dx^2} = -\frac{2}{x} \frac{dC}{dx} \qquad (6)$$

Once the profile of the concentration is calculated from this differential equation, it can be inserted into Eq. (5) to quantify the flux J. To solve Eq. (6) boundary conditions are imposed. The monomer concentration C_i on the surface of a crystal as well as the monomer concentration C_b in the bulk of the solution, i.e. far from any crystal, are considered. With these boundary conditions the general form of the concentration profile around a crystal is derived as

$$C(x) = C_b - \frac{r(C_b - C_i)}{x} \qquad (7)$$

and thus the derivative of C reads:

$$\frac{dC}{dx} = \frac{r(C_b - C_i)}{x^2} \qquad (8)$$

This derivative can be inserted into Eq. (5) and the flux towards a crystal of radius r is described as

$$J = 4\pi D r (C_b - C_i) \qquad (9)$$

Finally, with this result the growth rate of a crystal in Eq. (3) reads:

$$\frac{dr}{dt} = \frac{D}{r d_m} (C_b - C_i) \qquad (10)$$

To this point an infinite stability of the nanocrystals is assumed. This assumption cannot be maintained. The Gibbs–Thompson effect actually introduces a competing effect to the growth [40]. Crystals – or particles and droplets in general – have a higher vapour pressure the smaller they are, and thus monomers "evaporate" into solution more easily from smaller crystals than from larger ones. This can be understood on a molecular level when considering the higher curvature of smaller crystals. Due to the increased surface curvature the surface atoms are more exposed to the surrounding and at the same time experience a weaker binding strength to the smaller crystal core. Experimentally this effect is seen in the lower melting temperature of smaller nanocrystals [4, 42].

With the help of the Gibbs–Thompson equation the vapour pressures of a crystal of radius r can be calculated. Through the general gas equation these vapour pressures can be expressed as the concentrations of monomers in the vicinity of the surface:

$$C_i = C_\infty \, e^{\frac{2\sigma}{rd_m k_B T}} \approx C_\infty \left(1 + \frac{2\sigma}{rd_m k_B T}\right) \quad (11)$$

In this equation C_∞ is the vapour pressure of a flat surface, σ is the surface tension. Formally the same reasoning is applied to the concentration C_b. The radius of a crystal in equilibrium with the concentration of monomers in the bulk is introduced as the critical size r^* of the growth process [41]:

$$C_b = C_\infty \, e^{\frac{2\sigma}{r^* d_m k_B T}} \approx C_\infty \left(1 + \frac{2\sigma}{r^* d_m k_B T}\right) \quad (12)$$

with these two quantities the growth rate from Eq. (10) can be calculated as

$$\frac{dr}{dt} = \frac{2\sigma D C_\infty}{d_m^2 k_B T} \frac{1}{r} \left(\frac{1}{r^*} - \frac{1}{r}\right) \quad (13)$$

The critical size r^* is characterised by a zero growth rate, as a crystal of this size is in equilibrium with the solution. For crystals smaller than r^* the growth rate is negative, the dissociation of monomers is more important than the supply of fresh monomers, and therefore these crystals melt. Here it becomes evident that the quantity r^* is actually equal to the critical size r_c that characterises the position of the energy barrier in the nucleation event.

The general dependence of the growth rate on the radius of the crystal is illustrated in Fig. 3. It is interesting to note the presence of a maximum at a radius of $2r^*$. If all crystals present in the solution have a radius larger than this value, the smallest crystals grow fastest, and therefore the size distribution becomes narrower over time. The value of r^* depends mainly on the overall concentration of free monomers, but also on the reaction temperature and on the surface tension σ. Especially the latter effect will be discussed in Sect. 4 of this chapter. During the run of the reaction the concentration of monomers decreases and the critical size shifts to higher values. If the critical size is sufficiently small the system is said to be in the *narrowing* or *size focussing regime*. During the run of the synthesis the critical size increases and as soon as the size of maximal growth rate, $2r^*$, has reached a value situated in the lower end of the size distribution of the nanocrystals, the system enters in the *broadening regime*. Ultimately, when r^* is larger than the radius of the smallest nanocrystals

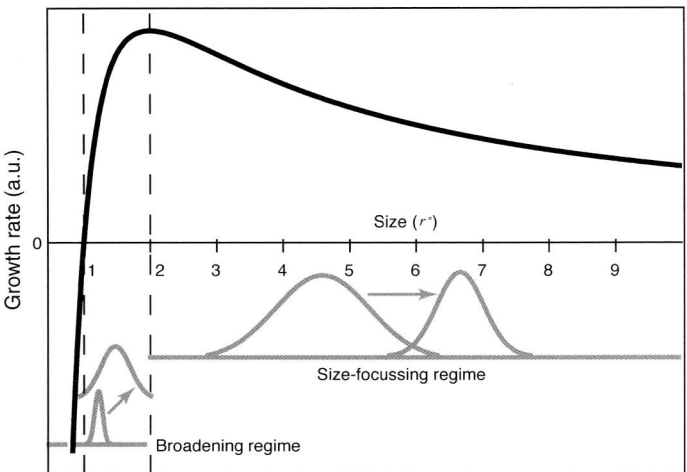

Fig. 3. Sketch of the growth rate dr/dt of the nanocrystals in units of the critical size r^*. As examples two size distributions and their development with time are displayed. Note that in the example for the broadening regime, the smallest particles are larger than r^*. Therefore the mean size of the particles still increases. The situation would be different if the size distribution would comprise nanocrystals smaller than r^*

present, the system enters into the *Ostwald ripening regime*, which is characterised by a large broadening of the size distribution and – more important – by a decrease of the total concentration of the nanocrystals. The smallest nanocrystals melt to free monomers that are incorporated into the larger nanocrystals.

In an experiment the effect of the size focusing is limited by the broadness of the initial size distribution, in other words, the nucleation event has a strong influence on the size distribution of the final sample [43]. Ideally the nucleation event should be finished well before the system enters into the diffusion controlled growth regime. The sharpness (in terms of time duration) of this event is therefore of great importance [43]. If on the contrary the nucleation event spans a considerable amount of time, those nanocrystals that nucleated first have grown already too far, resulting in a broad size distribution. In that case the effect of the size focusing might not be sufficient to obtain a reasonably narrow distribution at the desired average size of the nanocrystals. The limiting factor here is that the effect of the size focussing is always accompanied with an overall growth of all nanocrystals.

In the case of fluorescent nanocrystals, the width of the fluorescence spectrum is a good indicator for the quality of the size distribution. Samples of CdSe or CdTe nanocrystals generally have a band linewidth of ca. 30 nm or less, which in this case corresponds to ca. 100 meV. In Fig. 4 an example for the focussing of the size distribution is displayed. In this example the synthesis was carried out under conditions that favour size focussing. The nanocrystals were synthesised at a very high temperature, which reduces the nucleation event to a very short time span. Also, the concentration of monomers was sufficiently high to prevent the system to enter into the broadening regime.

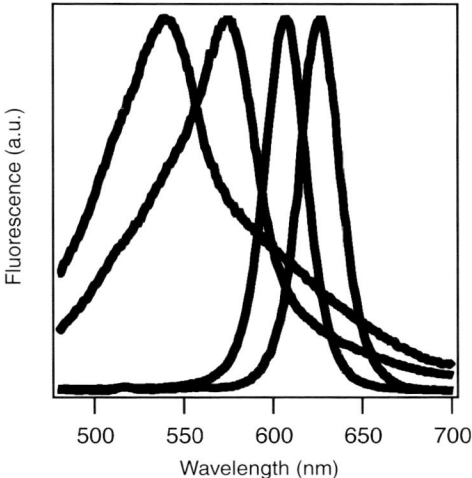

Fig. 4. Effect of size focussing in the synthesis of CdSe nanocrystals. The synthesis was performed at high temperature (370°C). Aliquots were taken every 20 s during the run of the synthesis and fluorescence spectra were recorded. The first aliquot (leftmost spectrum) shows a broad fluorescence spectrum and thus a wide size distribution, whereas the fluorescence spectrum of the latest aliquot (rightmost spectrum) is very narrow with a FWHM of 28 nm

In some synthesis schemes it is inevitable to consume the reservoir of monomers and thus obtain a broadening of the fluorescence band. In these cases the system can be maintained in focussing regime by repeated injections of fresh monomers [29, 44]. Experimentally it is not always possible to obtain a perfect size distribution. In some cases the distribution is broadened or shows several distinct peaks. A possible, but somewhat laborious way to improve the size distribution is to perform a size-selective precipitation after the synthesis is completed. In order to do so, a non-solvent is added slowly in a controlled way. As an example the addition of the polar solvent methanol to nanocrystals dissolved in a non-polar solvent such as chloroform or toluene induces the precipitation of the nanocrystals. The larger particles become unstable in the solution at lower concentrations of the non-solvent than the smaller particles, and thus they precipitate first [45]. This process has been successfully applied for instance in the synthesis of CdTe [46] and CdSe [29]. In Fig. 5 an example of a size-selective precipitation is shown.

3. Magic size clusters

For many applications it is of interest to have materials at hand that exhibit a high fluorescence yield and at the same time can be packed very densely. Through this many light-emitting units could be packed into a small volume and the overall fluorescence yield of the material would be enhanced as an eventually low fluorescence quantum yield might be overcome by a larger number of light-emitting sites. Also, in applications as markers the size of the fluorophore is a limiting factor. Most techniques to produce water-soluble nanocrystals involve the generation of a

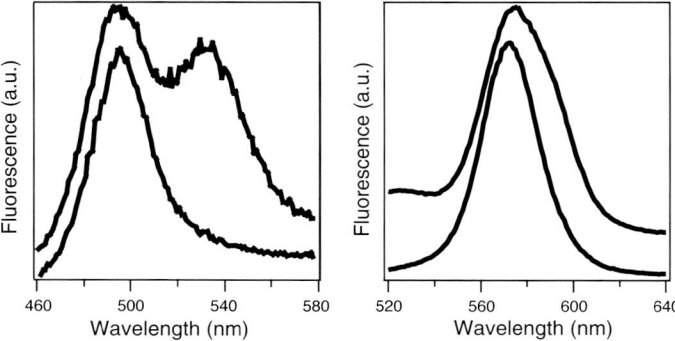

Fig. 5. Effect of the size-selective precipitation. By careful addition of a non-solvent to samples of CdSe nanocrystals with a bimodal (left) or simply broadened (right) size distribution the largest particles, i.e. those with an emission at higher wavelength, can be precipitated by centrifugation while the smaller particles remain in solution. The upper spectra show the fluorescence of the initial samples, whereas the lower spectra show the fluorescence of the supernatants. In the right example a further contamination of small, i.e. blue fluorescent, nanocrystals has been removed by a second precipitation

shell of organic molecules around the nanocrystals [47, 48]. This additional shell contributes to the total size of the nanocrystals and their overall size is more than doubled as compared to the bare inorganic core [49]. The hydrodynamic radius might vary also with other parameters such as salt concentration and lead to a further increase of the nanocrystals' effective size.

With published synthesis schemes for CdSe particles [29, 37] generally nanocrystals with a diameter of the inorganic core in the range of 2–8 nm are produced. To synthesise routinely nanocrystals with a smaller diameter, a different concept has to be employed. Already in the reported synthesis schemes of the CdSe nanocrystals the existence of highly stable and very small CdSe clusters was observed [50]. In the size regime below 2 nm the stability of these clusters can be explained with the concept of magic sizes. Several discrete cluster configurations exhibit a higher stability as compared to other configurations. Generally their high stability coincides with a high symmetry in their structure. The magic size clusters could be considered as those clusters that possess complete shells of atoms [51]. A classical example is the

Fig. 6. A series of different clusters formed by addition of closed shells in *fcc* structure. Each of the presented clusters represents the core of the subsequent cluster. Starting from a single atom, cubo octahedra of 13, 55 and 147 Au atoms can be build. Note that the percentage of surface atoms is far above 50% for all structures

series of truncated octahedra ("cubo octahedra"). The first member of this series is a single atom along with its 12 nearest neighbours in the *fcc* structure. The larger clusters can be constructed from the previous ones by covering each of the exposed facets with one layer of atoms. Thus the smallest cluster contains 13 atoms, the second 55 atoms and the third 147 atoms, as sketched in Fig. 6 [52]. These clusters differ only little from a sphere and therefore minimise the contribution of the surface atoms to the entire structure. For the larger clusters the facetted surface structure is obvious. Two types of facets are exposed: The (1 1 1) facets with a triangular shape and the (1 0 0) facets with quadratic shape. Clusters comprising 13 atoms have been shown for instance for Au and Rh [53, 54]. The cluster Au_{55} has been identified by Mößbauer spectroscopy [55] and it has been proven to be exceptionally stable against oxidation [56]. The same cluster has been prepared for several other metals [57]. Larger clusters with a fourth shell have been synthesised of palladium [58] and platinum [59]. Clusters with different geometry can be constructed with a similar approach [51].

In this section a synthesis scheme for the production of magic size clusters of CdSe is presented and discussed [60]. Due to their small size the magic-size clusters of CdSe emit blue light. Therefore these clusters extend the spectrum of the accessible fluorescence colours of CdSe nanocrystals. A method to exploit this for the production of blue-emitting diodes is presented by Rizzo et al. [61].

3.1 Synthesis of magic-size clusters.
In literature one can find protocols for the synthesis of several magic-size clusters [62–65]. In these reaction schemes for each different cluster size a different technique is employed. For this reason also the structure and nature of the surfactants is different for each of the different clusters. We have used a slightly varied reaction scheme [60] through which a series of magic-size clusters can be synthesised sequentially one after another with only one synthesis protocol. The size of the cluster is determined only by the reaction time.

The synthesis as described by Kudera et al. [60] is derived from the standard synthesis of green- to red-emitting CdSe nanocrystals [29, 36, 37], i.e. of nanocrystals with a size comprised in the range between 2 and 6 nm. The most important difference between this synthesis and the synthesis of the magic-size cluster is the reaction temperature of 80–100°C. In general the synthesis of semiconductor nanocrystals in organic solvents is performed at temperatures between 200 and 300°C. In addition the relatively weak ligands carboxylic acid [33, 66] and amine are used that do not suppress growth at these low temperatures.

A synthesis of CdSe nanocrystals under these conditions shows a quite remarkable result [60]. In the absorption spectra, instead of a continuously shifting peak one can find a series of peaks that appear one after another and only vary in their relative amplitude (see Fig. 8). The position of the individual peaks remains almost constant over time. As the position of the absorption peak is related to the size of the absorbing nanocrystal, this result points towards the sequential appearance of different sizes. The co-existence of clusters with different sizes in the solution can be proven by size-selective precipitation. It is possible to separate the fraction of the solution that is producing the absorption peak situated at red edge of the spectrum. This fraction is found to be the clusters that precipitated first, and thus corresponds to the largest

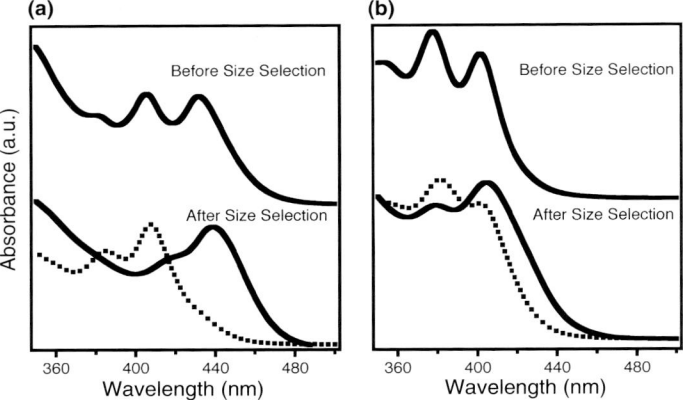

Fig. 7. Size-selective precipitation of two samples of magic size clusters. Methanol was added carefully to the cooled reaction solution (upper spectra) and the mixture was centrifuged. In the spectrum of the first precipitates (solid lines, lower spectra) the amplitude of the first peak is increased with respect to the other peaks. In the spectrum of the supernatants (dashed lines, lower spectra) the situation is inverted and the second peak is more pronounced than the first. From [60], © Wiley–VCH Verlag GmbH & Co. KGaA. Reproduced with permission

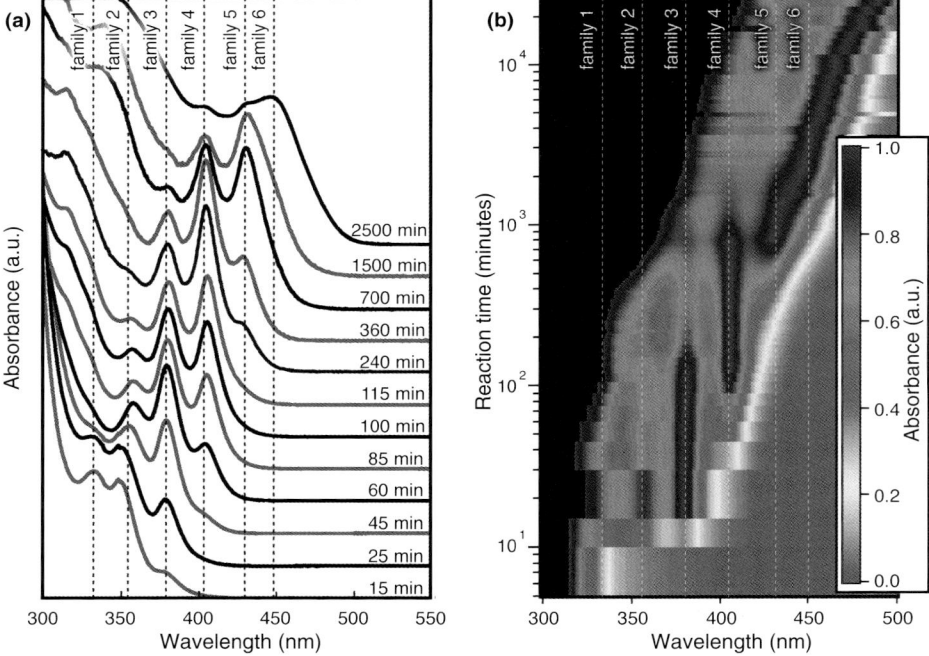

Fig. 8. Characteristic absorption spectra of magic size clusters of CdSe. Absorption spectra of aliquots taken during the synthesis of magic size clusters in two different representations (**a**, **b**). In both cases the spectra are normalised to the amplitude of the dominant peak. At the early stage of the reaction, only distinct peaks are visible at 330 (family 1), 360, 384, 406, 431 and 447 nm (family 6). In the upper right part of **b** the transition from the discrete growth dynamics to a continuous growth mode can be observed. From [60], © Wiley–VCH Verlag GmbH & Co. KGaA. Reproduced with permission

particles. By a careful treatment of the remaining supernatant it is even possible to eliminate the peak related to these large clusters almost completely, as shown in Fig. 7a. In the following the clusters are classified into six families, each of them characterised by a different peak in the absorption spectrum (see Fig. 8).

In the spectra of size-selected samples and generally in spectra recorded after a purification of magic-size clusters a red shift of the peaks can be noticed. This shift is most likely due to the partial stripping off of some surfactants from the clusters. The shift is of the order of ca. 10 nm. Upon addition of free surfactants to the solution the peaks can be shifted back. The recovery is not complete and bears also the disadvantage that fresh organic material is introduced into the solution as impurity.

Another indication for the discontinuous size distribution can be found in the fluorescence spectra. As shown in Fig. 9 the fluorescence spectra at late stages of the reaction, i.e. when the families 4–6 are present, are composed of a series of distinct peaks, similar to the behaviour of the absorption spectra. The observation of the discontinuity in the fluorescence spectra is experimentally limited by the low fluorescence efficiency of the bare clusters. Especially the smallest clusters (families 1–3) show almost no sharp band edge fluorescence. For applications (as shown by Rizzo et al. [61]) the fluorescence quantum yield is enhanced by covering the clusters with a shell of ZnS. Generally this process strongly influences the position of the absorption features.

Once a cluster has reached the size relative to family 6, the growth mode changes. The further growth proceeds in the classical continuous mode as normally observed in nanocrystal growth [44]. This behaviour is clearly evident in the plot shown

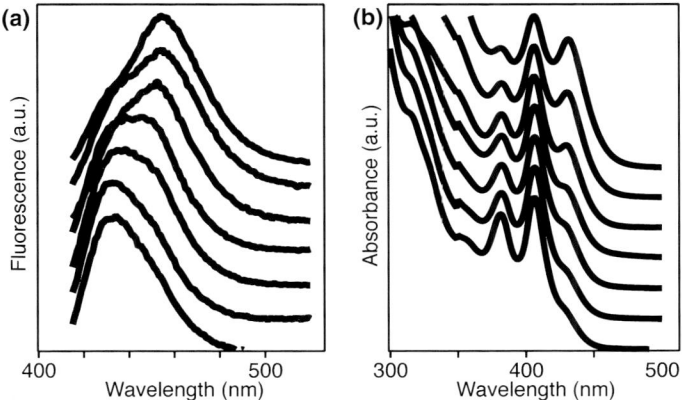

Fig. 9. Fluorescence spectra of the growing magic-size clusters. Fluorescence spectra (**a**) were recorded at different times during the growth process and for comparison the corresponding absorption spectra are shown in **b**. For clarity the spectra recorded at later stages of the growth are shifted upwards parallel the y axis. Even though it is not as clearly visible as in the case of absorption also the fluorescence spectra are composed of discrete peaks. The first (lowest) fluorescence spectrum shows a strong peak centred at ca. 430 nm with a small shoulder at higher wavelengths. The latest spectrum has a strong peak at ca. 450 nm and a shoulder at lower wavelengths. Intermediate spectra show a broadened peak. By comparison with the amplitude of the first two absorption peaks, one can infer that the quantum yield of the larger clusters, i.e. those fluorescing at 450 nm, is superior to that of the smaller clusters. From [60], © Wiley–VCH Verlag GmbH & Co. KGaA. Reproduced with permission

in Fig. 8b. At the first stage of the synthesis, the behaviour of the spectra is discontinuous. The vertical features represent the appearance and disappearance of the peak along with their stability in position. At later stages, i.e. after more than 1000 min of reaction and once family 6 is populated the principal absorption feature shifts continuously to larger wavelengths with the proceeding of the reaction, as can be discerned from the diagonal feature in the plot.

3.2 Growth mechanism of the magic-size clusters.

It is of great interest to understand the mechanism of formation of the magic-size clusters. In the preceding section it was shown that the individual peaks actually point towards a non-continuous size distribution. In this section the mechanism of growth will be discussed under this assumption. Certainly it is different from the standard models for the growth of nanocrystals. One can imagine the growth mechanism as a variation of a continuous growth model. The growing nanocrystal is exposed to a flux of monomers. In the same way as in a normal growth model, the monomers dynamically bind and unbind onto the surface of the nanocrystal. However, in the case of magic-size clusters the ratio between the binding- and unbinding-rate depends on the size of the crystal. The magic sizes represent sizes of extraordinary stability. If the crystal is slightly larger than a magic-size cluster, the release of the outer atoms is more probable than the deposition of other monomers onto the nanocrystal. Accordingly, nanocrystals slightly smaller than a magic-size cluster have a high affinity for monomers to reach the closed-shell structure. In the normal growth model in contrast, the binding rate dominates for all crystal sizes above the critical size, so that the crystals grow continuously.

A way to formally rationalise the stability of the magic-size clusters is to introduce a slight modification to the energy landscape presented in Sect. 2.1 as sketched in Fig. 10. In this modified landscape the magic sizes represent local minima. The sketch is a simplified model for the one-dimensional case. In principle the modified

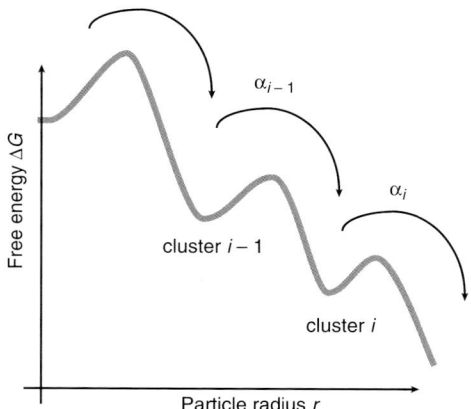

Fig. 10. Sketch of the potential landscape in the growth of CdSe magic size clusters. For simplicity it is sketched only in one-dimension. The model is to be understood as a modification to the classical nucleation potential as shown in Fig. 2

growth model can be understood as a subsequent nucleation. The growth proceeds only through a series of jumps over potential barriers. In the potential a preferential direction is inscribed. A given magic size cluster can evolve only into a larger magic size cluster. The melting, i.e. the evolution to a smaller magic size cluster, is virtually suppressed. Clusters that grow leave their own family i and after they have overcome the potential barrier, they become part of the subsequent family $i + 1$. In the same way the depopulation of the preceding family $i - 1$ feeds the population of family i. Therefore, the evolution of the concentration c_i of clusters of family i can be described by only two rate constants. The concentration of clusters of family i grows at the rate $\alpha_{i-1} c_{i-1}$ and decreases at the rate $\alpha_i c_i$:

$$\frac{dc_i}{dt} = \alpha_{i-1} c_{i-1} - \alpha_i c_i \qquad (14)$$

The differential equations for the individual families can be summarised in a vectorial form:

$$\frac{d}{dt} \begin{pmatrix} c_0 \\ c_1 \\ c_2 \\ \vdots \\ c_n \end{pmatrix} = \begin{pmatrix} -\alpha_0 & 0 & 0 & & 0 \\ \alpha_0 & -\alpha_1 & 0 & \cdots & 0 \\ 0 & \alpha_1 & -\alpha_2 & & 0 \\ \vdots & & & \ddots & \\ 0 & 0 & 0 & & -\alpha_n \end{pmatrix} \begin{pmatrix} c_0 \\ c_1 \\ c_2 \\ \vdots \\ c_n \end{pmatrix} \qquad (15)$$

It is interesting to note the singular role of the families 0 and n in the coefficient matrix. Family n is exceptional only for a trivial reason. It represents the last family described by the matrix. Clusters that leave this family enter into a different growth regime. From the data represented in Fig. 8b it can be deduced that this growth regime is continuous growth. More interesting is family 0. It is not fed by any other family, thus the clusters of this family must exist from the beginning of the reaction. At that moment only free monomers are found in the solution. Therefore the clusters of family 0 can at maximum be single atoms, which actually cannot be discerned from the other atoms. In this sense it is difficult to find a physical relevance for the number c_0. Only the rate of depopulation of family 0 can be interpreted nicely. It coincides with the rate at which clusters of family 1 appear. In other words the rate $\alpha_0 c_0$ can be interpreted as the nucleation rate of the system. This becomes evident when calculating the total concentration of magic size clusters:

$$c_{\text{total}} = \sum_{i=1}^{n} c_i \qquad (16)$$

As long as the largest family n is not populated, this number is constant over time. A reasonable boundary condition to solve Eq. (15) is to set all concentrations to zero at $t = 0$, except for c_0, which will take a finite value. If we excluded family 0 from the sum, the total concentration of clusters would be zero at the beginning of the reaction and then increase until it reaches a saturation value. It would remain stable until the first clusters appear in family n.

Kudera et al. [60] have shown that with the solutions of Eq. (15) it is possible to reproduce the general behaviour of the absorption spectra shown in Fig. 8. The exact shape of the individual entities that are deposited as well as the atomic structure of the clusters produced in this synthesis scheme are still unknown. One could imagine that the growth proceeds through the deposition of either individual atoms or very small molecular units, such as rings of the form $(CdSe)_3$. The magic size clusters are merely interesting for the possibility to observe the process of growth on a very low level. One can assume that between the individual steps of the growth single layers of atoms are deposited onto the clusters. Therefore it is very likely that the clusters have a pyramidal shape as seen also in similar clusters of CdSe [63, 67]. This structure also provides a highly symmetric shape in the cubic zinc blende phase of the presented clusters [60]. Generally this layer by layer growth is assumed to take place even at larger length scales. It has been observed in situ in a high-resolution TEM experiment in the growth of Al_2O_3 from a drop of liquid aluminium [68]. Also in that case the growth of layers of hundreds of atoms proceeds in a stepwise manner. Individual layers are deposited in time periods shorter than 0.04 s.

Magic-size clusters are of interest for various reasons. First, they extend the accessible spectrum of for instance CdSe [60] and CdTe [69] towards lower wavelengths. One can assume that by this one approaches a minimum. Smaller clusters would be in the range of atomic species. In that sense the clusters can also be understood as the molecular edge of the size spectrum of nanocrystals. A second reason for the interest in magic size clusters is their reproducibility. As the number of atoms in the clusters is well defined also the physical properties are defined with the same precision in comparison between two badges. This could be of great technological interest.

4. Shape control

The nanocrystals discussed so far were of almost spherical shape. However, in a more detailed analysis already in very small nanocrystals such as the magic-size clusters presented in the previous section different crystalline facets can be identified. On the other hand, due to the small size of these facets the deposition of only few atoms onto one of them changes its nature completely. With some nanocrystals the

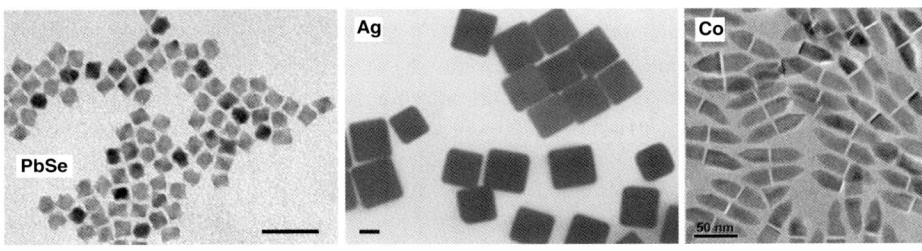

Fig. 11. Examples of faceting of nanocrystals. TEM images of differently shaped nanocrystals: octahedron-shaped PbSe, cube-shaped silver and pencil-shaped CoO. The scalebars represent 50 nm. Silver nanocrystals reproduced from [71]. Reprinted with permission from AAAS. CoO nanocrystals reprinted with permission from [72]. © 2006 American Chemical Society

faceted shape becomes obvious, as for instance on octahedral PbSe nanocrystals [70], silver nanocubes [71] or pencil-shaped Co nanocrystals [72] (see Fig. 11). Indeed, the growth of spherical crystals is more an exception than the rule. For instance macroscopic quartz is usually found in an elongated shape with six large facets that intersect under an angle of 120°. The tip of the crystal shows a series of facets, all inclined with respect to the long axis of the crystal. Generally the shape of a crystal depends on the relative speeds at which the individual facets grow. Here, the speed of growth of a facet is measured as the speed at which the distance of its centre to the centre of the entire crystal increases. The faster a crystal grows on one facet the more likely to disappear is this facet [73].

To rationalise the differences in the growth speeds let us reconsider the discussion of Sect. 2 on the growth process, where the exact nature of the surface or the existence of facets of the growing crystal was not taken into account. The introduction of the critical size in the growth process was actually based on the values of the vapour pressure of a flat surface (C_∞) and the surface tension σ, and in Eq. (13) these values were implicitly assumed constant for all facets. In fact these two quantities depend strongly on the exact nature of the facet under consideration. Especially the surface tension σ of the different facets can be influenced strongly by the choice of the surfactants [66, 74–76]. If the surfactants bind stronger to one facet than to its neighbouring facets, new monomers are more likely to be incorporated into these neighbouring facets. In other words, those facets onto which the surfactants bind stronger have a lower surface tension, as the area of this facet can be extended easier than that of a facet to which ligands bind weakly. In literature this relation between growth rate of the different facets and their individual surface tension is known as the Curie Gibbs Wulff theorem [41, 77]. Strictly speaking this theorem is valid only for a system in thermodynamic equilibrium. Generally in the synthesis of colloidal nanocrystals one works at high supersaturations of the monomers. In that case the relative speeds of growth of the different facets is changed sensitively [31]. In an analogous way the epitaxial growth regime depends very sensitively on the degree of supersaturation [78].

This effect is seen on many different nanomaterials, where nanocrystals of different shapes can be synthesised, as for instance cubes can be formed from Cu_2O, gold, silver or PbSe [71, 79, 80], disks of cobalt, silver and iron oxide [81–84], rods or wires of gold [26, 85].

In the wurtzite structure the choice of the growth axis is relatively straightforward. Through its high symmetry the unique c-axis is distinguished from the other axes. In CdE (E = S, Se, Te) nanocrystals it serves as the directional axis for the asymmetric growth [87]. In the case of Co nanocrystals in the ε-phase growth along this axis is suppressed [84]. But the asymmetric growth is not restricted to materials that expose a hexagonal structure [88]. CaF_2 nanorods grow in the cubic fluorite structure along the (1 1 1) axis [89]. PbSe grows in the highly symmetric cubic rock salt structure, but with the help of bulky surfactants a unique growth direction can be assigned and nanowires are formed [80]. For the materials of interest in this chapter (CdS, CdSe, CdTe) generally the crystalline structure is the wurtzite structure. For instance, nanorods of CdSe (as shown in Fig. 12) preferentially grow in wurtzite structure, as can be inferred from XRD measurements and high resolution TEM [90]. A nice

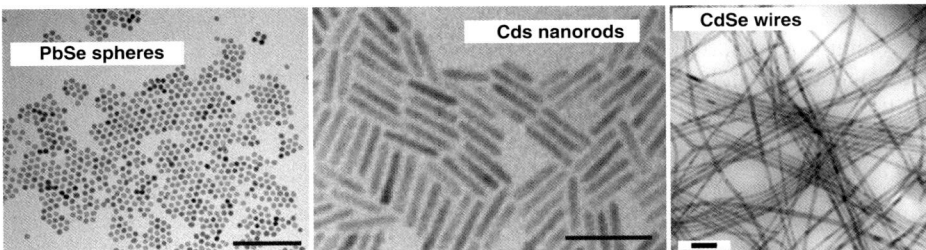

Fig. 12. TEM images of nanocrystals with different aspect ratios. Spherical PbSe nanocrystals (left), rod-shaped CdS and nanowires of CdSe (right). The scale bars correspond to 50 nm. CdSe nanowires reprinted from [86]. © 2003 American Chemical Society

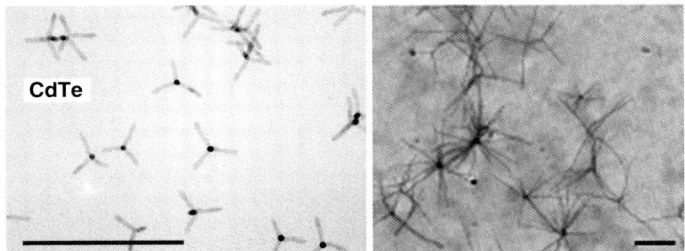

Fig. 13. TEM images of branched CdTe. (Left) The nanocrystals consist of four rods that are fused together in one point so that they span a tetrahedral angle between two legs. In the projection of the TEM, three legs can easily be identified, the forth legs points directly upwards. Due to the longer distance that the electron beam has to travel through the material this leg appears as a dark spot. (Right) Hyper-branched particles of CdTe. The scale bars represent 200 nm. Hyper-branched particles reprinted from [101]. © 2005 American Chemical Society

demonstration of the existence of the well-developed facets in the rods can be seen in honeycomb structures, in which the nanorods align vertically and in TEM images the rods can be seen along the long axis, or parallel to the lateral facets. In this case the hexagonal boundaries of each rod can be observed clearly [90]. It is interesting to notice that for thin nanorods the wurtzite structure is favoured even in conditions where the bulk material favours the cubic zinc blende structure due to the reduced number of dangling bonds on the surface [91, 92].

A higher level of complexity is observed in branched nanocrystals (Fig. 13). An intriguing example of such structures is the tetrapod. It is composed of four rods that are fused together in a central core. This shape is observed in several materials, such as ZnO [93], ZnSe [94, 95], ZnS [96], CdSe [97–99], CdTe [100–102], Pt [103] and Fe_2O_3 [104]. Generally, the growth mode of the tetrapods is driven by the same mechanism as the growth of the rods. During growth monomers are deposited onto the high energy facets of the arms, i.e. on their tips, whereas deposition of monomers onto the lateral facets of the arms is hindered. The main difference in the growth process resides in the nucleation event and the very early stage of the growth, when the core of the tetrapod is formed. In the following we will present two possible models that describe the crystalline structure of the core of a tetrapod.

In recent recipes for the preparation of nanorods their shape is still controlled by surfactants as described above, but in addition, the quality of the sample in terms of homogeneity is enhanced by the use of a seeded growth approach. This technique will be described in the section on hybrid materials.

4.1 Origin of branching: polymorphism versus twinning. To understand the dynamics of the growth of tetrapods, the intrinsic structure of these peculiar nanocrystals should be discussed. One can actually refer to two competing models to rationalise the four arm shape. One model – the *octa-twin model* – has been discussed and supported since the early 1990s for tetrapods of ZnO and ZnS [94]. This model is based on the controlled formation of twin planes. The second model – the *polymorphism model* – relies on the polymorphism of the materials and has been assumed to describe the formation of CdTe tetrapods [100]. In this model the core of the tetrapod is grown in zinc blende structure, whereas the arms grow in wurtzite structure.

4.1.1 Polymorphism model. Being a tetrahedrally coordinated binary compound, CdTe can grow either in the hexagonal wurtzite or the cubic zinc blende structure. The different structures have different energies of formation. Thus, at lower reaction temperatures growth in the zinc blende phase is favoured, whereas at higher temperatures the crystals preferentially grow in the wurtzite structure [105, 106]. The injection of precursors induces a drop in temperature and thus enables the formation of zinc blende nuclei. After the temperature has recovered the growth shifts to the wurtzite phase. Both crystalline phases have one set of facets in common. The {1 1 1} facets of the zinc blende structure are atomically identical to the $\pm(0\,0\,0\,1)$ facets of the wurtzite structure. Therefore the wurtzite arms will grow out of the {1 1 1} facets of the core. Apart from the temperature effect, the change in the crystalline phase is also supported by the affinity of the ligand – namely the phosphonic acids – to the lateral facets of the rods in wurtzite structure, i.e. to the arms of the tetrapod [107]. At the very early stage the surfactants are too bulky to efficiently bind to these facets and thus support their formation.

The fact that only four out of the eight {1 1 1} facets of zinc blende structure allow for the growth of the arms is explained by the atomic difference between the (1 1 1) and the opposed $(\bar{1}\,\bar{1}\,\bar{1})$ facets. In the same way as in the wurtzite structure on one of the facets the Cd atoms expose three dangling bonds, whereas on the other facet they expose only one. The later can be efficiently passivated by surfactants and thus the growth on this facet is hindered [107]. Therefore only on four out of the eight facets of the {1 1 1} family a wurtzite arm can grow.

4.1.2 Octa-twin model. A different model describes the tetrapod as composed of several domains that are exclusively in wurtzite structure. According to this twinning model, the core is composed of eight wurtzite domains. These domains have a tetrahedral shape and can be of two different types. In one type (type A) the basal facet is a (0 0 0 1) facet of the wurtzite structure, in the other type (type B) it is a $(0\,0\,0\,\bar{1})$ facet. The other three facets of the tetrahedra are either of the family $\{1\,1\,\bar{2}\,2\}$ (type A) or of the family $\{1\,1\,\bar{2}\,2\}$ (type B). Two tetrahedra of different type

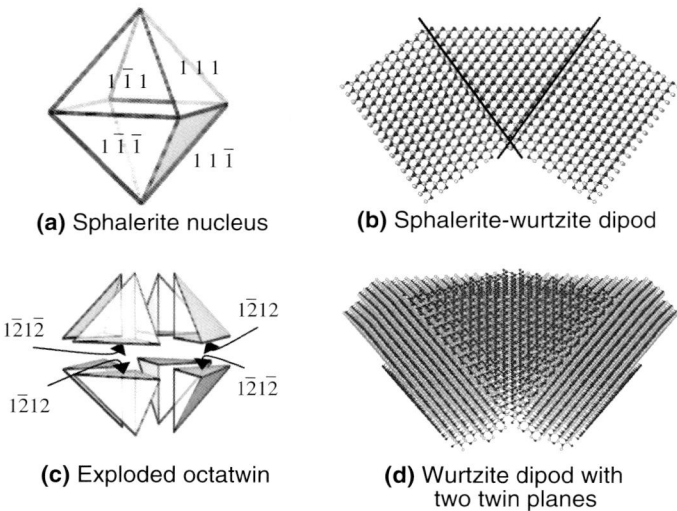

Fig. 14. Comparison between the polymorphism and the octa-twin model. An octahedral core can be formed either in cubic zinc blende structure (**a**) or out of eight tetrahedral domains in the hexagonal wurtzite phase (**c**). In both cases only four of the eight facets allow the growth of a rod, so that a tetrapod can be formed. The two models differ in the way the junction between two arms is described. Either the arms grow on top of a sphalerite core (**b**) or they are interconnected through a hexagonal domain (**d**). In the latter case the individual domains are fused through twin planes. Adopted from [3], with permission from Elsevier

are interconnected through a twin plane. This twin structure shows an inverted symmetry. As domains of type A are connected preferentially to domains of type B and vice versa, the octahedron formed by eight of these domains exposes an alternating pattern of $(0\,0\,0\,1)$ and $(0\,0\,0\,\bar{1})$ facets. These facets exhibit hexagonally organised atoms, identically to the $\pm(1\,1\,1)$ facets of the cubic structure (Fig. 14).

4.1.3 Comparison of the two models.

In both cases the Cd atoms on the individual facets are organised hexagonally and exhibit either one or three dangling bonds. Therefore both models describe a core, which exhibits four fast-growing and four slow-growing facets. Therefore both octahedra are chemically identical. As a result both models are able to describe the formation of tetrapods.

The *octa-twin* model ("octahedral multiple twin model") was first proposed to describe the architecture of ZnO and ZnS tetrapods [93, 108]. The structure of the core, as constructed according to this model, does not necessarily fill the volume of the octahedron. In fact it is very likely that the individual domains are formed such that, when assembling eight of them, not all of them can evolve three twin planes without introducing stress in the crystalline structure, so that slices of material are missing along the twin planes leading to a deviation from the ideal tetrapod shape, in the sense that the angles between the arms are different from the tetrahedral angle [94]. The successful measurement of this deviation on ZnO tetrapods supports the octa-twin model [109].

To date it is not yet ultimately decided which of the models describes the tetrapods the best. In the case of CdE tetrapods there is only little experimental evidence for either of the models. The polymorphism model is accepted widely in literature for the growth of CdE tetrapods [97, 100, 110]. Some XRD studies opt for the polymorphism model [32, 110]. On the other hand, some experiments seem to point towards the octa-twin model. It is possible to constantly tune the average number of branches in the tetrapod increasing the concentration of a single reactant [101, 102]. In Fig. 13 the case of a moderate and an excessive addition of this reactant is shown. This fact can not be explained by the polymorphism model. Therefore we can assume that at least in a synthesis scheme similar to that employed in [101, 102] a considerable fraction of the branched nanocrystals is formed according to the octa-twin model.

5. Hybrid materials

The relative ease to tune the properties of the nanocrystals actually makes them a very useful tool. Yet the range of accessible properties for one type of material is still relatively limited. Magnetic nanocrystals can be directed towards specific areas in e.g. tissues [111], but there is no simple and direct method to visualise their enrichment. It would be therefore of advantage to design nanostructures that consist of different domains, each of these domains contributing with its particular properties to the entire object. In the described case it would be interesting if the material could also emit light. There are several strategies to reach this target. One of the first approaches was to assemble different nanocrystals through pairs of complementary biomolecules such as single strands of DNA into larger structures [112]. By a more careful choice of the DNA sequences and some advanced treatment, even dimers or chains of nanocrystals can be formed [113, 114]. The disadvantage of these nanostructures is their instability. The linker is stable only under precisely defined conditions, otherwise it very likely releases the individual nanocrystals. A different approach involves the formation of polymer capsules that contain different materials [115]. In principle these capsules can contain any type of nanomaterials and are prone to serve for tasks like drug delivery. In this section we will discuss a third approach. It consists in fusing two materials together into a single crystalline structure.

It has been demonstrated that with this approach fluorescent and magnetic fractions can be combined into a single nanocrystal [116]. Heterodimers comprising gold have been prepared with magnetic materials, such as Au-CoPt$_3$ [117] or Au-Fe$_3$O$_4$ [118, 119], and also with semiconductor materials, e.g. CdSe-Au [120]. Here, gold is of interest as its ability to link functional molecules is well explored. Another possible use of gold domains could be to use them as local heat source [121, 122]. Furthermore, heterodimers of materials with different wettability can be prepared. In this case the interfacial tension between a hexane and water phase can be influenced [123].

Especially for the construction of hybrid nanocrystals based on spherical units there are different strategies [124]. In this section we will discuss an approach to synthesise hybrid nanocrystals. The techniques discussed here rely on the heterogeneous nucleation of some material on nanocrystals present in the growth solution. Nanorods are decorated with tips of a different material to form *nanodumbbells*.

Furthermore the technique for the production of hybrid structures can be exploited in order to obtain nanorods with a very sharp distribution of lengths [20].

5.1 Heterogeneous nucleation. It is known that in the presence of inhomogeneities or impurities the formation of crystals is favoured. Especially in the synthesis of metal nanocrystals this effect can be observed as a hindrance in the production of homogeneous samples. The presence of only a small amount of impurities might lead to a broad size distribution. The impurities act as seeds for the growth of material. In this case heterogeneous nucleation, i.e. nucleation on the impurities, is preferred as it effectively lowers the nucleation barrier [40]. Therefore nucleation is possible even at a relatively low supersaturation, i.e. at later stages of the synthesis, and the size distribution is broadened due to the prolonged nucleation event.

In the following paragraph we will introduce the concept of heterogeneous nucleation using the example of bubble formation in a liquid. In a glass of soda water, one commonly observes bubbles being formed on the walls and interestingly one finds only a limited number of nucleation sites, which then generate long strings of bubbles. In this example small gas filled cavities on the surface of the vessel serve as nucleation sites. The bubbles grow in these cavities and detach once they reached a certain size. After the detachment, some gas remains in the cavity as nucleus for the subsequent bubble. If the radius of the meniscus of the gas bubbles in the cavities is superior to the critical radius of the nucleation event (see Fig. 2) the nucleation barrier is basically by-passed and the visible formation of free bubbles is controlled only by the growth of the bubbles in and on the cavity to a size sufficient for their detachment [125]. In this argumentation we must thus understand the radius r used in the discussion of the nucleation event (Eq. 1) as the curvature of the surface onto which material is deposited. Growth is possible even if the volume of the seed is not sufficient to establish a spherical bubble larger than the critical size. This problem is overcome by the use of seeds that might contain less material than a free nucleus of the critical size, but due to some constraints – in this case being enclosed in a cavity – the exposed surface shows a curvature related to bubbles of size larger than the critical size.

In the case of crystal formation in solution the effect of the seed is to provide a substrate onto which the second material can nucleate. If this material perfectly wets the (spherical) seed, the growth practically proceeds as if the system was directly transferred to a situation in which the growth rate as described in Eq. (13) is determined by parameters such as concentration of monomers in solution and temperature, whereas the size distribution is determined by the size distribution of the seeds. In this case the nucleation barrier is simply by-passed as seen also in the case of the bubble formation.

The case when the second material does not perfectly wet the support can be qualitatively understood when considering the nucleation of this material on a flat surface. Here the relative surface tensions determine the contact angle γ between the two materials. To nucleate an island only a fraction of the volume of a sphere of the critical size r^* is necessary. The curvature of the surface of the island is determined by r^* as shown in Fig. 2, and the volume of this section depends on the contact angle, as

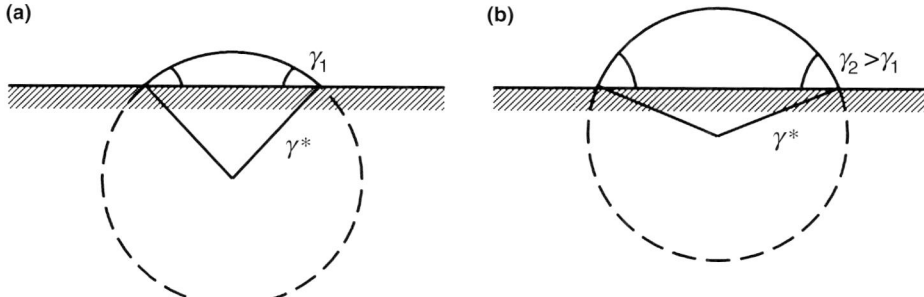

Fig. 15. Sketch of a stable nucleus on a substrate. The curvature r^* of the nucleus is determined by the critical size as in the homogeneous nucleation. Only the total volume of the nucleus is reduced, as the material grows with the constraint of forming a contact angle γ with the substrate

illustrated in Fig. 15. The larger the contact angle is, the more material needs to be assembled to form a stable nucleus. A detailed calculation shows that through the interplay between the reductions of the surface and the volume the energetic barrier can be reduced [40]. In simple words the effect of the reduced volume renders more probable the formation of a stable nucleus through thermodynamic fluctuations.

5.2 Growth strategies. The general procedure for the growth of hybrid materials is to initiate the growth of some material in the presence of another preformed material. First, we consider the case of a shell grown onto spherical nanocrystals. This technique has been developed only shortly after a stable synthesis scheme for semiconductor nanocrystals was established. In the process of a shell growth ideally the conditions are chosen such that the shell material perfectly wets the seed. One major parameter to attain this wettability in the growth of nanocrystals is to select materials with only slightly different lattice constants. The technique employed to grow a thin shell around nanocrystals involves low concentrations of the shell material in the reaction solution. Under normal conditions this would result in a very large critical size and thus Ostwald ripening. This is usually prevented by performing the synthesis at relatively low temperatures [126–130] (Fig. 16).

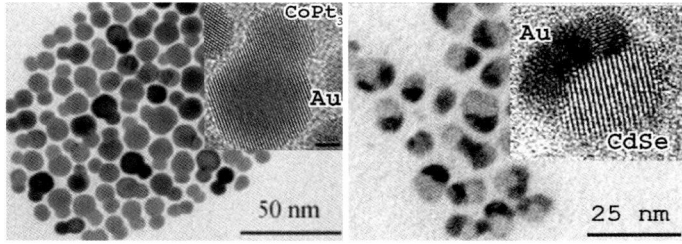

Fig. 16. Colloidal heterodimers. The nanocrystals presented in this figure are produced by nucleating gold on spherical nanocrystals of $CoPt_3$ (left) and CdSe (right). The insets show individual structures of the respective material, reproduced from [124] with permission of The Royal Society of Chemistry (RSC). Left figure reprinted with permission from [117]. © 2006 American Chemical Society. Right figure reprinted by permission from Macmillan Publishers Ltd: Nature Materials [120], © 2005

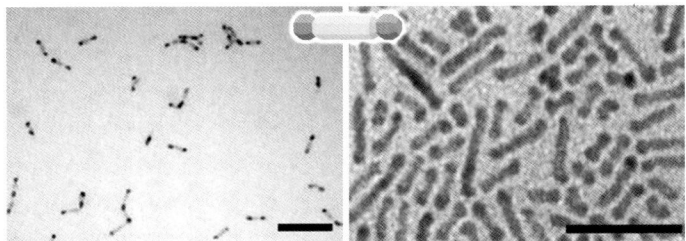

Fig. 17. Examples of dumbbell structures. (Left) CdSe nanorods are decorated with Au tips. Reprinted by permission from Macmillan Publishers Ltd: Nature Materials [120], © 2005. (Right) Dumbbells of CdTe–CdS–CdTe. Reproduced from [137] with permission of the Royal Society of Chemistry (RSC). The scale bars represent 50 nm

In case the second material does not wet the seed it might be deposited only on a precisely defined facet of the seed so that the lattice mismatch is minimised at the interface. In this case one usually finds nanocrystals that consist of two joint spheres, as observed in the case of CdS–FePt [116], γ-Fe$_2$O$_3$-MS (M = Cd, Zn, Hg) [131], Au-Fe$_3$O$_4$ [118, 119], CdSe-Fe$_2$O$_3$ [132], Au-CoFe$_2$O$_4$ [133], Au-CoPt$_3$ [117] and Au-CdSe [120]. Some examples of such structures are shown in Fig. 17. In some of these cases the peculiar shape is not necessarily obtained by a simple deposition of the second material on top of the seed, but the growth proceeds in two steps. First an amorphous and isotropic shell is formed, which then retracts to only one side of the seed upon annealing of the crystalline structure [116, 131].

When shape controlled nanocrystals are used as seeds, the selectivity of the interface can be nicely visualised. When some material is deposited on nanorods, it is very likely that it nucleates only at the tips of the rods [120, 134–136]. This can be explained with the matching of the lattice constants and also with the reduced passivation of the tips. As explained in the section on the shape control, surfactants bind stronger to the lateral facets of the rods and therefore enable the anisotropic growth. In the same way also when the second material is deposited onto the rods, these are likely to nucleate at unpassivated sites. With this technique it is even possible to distinguish between the two basal opposed facets of nanorods grown in wurtzite structure. As explained in Sect. 4 in the case of the growth of tetrapods, the opposing hexagonal facets in both the wurtzite $((0\,0\,0\,1)$ and $(0\,0\,0\,\bar{1}))$ and the zinc blende phase $((1\,1\,1)$ and $(\bar{1}\,\bar{1}\,\bar{1}))$ show chemically different behaviour. On one side the surface atoms of one type exhibit three dangling bonds, which cannot be efficiently passivated, on the other side atoms of the same type show only one passivated dangling bond.

A different method to produce hybrid materials involves the partial substitution of material. It was shown that in dumbbells tips of CdTe or PbSe can be transformed into Au tips [137]. In a more elaborate scheme a superlattice of AgS$_2$ domains can be formed in CdS nanorods [138].

5.3 Improved synthesis of nanorods. In this section we will present a seeded growth approach that helps to improve the size distribution of nanorods. A general problem in the growth of nanorods is the duration of the nucleation event, even more than in the case of spherical nanocrystals. The synthesis is usually carried out at high supersaturations [31], so that there is always a considerable probability to nucleate new nanocrystals. Therefore, in classical schemes for the growth of nanorods the particles nucleate during a period of time that spans a considerable fraction of the entire duration of the synthesis. As a result samples with a wide distribution of lengths are produced. In certain limits the distribution can be narrowed slightly by size-selective precipitations but still improvements would be desirable. As a consequence, a viable technique for the production of samples with narrow length distribution consists in the drastic shortening of the nucleation event. This might be performed in a seeded growth approach. For the shape control the seeded growth approach was first employed by Jana et al. in the synthesis of nanorods of gold [139] and silver [140]. Their approach makes use of two effects. First preformed nanocrystals are used as seeds, and second a rod-shaped micellar template is used to control the shape of the nanocrystals.

The seeded growth approach for the production of nanorods in a template-free surrounding was employed by Talapin et al. in a synthesis of anisotropic shells of CdS on CdSe seeds [141]. They present a variation of the shell growth, in which the shell adopts an anisotropic form due to the different lattice mismatches on the individual facets of the core.

In more recently presented recipes preformed spherical nanocrystals with a narrow size distribution are exploited as seeds for anisotropic growth [20, 142]. In this case the monomer concentration has to be much higher as compared to the shell growth, in order to favour anisotropic growth. In fact these synthesis schemes can be considered as variations of nanorod growth. The conditions are the ones that favour nucleation and growth of the rods, but the nucleation event is influenced by the presence of adequate seeds. As a result the nanorods grow with a very narrow size distribution as can be inferred from Fig. 18.

This method might be exploited for a variety of purposes. The possibility to form monodisperse nanorods enables the formation of ordered arrays of these nanorods on surfaces, as can be inferred from the comparison between the quality of the alignments presented by Nobile et al. [19] and Carbone et al. [20].

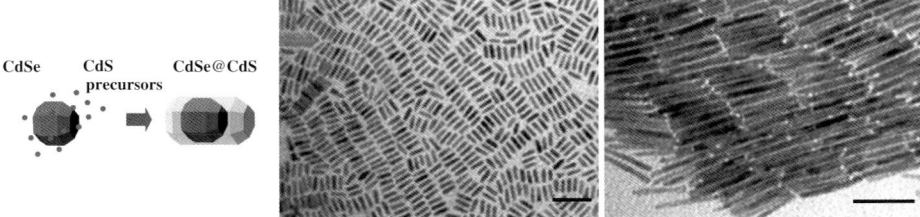

Fig. 18. TEM image of nanorods with ultra narrow length distribution. Due to van der Waals interactions the nanorods arrange in close packages when the solvent evaporates. In the present case the nanorods form ribbons that again assemble into parallel structures. From the straightness of the lines that separate the ribbons one can deduce the narrowness of the length distribution. The scale bars represent 50 nm.
Reprinted with permission from [20]. © 2007 American Chemical Society

As concerns the actual growth process, it is interesting to mention the possibility to exploit also intrinsic properties of the seed. In the discussed case the relative crystalline structures and the chemical surrounding were chosen in a way to favour the growth of (truncated) cubes [142] or nanorods [20, 142]. Additionally it is also possible to use a combination of core and shell material that favours a more complex shape. If the core is of cubic structure and the shell of hexagonal structure it is possible to grow tetrapods [20, 143, 144]. In that case the model describing their structure is indeed the polymorphism model as presented in the last section. These structures are again of interest for their expected electronic properties. One could imagine to sensitively influence the gating behaviour of the tetrapod [23] by altering the composition of the core and the arm material.

6. Conclusion

In this chapter we have discussed the control on three major features of nanocrystals: size, shape and composition. The topic of size control is the foundation of modern research on colloidal nanocrystals. In this field most of the current synthesis techniques have been explored and developed. It has soon turned out that even small variations of the standard synthesis schemes could lead to interesting effects, such as the formation of nanocrystals with peculiar shapes. It is now possible to attain well-defined sizes and shapes so that applications can be addressed. The attraction of the field resides in the possibility to easily adjust the characteristics of the materials to a desired value.

Among the three fields introduced in this chapter, composition control represents the newest trend in nanocrystal research. It has a high potential to extend the accessible range of applications, as it is based on the promise to render possible combination of complementary properties in single objects on the nanoscale. Core/shell nanocrystals of semiconductor materials with staggered bandgaps show bright fluorescence in a range that is not accessible to either of the materials. The simple idea of combining different properties in just one material is certainly exciting. As an example, gold tips on nanorods can be used to specifically bind functional molecules to these tips and with the help of these assemble nanorods to chains [135] or facilitate the connection between nanocrystals and macroscopic electrodes.

To attain such aims, still many questions need to be resolved. The core/shell systems as the simplest model of hybrid materials are well explored. More complex systems are only uprising. In the synthesis of these materials still many issues need to be addressed. First of all, it is still open for discussion, in how far it is physically possible and desirable to combine different properties in a single nanocrystal. A second issue could be to attain a more precise control on the size of individual domains. Generally the size of the second domain is much less controlled than it is possible for free nanocrystals. We are convinced that answering these questions will pave the way for many future applications.

Acknowledgements

This publication was supported by the EU-STREP SA-NANO and the Max Planck Society.

References

[1] Parak WJ, Manna L, Simmel FC, Gerion D, Alivisatos P (2004) Quantum dots. In: Nanoparticles: From Theory to Application (Schmid G. ed), pp 4–49. Wiley–VCH: Weinheim
[2] Ashcroft NW, Mermin ND (2001) Festkörperphysik. München: Oldenbourg
[3] Kudera S, Carbone L, Carlino E, Cingolani R, Cozzoli PD, Manna L (2007) Synthesis routes for the growth of complex nanostructures. Phys E Low-dimensional Syst Nanostruct 37(1–2): 128–133
[4] Alivisatos AP (1996) Perspectives on the physical chemistry of semiconductor nanocrystals. J Phys Chem 100(31): 13226–13239
[5] Hu J, Li L-S, Yang W, Manna L, Wang L-W, Alivisatos AP (2001) Linearly polarized emission from colloidal semiconductor quantum rods. Science 292(15 June): 2060–2063
[6] Kan S, Mokari T, Rothenberg E, Banin U (2003) Synthesis and size-dependent properties of zinc-blend semiconductor quantum rods. Nat Mater 2: 155–158
[7] Millo O, Katz D, Steiner D, Rothenberg E, Mokari T, Kazes M, Banin U (2004) Charging and quantum size effects in tunnelling and optical spectroscopy of CdSe nanorods. Nanotechnology 15(1): R1
[8] Huang Y, Duan XF, Cui Y, Lauhon LJ, Kim KH, Lieber CM (2001) Logic gates and computation from assembled nanowire building blocks. Science 294(5545): 1313–1317
[9] Duan XF, Huang Y, Cui Y, Wang JF, Lieber CM (2001) Indium phosphide nanowires as building blocks for nanoscale electronic and optoelectronic devices. Nature 409(6816): 66–69
[10] Gudiksen MS, Maher KN, Ouyang L, Park H (2005) Electroluminescence from a single-nanocrystal transistor. Nano Lett 5(11): 2257–2261
[11] Li LS, Walda J, Manna L, Alivisatos AP (2002) Semiconductor nanorod liquid crystals. Nano Lett 2(6): 557–560
[12] Dumestre F, Chaudret B, Amiens C, Respaud M, Fejes P, Renaud P, Zurcher P (2003) Unprecedented crystalline super-lattices of monodisperse cobalt nanorods. Angew Chem Int Ed 42(42): 5213–5216
[13] Talapin DV, Shevchenko EV, Murray CB, Kornowski A, Forster S, Weller H (2004) CdSe and CdSe/CdS nanorod solids. J Am Chem Soc 126(40): 12984–12988
[14] Li L-S, Alivisatos AP (2003) Semiconductor nanorod liquid crystals and their assembly on a substrate. Adv Mater 15(5): 408–411
[15] Harnack O, Pacholski C, Weller H, Yasuda A, Wessels JM (2003) Rectifying behavior of electrically aligned ZnO nanorods. Nano Lett 3(8): 1097–1101
[16] Ryan KM, Mastroianni A, Stancil KA, Liu HT, Alivisatos AP (2006) Electric-field-assisted assembly of perpendicularly oriented nanorod superlattices. Nano Lett 6(7): 1479–1482
[17] Gupta S, Zhang QL, Emrick T, Russell TP (2006) "Self-corralling" nanorods under an applied electric field. Nano Lett 6(9): 2066–2069
[18] Hu ZH, Fischbein MD, Querner C, Drndic M (2006) Electric-field-driven accumulation and alignment of CdSe and CdTe nanorods in nanoscale devices. Nano Lett 6(11): 2585–2591
[19] Nobile C, Fonoberov VA, Kudera S, DellaTorre A, Ruffino A, Chilla G, Kipp T, Heitmann D, Manna L, Cingolani R, Balandin AA, Krahne R (2007) Confined optical phonon modes in aligned nanorod arrays detected by resonant inelastic light scattering. Nano Lett 7(2): 476–479
[20] Carbone L, Nobile C, De Giorg M, Sala FD, Morello G, Pompa P, Hytch M, Snoeck E, Fiore A, Franchini IR, Nadasan M, Silvestre AF, Chiodo L, Kudera S, Cingolani R, Krahne R, Manna L (2007) Synthesis and micrometer-scale assembly of colloidal CdSe/CdS nanorods prepared by a seeded growth approach. Nano Lett 7(10): 2942–2950
[21] Huynh WU, Dittmer JJ, Alivisatos AP (2002) Hybrid nanorod-polymer solar cells. Science 295 (5564): 2425–2427
[22] Gur I, Fromer NA, Geier ML, Alivisatos AP (2005) Air-stable all-inorganic nanocrystal solar cells processed from solution. Science 310(5747): 462–465
[23] Cui Y, Banin U, Bjork MT, Alivisatos AP (2005) Electrical transport through a single nanoscale semiconductor branch point. Nano Lett 5(7): 1519–1523
[24] Gur I, Fromer NA, Chen CP, Kanaras AG, Alivisatos AP (2007) Hybrid solar cells with prescribed nanoscale morphologies based on hyperbranched semiconductor nanocrystals. Nano Lett 7(2): 409–414
[25] Li JB, Wang LW (2003) Shape effects on electronic states of nanocrystals. Nano Lett 3(10): 1357–1363
[26] Jana NR, Gearheart L, Murphy CJ (2001) Seed-mediated growth approach for shape-controlled synthesis of spheroidal and rod-like gold nanoparticles using a surfactant template. Adv Mater 13 (18): 1389–1393

[27] Nikoobakht B, El-Sayed MA (2003) Preparation and growth mechanism of gold nanorods (nrs) using seed-mediated growth method. Chem Mater 15(10): 1957–1962
[28] Rogach AL, Franzl T, Klar TA, Feldmann J, Gaponik N, Lesnyak V, Shavel A, Eychmüller A, Rakovich YP, Donegan JF (2007) Aqueous synthesis of thiol-capped CdTe nanocrystals: state-of-the-art. J Phys Chem C 111(40): 14628–14637
[29] Murray CB, Norris DJ, Bawendi MG (1993) Synthesis and characterization of nearly monodisperse CdE (E = S, Se, Te) semiconductor nanocrystallites. J Am Chem Soc 115: 8706–8715
[30] Manna L, Scher EC, Alivisatos AP (2000) Synthesis of soluble and processable rod-, arrow-, teardrop-, and tetrapod-shaped CdSe nanocrystals. J Am Chem Soc 122(51): 12700–12706
[31] Peng ZA, Peng XG (2001) Mechanisms of the shape evolution of CdSe nanocrystals. J Am Chem Soc 123: 1389–1395
[32] Yu WW, Wang YA, Peng X (2003) Formation and stability of size-, shape-, and structure-controlled CdTe nanocrystals: ligand effects on monomers and nanocrystals. Chem Mater 15: 4300–4308
[33] Qu L, Peng Z, Peng X (2001) Alternative routes toward high quality CdSe nanocrystals. Nano Lett 1(6): 333–337
[34] Pan B, He R, Gao F, Cui D, Zhang Y (2006) Study on growth kinetics of CdSe nanocrystals in oleic acid/dodecylamine. J Cryst Growth 286(2): 318
[35] Sapra S, Rogach AL, Feldmann J (2006) Phosphine-free synthesis of monodisperse CdSe nanocrystals in olive oil. J Mater Chem 16(33): 3391–3395
[36] Peng ZA, Peng X (2001) Formation of high-quality CdTe, CdSe, and CdS nanocrystals using CdO as precursor. J Am Chem Soc 123(1): 183–184
[37] Reiss P, Bleuse J, Pron A (2002) Highly luminescent CdSe/ZnSe core/shell nanocrystals of low size dispersion. Nano Lett 2(7): 781–784
[38] Yu WW, Peng X (2002) Formation of high-quality CdS and other (II–VI) semiconductor nanocrystals in noncoordinating solvents: tunable reactivity of monomers. Angew Chem Int Ed 41(13): 2368–2371
[39] Benson GC, Shuttleworth R (1951) The surface energy of small nuclei. J Chem Phys 19(1): 130–131
[40] Markov IV (2003) Crystal Growth for Beginners. 2nd ed., Singapore: World Scientific. p 546
[41] Sugimoto T (1987) Preparation of monodispersed colloidal particles. Adv Colloid Interf Sci 28: 65
[42] Buffat P, Borel J-P (1976) Size effect on the melting temperature of gold particles. Phys Rev A 13(6): 2287–2298
[43] LaMer VK, Dinegar RH (1950) Theory, production and mechanism of formation of monodispersed hydrosols. J Am Chem Soc 72(11): 4847
[44] Peng X, Wickham J, Alivisatos AP (1998) Kinetics of II–VI and III–V colloidal semiconductor nanocrystal growth: "focusing" of size distributions. J Am Chem Soc 120(21): 5343–5344
[45] Shah PS, Holmes JD, Johnston KP, Korgel BA (2002) Size-selective dispersion of dodecanethiol-coated nanocrystals in liquid and supercritical ethane by density tuning. J Phys Chem B 106(10): 2545–2551
[46] Talapin DV, Haubold S, Rogach AL, Kornowski A, Haase M, Weller H (2001) A novel organometallic synthesis of highly luminescent CdTe nanocrystals. J Phys Chem B 105(12): 2260–2263
[47] Parak WJ, Pellegrino T, Plank C (2005) Labelling of cells with quantum dots. Nanotechnology 16(2): R9–R25
[48] Klostranec JM, Chan WCW (2006) Quantum dots in biological and biomedical research: recent progress and present challenges. Adv Mater 18(15): 1953–1964
[49] Sperling RA, Liedl T, Duhr S, Kudera S, Zanella M, Lin C-AJ, Chang WH, Braun D, Parak WJ (2007) Size determination of (bio)conjugated water-soluble colloidal nanoparticles: a comparison of different techniques. J Phys Chem C 111(31): 11552–11559
[50] Peng ZA, Peng X (2002) Nearly monodisperse and shape-controlled CdSe nanocrystals via alternative routes: nucleation and growth. J Am Chem Soc 124(13): 3343–3353
[51] Martin TP (1996) Shells of atoms. Phys Rep 273: 199–241
[52] Schmid G (1987) Von Metallclustern und Clustermetallen. Nachr Chem Tech Lab 35(3): 249–254
[53] Albano VG, Ceriotti A, Chini P, Ciani G, Martinengo S, Anker WM (1975) Hexagonal close packing of metal atoms in new polynuclear anions $[Rh_{13}(CO)_{24}H_{5-n}]^{n-}$ ($n = 2$ or 3) – X-ray structure of $[(Ph_3P)_2N]_2[Rh_{13}(CO)_{24}H_3]$. J Chem Soc Chem Commun (20): 859–860
[54] Briant CE, Theobald BRC, White JW, Bell LK, Mingos DMP, Welch AJ (1981) Synthesis and X-ray structural characterization of the centered icosahedral gold cluster compound $[Au_{13}(PMe_2Ph)_{10}Cl_2]$ $(Pf_6)_3$ – the realization of a theoretical prediction. J Chem Soc Chem Commun (5): 201–202
[55] Schmid G, Pfeil R, Boese R, Bandermann F, Meyer S, Calis GHM, Velden JWAvd (1981) $Au_{55}[P(C_6H_5)_3]_{12}Cl_6$ – ein Goldcluster ungewöhnlicher Grösse. Chem Ber 114(11): 3634–3642

[56] Boyen HG, Kästle G, Weigl F, Koslowski B, Dietrich C, Ziemann P, Spatz JP, Riethmuller S, Hartmann C, Möller M, Schmid G, Garnier MG, Oelhafen P (2002) Oxidation-resistant gold-55 clusters. Science 297(5586): 1533–1536
[57] Schmid G, Huster W (1986) Large transition-metal clusters, IV – Ru_{55}-, Rh_{55}- and Pt_{55}-clusters. Z Naturforsch Sect B 41(8): 1028–1032
[58] Vargaftik MN, Zagorodnikov VP, Stolyarov IP, Moiseev II, Likholobov VA, Kochubey DI, Chuvilin AL, Zaikovsky VI, Zamaraev KI, Timofeeva GI (1985) A novel giant palladium cluster. J Chem Soc Chem Commun (14): 937–939
[59] Schmid G, Morun B, Malm JO (1989) $Pt_{309}Phen_{36}{}^{*}O_{30\pm10}$, a 4-shell platinum cluster. Angew Chem Int Ed 28(6): 778–780 (in English)
[60] Kudera S, Zanella M, Giannini C, Rizzo A, Li Y, Gigli G, Cingolani R, Ciccarella G, Parak WJ, Manna L (2007) Sequential growth of magic-size CdSe nanocrystals. Adv Mater 19(4): 548–552
[61] Rizzo A, Li Y, Kudera S, Sala FD, Zanella M, Parak WJ, Cingolani R, Manna L, Gigli G (2007) Blue light emitting diodes based on fluorescent CdSe/ZnS nanocrystals. Appl Phys Lett 90(5): 051106
[62] Behrens S, Bettenhausen M, Eichhöfer A, Fenske D (1997) Synthesis and crystal structure of $[Cd_{10}Se_4(SePh)_{12}(PPh_3)_4]$ and $[Cd_{16}(SePh)_{32}(PPh_3)_2]$. Angew Chem Int Ed 36(24): 2797–2799
[63] Soloviev VN, Eichhofer A, Fenske D, Banin U (2001) Size-dependent optical spectroscopy of a homologous series of CdSe cluster molecules. J Am Chem Soc 123(10): 2354–2364
[64] Kasuya A, Sivamohan R, Barnakov YA, Dmitruk IM, Nirasawa T, Romanyuk VR, Kumar V, Mamykin SV, Tohji K, Jeyadevan B, Shinoda K, Kudo T, Terasaki O, Liu Z, Belosludov RV, Sundararajan V, Kawazoe Y (2004) Ultra-stable nanoparticles of CdSe revealed from mass spectrometry. Nat Mater 3(2): 99–102
[65] Zheng N, Bu X, Lu H, Zhang Q, Feng P (2005) Crystalline superlattices from single-sized quantum dots. J Am Chem Soc 127(34): 11963–11965
[66] Puzder A, Williamson AJ, Zaitseva N, Galli G, Manna L, Alivisatos AP (2004) The effect of organic ligand binding on the growth of CdSe nanoparticles probed by ab initio calculations. Nano Lett 4(12): 2361–2365
[67] Soloviev VN, Eichhofer A, Fenske D, Banin U (2000) Molecular limit of a bulk semiconductor: size dependence of the "Band gap" in CdSe cluster molecules. J Am Chem Soc 122(11): 2673–2674
[68] Oh SH, Kauffmann Y, Scheu C, Kaplan WD, Rühle M (2005) Ordered liquid aluminum at the interface with sapphire. Science 310(5748): 661–663
[69] Dai QQ, Li DM, Chang JJ, Song YL, Kan SH, Chen HY, Zou BQ, Xu WP, Xu SP, Liu BB, Zou GT (2007) Facile synthesis of magic-sized CdSe and CdTe nanocrystals with tunable existence periods. Nanotechnology 18(40): 405603
[70] Cho K-S, Talapin DV, Gaschler W, Murray CB (2005) Designing PbSe nanowires and nanorings through oriented attachment of nanoparticles. J Am Chem Soc 127: 7140–7147
[71] Sun YG, Xia YN (2002) Shape-controlled synthesis of gold and silver nanoparticles. Science 298 (5601): 2176–2179
[72] An K, Lee N, Park J, Kim SC, Hwang Y, Park JG, Kim JY, Park JH, Han MJ, Yu JJ, Hyeon T (2006) Synthesis, characterization, and self-assembly of pencil-shaped CoO nanorods. J Am Chem Soc 128(30): 9753–9760
[73] Kleber W, Bautsch H-J, Bohm J (1998) Einführung in die Kristallographie. 18th ed., Berlin: Verlag Technik
[74] Manna L, Wang LW, Cingolani R, Alivisatos AP (2005) First-principles modeling of unpassivated and surfactant-passivated bulk facets of wurtzite CdSe: a model system for studying the anisotropic growth of CdSe nanocrystals. J Phys Chem B 109(13): 6183–6192
[75] Rempel JY, Trout BL, Bawendi MG, Jensen KF (2005) Properties of the CdSe (0 0 0 1), (0 0 0 $\bar{1}$), and (1 1 2 0) single crystal surfaces: relaxation, reconstruction, and adatom and admolecule adsorption. J Phys Chem B 109(41): 19320–19328
[76] Rempel JY, Trout BL, Bawendi MG, Jensen KF (2006) Density functional theory study of ligand binding on CdSe (0 0 0 1), (0 0 0 $\bar{1}$), and (1 1 2 0) single crystal relaxed and reconstructed surfaces: implications for nanocrystalline growth. J Phys Chem B 110(36): 18007–18016
[77] Wulff G (1901) Zur Frage der Geschwindigkeit des Wachstums und der Auflösung der Krystallflächen. Z Krystallogr Mineral 34: 449–530
[78] Scheel HJ, Fukuda T, eds (2004) Crystal Growth Technology. 1st ed., John Wiley & Sons
[79] Gou LF, Murphy CJ (2003) Solution-phase synthesis of Cu_2O nanocubes. Nano Lett 3(2): 231–234
[80] Lifshitz E, Bashouti M, Kloper V, Kigel A, Eisen MS, Berger S (2003) Synthesis and characterization of PbSe quantum wires, multipods, quantum rods, and cubes. Nano Lett 3(6): 857–862
[81] Casula MF, Jun YW, Zaziski DJ, Chan EM, Corrias A, Alivisatos AP (2006) The concept of delayed nucleation in nanocrystal growth demonstrated for the case of iron oxide nanodisks. J Am Chem Soc 128(5): 1675–1682

[82] Maillard M, Giorgio S, Pileni MP (2002) Silver nanodisks. Adv Mater 14(15): 1084–1086
[83] Chen SH, Fan ZY, Carroll DL (2002) Silver nanodisks: synthesis, characterization, and self-assembly. J Phys Chem B 106(42): 10777–10781
[84] Puntes VF, Zanchet D, Erdonmez CK, Alivisatos AP (2002) Synthesis of hcp-{c}o nanodisks. J Am Chem Soc 124(43): 12874–12880
[85] Jiang XC, Pileni MP (2007) Gold nanorods: influence of various parameters as seeds, solvent, surfactant on shape control. Colloids Surf A: Physicochem Eng Asp 295(1–3): 228–232
[86] Yu H, Li J, Loomis RA, Gibbons PC, Wang L-W, Buhro WE (2003) Cadmium selenide quantum wires and the transition from 3d to 2d confinement. J Am Chem Soc 125(52): 16168–16169
[87] Dalpian GM, Tiago ML, del Puerto ML, Chelikowsky JR (2006) Symmetry considerations in CdSe nanocrystals. Nano Lett 6(3): 501–504
[88] Allen PB (2007) Nanocrystalline nanowires: I. Structure. Nano Lett 7(1): 6–10
[89] Mao YB, Zhang F, Wong SS (2006) Ambient template-directed synthesis of single-crystalline alkaline-earth metal fluoride nanowires. Adv Mater 18(14): 1895–1899
[90] Peng X, Manna L, Yang W, Wickham J, Scher E, Kadavanich A, Alivisatos AP (2000) Shape control of CdSe nanocrystals. Nature 404(6773): 59–61
[91] Lawaetz P (1972) Stability of the wurtzite structure. Phys Rev B: Cond Matt 5(10): 4039–4045
[92] Akiyama T, Nakamura K, Ito T (2006) Structural stability and electronic structures of inp nanowires: role of surface dangling bonds on nanowire facets. Phys Rev B 73(23): 235308
[93] Nishio K, Isshiki T, Kitano M, Shiojiri M (1997) Structure and growth mechanism of tetrapod-like ZnO particles. Philos Mag A 76(4): 889–904
[94] Hu J, Bando Y, Golberg D (2005) Sn-catalyzed thermal evaporation synthesis of tetrapod-branched ZnSe nanorod architectures. Small 1(1): 95–99
[95] Cozzoli PD, Manna L, Curri ML, Kudera S, Giannini C, Striccoli M, Agostiano A (2005) Shape and phase control of colloidal ZnSe nanocrystals. Chem Mater 17(6): 1296–1306
[96] Zhu Y-C, Bando Y, Xue D-F, Golberg D (2003) Nanocable-aligned ZnS tetrapod nanocrystals. J Am Chem Soc 125(52): 16196–16197
[97] Pang Q, Zhao LJ, Cai Y, Nguyen DP, Regnault N, Wang N, Yang SH, Ge WK, Ferreira R, Bastard G, Wang JN (2005) CdSe nano-tetrapods: controllable synthesis, structure analysis, and electronic and optical properties. Chem Mater 17(21): 5263–5267
[98] Asokan S, Krueger KM, Colvin VL, Wong MS (2007) Shape-controlled synthesis of CdSe tetrapods using cationic surfactant ligands. Small 3(7): 1164–1169
[99] Nobile C, Kudera S, Fiore A, Carbone L, Chilla G, Kipp T, Heitmann D, Cingolani R, Manna L, Krahne R (2007) Confinement effects on optical phonons in spherical, rod- and tetrapod-shaped nanocrystals detected by raman spectroscopy. Phys Stat Sol (A) 204(2): 483–486
[100] Manna L, Milliron DJ, Meisel A, Scher EC, Alivisatos AP (2003) Controlled growth of tetrapod-branched inorganic nanocrystals. Nat Mater 2: 382–385
[101] Kanaras AG, Sonnichsen C, Liu H, Alivisatos AP (2005) Controlled synthesis of hyperbranched inorganic nanocrystals with rich three-dimensional structures. Nano Lett 5(11): 2164–2167
[102] Carbone L, Kudera S, Carlino E, Parak WJ, Giannini C, Cingolani R, Manna L (2006) Multiple wurtzite twinning in CdTe nanocrystals induced by methylphosphonic acid. J Am Chem Soc 128(3): 748–755
[103] Teng X, Yang H (2005) Synthesis of platinum multipods: an induced anisotropic growth. Nano Lett 5(5): 885–891
[104] Cozzoli PD, Snoeck E, Garcia MA, Giannini C, Guagliardi A, Cervellino A, Gozzo F, Hernando A, Achterhold K, Ciobanu N, Parak FG, Cingolani R, Manna L (2006) Colloidal synthesis and characterization of tetrapod-shaped magnetic nanocrystals. Nano Lett 6(9): 1966–1972
[105] Yeh C-Y, Lu ZW, Froyen S, Zunger A (1992) Zinc-blende-wurtzite polytypism in semiconductors. Phys Rev B: Cond Matt 46: 10086–10097
[106] Wei S, Zhang S (2000) Structure stability and carrier localization in CdX (X = S, Se, Te) semiconductors. Phys Rev B 62(11): 6944
[107] Manna L, Scher EC, Alivisatos AP (2002) Shape control of colloidal semiconductor nanocrystals. J Cluster Sci 13(4): 521–532
[108] Takeuchi S, Iwanaga H, Fujii M (1995) Octahedral multiple-twin model of tetrapod ZnO crystals. Philos Mag A 69(6): 1125–1129
[109] Iwanaga H, Fujii M, Takeuchi S (1998) Inter-leg angles in tetrapod ZnO particles. J Cryst Growth 183(1–2): 190–195
[110] Zhang JY, Yu WW (2006) Formation of CdTe nanostructures with dot, rod, and tetrapod shapes. Appl Phys Lett 89(12): 123108

[111] Alexiou C, Arnold W, Klein RJ, Parak FG, Hulin P, Bergemann C, Erhardt W, Wagenpfeil S, Lubbe AS (2000) Locoregional cancer treatment with magnetic drug targeting. Cancer Res 60(23): 6641–6648
[112] Mirkin CA, Letsinger RL, Mucic RC, Storhoff JJ (1996) A DNA-based method for rationally assembling nanoparticles into macroscopic materials. Nature 382: 607–609
[113] Alivisatos AP, Johnsson KP, Peng XG, Wilson TE, Loweth CJ, Bruchez MP, Schultz PG (1996) Organization of 'nanocrystal molecules' using DNA. Nature 382(6592): 609–611
[114] Parak WJ, Gerion D, Pellegrino T, Zanchet D, Micheel C, Williams SC, Boudreau R, Gros MAL, Larabell CA, Alivisatos AP (2003) Biological applications of colloidal nanocrystals. Nanotechnology 14: R15–R27
[115] Sukhorukov GB, Rogach AL, Garstka M, Springer S, Parak WJ, Munoz-Javier A, Kreft O, Skirtach AG, Susha AS, Ramaye Y, Palankar R, Winterhalter M (2007) Multifunctionalized polymer microcapsules: novel tools for biological and pharmacological applications. Small 3(6): 944–955
[116] Gu H, Zheng R, Zhang X, Xu B (2004) Facile one-pot synthesis of bifunctional heterodimers of nanoparticles: a conjugate of quantum dot and magnetic nanoparticles. J Am Chem Soc 126(18): 5664–5665
[117] Pellegrino T, Fiore A, Carlino E, Giannini C, Cozzoli PD, Ciccarella G, Respaud M, Palmirotta L, Cingolani R, Manna L (2006) Heterodimers based on CoPt$_3$-Au nanocrystals with tunable domain size. J Am Chem Soc 128(20): 6690–6698
[118] Yu H, Chen M, Rice PM, Wang SX, White RL, Sun S (2005) Dumbbell-like bifunctional Au-Fe$_3$O$_4$ nanoparticles. Nano Lett 5(2): 379–382
[119] Gu H, Yang Z, Gao J, Chang CK, Xu B (2005) Heterodimers of nanoparticles: formation at a liquid–liquid interface and particle-specific surface modification by functional molecules. J Am Chem Soc 127(1): 34–35
[120] Mokari T, Sztrum CG, Salant A, Rabani E, Banin U (2005) Formation of asymmetric one-sided metal-tipped semiconductor nanocrystal dots and rods. Nat Mater 4(11): 855–863
[121] Govorov AO, Zhang W, Skeini T, Richardson H, Lee J, Kotov NA (2006) Gold nanoparticle ensembles as heaters and actuators: melting and collective plasmon resonances. Nanoscale Res Lett 1(1): 84–90
[122] Govorov AO, Richardson HH (2007) Generating heat with metal nanoparticles. Nano Today 2(1): 30–38
[123] Glaser N, Adams DJ, Böker A, Krausch G (2006) Janus particles at liquid–liquid interfaces. Langmuir 22(12): 5227–5229
[124] Cozzoli PD, Pellegrino T, Manna L (2006) Synthesis, properties and perspectives of hybrid nanocrystal structures. Chem Soc Rev 35(11): 1195–1208
[125] Jones SF, Evans GM, Galvin KP (1999) Bubble nucleation from gas cavities – a review. Adv Colloid Interf Sci 80(1): 27–50
[126] Hines MA, Guyot-Sionnest P (1996) Synthesis and characterization of strongly luminescing ZnS-capped CdSe nanocrystals. J Phys Chem 100(2): 468–471
[127] Dabbousi BO, Rodriguez-Viejo J, Mikulec FV, Heine JR, Mattoussi H, Ober R, Jensen KF, Bawendi MG (1997) (CdSe)ZnS core-shell quantum dots: synthesis and characterization of a size series of highly luminescent nanocrystallites. J Phys Chem B 101(46): 9463–9475
[128] Peng XG, Schlamp MC, Kadavanich AV, Alivisatos AP (1997) Epitaxial growth of highly luminescent CdSe/CdS core/shell nanocrystals with photostability and electronic accessibility. J Am Chem Soc 119(30): 7019–7029
[129] Kim S, Fisher B, Eisler HJ, Bawendi M (2003) Type-II quantum dots: CdTe/CdSe(core/shell) and CdSe/{ZnTe}(core/shell) heterostructures. J Am Chem Soc 125(38): 11466–11467
[130] Yu K, Zaman B, Romanova S, Wang DS, Ripmeester JA (2005) Sequential synthesis of type II colloidal CdTe/CdSe core-shell nanocrystals. Small 1(3): 332–338
[131] Kwon KW, Shim M (2005) Gamma-Fe$_2$O$_3$/II–VI sulfide nanocrystal heterojunctions. J Am Chem Soc 127(29): 10269–10275
[132] Selvan ST, Patra PK, Ang CY, Ying JY (2007) Synthesis of silica-coated semiconductor and magnetic quantum dots and their use in the imaging of live cells. Angew Chem Int Ed 46(14): 2448–2452
[133] Li YQ, Zhang G, Nurmikko AV, Sun SH (2005) Enhanced magnetooptical response in dumbbell-like Ag-CoFe$_2$O$_4$ nanoparticle pairs. Nano Lett 5(9): 1689–1692
[134] Pacholski C, Kornowski A, Weller H (2004) Site-specific photodeposition of silver on ZnO nanorods. Angew Chem Int Ed 43(36): 4774–4777
[135] Mokari T, Rothenberg E, Popov I, Costi R, Banin U (2004) Selective growth of metal tips onto semiconductor quantum rods and tetrapods. Science 304: 1787–1790

[136] Kudera S, Carbone L, Casula MF, Cingolani R, Falqui A, Snoeck E, Parak WJ, Manna L (2005) Selective growth of PbSe on one or both tips of colloidal semiconductor nanorods. Nano Lett 5(3): 445–449
[137] Carbone L, Kudera S, Giannini C, Ciccarella G, Cingolani R, Cozzoli PD, Manna L (2006) Selective reactions on the tips of colloidal semiconductor nanorods. J Mater Chem 16(40): 3952–3956
[138] Robinson RD, Sadtler B, Demchenko DO, Erdonmez CK, Wang L-W, Alivisatos AP (2007) Spontaneous superlattice formation in nanorods through partial cation exchange. Science 317 (5836): 355–358
[139] Jana NR, Gearheart L, Murphy CJ (2001) Wet chemical synthesis of high aspect ratio cylindrical gold nanorods. J Phys Chem B 105(19): 4065–4067
[140] Jana NR, Gearheart L, Murphy CJ (2001) Wet chemical synthesis of silver nanorods and nanowires of controllable aspect ratio. Chem Commun (7): 617–618
[141] Talapin DV, Koeppe R, Gotzinger S, Kornowski A, Lupton JM, Rogach AL, Benson O, Feldmann J, Weller H (2003) Highly emissive colloidal CdSe/CdS heterostructures of mixed dimensionality. Nano Lett 3(12): 1677–1681
[142] Habas SE, Lee H, Radmilovic V, Somorjai GA, Yang P (2007) Shaping binary metal nanocrystals through epitaxial seeded growth. Nat Mater 6(9): 692–697
[143] Xie RG, Kolb U, Basche T (2006) Design and synthesis of colloidal nanocrystal heterostructures with tetrapod morphology. Small 2(12): 1454–1457
[144] Talapin DV, Nelson JH, Shevchenko EV, Aloni S, Sadtler B, Alivisatos AP (2007) Seeded growth of highly luminescent CdSe/CdS nanoheterostructures with rod and tetrapod morphologies. Nano Lett 7(10): 2951–2959

Synthesis of semiconductor nanocrystals in organic solvents

By

Peter Reiss

CEA Grenoble DSM/INAC/SPrAM (UMR 5819 CEA-CNRS-Université Joseph Fourier),
Laboratoire d'Electronique Moléculaire, Organique et Hybride, Grenoble, France

1. Introduction

Colloidal semiconductor nanocrystals (NCs) are crystalline particles with diameters ranging typically from 1 to 10 nm, comprising some hundreds to a few thousands of atoms. The inorganic core consisting of the semiconductor material is capped by an organic outer layer of surfactant molecules ("ligands"), which provide sufficient repulsion between the crystals to prevent them from agglomeration. In the nanometer size regime many physical properties of the semiconductor particles change with respect to the bulk material. Examples of this behavior are melting points and charging energies of NCs, which are, to a first approximation, proportional to the reciprocal value of their radii. At the origin of the great interest in NCs was yet another observation, namely the possibility of changing the semiconductor band gap – that is the energy difference between the electron-filled valence band and the empty conduction band – by varying the particle size. In a bulk semiconductor an electron e^- can be excited from the valence to the conduction band by absorption of a photon with an appropriate energy, leaving a hole h^+ in the valence band. Feeling each other's charge, the electron and hole do not move independently from each other because of the Coulomb attraction. The formed e^-–h^+ bound pair is called an exciton and has its lowest energy state slightly below the lower edge of the conduction band. At the same time its wave function is extended over a large region (several lattice spacings), i.e. the exciton radius is large, since the effective masses of the charge carriers are small and the dielectric constant is high [1]. To give examples, the Bohr exciton radii in bulk CdS and CdSe are approximately 3 and 5 nm. Reduction of the particle size to a few nanometers produces the unusual situation that the exciton size can exceed the crystal dimensions. To "fit" into the NC, the charge carriers have to assume higher kinetic energies leading to an increasing band gap and quantization of the energy levels to discrete values. This phenomenon is commonly called *"quantum confinement effect"* [2], and its theoretical treatment is usually based on the quantum mechanical particle-in-a-box model [3]. With decreasing particle size, the energetic structure of the NCs (also termed *quantum dots*) changes from a band-like one to discrete levels. Therefore, in some cases a description by means of molecular orbital theory may be more appropriate, applying the terms HOMO (highest occupied molecular orbital) and LUMO (lowest unoccupied molecular orbital) instead of conduction band and valence band. This ambiguity in terminology reflects the fact that the properties of semiconductor NCs lie in between those of the corresponding bulk material and molecular compounds. The unique optical properties of semiconductor NCs are exploited in a large variety of applications essentially in the fields of biological labeling and (opto-)electronics.

Initiated by pioneering work in the early 1980s [4–7], the research on semiconductor NCs went through a remarkable progress after the development of a novel chemical synthesis method in 1993, which allowed for the preparation of samples with a low size dispersion [8]. The physical properties of NCs, and in particular the optical ones, are strongly size-dependent. To give an example, the linewidth of the photoluminescence (PL) peak is directly related to the size dispersion of the NCs, and thus a narrow size distribution is necessary to obtain a pure emission colour. It is common practice to term samples whose deviation from the mean size is inferior or equal to 5–10% as "monodisperse". The method developed by Murray et al. [8] was the first one to allow for the synthesis of monodisperse cadmium chalcogenide NCs in a size range of 2–12 nm. It relies on the rapid injection of organometallic precursors into a hot organic solvent. Numerous synthesis methods deriving from the original one have been reported in literature in the last 15 years. Nowadays a much better understanding of the influence of the different reaction parameters has been achieved, allowing for the rational design of synthesis protocols.

The goal of this chapter is to give a representative overview of the state of the art concerning the chemical synthesis of semiconductor NCs/quantum dots in organic solvents with a special emphasis on recent developments. Surface functionalization as well as applications of semiconductor NCs go beyond the scope of this chapter. Several reviews on NCs' synthesis have appeared in the literature during the last 5 years [9–12]. In addition, the synthesis of core/shell (CS) structures, comprising a NC of a first semiconductor (*core*), surrounded by an epitaxial, generally 0.5–2-nm thick layer of another semiconductor (*shell*) is discussed. Core/shell systems are intensively studied due to their superior optical properties and higher stability as compared to core NCs. On the other hand, differing synthesis methods, such as the preparation of NCs in aqueous media [13, 14], in microemulsions [15], at the oil-water interface [16], in ionic liquids [17] or in supercritical fluids [18] are not treated here. Concerning the shape control of NCs, the preparation of core/multishell structures, or the synthesis of NCs or heterostructures comprising other types of materials than semiconductors (e.g. oxides or metals), the interested reader is referred to other chapters of this book.

The text is organized as follows:

Succeeding this introduction (Sect. 1), a brief description of the basic optical and structural properties of semiconductor NCs is given (Sect. 2). In addition, the general principles of the synthesis of core and core/shell NCs in organic solvents at elevated temperature are outlined.

In Sect. 3, the current state of the art concerning the synthesis of core NCs of different families of elemental and compound (binary and tertiary) semiconductors is reviewed. Section 4 deals with the preparation of core/shell structures comprising two different semiconductors. At the same time, the consequences of the band alignment, i.e. the relative positions of the electronic energy levels of both constituents, on the optical properties of the NCs are discussed. Finally, it is attempted to sketch some perspectives concerning the development of this domain in the near future (Sect. 5).

2. Basic properties of nanocrystals

2.1 Optical properties of semiconductor nanocrystals

2.1.1 *Absorption.* As it has already been stated in the introduction, absorption of a photon by the NC occurs if its energy exceeds the band gap. Due to quantum confinement, decreasing the particle size results in a hypsochromic (blue-) shift of the absorption onset [19]. A relatively sharp absorption feature near the absorption onset corresponds to the excitonic peak, i.e. the lowest excited state exhibiting a large oscillator strength (cf. Fig. 1). While its position depends on the band gap and, consequently, on the particle size, its form and width is strongly influenced by the

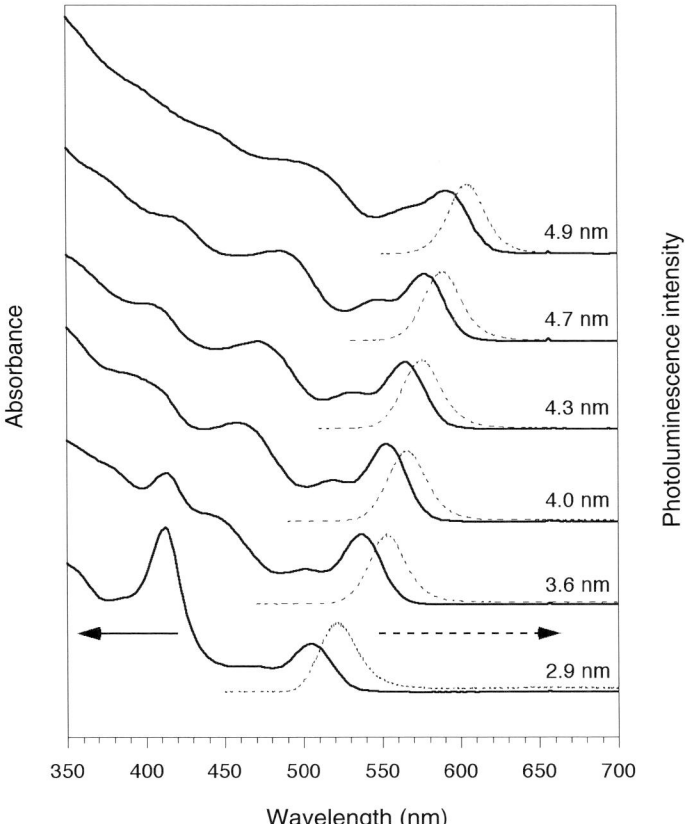

Fig. 1. Absorption and normalized photoluminescence spectra of a size series of CdSe NCs (spectra vertically shifted for clarity) [21]

distribution in size, as well as the form and stoichiometry of the NCs. Therefore polydisperse samples typically exhibit only a shoulder in the absorption spectrum at the position of the excitonic transition. Less pronounced absorption features in the shorter wavelength range correspond to excited states of higher energy [20]. As a rule of thumb, it can be asserted that the larger the number of such spectral features and the more distinctly they are resolved in the absorption spectrum, the smaller is the size dispersion of the sample. In Fig. 1 the absorption and photoluminescence spectra of a series of CdSe NCs differing in size are depicted.

2.1.2 Photoluminescence. Photoluminescence, i.e. the generation of luminescence through excitation by photons, is formally divided into two categories, fluorescence and phosphorescence, depending upon the electronic configuration of the excited state and the emission pathway. Fluorescence is the property of a semiconductor to absorb photons with an energy $h\nu_e$ superior to its band gap, and – after charge carrier relaxation via phonons to the lowest excited state – to emit light of a higher wavelength (lower energy $h\nu_f$) after a brief interval, called the fluorescence lifetime (Fig. 2). The process of phosphorescence occurs in a similar manner, but with a much longer excited state lifetime, due to the symmetry of the state.

The emitted photons have an energy corresponding to the band gap of the NCs and for this reason the emission colour can be tuned by changing the particle size. It should be noted here that efficient room temperature band edge emission is only observed for NCs with proper surface passivation because otherwise charge carriers are very likely to be trapped in surface states, enhancing non-radiative recombination. Due to spectral diffusion and the size distribution of NCs, the room temperature luminescence linewidths of ensembles lie for the best samples of CdSe NCs in the range of 20–25 nm (full width at half maximum, FWHM). As can be seen in Fig. 1, the maxima of the emission peak are red-shifted by ca. 10–20 nm as compared to the excitonic peak in the absorption spectra. This phenomenon is usually referred to as Stokes-shift and has its origin in the particular structure of the exciton energy levels inside the NC. Models using the effective mass approximation show that in bulk wurtzite CdSe the exciton state ($1S_{3/2}1S_e$) is eight-fold degenerate [22]. In CdSe NCs,

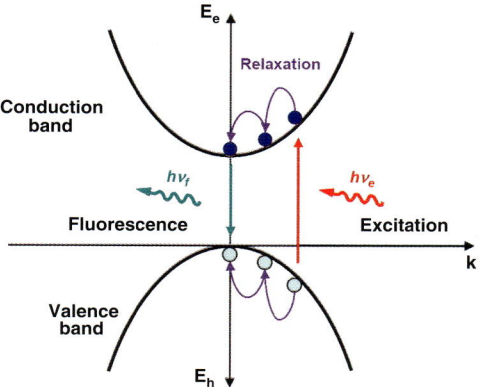

Fig. 2. Fluorescence in a bulk semiconductor

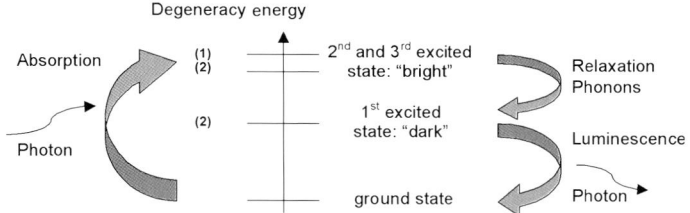

Fig. 3. Schematic representation of the exciton states of CdSe NCs involved in absorption and emission processes

this degeneracy is partially lifted and the band edge state is split into five states, due to the influence of the internal crystal field, effects arising from the non-spherical particle shape and the electron–hole exchange interaction (see Fig. 3). The latter term is strongly enhanced by quantum confinement [23].

Two states, one singlet state and one doublet state, are optically inactive for symmetry reasons. The energetic order of the three remaining states depends on the size and form of the NC. In the case of weak excitation on a given state, absorption depends exclusively on its oscillator strength. As the oscillator strength of the second and third excited ("bright") states is significantly higher than that of the first ("dark") one, excitation by photon absorption occurs to the bright states. On the opposite, photoluminescence depends on the product of oscillator strength and population of the concerned state. Relaxation via acoustic phonon emission from bright states to the dark band edge state causes strong population of the latter and enables radiative recombination (Fig. 3). This model is corroborated by the experimental room temperature values of the Stokes-shift, which are consistent with the energy differences between the related bright and dark states.

2.1.3 Emission of single nanocrystals, blinking phenomenon.

Spectroscopic investigation of single semiconductor NCs revealed that their emission under continuous excitation turns on and off intermittently. This blinking is a common feature also for other nanostructured materials, involving porous Si [24] and epitaxially grown InP quantum dots [25], as well as chromophores at the single molecule level such as polymer segments [26], organic dye molecules [27] and green fluorescent protein (GFP) [28]. However, the origin of the intermittence is completely different for NCs and single dye molecules: in the latter resonant excitation into a single absorbing state takes place. Due to spectral shifting events the excitation is no longer in resonance and a dark period begins. NCs, on the other hand, are excited non-resonantly into a large density of states above the band edge. While their emission statistics and its modelling is the issue of a large number of publications, detailed understanding of the blinking phenomenon has not yet been achieved. The sequence of "on" and "off" periods resembles a random telegraph signal on a time scale varying over several orders of magnitude up to minutes and follows a temporal statistics described by an inverse power law [29]. Transition from an "on" to an "off" state of the NC occurs by photo-ionization, which implies the trapping of a charge carrier in the surrounding matrix (dangling bonds on the surface, solvent, etc.). A single delocalized electron or hole rests in the NC core. Upon further excitation this gives rise to fast (nanosecond order) non-radiative relaxation through Auger processes, i.e. energy transfer from the created exciton to the delocalized charge carrier [30]. Mechanisms for a return to the "on" state are the recapture of the localized charge carrier into the core or the capture of an opposite charge carrier from traps in the proximity. Both pathways can be accompanied by a reorganization of the charge distribution around the NCs. As a consequence the local electric field changes leading to a Stark shift of the photoluminescence peak [31, 32].

2.1.4 *Efficiency of the emission: fluorescence quantum yield.*

The emission efficiency of an ensemble of NCs is expressed in terms of the fluorescence quantum yield (Q.Y.), i.e. the ratio between the number of absorbed photons and the number of emitted photons. As a consequence of the blinking phenomenon (*vide supra*) the theoretical value of 1 is hard to observe because a certain number of NCs are in "off" states. Furthermore the Q.Y. may be additionally reduced as a result of quenching caused by surface trap states. As both of these limiting factors are closely related to the quality of the NC surface, they can be considerably diminished by its improved passivation. This can be achieved by changing the nature of the organic ligands, capping the NCs after their synthesis. To give an example, after substitution of the trioctylphosphine oxide (TOPO) cap on CdSe NCs by hexadecylamine (HDA) or allylamine, an increase of the Q.Y. from about 10% to values of 40–50% has been reported [33]. In this case better surface passivation probably results from an increased capping density of the sterically less-hindered amines as compared to TOPO. Very recently the influence of thiol and amine ligands on the PL properties of CdSe-based NCs has been studied in further details by Munro et al. [34]. Jang et al. obtained cadmium chalcogenide NCs with Q.Y.s up to 75% after treatment with $NaBH_4$ and explained the better surface passivation by the formation of a cadmium oxide layer [35]. In the case of III–V semiconductors, Mićić et al. and Talapin et al. reported an enhancement of the PL Q.Y. of InP NCs from less than 1% to 25–40% upon treatment with HF [36, 37]. Here the improved emission properties have been attributed to the removal of phosphorus dangling bonds under PF_3 elimination from the NC surface.

However, in view of further NC functionalization, it is highly desirable to provide a surface passivation, which is insensitive to subsequent ligand exchange. This is obviously not the case with the described procedures for Q.Y. enhancement. A suitable and widely applied method consists of the growth of an inorganic shell on the surface of the NCs. The resulting core/shell systems will be described in detail in Sect. 2.3.

2.2 Structural properties of nanocrystals.

Most binary octet semiconductors crystallize either in the cubic zinc blende (ZB) or in the hexagonal wurtzite (W) structure, both of which are four-coordinate and vary in the layer stacking along (111), showing an ABCABC or an ABAB sequence, respectively (Fig. 4).

The room temperature ground state structures of selected II–VI, III–V and IV–VI semiconductors are given in Table 1 (Sect. 2.4.1). In cases of relatively low difference in the total energy between the ZB and the W structure (e.g. CdTe, ZnSe), the materials exhibit the so-called W–ZB polytypism [38]. Depending on the

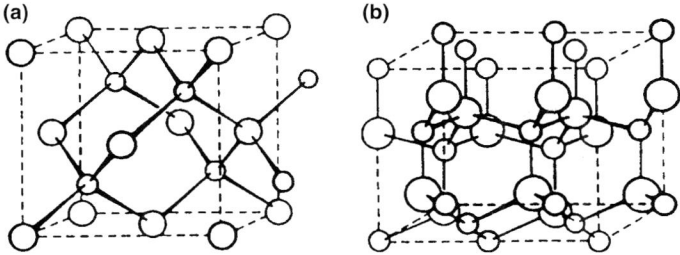

Fig. 4. **a** Zinc blende, **b** wurtzite crystal structure

experimental conditions, nucleation and growth of the NCs can take place in either structure and also the coexistence of both structures in the same nanoparticle is possible. Lead chalcogenide NCs crystallize in the six-coordinate rocksalt structure (cf. Table 1, p 46), and it has been shown that also CdSe NCs can exist in this crystal structure at ambient pressure, provided that their diameter exceeds a threshold size of 11 nm, below which they transform back to the four-coordinate structure [39].

2.3 Core/shell structures. Surface engineering is an important tool to control the properties of the NCs and in particular the optical ones. One important strategy is the overgrowth of NCs with a shell of a second semiconductor, resulting in CS systems. This method has been applied to improve the fluorescence Q.Y. and the stability against photo-oxidation but also, by proper choice of the core and shell materials, to tune the emission wavelength in a large spectral window. After pioneering work in the 1980s and the development of powerful chemical synthesis routes in the end of the 1990s [40–42], a strongly increasing number of articles have been devoted to CS NCs in the last 5 years. Nowadays, almost any type of core NC prepared by a robust chemical synthesis method has been overgrown with shells of other semiconductor materials.

Depending on the band gaps and the relative position of electronic energy levels of the involved semiconductors, the shell can have different functions in CS NCs.

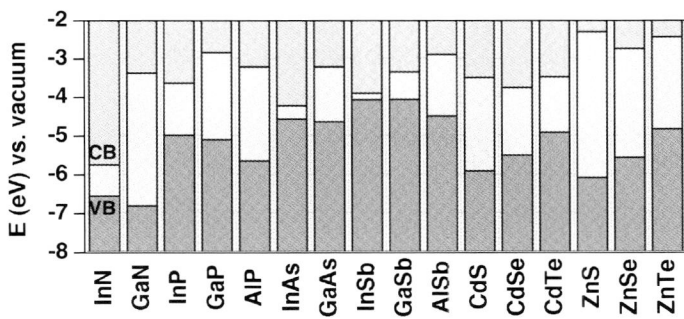

Scheme 1. Electronic energy levels of selected III–V and II–VI semiconductors using the valence band offsets from [43] (VB: valence band, CB: conduction band)

Scheme 1 gives an overview of the band alignment of the bulk materials, which are mostly used in NC synthesis. Two main cases can be distinguished, denominated type I and type II band alignment, respectively. In the former, the band gap of the shell material is larger than that of the core one, and both electrons and holes are confined in the core. In the latter, either the valence band edge or the conduction band edge of the shell material is located in the band gap of the core. The resulting staggered band alignment leads upon excitation of the NC to a spatial separation of the hole and the electron in different regions of the CS structure.

At this point, it is important to stress that the type I/type II distinction must not be mistaken for the difference between direct and indirect semiconductors. In the first case, type I heterostructures experience optical transitions between electron and hole states, whose wavefunctions are localized in the same region in the *real space*, whereas for type II heterostructures, the electron and hole lie in different regions (here, the core and shell of the NCs). In the second case, the distinction between direct and indirect semiconductors concerns transitions between electron and hole states located at the same (direct) or at different (indirect) points in the *reciprocal (wavevector) space*. Indirect optical transitions involve the simultaneous emission or absorption of a phonon, thus having a much lower probability than direct transitions. A typical example of an indirect semiconductor is silicon, consequently exhibiting a very low fluorescence Q.Y. in its bulk form.

In type I CS NCs, the shell is used to "passivate" the surface of the core with the goal to improve its optical properties. The shell separates physically the surface of the optically active core NC from its surrounding medium. As a consequence, the sensitivity of the optical properties to changes in the local environment of the NCs' surface, induced for example by the presence of oxygen or water molecules, is reduced. With respect to core NCs, CS systems exhibit generally enhanced stability against photo-degradation. At the same time, shell growth reduces the number of surface dangling bonds, which can act as trap states for charge carriers and reduce the fluorescence Q.Y. The first published prototype system was CdSe/ZnS [40]. The ZnS shell significantly improves the fluorescence Q.Y. and stability against photo-bleaching. Shell growth is accompanied by a *small* red shift (5–10 nm) of the excitonic peak in the UV–vis absorption spectrum and the PL wavelength. This observation is attributed to a partial leakage of the exciton into the shell material.

In type II systems, shell growth aims at a *significant* red shift of the emission wavelength of the NCs. The staggered band alignment leads to a smaller effective band gap than each one of the constituting core and shell materials. The interest of these systems is the possibility to tune the emission colour with the shell thickness towards spectral ranges, which are difficult to attain with other materials. Type II NCs have been developed in particular for near infrared emission, using for example CdTe/CdSe or CdSe/ZnTe. In contrast to type I systems, the PL decay times are strongly prolonged in type II NCs due to the lower overlap of the electron and hole wavefunctions. As one of the charge carriers (electron or hole) is located in the shell, an overgrowth of type II CS NCs with an outer shell of an appropriate material can be used in the same way as in type I systems to improve the fluorescence Q.Y. and photo-stability.

2.4 Chemical synthesis of semiconductor nanocrystals

2.4.1 *Synthesis methods*. Historically the synthesis in aqueous media was the first successful preparation method of colloidal semiconductor NCs. Therefore, it is briefly described in this paragraph, even though the present chapter is dedicated to the synthesis in organic solvents. The topic is addressed in details in the Chapter of Gaponik and Rogach. Initially developed procedures comprise NC formation in homogenous aqueous solutions containing appropriate reagents and surfactant-type or polymer-type stabilizers [44, 45]. The latter bind to the NC surface and stabilize the particles by steric hinderance and/or electrostatic repulsion in the case of charged stabilizers. In parallel to this monophase synthesis, a bi-phase technique has been developed, which is based on the arrested precipitation of NCs within inverse micelles [5, 7, 46]. Here nanometer-sized water droplets (dispersed phase) are stabilized in an organic solvent (continuous phase) by an amphiphilic surfactant. They serve as nanoreactors for the NC growth and prevent at the same time from particle agglomeration. Both methods provide relatively simple experimental approaches using standard reagents as well as room temperature reactions and were of great importance for the development of NC synthesis. Furthermore, for some materials (e.g. mercury chalcogenides) [47–49] the aqueous synthetic technique is the only successful preparation method reported today. On the other hand, the samples prepared by these synthetic routes usually exhibit size dispersions at least of the order of 15% and therefore fastidious procedures of NCs separation into "sharp" fractions have to be applied in order to obtain monodisperse samples.

The introduction of a high temperature preparation method using organic solvents in 1993 [8] constituted an important step towards the fabrication of monodisperse CdS, CdSe and CdTe NCs. As demonstrated in classical studies by LaMer and Dinegar [50], the synthesis of monodisperse colloids via homogeneous nucleation requires a temporal separation of nucleation and growth of the seeds. The LaMer plot (Fig. 5) is very useful to illustrate the separation of nucleation and growth by means of a nucleation burst.

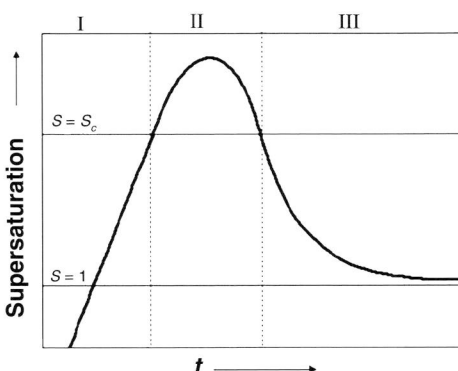

Fig. 5. LaMer plot depicting the degree of supersaturation as a function of reaction time [50]

Initially the concentration of monomers, i.e. the minimum subunits of the crystal, constantly increases by addition from exterior or by in situ generation within the reaction medium. It should be noted that in stage I no nucleation occurs even in supersaturated solution ($S>1$), due to the extremely high energy barrier for spontaneous homogeneous nucleation. The latter is overcome in stage II for a yet higher degree of supersaturation ($S>S_c$), where nucleation and formation of stable nuclei take place. As the rate of monomer consumption induced by the nucleation and growth processes exceeds the rate of monomer supply, the monomer concentration and hence the supersaturation decreases below S_c, the level at which the nucleation rate becomes zero. In the following stage III, the particle growth continues under further monomer consumption as long as the system is in the supersaturated regime.

Experimentally, the separation of nucleation and growth can be achieved by rapid injection of the reagents into the hot solvent, which raises the precursor concentration in the reaction flask above the nucleation threshold (*"hot-injection method"*) [51]. The hot-injection leads to an instantaneous nucleation, which is quickly quenched by the fast cooling of the reaction mixture (the solution to be injected is at room temperature) and by the decreased supersaturation after the nucleation burst. Another possibility relies on attaining the degree of supersaturation necessary for homogeneous nucleation via the in situ formation of reactive species upon supply of thermal energy (*"heating-up method"*) [12]. This method is widely used in the synthesis of metallic nanoparticles, but recently an increasing number of examples of semiconductor NCs prepared by this approach can be found, as will be shown in Sect. 3. In an ideal case all crystallization nuclei are created at the same time and undergo identical growth. During the growth stage it is possible to carry out subsequent injections of precursors in order to increase the mean particle size without deterioration of the narrow size distribution as long as the concentration corresponding to the critical supersaturation S_c is not exceeded. Crystal growth from solution is in many cases followed by a second distinct growth process, which is referred to as Ostwald ripening [52, 53]. It consists of the dissolution of the smallest particles because of their high surface energy and subsequent redeposition of the dissolved matter onto the bigger ones. Thereby the total number of NCs decreases, whereas their mean size increases. As shown in early studies [54], Ostwald ripening can lead to reduced size dispersions of micron-sized colloids. In the case of nanometer-sized particles, however, Ostwald ripening generally yields size dispersions of the order of 15–20% [13], and therefore the reaction should be stopped before this stage.

From a thermodynamic point of view, if diffusion is the rate-limiting step, the size-dependent growth rate can be expressed by means of the *Gibbs–Thompson equation* (Eq. 1) [55]:

$$L(r) = L_{\text{bulk}} \exp(2\gamma V_m / rRT) \qquad (1)$$

$L(r)$ and L_{bulk} are the solubilities of a NC with radius r and of the bulk solid, respectively. g is the specific surface energy, V_m the molar volume of the solid, R the gas constant and T the temperature. The validity of Eq. (1) and the focusing of size distributions during the diffusion-controlled NC growth have been confirmed experimentally [56, 57]. Using the hot-injection or heating-up synthesis methods, it is possible to obtain samples with 5–10% standard deviation from the mean size without post-preparative size fractionation.

One of the main disadvantages of the initially reported preparation methods lies in the fact that pyrophoric organometallic precursors were applied. Their use requires special experimental precautions and their extremely high reactivity restricts the batches to laboratory scale quantities. As a result, in recent years the development of NC high temperature synthesis focused on the replacement of these organometallic precursors by easy to handle standard reagents. To give an example, in the preparation of cadmium chalcogenide NCs, dimethylcadmium has been successfully substituted by cadmium oxide or cadmium salts of weak acids (cadmium acetate, cadmium carbonate) after complexation with long chain phosphonic or carboxylic acids [58, 59]. A further modification of the high temperature methods consisted of the appropriate selection of coordinating and non-coordinating solvents, with the goal to determine their influence on the nucleation and growth kinetics of the NCs and to fine-tune the reactivity of the precursors aiming at obtain narrow size distributions of the order of 5% [60]. More recently, studies have been undertaken by means of NMR spectroscopy, giving important insight into the chemical reactions occurring during the formation of CdSe NCs using the hot-injection method [61]. Concerning other materials, it can be concluded that the synthesis methods initially developed for CdSe NCs have been adapted to the majority of II–VI and a few III–V semiconductors in the last 5 years. The exponentially increasing number of publications proves that this research field is worldwide highly active.

2.4.2 Synthesis of core/shell systems.
A general requirement for the synthesis of CS NCs with satisfactory optical properties is *epitaxial-type shell growth*. Therefore an *appropriate band alignment* is not the sole criterion for materials' choice but, in addition, the core and shell materials should *crystallize in the same structure and exhibit a small lattice mismatch*. In the opposite case, the growth of the shell results in strain and the formation of defect states at the core/shell interface or within the shell. These can act as trap states for photo-generated charge carriers and diminish the fluorescence Q.Y. [62]. Table 1 lists structural parameters of selected semiconductor materials.

Good precursors for shell growth should fulfill the criteria of high reactivity and selectivity (no side reactions). For practical reasons, and in particular if an upscaling of the reaction or industrialization of the production process is aimed, additional properties of the precursors come into play. Pyrophoric and/or highly toxic compounds require special precautions for their manipulation, especially if used in large quantities. To give an example, for the synthesis of zinc sulfide shells on various core NCs, first diethylzinc (pyrophoric) and hexamethyldisilathiane (toxic) have been proposed. Even though widely used in laboratory scale syntheses, these compounds are not very suitable for a large scale production of ZnS overcoated NCs. Further criteria, which have to be taken into account for the choice of the precursors concern the environmental risks related to the use of these compounds and eventually their

Table 1. Structural parameters of selected bulk semiconductors [63, 64]

Material	Structure (300 K)	Type	E_{gap} (eV)	Lattice parameter (Å)	Density (kg/m^3)
ZnS	Zinc blende	II–VI	3.61	5.41	4090
ZnSe	Zinc blende	II–VI	2.69	5.668	5266
ZnTe	Zinc blende	II–VI	2.39	6.104	5636
CdS	Wurtzite	II–VI	2.49	4.136/6.714	4820
CdSe	Wurtzite	II–VI	1.74	4.3/7.01	5810
CdTe	Zinc blende	II–VI	1.43	6.482	5870
GaN	Wurtzite	III–V	3.44	3.188/5.185	6095
GaP	Zinc blende	III–V	2.27 i[a]	5.45	4138
GaAs	Zinc blende	III–V	1.42	5.653	5318
GaSb	Zinc blende	III–V	0.75	6.096	5614
InN	Wurtzite	III–V	0.8	3.545/5.703	6810
InP	Zinc blende	III–V	1.35	5.869	4787
InAs	Zinc blende	III–V	0.35	6.058	5667
InSb	Zinc blende	III–V	0.23	6.479	5774
PbS	Rocksalt	IV–VI	0.41	5.936	7597
PbSe	Rocksalt	IV–VI	0.28	6.117	8260
PbTe	Rocksalt	IV–VI	0.31	6.462	8219

[a] Indirect band gap

degradation products, their price and their commercial availability. As it is rather difficult to satisfy all of these factors, the development of shell synthesis methods is currently an active area of research.

The control of the shell thickness is a delicate point in the fabrication of CS NCs and deserves special attention. If the shell is too thin, the passivation of the core NCs is inefficient resulting in reduced photo-stability. In the opposite case, in turn, the optical properties of the resulting CS NCs deteriorate as, driven by the lattice mismatch of the core and shell materials, defects are created with increasing shell thickness. CS systems are generally fabricated in a two-step procedure, consisting of core NCs' synthesis, followed by a purification step, and the subsequent shell growth reaction, during which a small number of monolayers (typically 1–5) of the shell material are deposited on the cores. The temperature used for the core NC synthesis is generally higher than that used for the shell growth and the shell precursors are slowly added, for example by means of a syringe pump. The major advantages over a so-called one-pot approach without intermediate purification step is the fact that unreacted precursors or side-products can be eliminated before the shell growth. The core NCs are purified by precipation and redispersion cycles, and finally they are redispersed in the solvent used for the shell growth. In order to calculate the required amount of shell precursors to obtain the desired shell thickness, the knowledge of the concentration of the core NCs is indispensable. In the case of cadmium chalcogenide NCs, the concentration of a colloidal solution can be determined in good approximation by means of UV–vis absorption spectroscopy thanks to tabulated relationships between the excitonic peak, the NC size and the molar exctinction coefficient [65]. An advanced approach for shell growth derived from chemical bath deposition techniques and aiming at the precise control of the shell thickness, is the so-called

SILAR (successive ion layer adsorption and reaction) method [66]. It is based on the formation of one monolayer at a time by alternating the injections of cationic and anionic precursors and has firstly been applied for the synthesis of CdSe/CdS CS NCs. Monodispersity of the samples was maintained for CdS shell thicknesses up to five monolayers on 3.5 nm core CdSe NCs, as reflected by the narrow PL linewidths obtained in the range of 23–26 nm (FWHM).

3. Chemical synthesis of core nanocrystals in organic solvents

3.1 II–VI semiconductor nanocrystals. Table 2 gives an overview of the combinations of precursors, stabilizers and solvents mainly used in the synthesis of II–VI semiconductor NCs. As already stated in Sect. 2, the synthesis of NCs in organic solvents can be divided into two principal groups, the *hot-injection method* and the *heating-up method*. Due to the less restrictive experimental requirements, the latter has significant advantages if the large scale production of monodisperse NCs is aimed. Therefore, a gain of importance of this method is observed within the last years.

3.1.1 *Binary systems.* The pioneering work of Murray et al. [8] on the synthesis of monodisperse CdSe, CdS, CdTe NCs via the hot-injection technique experienced a number of modifications since its publication in 1993. The latter can be classified into two groups, the first one aiming at the further improvement of the particles' size and shape control, and the second one at the simplification of the experimental protocol. The addition of HDA to the TOPO/trioctylphosphine(TOP) mixed solvent proposed by Talapin et al. [33] can be counted in the first group and led to an unprecedented low size distribution of the as-prepared CdSe NCs of the order of 5%. Concerning the second group, several contributions of Peng's group in the last 7 years have to be outlined. In a first time, the replacement of the pyrophoric Cd precursor dimethylcadmium by much easier to handle compounds such as cadmium oxide, cadmium acetate or cadmium nitrate was proposed [58, 59]. The latter are transformed by phosphonic or fatty acids into reactive species prior to the injection of the Se precursor. The next step concerned the substitution of the coordinating solvent TOPO by the non-coordinating one 1-octadecene (ODE) [60]. This solvent offers a number of practical advantages (liquid at room temperature, lower price and more environmentally benign than TOPO). Moreover, the use of ODE allows for a better fine-tuning of the reactivity of the Cd precursor, as the solvent does not have the additional function of being the stabilizing ligand. Fatty acids accomplish this role and it has been demonstrated that the mean size and size dispersion depend on the length of the carbonaceous chain. While also applicable to CdSe and CdTe NCs, this approach was the first one to yield monodisperse CdS NCs without additional size sorting procedures. Very recently, the use of different alkylamines as stabilizing ligands has been studied in detail, which allowed for the decrease of the reaction temperature to 150°C, 100–200°C lower than in previously reported procedures [91]. The last ingredient of the original CdSe synthesis to be replaced was the Se precursor TOP-Se. Motivated by the fact that trialkylphosphines such as TOP or TBP

Table 2. Precursors, stabilizers and solvents used in the synthesis of various II–VI semiconductor NCs

Material	Precursors and stabilizers	Solvent(s)	Method[a]	References
CdS, CdSe, CdTe	CdMe$_2$/TOP, (TMS)$_2$Se or (TMS)$_2$S or (BDMS)$_2$Te	TOPO	HI	[8]
CdSe	CdMe$_2$/TOP, TOP-Se	TOPO, HDA	HI	[33, 67]
CdSe, CdTe	CdO, TDPA, TOP-Se or TOP-Te	TOPO	HI	[58]
CdSe	CdO or Cd(ac)$_2$ or CdCO$_3$, TOP-Se, TDPA or SA or LA	TOPO	HI	[59]
CdS, CdSe	CdO, S/ODE or TBP-Se/ODE, OA	ODE	HI	[60]
CdSe	Cd(st)$_2$, TOP-Se, HH or BP	HDA, octadecane	HI	[68]
CdSe	Cd(my)$_2$, Se, OA/ODE	ODE	HU	[69]
CdSe	CdO, Se/ODE, OA	ODE	HI	[70]
CdSe	CdO, Se, OA	Olive oil	HI	[71]
CdSe	Cd(st)$_2$, TBP-Se, SA, DDA	ODE	HI	[72]
CdS	Cd(ac)$_2$, S, MA	ODE	HU	[73]
CdS, ZnS	Cd(hdx)$_2$ or Cd(ex)$_2$ or Cd(dx)$_2$ or Zn(hdx)$_2$	HDA	HU	[74, 75]
CdS, ZnS	CdCl$_2$/OAm or ZnCl$_2$/OAm/TOPO, S/OAm	OAm	HU	[76]
CdTe	CdMe$_2$, TOP-Te	DDA	HI	[77]
CdTe	CdO, TBP-Te/ODE or TOP-Te/ODE, OA	ODE	HI	[78, 79]
CdTe	CdO, TBP-Te, ODPA	ODE	HU	[69]
ZnS	Zn(st)$_2$, S/ODE	ODE, Tetracosane	HI	[80]
ZnS	ZnEt$_2$, S	HDA/ODE	HU	[81]
ZnSe	ZnEt$_2$, TOP-Se	HDA	HI	[82]
ZnSe	Zn(st)$_2$, TOP-Se	Octadecane	HI	[83]
ZnTe	Te and ZnEt$_2$ in TOP	ODA, ODE	HI	[84]
HgTe	HgBr$_2$, TOP-Te	TOPO	HI	[85]
Cd$_{1-x}$Zn$_x$Se	ZnEt$_2$/TOP, CdMe$_2$/TOP	TOPO, HDA	HI	[86, 87]
Cd$_{1-x}$Zn$_x$Se	Zn(st)$_2$, Cd(st)$_2$, TOP-Se	ODE	HI	[88]
Cd$_{1-x}$Zn$_x$S	CdO, ZnO, S/ODE, OA	ODE	HI	[89]
CdSe$_{1-x}$Te$_x$	CdO, TOP-Se, TOP-Te	TOPO, HDA	HI	[90]

[a] *HI* Hot-injection method; *HU* Heating-up method
CdMe$_2$ dimethylcadmium; *ZnEt$_2$* diethylzinc; *TMS* trimethylsilyl; *(BDMS)$_2$Te* bis(*tert*-butyldimethylsilyl) telluride; *TDPA* tetradecylphosphonic acid; *ODPA* octadecylphosphonic acid; *SA* stearic acid; *LA* lauric acid; *OA* oleic acid; *MA* myristic acid; *ac* acetate; *my* myristate; *st* stearate; *hdx* hexadecylxanthate; *ex* ethylxanthate; *dx* decylxanthate; *TOPO* trioctylphosphine oxide; *HAD* hexadecylamine; *DDA* dodecylamine; *ODA* octadecylamine; *TOP* trioctylphosphine; *TBP* tributylphosphine; *ODE* 1-octadecene; *HH* hexadecyl hexadecanoate; *BP* benzophenone

are air- and moisture-sensitive and relatively expensive compounds, the groups of Cao and of Mulvaney proposed phosphine-free synthesis methods for CdSe NCs in 2005 [69, 70]. In the first case, elemental selenium was used as the precursor in a one-pot reaction. It becomes soluble in ODE above 190°C and starts to react with the Cd precursor cadmium myristate at higher temperatures. This procedure does not require the injection of precursors into the hot reaction medium and is therefore a typical example of the heating-up method. Remarkably, the as-prepared NCs exhibit a size distribution of the order of 5% and zinc blende crystal structure. The

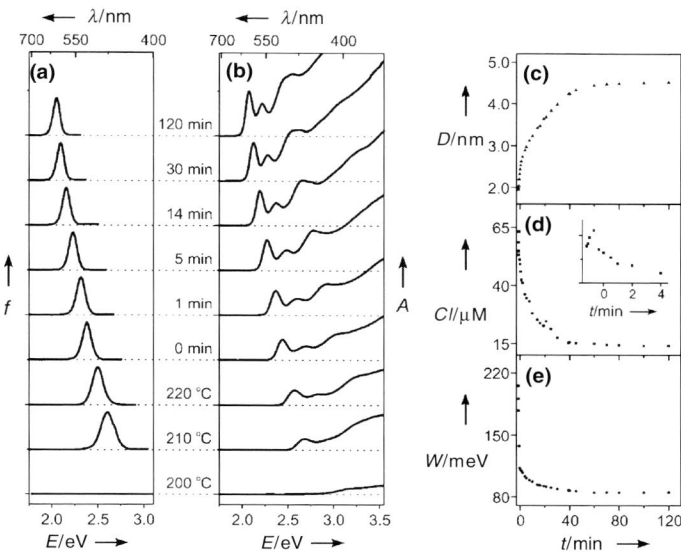

Fig. 6. Temporal evolution of **a** the fluorescence (f) spectrum, **b** the absorption (A) spectrum, **c** the diameter (D) of the nanocrystals, **d** (and inset) the concentration (C) of the nanocrystals, **e** FWHM (W) of the fluorescence spectrum during the phosphine-free synthesis of CdSe NCs using the heating-up method. Reproduced with permission from [69]. © 2005, Wiley–VCH Verlag GmbH & Co. KGaA

evolution of the optical properties of the CdSe NCs during synthesis is shown in Fig. 6.

In the protocol of Mulvaney et al. [70], Se powder is dissolved in ODE by 2 h heating at 200°C under inert atmosphere. After cooling to room temperature, this mixture is injected to the cadmium precursor containing solution at 300°C. Another recent modification of the CdSe synthesis was proposed by Sapra et al. who used olive oil as the reaction medium in combination with CdO and elemental Se [71]. As can be seen from these ingredients, this method is most probably the ultimate simplification of the original protocol. The possibility to dissolve CdO in olive oil comes from the fact that oleic acid (OA) is one of its natural components. The authors used additional OA to synthesize larger-sized NCs and the obtained size range was 2.3–6.0 nm. CdTe NCs with a size of 2.5–7.0 nm have been synthesized in dodecylamine (DDA) using the classical approach based on $CdMe_2$ and TOP-Te [77], before protocols using non-coordinating solvents came up [69, 78, 79].

The synthesis of II-sulfide NCs (CdS, ZnS) differs somehow from those of the selenides and tellurides in terms of the applied chalcogenide sources. Elemental yellow sulfur occurs in discrete S_8 molecules, while its heavier homologues form Se_x and Te_x rings and chains, which are more difficult to solubilize. Therefore, in most syntheses, elemental sulfur, dissolved for example in ODE or oleylamine, is used as the S precursor. An illustrative example is the synthesis of a series of transition metal sulfide NCs comprising CdS and ZnS described by Joo et al. (cf. Fig. 7) [76]. At the same time, a number of appropriate monomolecular precursors, containing both the metal and the S source, are commercially available or easy to prepare. In particular

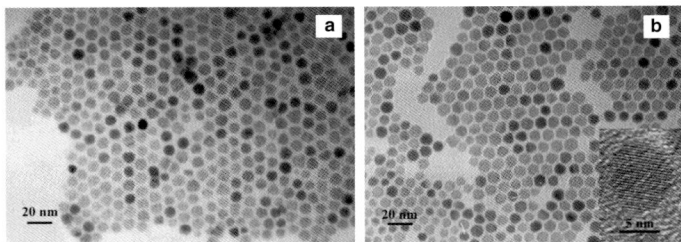

Fig. 7. TEM images of ZnS NCs synthesized via the heating-up method: **a** before size-selection process; **b** after size-selective precipitation. Inset: HRTEM image of a single ZnS NC. Reprinted with permission from [76]. © 2003 American Chemical Society

Zn or Cd xanthates and dithiocarbamates have to be mentioned in this context: most of these compounds decompose at temperatures below 200°C and are therefore suitable precursors for the preparation of transition metal sulfide NCs. Efrima and coworkers reported the use of Zn- and Cd-alkylxanthates for the synthesis of the corresponding sulfide NCs [74, 75].

Concerning zinc selenide, a promising compound for emission in the near UV/blue range, Hines and Guyot-Sionnest first reported the synthesis of ZnSe NCs showing strong band edge fluorescence in the range from 365 to 445 nm with PL Q.Y.s of 20–50% [82]. Adapting the work on CdSe NCs by Murray et al. [8] they used an organometallic zinc precursor (diethylzinc) and selenium powder dissolved in TOP (TOPSe). The major difference, with respect to the case of CdSe was the use of HDA as a solvent. TOPO turned out to bind too strongly to Zn and therefore an appropriate balance between the nucleation and the growth crucial for good size control could not be achieved. At the same time, the combination of HDA and TOP efficiently passivated the NCs' surface by removing trap states/dangling bonds, which resulted in pure band edge fluorescence.

A very simple method for the preparation of a size series of monodisperse ZnSe NCs was reported in 2004 by our group [83]. Here, pyrophoric diethylzinc was replaced by an air-stable and easy to manipulate precursor (zinc stearate), while keeping the established Se precursor (TOPSe) and using the saturated non-coordinating solvent octadecane. In the growth process zinc stearate served not only as the cation precursor but also as a source of stabilizing stearate ligands. The size of the obtained ZnSe could be varied from 3 to 7 nm by adjusting the concentration of the zinc precursor in the reaction mixture and/or varying the temperature, yielding samples emitting in the range of 390–440 nm (FWHM 12–17 nm).

Li et al. reported a similar route implying the use of a mixture of 1-octadecene and tetracosane combined with the addition of a small amount of octadecylamine (ODA) to the reaction mixture [80]. According to these authors, ODA was necessary to activate the zinc carboxylate precursor, otherwise no NCs with low size distributions could be obtained. Finally, Chen et al. used a shorter chain zinc carboxylate, namely zinc laurate generated in situ from ZnO and lauric acid in HDA. Size control of the obtained ZnSe NCs was achieved in a range of 2.5–6 nm by varying the reaction time [92]. ZnTe has been synthesized by injection of a solution containing $ZnEt_2$ and elemental Te in TOP into a mixture of ODA and ODE at 270°C [84]. Concerning the mercury chalcogenides, HgTe is the only example to be prepared in organic solvents to date via the reaction of $HgBr_2$ and TOP-Te in a mixture of TOPO and ODA [85].

3.1.2 Ternary systems ("alloys").
As aforementioned, by taking advantage of the quantum confinement effect it is possible to tune, through size control, the fluorescence wavelength of semiconductor NCs. The formation of a ternary structure is an alternative way to influence the band gap of the NCs, not by changing their *size* but their *composition*. Although the term "alloy" is, in its strict sense, limited to solid solutions of two or more metals, it is also widely applied in literature for the description of ternary systems comprising chalcogenide ions. A typical example is $Cd_{1-x}Zn_xSe$, corresponding to ZnSe NCs, in which a fraction of the Zn atoms is substituted by Cd ones in the crystal lattice. The band gap of the resulting ternary alloy is in between those of pure ZnSe and of pure CdSe NCs of the same size. In contrast to the crystal lattice parameters, which show, according to Végard's law, a linear evolution with composition, the curve describing the corresponding evolution of the band gap shows a deviation from linear behavior. However, this difference is not very pronounced for common anion systems such as $Cd_{1-x}Zn_xSe$ or $Cd_{1-x}Zn_xS$ [88, 89]. A measure of this deviation is the so-called bowing parameter, which depends on the difference in electronegativity of the two end components, here CdSe and ZnSe [93, 94]. Consequently, common cation alloys ($CdS_{1-x}Se_x$, $CdSe_{1-x}Te_x$) generally present larger bowing parameters than common anion ones. Different methods of the $Cd_{1-x}Zn_xSe$ NCs synthesis were developed, deriving directly from those used for the binary compounds. While the Se precursor was generally TOPSe, the use of different Cd and Zn precursors has been explored. Organometallic approaches based on dimethylcadmium and diethylzinc were first applied [86, 87] before the recent development of a method in which only air-stable precursors (cadmium stearate and zinc stearate) were used [88]. In the case of $CdSe_{1-x}S_x$ NCs, the influence of the nature of the solvent on the sample properties was recently studied by Al-Salim et al. [95]. The narrowing of the band gap of CdSe NCs by forming a solid solution with CdTe, further enhanced by the bowing effect, made the near IR emission range accessible [90].

3.1.3 Doped nanocrystals.
Doping – the introduction of a small amount of "impurities" into the crystal lattice – is an attractive way to change the NCs' physical properties. An important example is the doping of II–VI semiconductors with paramagnetic Mn^{2+} ions ($S = ^5/_2$), yielding materials denominated *dilute magnetic semiconductors* (DMS), which exhibit interesting magnetic and magneto-optical properties [96]. At the same time, the host NC can act as an antenna for the absorption of energy (e.g. light) and excitation of the dopant ions via energy transfer. In this case, mostly UV-absorbing NCs are chosen as the hosts, such as ZnS or ZnSe. Mn-doped ZnSe is an instructive example for the development of doped II–VI semiconductor NCs. Bulk ZnSe:Mn exhibits PL at 582 nm (2.13 eV), commonly assigned to an optically forbidden *d–d* transition of Mn^{2+} ($^4T_1$ to $^6A_1$) [97, 98]. This emission is

sensitive to the crystal field splitting being itself dependent on the local chemical environment.

A general problem encountered in essentially all attempts of NCs doping is the fact that the host matrix tends to expel the dopant ions to the surface, in some sort of "self-purification" process. Therefore, even in the favourable case of dopant ions, having the same valence state and similar ionic radius as the corresponding host ions, successful (volume) doping is difficult to achieve in a straightforward approach by simply adding a small amount of dopant precursor during the synthesis of the host NCs. Nevertheless, by adding dimethylmanganese [99] or manganese cyclohexanebutyrate [100] to the zinc precursor in the organometallic ZnSe synthesis [82], successful Mn-doping has been achieved, even though some residual blue emission from the ZnSe host matrix at low dopant concentrations indicated the co-existence of both doped and undoped NCs. Similarly, the injection of TBPSe into a solution of Zn- and Co-acetate in a mixture of ODE, HDA and oleic acid at 310°C led to the formation of Co^{2+}-doped ZnSe NCs [101]. However, the obtained DMS NCs contained no dopant ions in the central cores and therefore exhibited excitonic Zeeman splitting energies significantly (40%) smaller than expected from the bulk ZnSe:Co data. The same group reported recently the Co^{2+} and Mn^{2+} doping of CdSe, CdS and CdSe/CdS NCs using well-defined Cd-chalcogenide clusters [102] as starting materials [103].

A decisive step towards the understanding of the doping process was achieved by Erwin et al. who first succeeded in doping CdSe NCs with Mn^{2+} [104]. They introduced a model of doping based on kinetics and concluded that the doping mechanism is controlled by the initial adsorption of impurities on the surface of growing NCs. Only impurities remaining adsorbed on the surface for a time comparable to the reciprocal growth rate are incorporated into the NC. Three main factors influencing this residence time were determined, namely the surface morphology, NC shape and surfactants present in the growth solution. It has been shown that (0 0 1) surfaces of ZB crystals exhibit much higher impurity binding energies than the other two ZB orientations and than any facet of crystals with W or rock-salt (RS) structures. These findings were fully corroborated by the state-of-the-art, as all NCs successfully doped with Mn^{2+} ions exhibited the ZB crystal structure.

A new approach with the goal to achieve the doping of *all* NCs in a given sample was explored by Peng and coworkers. In the so-called *nucleation-doping* strategy, MnSe nuclei, formed from manganese stearate and TBPSe in octadecylamine at 280°C, were overcoated with ZnSe using zinc stearate or zinc undecylenate. No residual ZnSe emission was observed and the doped NCs exhibited thermally stable (up to 300°C) highly efficient (Q.Y. 40–70%) PL in a spectral window of 545–610 nm, depending on the ZnSe shell thickness and on the nature of the surface ligands (charged or neutral) [105]. The same approach was extended to the doping of ZnSe with Cu ions (Fig. 8) [106]. In conclusion, with exception of the comparably broad PL peaks (>50 nm at FWHM), their otherwise very interesting optical properties make transition metal-doped ZnSe NCs promising "green" alternatives to the widely studied II–VI semiconductor NCs for a number of applications including biological labeling [107].

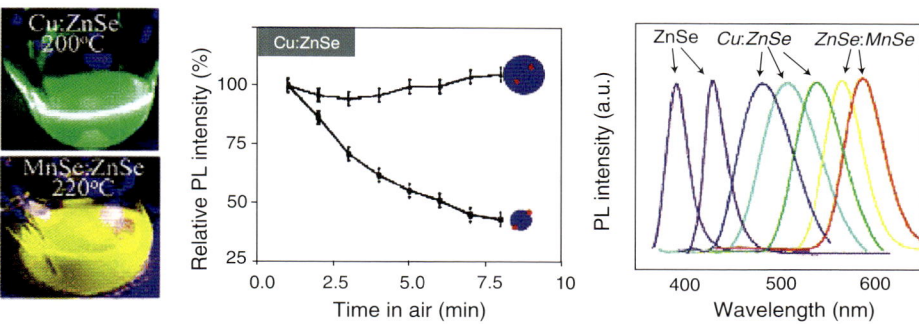

Fig. 8. Photoluminescence of Cu- and Mn-doped ZnSe NCs at high temperature (left); stability of Cu-doped ZnSe NCs in air (middle); PL spectra of ZnSe-based doped NCs. Reprinted with permission from [105]. © 2005 American Chemical Society

3.2 III–V semiconductor nanocrystals. Compared to most of the II–VI and IV–VI NCs, III–V compounds are generally referred to as "greener" NCs because the group III elements such as In or Ga are more environmentally friendly than Cd, Pb or Hg. Nevertheless, the studies and applications of III–V NCs are rather sparse as compared to their II–VI analogues, principally caused by significant difficulties in their synthesis. In fact, due to the stronger covalent bonding of the precursors generally a higher reaction temperature and a longer reaction time are necessary. These conditions favourize Ostwald ripening, leading to an increased size dispersion. Therefore, highly reactive organometallic precursors or monomolecular precursors, containing both the cation and anion already chemically bound in the same molecule, are applied in a large number of protocols.

The following section focuses on recent developments in the synthesis of monodisperse III–V semiconductor NCs. For a more detailed description of the methods published prior to 2002 the interested reader is referred to the reviews of Green [108] and Wells and Gladfelter [109]. An overview of the different synthetic procedures is given by means of Table 3.

Table 3. Precursors, stabilizers and solvents used in the synthesis of various III–V semiconductor NCs

Material	Precursors and stabilizers	Solvent(s)	References
InP	$InCl_3$ or $InCl_3/Na_2C_2O_4$, $P(TMS)_3$	TOPO or TOPO/TOP	[110, 111]
InP, InAs	$In(ac)_3$, $P(TMS)_3$ or $As(TMS)_3$, MA	ODE	[112]
InP	$InMe_3$, $P(TMS)_3$, MA	MM or DBS,	[113]
InAs	$InCl_3$, $As(TMS)_3$	TOP	[114]
GaP	$[Cl_2GaP(SiMe_3)_2]_2$	TOPO/TOP	[115]
GaP	$Ga(PtBu_2)_3$	TOA, HDA	[116]
GaP	$GaCl_3$, $P(TMS)_3$	TOPO	[117]
GaN, AlN, InN	$[M(H_2NCONH_2)_6]Cl_3$ (M=Ga, Al, In)	TOA	[118]

TMS trimethylsilyl; *ac* acetate; *MA* myristic acid; *OA* oleic acid; *TOPO* trioctylphosphine oxide; *TOP* trioctylphosphine; *ODE* 1-octadecene; *MM* methyl myristate; *DBS* dibutyl sebacate; *TOA* trioctylamine; *HDA* hexadecylamine

Most of the reports concern the synthesis of indium phosphide NCs. These are potentially an attractive alternative to CdSe or CdTe ones, due to their size-dependent emission in the visible and near infrared spectral range combined with the lower toxicity of indium with respect to cadmium. In initial synthetic routes [110, 111], the method established for cadmium chalcogenide NCs [8] was adapted to InP, but longer reaction times (3–7 days) were necessary to yield particles of good crystallinity. Interestingly, these approaches follow the heating-up method. Peng and coworkers later reported a new protocol, which is based on fatty acids as stabilizers in combination with the non-coordinating solvent ODE instead of TOPO/TOP [112]. The use of this medium provided a fast and controllable reaction, yielding high quality InP NCs. Similar results were obtained when organometallic In precursors were used in combination with ester type solvents [113]. However, to date the as-prepared NCs exhibit in all cases size dispersions exceeding 10% and the use of the expensive and pyrophoric phosphorus precursor $P(TMS)_3$ (tris(trimethylsilyl)phosphine) is mandatory. NCs of the narrow band gap semiconductor InAs can be synthesized using similar approaches as in the case of InP. Both the synthesis in coordinating (TOP) [114] and non-coordinating (ODE) [112] solvent have been reported.

The synthesis of GaP NCs was reported by Mićić et al. who decomposed in a TOPO/TOP mixture a monomolecular precursor complex, $[Cl_2GaP(SiMe_3)_2]_2$, in situ generated from $GaCl_3$ and $P(TMS)_3$ in toluene [115]. Monodispersed 8 nm GaP NCs have been synthesized from the monomolecular precursor $Ga(PtBu_2)_3$ in a mixture of trioctylamine (TOA) and HDA [116]. Depending on the concentration of HDA in the reaction medium, a shape transition from spherical to rod-like NCs has been observed. Green and O'Brien used the same monomolecular precursor in 4-ethylpyridine [119]. Furis et al. adapted the II–VI NCs hot-injection method, using $GaCl_3$, $P(TMS)_3$ as the gallium and phosphorus source, respectively, and TOPO as the solvent [117].

III-nitride NCs are extremely difficult to synthesize due to the absence of appropriate (i.e. highly reactive) nitrogen precursors. Further restrictions are the high growth temperature in the case of AlN and the low decomposition temperature of InN. Recent advances in this field have been achieved by Rao and coworkers [118]. They prepared AlN, InN and in particular GaN NCs of low size dispersion by the thermal decomposition of the metal–urea complexes in refluxing TOA under N_2 atmosphere.

3.3 IV–VI semiconductor nanocrystals.
The IV–VI semiconductor family comprises materials of high interest for applications relying on emission in the near infrared spectral range. While only very sparse information exists concerning tin chalcogenides, their lead homologues have been intensively studied in the last years. The latter are narrow band gap semiconductors as can be seen from Table 1 (Sect. 2.4.2) and exhibit some other unique properties as compared to II–VI or III–V compounds. In particular, they show high dielectric constants, large Bohr exciton radii and the electron and hole masses are approximately equal. Two recent reviews deal with the synthesis and properties of infrared-emitting NCs, and in particular with lead chalcogenides [120, 121]. While initially significant efforts were

Fig. 9. TEM image of as-prepared PbSe NCs synthesized in the non-coordinating solvent ODE. The mean size is 6.8 nm and the size distribution is 6.2%. Reprinted with permission from [124]. © 2004, American Chemical Society

made to synthesize them in aqueous media, today it seems clear that the high temperature methods in organic solvents yields superior results in terms of monodispersity of the prepared NCs. In all cases the hot-injection method has been applied, and the first example was the synthesis of PbSe by Murray et al. [122]. Already successfully used in the synthesis of metal nanoparticles (e.g. Co [122], FePt [123]), a high boiling point ether (diphenylether) was applied as the solvent in combination with lead oleate (prepared in situ from lead acetate and oleic acid), while maintaining the traditional Se precursor TOPSe. The size of the NCs could be tuned within a large range (3.5–15 nm) and the monodisperse fractions obtained after size-selective precipitation exhibited size-dependent absorption spectra. The latter comprised in addition to the excitonic peak, located at 1200–2200 nm depending on the NC size, several well-defined features at higher energies. Similar as in the case of II–VI compounds, the synthetic scheme was later modified by using the non-coordinating solvent ODE [124]. Figure 9 shows a TEM image of the PbSe NCs obtained with this method. They exhibit a narrow size dispersion (5–7%) without the necessity of fractionation procedures such as size-selective precipitation.

In 2003, the synthesis of PbS NCs by the hot-injection method was first published by Hines and Scholes [125]. In this case the non-coordinating solvent ODE was applied in combination with OA as the stabilizer and PbO and $(TMS)_2S$ as the Pb and S precursors, respectively. The pronounced excitonic peak visible in the presented absorption spectra spans a wavelength range from 800 to 1800 nm as a function of the NC size and narrow emission linewidths have been reported. In the meantime, a number of derived synthetic protocols have been published. Table 4 gives an overview of the explored experimental parameters. Concerning PbTe, essentially the same procedures as those developed for PbSe have been successfully adapted to yield NCs with narrow size distributions of 5–7% [126, 127]. Depending on the reaction parameters, instead of spherical particles a variety of different shapes can be

Table 4. Precursors, stabilizers and solvents used in the synthesis of various IV–VI semiconductor NCs

Material	Precursors and stabilizers	Solvent	References
PbSe	Pb(ac)$_2$, OA, TOP-Se	DPE	[122, 128–132]
PbSe	Pb(chbt)$_2$, TBP-Se	TOPO	[133]
PbSe	PbO, OA, TOP-Se	ODE	[124]
PbS	PbO, OA, (TMS)$_2$S	ODE	[125]
PbS	PbCl$_2$, S/OAm	OAm	[76]
PbTe	Pb(ac)$_2$, OA, TOP-Te	DPE	[126]
PbTe	PbO, OA, TOP-Te	ODE	[127]

TMS trimethylsilyl; *ac* acetate; *chbt* cyclohexylbutyrate; *OA* oleic acid; *OAm* Oleylamine; *DPE* diphenylether

obtained, such as cubes or stars. In this context, the paper of Houtepen et al. has to be cited, which revealed the crucial role of the concentration of acetate in the reaction medium on the NC shape [128].

3.4 Nanocrystals of other semiconductors.

Apart from III–V QDs, ternary semiconductor NCs such as I–III–VI$_2$ type chalcopyrites (CuInSe$_2$-CISe, CuInS$_2$-CIS) represent further potential alternative materials to cadmium-based systems. They are direct semiconductors and exhibit a relatively low band gap (1.05 eV for CISe, 1.5 eV for CIS). CIS and CISe NCs were intensively studied because of their high potential for use in photovoltaics [134–136]. To the contrast, their PL properties were rarely investigated in previous reports [137, 138]. Recently Castro et al. reported a new synthesis method for CIS via the decomposition of the single source precursor (PPh$_3$)$_2$CuIn(SEt)$_4$, yielding luminescent CIS samples with a PL Q.Y. of ca. 5% [139, 140]. The Hyeon group prepared large-sized CIS NCs of anisotropic shape from Cu- and In-oleate in a mixture of oleylamine and dodecanethiol via the heating-up method [141]. Nakamura et al. doped CIS NCs with Zn and were able to vary their PL wavelength from 570 to 800 nm, with Q.Y.s in the range of 5% [142]. Increased fluorescence Q.Y. upon addition of Zn has also recently been observed in another example of a I–III–VI$_2$ semiconductor: a solid solution of ZnS and AgInS$_2$ exhibited strong, tunable emission in the visible range [143]. An early approach for the synthesis of CISe comprised the use of copper(I) and indium chloride in TOPO, and the injection of TOP-Se [144]. More recently, CISe NCs have been prepared in the non-coordinating solvent ODE [145].

The synthesis of the elemental semiconductors Si and Ge is by far less-developed than that of the other semiconductor families discussed. This fact stands out against their technological importance. An inspection of the relevant literature seems to indicate that the synthesis in supercritical solvents [18, 146–149] or the use of the microemulsion technique [150, 151] are more appropriate than the preparation of these materials in organic solvents using the hot-injection or the heating-up method. An exception from this rule is the approach of Kauzlarich and coworkers, who used the metathesis reaction of the Zintl salts NaGe and KGe or of Mg$_2$Ge with excess GeCl$_4$ in refluxing glyme (ethylene glycol dimethyl ether), diglyme or triglyme for the preparation of Ge NCs [152]. The same group applied a similar procedure comprising Mg_2Si, $SiCl_4$ and glyme for the synthesis of Si NCs [153]. In an earlier report, Heath et al. prepared Ge NCs via the reduction of chlorogermanes and organochlorogermanes by a K/Na alloy in heptane, followed by thermal annealing in an autoclave at 270°C [154].

4. Core/shell systems

The following section is dedicated to the description of the synthesis of CS NCs. Core multiple shell structures such as core/shell/shell NCs or quantum-dot-quantum-well onion-like systems are considered in a separate Chapter of Dorfs and Eychmüller.

4.1 Type I systems. As already mentioned in Sect. 2, type I systems are generally synthesized with the goal to increase the fluorescence Q.Y. and stability against photo-bleaching by improving NCs' surface passivation. Unlike otherwise stated, the shell precursors are slowly injected to a dispersion of the purified core NCs.

4.1.1 Synthesis of core/shell nanocrystals of II–VI semiconductors.

One of the earliest CS structures reported was CdSe/ZnS, which is at the same time the most intensively studied system to date. Its synthesis was first described by Hines and Guyot-Sionnest who overcoated 3 nm CdSe NCs with 1–2 monolayers of ZnS, resulting in a Q.Y. of 50% [40]. ZnS shell growth has been achieved by the injection of a mixture of the organometallic precursors diethylzinc and hexamethyldisilathiane, also known as bis(trimethylsilyl)sulfide, $S(TMS_2)$ at high temperature (300°C). A whole size series of CdSe/ZnS NCs and their in depth characterization was published shortly afterwards by Bawendi's group [41]. A similar approach has later been applied for the ZnS capping of CdSe nanorods with lengths up to 30 nm [155]. The addition of HDA to the traditionally used solvent system TOPO/TOP led to a better control of the growth kinetics during both the CdSe core and ZnS shell synthesis resulting in a lower size distribution and Q.Y.s of the order of 60% (cf. Fig. 10) [33]. Very recently extremely small CdSe/ZnS CS NCs have been synthesized by Kudera et al., making the blue spectral region accessible with this system [156]. The synthetic approach was based on the sequential growth of CdSe magic size clusters in a mixture of trioctylphosphine, dodecylamine, and nonanoic acid at temperatures of 80°C and their subsequent overcoating with ZnS. Another procedure to access this spectral region was suggested by Jun and Jang [157]: by carrying out the ZnS overgrowth process at 300°C using precursors (zinc acetate and octanethiol) of relatively low reactivity, shell material diffused into the core resulting in a significant hypsochromic shift of the emission wavelength. The obtained NCs emit at 470 nm with a Q.Y. of 60%. Multimodal CS NCs appropriate for both optical and magnetic resonance imaging techniques have been obtained by doping of the ZnS shell with manganese ions in the range of 0.6–6.2% [158]. In this case, the growth of the $Zn_{1-x}Mn_xS$ shells of 1–6 monolayer thickness was achieved by injection of a mixture of diethylzinc, dimethylmanganese and H_2S gas to a dispersion of the CdSe core NCs at 170°C.

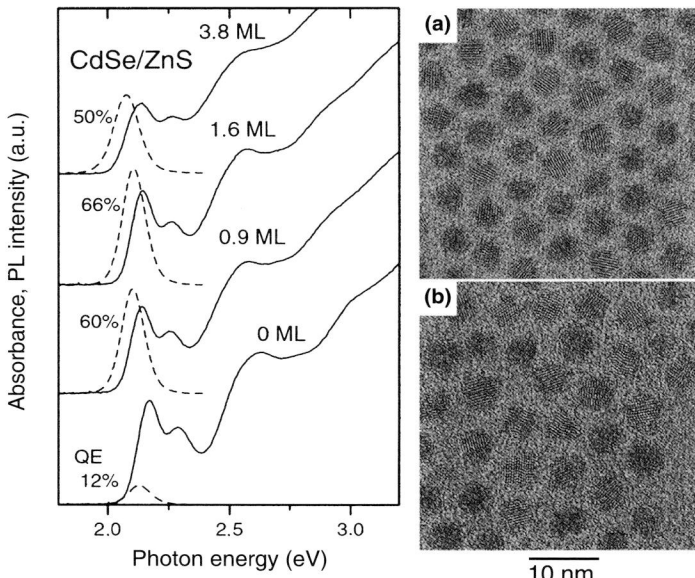

Fig. 10. Left panel: UV–vis absorption and PL spectra during the growth of a ZnS shell (ML = monolayer) on 4 nm CdSe NCs. Right panel: TEM images of 4 nm core NCs (**a**) and of CdSe/ZnS CS NCs (**b**; 1.6 ZnS ML). Reprinted with permission from [33]. © 2001, American Chemical Society

A further CdSe-based system exhibiting a different band alignment is CdSe/CdS. In this common cation heterostructure, a large band offset for the holes is combined with a relatively small one for the electrons. Epitaxial growth is favoured by the comparably small lattice mismatch of around 4% between the core and shell material. Peng et al. reported the synthesis and detailed characterization of series of CdSe/CdS NCs with core diameters ranging from 2.3 to 3.9 nm and Q.Y.s above 50% [42]. In contrast to the CdSe/ZnS system, exhibiting a rather small bathochromic shift (5–10 nm) of the excitonic and PL peak upon shell growth, here these features are continuously shifting throughout the shell growth, indicating a delocalization of the electron over the entire CS structure. While in this early report organometallic precursors (dimethylcadmium, bis(trimethylsilyl)sulfide) had been used, more recently the synthesis of this CS system was carried out using air-stable precursors, i.e. cadmium oleate and elemental sulfur dissolved in ODE [66]. O'Brien and coworkers extended their work on monomolecular precursors to the CdSe/CdS CS system using bis(hexyl(methyl)dithiocarbamato) and bis(hexyl(methyl)diselenocarbamato) cadmium compounds for the core NC and shell growth, respectively [159, 160]. Another modification of the original protocol [40] concerned the omission of the intermediate purification step of the core NCs and growth of the CS structure in a so-called one-pot synthesis, yielding NCs with Q.Y.s in the range of 50–85% [161]. In addition, alternative precursors for the shell growth have been proposed in the same article, namely in situ generated H_2S gas and cadmium acetate. The preparation of particularly small CdSe/CdS CS NCs with core sizes in the range of 1.2–1.5 nm, emitting in the range of 445–517 nm with Q.Y.s of 60–80%, has been described by Pan et al. [162]. Both core and shell synthesis were carried out at 180/ 140°C in an autoclave using cadmium myristate and selenourea/thiourea as precursors, oleic acid as a stabilizer and a toluene/water two phase solvent.

CdSe/ZnSe is a CS system exhibiting, in contrast to CdSe/CdS, efficient confinement of the electrons in the NC core due to the large conduction band offset, while only a relatively small barrier exists for the holes. Although the lattice mismatch is slightly larger than in CdSe/CdS (6.3 vs. 3.9%), the common anion structure is particularly favourable for epitaxial-type shell growth. While in earlier work on CdSe/ZnSe NCs rather low values of the PL Q.Y. (<1%) were published [163], our group reported more recently a modified synthesis method using for the first time the air-stable precursor zinc stearate instead of diethylzinc as the zinc source in combination with selenium dissolved in trioctylphosphine (TOPSe) as the Se source [164]. The obtained CS NCs exhibited Q.Y.s in ranging from 60 to 85% and narrow emission linewidths. Lee et al. studied the effect of lattice distortion in the CdSe/ZnSe CS system on the optical spectra by varying the concentration of the ZnSe precursor solution used for the shell growth [165] as well as the ripening kinetics upon thermal annealing [166].

Both CdSe/CdS and CdSe/ZnSe heterostructures exhibit high fluorescence Q.Y.s and can have specific interests due to the "accessibility" of the weakly confined electrons or holes, respectively. On the other hand, if purely high stability of the optical properties against photo-degradation and chemical inertness of the shell material are desired, zinc sulfide is the shell material of choice. Although it is in principle possible to obtain green or even blue emission with CdSe NCs of small size, their capping with ZnS has only very recently been achieved [156]. As a matter of fact, an analysis of the present state of the art suggests that for a large variety of materials the preparation of CS NCs of low size dispersion and satisfying optical properties is facilitated when core NCs with diameters in the range of approximately 2.5 and 5 nm are used. Consequently the synthesis of other types of core NCs than CdSe has been developed with the goal to better cover the green/blue/UV (and the near infrared) spectral region. In this context, an attractive alternative to the tuning of the emission colour with size is the formation of alloy structures, allowing for the colour variation by changing the composition of the NCs. An example are $Cd_{1-x}Zn_xSe$ NCs, whose band gap can be varied by changing x between the values of pure CdSe and pure ZnSe NCs of the same size. This system is particularly interesting for the fabrication of efficient green emitters for use in display applications. To do so, the alloy core NCs have been overgrown with a ZnS shell either using the established diethylzinc/bis(trimethylsilyl)sulfide method [87] or, more recently, by means of the air-stable monomolecular precursor zinc diethylxanthate [88]. In the latter case, after the growth of three ZnS monolayers, the obtained CS NCs emitted at 530 nm with a linewidth of 35 nm (FWHM) and a Q.Y. of 65%.

Emission wavelengths in the blue and near UV spectral region have been obtained by using larger band gap core materials, in particular CdS and ZnSe. CdS NCs have been overgrown with a ZnS shell using the classical organometallic approach with dimethylcadmium and $S(TMS)_2$ as precursors, yielding an emission in the range of 460–480 nm (FWHM 24–28 nm) with Q.Y.s in of 20–30% [167]. Recently these reagents were replaced by a combination of the air-stable monomolecular precursor

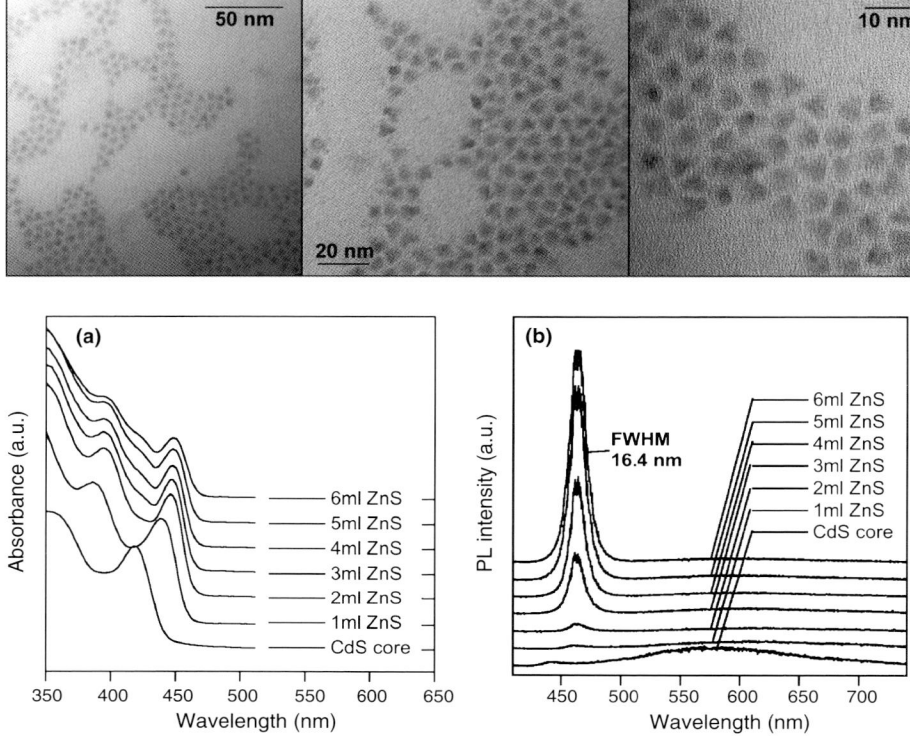

Fig. 11. Top: TEM images at different magnifications of CdS/ZnS NCs using zinc ethylxanthate as precursor for the ZnS shell growth. Bottom: **a** UV–vis absorption spectra; **b** PL spectra recorded during the addition of 6 mL of the ZnS precursor solution corresponding to the growth of a five monolayer-thick ZnS shell on 4 nm CdS core NCs [168]

zinc ethylxanthate and zinc stearate, resulting in monodisperse CdS/ZnS CS NCs emitting in the range of 440–480 nm (15–18 nm FWHM) with Q.Y.s of 35–45% (Fig. 11) [168].

The CdS core NCs have further been used as a host matrix for Mn dopant ions in CdS:Mn/ZnS CS systems [169]. ZnSe has equally been overcoated by ZnS using organometallic precursors, leading to emission wavelengths in the range of 400 nm, PL linewidths of 20 nm and Q.Y.s of the order of 15% [170]. In a newer approach based on the use of alternative precursors, i.e. ZnO and TOPSe for the core NCs and zinc laurate and TOPS for the shell growth carried out at 180°C in HDA, Q.Y.s up to 30% were reached [171].

Cadmium telluride, exhibiting a smaller bulk band gap than cadmium selenide (1.5 vs. 1.75 eV at 300 K), is in principle a good candidate for the fabrication of red or near infrared-emitting quantum dots. However, due to the high oxidation sensitivity of CdTe NCs prepared in organic solvents, comparably few reports exist concerning the preparation of related CS systems in organic solvents, such as CdTe/ZnS [172]. In this work, CdTe core NCs were synthesized in water following the recipe of [14] and transferred to organic solvent by ligand exchange before overgrowing them with the ZnS shell.

4.1.2 *Synthesis of core/shell nanocrystals of III–V semiconductors.* Literature on III–V semiconductor based CS systems is much more sparse than in the case of II–VI compounds, as a consequence of the lack of robust synthesis methods for most core NCs of this family (cf. Sect. 3.2). As-prepared InP NCs exhibit rather poor optical properties as compared to CdSe. The PL linewidth is significantly broader, of the order of 50 nm (FWHM), and in addition to band edge emission peaks related to defect state emission occur in the spectrum. Furthermore, the Q.Y. is low, typically inferior to 1%. Talapin et al. described an efficient way to increase the Q.Y. of InP NCs to values approaching 40% by photo-assisted etching of their surface with HF [37]. This process resulted in the removal of surface phosphorous atoms being at the origin of trap states, which provided non-radiative recombination pathways. Concerning InP-based CS systems, Haubold et al. used organometallic precursors to grow a ZnS shell on InP and observed a subsequent increase of the Q.Y. in a slow room temperature process to 15% after 3 days and to 23% after 3 weeks [173]. In order to adjust the lattice parameters of the core and shell materials and to reduce strain-induced defects, Mićić et al. developed a CdZnSe$_2$ shell leading to a fluorescence Q.Y. of 5–10% [174]. Cao and Banin reported the preparation of several CS systems based on near infrared-emitting InAs core NCs, including InP, GaAs, CdSe, ZnSe and ZnS using high-temperature pyrolysis of organometallic precursors in TOPO [175, 176]. The obtained fluorescence Q.Y.s depended on the shell material. In the case of InP, PL quenching was observed, whereas ZnS led to 8% and for CdSe and ZnSe an enhancement up to 20% was detected. More recently the synthesis of a series of small InAs/ZnSe CS NCs has been described, aiming at emission wavelengths in the range of 700–900 nm [177]. This spectral region is especially well-adapted for in vivo biological imaging due to the reduced light scattering by the tissue. The obtained NCs exhibited a Q.Y. of 6–9% after transfer to the aqueous phase and were successfully applied for the in vivo imaging of lymph nodes.

4.1.3 *Synthesis of core/shell nanocrystals of IV–VI semiconductors.* In contrast to the discussed II–VI and III–V semiconductors exhibiting either the hexagonal wurtzite or the cubic zinc blende crystal structure, the IV–VI family is characterized by the rocksalt structure (cf. Table 1). Only lead-based NCs (PbS, PbSe) have been studied in form of CS systems. Lifshitz and coworkers reported the synthesis of PbSe/PbS and PbSe/PbSe$_x$S$_{1-x}$ CS NCs emitting in the range of 1–2 µm with Q.Y.s of 40–50% and 65%, respectively, by means of the use of lead(II)acetate, TOPSe and TOPS as precursors, oleic acid as stabilizer and diphenylether as the solvent (Fig. 12) [178–180]. The SILAR method mentioned before in the case of CdSe/CdS has recently been applied for the synthesis of PbSe/PbS NCs [181].

4.2 Type II systems. Research on colloidal type II systems was triggered by the seminal work of Bawendi et al. [182] who described the synthesis and optical properties of CdTe/CdSe and CdSe/ZnTe CS NCs [182]. The emission wavelength of

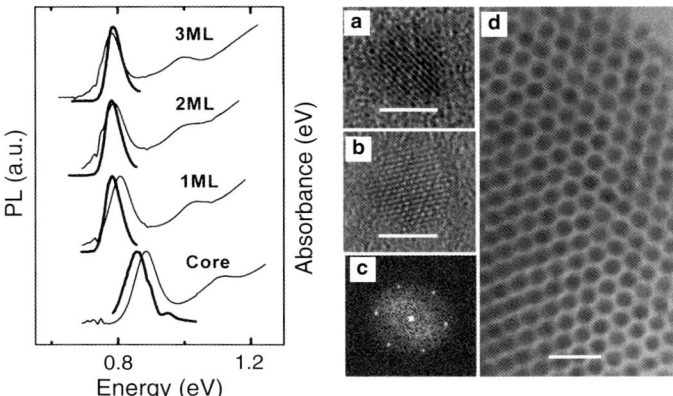

Fig. 12. Left panel: Evolution of the UV–vis absorption and PL spectra of 4.9 nm PbSe core NCs during the growth of a PbS shell of indicated thickness. Right panel: **a** HRTEM image of a PbSe/PbS CS NC comprising a 4.8 nm core and a 1.2 nm shell; **b** HRTEM image of a PbSe/PbSe$_{0.5}$S$_{0.5}$ core/alloyed shell NC; **c** FFT image of the particle in image A; **d** TEM image of self-assembled 6.7 nm core/alloyed shell NCs. The scale bars are 5 nm in **a** and **b**, and 20 nm in **d**. Reprinted with permission from [180]. © 2006 American Chemical Society

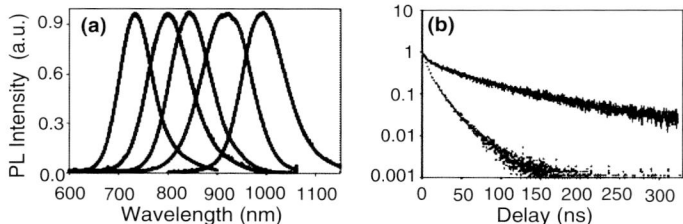

Fig. 13. **a** Normalized PL spectra of CdTe/CdSe CS NCs having the core/shell radii of 1.6/1.9, 1.6/3.2, 3.2/1.1, 3.2/2.4, and 5.6/1.9 nm (from left to right). **b** Normalized PL decays of 3.2/1.1 nm CdTe/CdSe CS NCs and of the corresponding 3.2 nm CdTe core NCs (dotted line). Reprinted with permission from [182]. © 2003 American Chemical Society

CdTe/CdSe NCs could be tuned by changing the shell thickness and the core NC size from 700 to 1000 nm (Fig. 13). This approach is an alternative possibility to shift the emission peak to higher wavelengths, which would not be attainable by simply increasing the size of the core NC in a type I CS system. On the other hand the observed mean decay lifetime (57 ns) was significantly larger than that of the corresponding core CdTe NCs (9.6 ns), and the Q.Y. was low (4%) with respect to type I CS systems. CdTe/CdSe CS NCs were further synthesized without the use of organometallic precursors applying CdO, TOPTe and TOPSe, leading to Q.Y.s approaching 40% for small shell thicknesses below 0.5 nm [183]. Similarly, CdSe/ZnTe NCs have been prepared with CdO as the cadmium precursor and femtosecond dynamics measurements revealed that the rate of photo-induced electron/hole spatial separation decreased with increasing core size and is independent of the shell thickness [184].

Klimov and coworkers first studied the optical properties of so-called "inverted" CS NCs, i.e. the band gap of the core material (ZnSe) is larger than that of the shell material (CdSe) [185]. On the basis of the radiative recombination lifetimes recorded for NCs with a fixed core size and increasing shell thickness, a continuous transition from type I (both electron and hole wave functions are distributed over the entire NC) to type II (electron and hole are spatially separated between the shell and the core) and back to type I (both electron and hole primarily reside in the shell) localization regimes was observed. The samples exhibited emission in the range of 430–600 nm and Q.Y.s of 60–80%. The same CS system was also synthesized applying CdO dissolved with oleic acid in octadecane and TOPSe as the shell precursors [186]. By varying the CdSe shell thickness on 2.8 core ZnSe NCs, the emission wavelength could be tuned in a broad spectral range (417–678 nm) with Q.Y.s of 40–85%. Furthermore size focusing during the shell growth has been observed in contrast to the usually occurring broadening of the size distribution at this stage of the reaction. ZnTe/CdTe represents the first common anion type II system, reported by Basché coworkers [187]. The same article also comprises the synthesis of ZnTe NCs with CdS and CdSe shells. These heterostructures were obtained by addition of precursors (cadmium oleate, TOPTe, TOPSe or sulfur dissolved in octadecene) to the crude dispersion of ZnTe core NCs. Q.Y.s of up to 30% have been observed and the emission could be tuned in the range of 500–900 nm. Interestingly, the same group observed in the case of the ZnTe/CdSe CS system, also described in [188], a transition from the concentric CS structure via pyramidal to tetrapod-shaped heterostructures when the shell growth was carried out at 215°C instead of 240°C (cf. Fig. 14) [189].

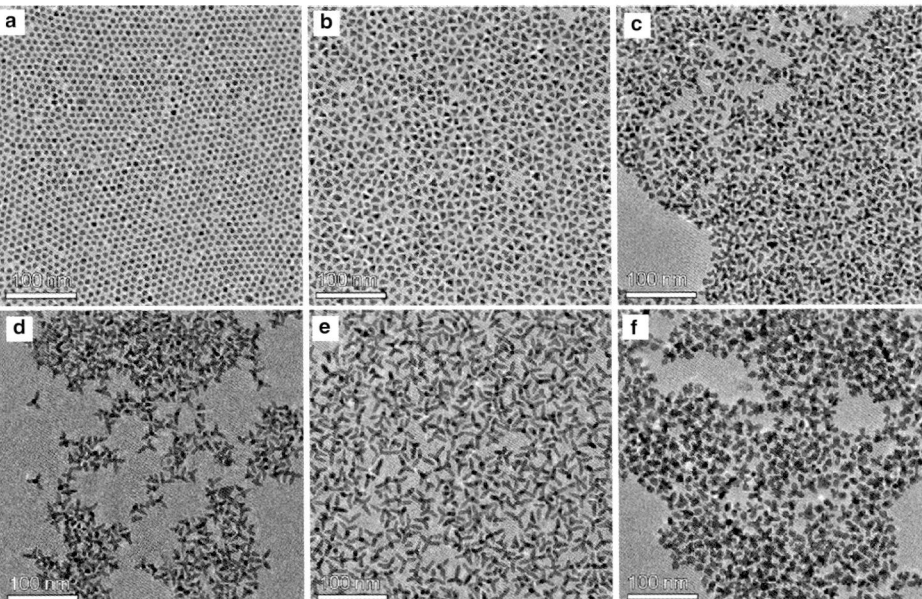

Fig. 14. TEM images of ZnTe/CdSe NCs: transition from a slightly anisotropic core/shell to b pyramidal to c, d, e tetrapod-shaped heterostructures with different arm lengths. f TEM image of tetrapod-shaped ZnTe/CdS NCs. Reproduced with permission from [189]. © 2006 Wiley–VCH Verlag GmbH & Co. KGaA

5. Conclusions and outlook

The synthesis of cadmium chalcogenide NCs via the hot-injection method reported in 1993 [8] was the starting shot for an impressive evolution of this research field. Within the last 15 years the synthesis of monodisperse spherical NCs of a large variety of semiconductors has been achieved. Moreover, the same synthetic scheme allows for the preparation of shape-controlled anisotropic NCs (e.g. rod-like or branched structures) [190, 191] and of composition-controlled ternary compounds ("alloy" semiconductors). For all materials, an ongoing trend toward the simplification of the synthesis procedure can be observed in view of an enhanced reproducibility and scale-up of the reactions. As a consequence, recent advances comprise the substitution of pyrophoric precursors by air-stable ones and the use of the heating-up technique instead of the hot-injection method. However, most of the published work to date is dedicated to the II–VI semiconductor family, and in particular to cadmium chalcogenides. The latter have an extremely low acceptability for technological applications due to the toxicity of cadmium and related problems with the handling of these compounds. To give an example, the use of toxic elements such as Cd, Pb and Hg has been severely limited by recent restrictions established in the European Union[1]. Therefore it is necessary to develop robust synthesis methods for NCs of alternative semiconductor materials. Ternary semiconductors such as chalcopyrites or doped NCs (e.g. ZnSe:Mn) may gain importance in this context. Still important challenges remain for the synthetic chemist. When moving for example from the II–VI family to the more covalently bound III–V compounds, it is much more difficult to produce monodisperse samples. Furthermore, hot-injection or heating-up synthesis methods for NCs of several technologically highly important materials are completely lacking up till now, such as gallium nitride or silicon, to name a few. The successful rational design of new synthesis strategies implies a better understanding of the NCs' formation mechanisms, which can be achieved by the precise investigation of their nucleation and growth processes.

In view of the developments in the domain of CS semiconductor NCs in the last decade, it can be speculated that a large variety of new heterostructures with exciting and, in some cases unprecedented features will be synthesized by chemical routes in the next years. The ability to precisely control the shell thickness will further boost advances in the preparation of core/shell/shell or other complex structures, such as quantum-dot-quantum-wells. Significant progress has also been achieved in the field of anisotropic shell growth on spherical CdSe NCs, leading to new rod-shaped CS nanostructures, which combine unique optical properties (high Q.Y., polarized emission) with appealing self-assembly properties [192, 193]. The reported synthetic procedure using small core NCs as seeds in the so-called seeded-growth approach, opens up the way for the generation of new heterostructures, including nanorods and branched ones, such as bi-, tri-, tetra-pods or multipods. Similar as in the case of II–VI compounds, it can be expected that a growing number of CS systems based on III–V semiconductors will be developed, inspired by the huge amount of research carried out in the field of III–V nanostructures grown by molecular beam epitaxy

---

[1] cf. the Reduction of Hazardous Substances (RoHS) mandate, banning the use of six chemicals – including Cd, Hg, Cr^{VI} and Pb – in almost all electronics products (date of effect: 01/07/2006).

(MBE) techniques. Finally, the association of semiconductors with other materials such as metals or oxides in the same CS heterostructure allows for the design of NCs combining different physical properties, e.g. fluorescence, magnetism, etc. In such a manner, novel functional building blocks can be generated for applications in fields ranging from (opto-)electronics via information technology to healthcare. In most of these examples, the NCs are used as a platform for further surface functionalization, enabling their integration into devices or materials, or their binding to other molecules or macromolecules.

References

[1] Brus LE (1984) Electron–electron and electron–hole interactions in small semiconductor crystallites – the size dependence of the lowest excited electronic state. Journal of Chemical Physics 80: 4403–4409
[2] Bawendi MG, Steigerwald ML, Brus LE (1990) The quantum-mechanics of larger semiconductor clusters um dots). Annual Review of Physical Chemistry 41: 477–496
[3] Efros AL, Rosen M, Kuno M, Nirmal M, Norris DJ, Bawendi M (1996) Band-edge exciton in quantum dots of semiconductors with a degenerate valence band: dark and bright exciton states. Physical Review B 54: 4843–4856
[4] Efros AL, Efros AL (1982) Interband absorption of light in a semiconductor sphere. Soviet Physics Semiconductors—USSR 16: 772–775
[5] Henglein A (1982) Photo-degradation and fluorescence of colloidal-cadmium sulfide in aqueous-solution. Berichte der Bunsen-Gesellschaft-Physical Chemistry Chemical Physics 86: 301–305
[6] Rossetti R, Brus L (1982) Electron–hole recombination emission as a probe of surface-chemistry in aqueous cds colloids. Journal of Physical Chemistry 86: 4470–4472
[7] Rossetti R, Nakahara S, Brus LE (1983) Quantum size effects in the redox potentials, resonance Raman-spectra, and electronic-spectra of CdS crystallites in aqueous-solution. Journal of Chemical Physics 79: 1086–1088
[8] Murray CB, Norris DJ, Bawendi MG (1993) Synthesis and characterization of nearly monodisperse CdE (E = S, Se, Te) semiconductor nanocrystallites. Journal of the American Chemical Society 115: 8706–8715
[9] Murray CB, Kagan CR, Bawendi MG (2000) Synthesis and characterization of monodisperse nanocrystals and close-packed nanocrystal assemblies. Annual Review of Materials Science 30: 545–610
[10] Yin Y, Alivisatos AP (2005) Colloidal nanocrystal synthesis and the organic-inorganic interface. Nature 437: 664–670
[11] Green M (2005) Organometallic based strategies for metal nanocrystal synthesis. Chemical Communications 24: 3002–3011
[12] Park J, Joo J, Kwon SG, Jang Y, Hyeon T (2007) Synthesis of monodisperse spherical nanocrystals. Angewandte Chemie-International Edition 46: 4630–4660
[13] Rogach AL, Talapin DV, Shevchenko EV, Kornowski A, Haase M, Weller H (2002) Organization of matter on different size scales: monodisperse nanocrystals and their superstructures. Advanced Functional Materials 12: 653–664
[14] Rogach AL, Franzl T, Klar TA, Feldmann J, Gaponik N, Lesnyak V et al (2007) Aqueous synthesis of thiol-capped CdTe nanocrystals: state-of-the-art. Journal of Physical Chemistry C 111: 14628–14637
[15] Pileni MP (1997) Nanosized particles made in colloidal assemblies. Langmuir 13: 3266–3276
[16] Wang X, Peng Q, Li YD (2007) Interface-mediated growth of monodispersed nanostructures. Accounts of Chemical Research 40: 635–643
[17] Biswas K, Rao CNR (2007) Use of ionic liquids in the synthesis of nanocrystals and nanorods of semiconducting metal chalcogenides. Chemistry-A European Journal 13: 6123–6129
[18] Shah PS, Hanrath T, Johnston KP, Korgel BA (2004) Nanocrystal and nanowire synthesis and dispersibility in supercritical fluids. Journal of Physical Chemistry B 108: 9574–9587
[19] Gaponenko SV (1998) Optical properties of semiconductor nanocrystals. Cambridge University Press, Cambridge (UK)
[20] Norris DJ, Sacra A, Murray CB, Bawendi MG (1994) Measurement of the size-dependent hole spectrum in CdSe quantum dots. Physical Review Letters 72: 2612–2615

[21] Reiss P, Bleuse J (unpublished)
[22] Ekimov AI, Hache F, Schanne-Klein MC, Ricard D, Flytzanis C, Kudryavtsev IA et al (1993) Absorption and intensity-dependent photoluminescence measurements on CdSe quantum dots – assignment of the 1st electronic-transitions. Journal of the Optical Society of America B-Optical Physics 10: 100–107
[23] Nirmal M, Norris DJ, Kuno M, Bawendi MG, Efros AL, Rosen M (1995) Observation of the dark exciton in CdSe quantum dots. Physical Review Letters 75: 3728–3731
[24] Mason MD, Credo GM, Weston KD, Buratto SK (1998) Luminescence of individual porous Si chromophores. Physical Review Letters 80: 5405–5408
[25] Pistol ME, Castrillo P, Hessman D, Prieto JA, Samuelson L (1999) Random telegraph noise in photoluminescence from individual self-assembled quantum dots. Physical Review B 59: 10725–10729
[26] VandenBout DA, Yip WT, Hu DH, Fu DK, Swager TM, Barbara PF (1997) Discrete intensity jumps and intramolecular electronic energy transfer in the spectroscopy of single conjugated polymer molecules. Science 277: 1074–1077
[27] Xie XS, Trautman JK (1998) Optical studies of single molecules at room temperature. Annual Review of Physical Chemistry 49: 441–480
[28] Dickson RM, Cubitt AB, Tsien RY, Moerner WE (1997) On/off blinking and switching behaviour of single molecules of green fluorescent protein. Nature 388: 355–358
[29] Kuno M, Fromm DP, Hamann HF, Gallagher A, Nesbitt DJ (2000) Nonexponential "blinking" kinetics of single CdSe quantum dots: a universal power law behavior. Journal of Chemical Physics 112: 3117–3120
[30] Chepic DI, Efros AL, Ekimov AI, Vanov MG, Kharchenko VA, Kudriavtsev IA et al (1990) Auger ionization of semiconductor quantum drops in a glass matrix. Journal of Luminescence 47: 113–127
[31] Empedocles S, Bawendi M (1999) Spectroscopy of single CdSe nanocrystallites. Accounts of Chemical Research 32: 389–396
[32] Neuhauser RG, Shimizu KT, Woo WK, Empedocles SA, Bawendi MG (2000) Correlation between fluorescence intermittency and spectral diffusion in single semiconductor quantum dots. Physical Review Letters 85: 3301–3304
[33] Talapin DV, Rogach AL, Kornowski A, Haase M, Weller H (2001) Highly luminescent monodisperse CdSe and CdSe/ZnS nanocrystals synthesized in a hexadecylamine–trioctylphosphine oxide–trioctylphospine mixture. Nano Letters 1(4): 207–211
[34] Munro AM, Plante IJL, Ng MS, Ginger DS (2007) Quantitative study of the effects of surface ligand concentration on CdSe nanocrystal photoluminescence. Journal of Physical Chemistry C 111: 6220–6227
[35] Jang E, Jun S, Chung YS, Pu LS (2004) Surface treatment to enhance the quantum efficiency of semiconductor nanocrystals. Journal of Physical Chemistry B 108: 4597–4600
[36] Mićić OI, Cheong HM, Fu H, Zunger A, Sprague JR, Mascarenhas A, Nozik AJ (1997) Size-dependent spectroscopy of InP quantum dots. Journal of Physical Chemistry B 101(25): 4904–4912
[37] Talapin DV, Gaponik N, Borchert H, Rogach AL, Haase M, Weller H (2002) Etching of colloidal InP nanocrystals with fluorides: photochemical nature of the process resulting in high photoluminescence efficiency. Journal of Physical Chemistry B 106(49): 12659–12663
[38] Yeh CY, Lu ZW, Froyen S, Zunger A (1992) Zinc-blende-wurtzite polytypism in semiconductors. Physical Review B 46: 10086–10097
[39] Jacobs K, Wickham J, Alivisatos AP (2002) Threshold size for ambient metastability of rocksalt CdSe nanocrystals. Journal of Physical Chemistry B 106: 3759–3762
[40] Hines MA, Guyot-Sionnest P (1996) Synthesis and characterization of strongly luminescing ZnS-capped CdSe nanocrystals. Journal of Physical Chemistry 100: 468–471
[41] Dabbousi BO, Rodriguez Viejo J, Mikulec FV, Heine JR, Mattoussi H, Ober R et al (1997) (CdSe)ZnS core-shell quantum dots: synthesis and characterization of a size series of highly luminescent nanocrystallites. Journal of Physical Chemistry B 101: 9463–9475
[42] Peng XG, Schlamp MC, Kadavanich AV, Alivisatos AP (1997) Epitaxial growth of highly luminescent CdSe/CdS core/shell nanocrystals with photostability and electronic accessibility. Journal of the American Chemical Society 119: 7019–7029
[43] Wei SH, Zunger A (1998) Calculated natural band offsets of all II–VI and III–V semiconductors: chemical trends and the role of cation d orbitals. Applied Physics Letters 72: 2011–2013
[44] Fojtik A, Weller H, Koch U, Henglein A (1984) Photo-chemistry of colloidal metal sulfides. 8. Photophysics of extremely small CdS particles – Q-state CdS and magic agglomeration numbers. Berichte der Bunsen-Gesellschaft für Physikalische Chemie-Physical Chemistry Chemical Physics 88: 969–977

[45] Spanhel L, Haase M, Weller H, Henglein A (1987) Photochemistry of colloidal semiconductors. 20. Surface modification and stability of strong luminescing CdS particles. Journal of the American Chemical Society 109: 5649–5655
[46] Lianos P, Thomas JK (1986) Cadmium-sulfide of small dimensions produced in inverted micelles. Chemical Physics Letters 125: 299–302
[47] Rogach A, Kershaw S, Burt M, Harrison M, Kornowski A, Eychmüller A et al (1999) Colloidally prepared HgTe nanocrystals with strong room-temperature infrared luminescence. Advanced Materials 11: 552–556
[48] Harrison MT, Kershaw SV, Rogach AL, Kornowski A, Eychmüller A, Weller H (2000) Wet chemical synthesis of highly luminescent HgTe/CdS core/shell nanocrystals. Advanced Materials 12: 123–125
[49] Kuno M, Higginson KA, Qadri SB, Yousuf M, Lee SH, Davis BL, Mattoussi H (2003) Molecular clusters of binary and ternary mercury chalcogenides: colloidal synthesis, characterization, and optical spectra. Journal of Physical Chemistry B 107: 5758–5767
[50] LaMer VK, Dinegar RH (1950) Theory, production and mechanism of formation of monodispersed hydrosols. Journal of the American Chemical Society 72: 4847–4854
[51] de Mello Donegá C, Liljeroth P, Vanmaekelbergh D (2005) Physicochemical evaluation of the hot-injection method, a synthesis route for monodisperse nanocrystals. Small 1: 1152–1162
[52] Ostwald W (1901) Z Physical Chemistry 37: 385
[53] Voorhees PW (1985) The theory of Ostwald ripening. Journal of Statistical Physics 38: 231–252
[54] Reiss H (1951) The growth of uniform colloidal dispersions. Journal of Chemical Physics 19: 482–487
[55] Sugimoto T (1987) Preparation of monodispersed colloidal particles. Advances in Colloid and Interface Science 28: 65–108
[56] Sugimoto T, Shiba F (1999) A new approach to interfacial energy. 3. Formulation of the absolute value of the solid–liquid interfacial energy and experimental collation to silver halide systems. Journal of Physical Chemistry B 103: 3607–3615
[57] Peng XG, Wickham J, Alivisatos AP (1998) Kinetics of II–VI and III–V colloidal semiconductor nanocrystal growth: "Focusing" of size distributions. Journal of the American Chemical Society 120: 5343–5344
[58] Peng ZA, Peng XG (2001) Formation of high-quality CdTe, CdSe, and CdS nanocrystals using CdO as precursor. Journal of the American Chemical Society 123: 183–184
[59] Qu L, Peng ZA, Peng X (2001) Alternative routes toward high quality CdSe nanocrystals. Nano Letters 1: 333–336
[60] Yu W, Peng X (2002) Formation of high quality CdS and other II–VI semiconductor nanocrystals in noncoordinating solvents: tunable reactivity of monomers. Angewandte Chemie-International Edition 41: 2368–2371
[61] Liu HT, Owen JS, Alivisatos AP (2007) Mechanistic study of precursor evolution in colloidal group II–VI semiconductor nanocrystal synthesis. Journal of the American Chemical Society 129: 305–312
[62] Chen XB, Lou YB, Samia AC, Burda C (2003) Coherency strain effects on the optical response of core/shell heteronanostructures. Nano Letters 3: 799–803
[63] Madelung O, Schulz M, Weiss H (1982) Landolt-Börnstein: numerial data and functional relationships in science and technology, new series, group III: crystal and solid state physics, vol. III/17b. Springer, Berlin
[64] Singh J (1993) Physics of semiconductors and their heterostructures. McGraw-Hill, New York
[65] Yu WW, Qu LH, Guo WZ, Peng XG (2003) Experimental determination of the extinction coefficient of CdTe, CdSe, and CdS nanocrystals. Chemistry of Materials 15: 2854–2860
[66] Li JJ, Wang YA, Guo WZ, Keay JC, Mishima TD, Johnson MB et al (2003) Large-scale synthesis of nearly monodisperse CdSe/CdS core/shell nanocrystals using air-stable reagents via successive ion layer adsorption and reaction. Journal of the American Chemical Society 125: 12567–12575
[67] Talapin DV, Rogach AL, Shevchenko EV, Kornowski A, Haase M, Weller H (2002) Dynamic distribution of growth rates within the ensembles of colloidal II–VI and III–V semiconductor nanocrystals as a factor governing their photoluminescence efficiency. Journal of the American Chemical Society 124: 5782–5790
[68] Wu DG, Kordesch ME, Van Patten PG (2005) A new class of capping ligands for CdSe nanocrystal synthesis. Chemistry of Materials 17: 6436–6441
[69] Yang YA, Wu H, Williams KR, Cao YC (2005) Synthesis of CdSe and CdTe nanocrystals without precursor injection. Angewandte Chemie-International Edition 44(41): 6712–6715
[70] Jasieniak J, Bullen C, van Embden J, Mulvaney P (2005) Phosphine-free synthesis of CdSe nanocrystals. Journal of Physical Chemistry B 109: 20665–20668

[71] Sapra S, Rogach AL, Feldmann J (2006) Phoshine-free synthesis of monodisperse CdSe nanocrystals in olive oil. Journal of Materials Chemistry 16: 3391–3395
[72] Pradhan N, Reifsnyder D, Xie RG, Aldana J, Peng XG (2007) Surface ligand dynamics in growth of nanocrystals. Journal of the American Chemical Society 129: 9500–9509
[73] Cao YC, Wang JH (2004) One-pot synthesis of high-quality zinc-blende CdS nanocrystals. Journal of the American Chemical Society 126: 14336–14337
[74] Pradhan N, Efrima S (2003) Single-precursor, one-pot versatile synthesis under near ambient conditions of tunable, single and dual band fluorescing metal sulfide nanoparticles. Journal of the American Chemical Society 125: 2050–2051
[75] Pradhan N, Katz B, Efrima S (2003) Synthesis of high-quality metal sulfide nanoparticles from alkyl xanthate single precursors in alkylamine solvents. Journal of Physical Chemistry B 107: 13843–13854
[76] Joo J, Na HB, Yu T, Yu JH, Kim YW, Wu FX, Zhang JZ, Hyeon T (2003) Generalized and facile synthesis of semiconducting metal sulfide nanocrystals. Journal of the American Chemical Society 125: 11100–11105
[77] Talapin DV, Haubold S, Rogach AL, Kornowski A, Haase M, Weller H (2001) A novel organometallic synthesis of highly luminescent CdTe nanocrystals. Journal of Physical Chemistry B 105: 2260–2263
[78] Yu WW, Wang YA, Peng X (2003) Formation and stability of size-, shape-, and structure-controlled CdTe nanocrystals: ligand effects on monomers and nanocrystals. Chemistry of Materials 15(22): 4300–4308
[79] Kloper V, Osovsky R, Kolny-Olesiak J, Sashchiuk A, Lifshitz E (2007) The growth of colloidal cadmium telluride nanocrystal quantum dots in the presence of Cd-0 nanoparticles. Journal of Physical Chemistry C 111: 10336–10341
[80] Li LS, Pradhan N, Wang Y, Peng X (2004) High quality ZnSe and ZnS nanocrystals formed by activating zinc carboxylate precursors. Nano Letters 4(11): 2261–2264
[81] Yu JH, Joo J, Park HM, Baik SI, Kim YW, Kim SC, Hyeon T (2005) Synthesis of quantum-sized cubic ZnS nanorods by the oriented attachment mechanism. Journal of the American Chemical Society 127: 5662–5670
[82] Hines MA, Guyot-Sionnest P (1998) Bright UV-blue luminescent colloidal ZnSe nanocrystals. Journal of Physical Chemistry B 102(19): 3655–3657
[83] Reiss P, Quemard G, Carayon S, Bleuse J, Chandezon C, Pron A (2004) Luminescent ZnSe nanocrystals of high color purity. Materials Chemistry and Physics 84: 10–13
[84] Xie R, Zhong X, Basché T (2005) Synthesis, characterization, and spectroscopy of type-II core/shell semiconductor nanocrystals with ZnTe cores. Advanced Materials 17: 2741–2746
[85] Green M, Wakefield G, Dobson PJ (2003) A simple metalorganic route to organically passivated mercury telluride nanocrystals. Journal of Materials Chemistry 13: 1076–1078
[86] Zhong X, Zhang Z, Liu S, Han M, Knoll W (2004) Embryonic nuclei-induced alloying process for the reproducible synthesis of blue-emitting $Zn_xCd_{1-x}Se$ nanocrystals with long-time thermal stability in size distribution and emission wavelength. Journal of Physical Chemistry B 108(40): 15552–15559
[87] Steckel JS, Snee P, Coe-Sullivan S, Zimmer JR, Halpert JE, Anikeeva P et al (2006) Color-saturated green-emitting QD-LEDs. Angewandte Chemie-International Edition 45: 5796–5799
[88] Protière M, Reiss P (2007) Highly luminescent $Cd_{1-x}Zn_xSe/ZnS$ core shell nanocrystals emitting in the blue-green spectral range. Small 3: 399–403
[89] Zhong X, Feng Y, Knoll W, Han M (2003) Alloyed $Zn_xCd_{1-x}S$ nanocrystals with highly narrow luminescence spectral width. Journal of the American Chemical Society 125(44): 13559–13563
[90] Bailey RE, Nie S (2003) Alloyed semiconductor quantum dots: tuning the optical properties without changing the particle size. Journal of American Chemical Society 125(23): 7100–7106
[91] Pradhan N, Reifsnyder D, Xie R, Aldana J, Peng X (2007) Surface ligand dynamics in growth of nanocrystals. Journal of American Chemical Society 129: 9500–9509
[92] Chen HS, Lo B, Hwang JY, Chang GY, Chen CM, Tasi SJ et al (2004) Colloidal ZnSe, ZnSe/ZnS, and ZnSe/ZnSeS quantum dots synthesized from ZnO. Journal of Physical Chemistry B 108: 17119–17123
[93] Bernard JE, Zunger A (1986) Optical bowing in zinc chalcogenide semiconductor alloys. Physical Review B 34: 5992–5995
[94] Bernard JE, Zunger A (1987) Electronic-structure of ZnS, ZnSe, ZnTe, and their pseudobinary alloys. Physical Review B 36: 3199–3228
[95] Al-Salim N, Young AG, Tilley RD, McQuillan AJ, Xia J (2007) Synthesis of CdSeS nanocrystals in coordinating and noncoordinating solvents: solvent's role in evolution of the optical and structural properties. Chemistry of Materials 19: 5185–5193

[96] Furdyna JK (1988) Diluted magnetic semiconductors. Journal of Applied Physics 64: R29–R64
[97] Oczkiewicz B, Twardowski A, Demianiuk M (1987) Intra-manganese absorption and luminescence in $Zn_{1-x}Mn_xSe$ semimagnetic semiconductor. Solid State Communications 64: 107–111
[98] Xue J, Ye Y, Medina F, Martinez L, Lopez-Rivera SA, Giriat W (1998) Temperature evolution of the 2.1 eV band in the $Zn_{1-x}Mn_xSe$ system for low concentration. Journal of Luminescence 78: 173–178
[99] Norris DJ, Yao N, Charnock FT, Kennedy TA (2001) High-quality manganese-doped ZnSe nanocrystals. Nano Letters 1: 3–7
[100] Suyver JF, Wuister SF, Kelly JJ, Meijerink A (2000) Luminescence of nanocrystalline ZnSe: Mn^{2+}. Physical Chemistry Chemical Physics 2: 5445–5448
[101] Norberg NS, Parks GL, Salley GM, Gamelin DR (2006) Giant excitonic Zeeman splittings in colloidal Co^{2+}-doped ZnSe quantum dots. Journal of the American Chemical Society 128: 13195–13203
[102] Dance IG, Choy A, Scudder ML (1984) Syntheses, properties, and molecular and crystal-structures of $(Me_4N)_4[S_4Zn_{10}(SPh)_{16}]$ $(Me_4N)_n[Se_4Cd_{10}(SPh)_{16}]$ – molecular supertetrahedral fragments of the cubic metal chalcogenide lattice. Journal of the American Chemical Society 106: 6285–6295
[103] Archer PI, Santangelo SA, Gamelin DR (2007) Inorganic cluster syntheses of TM^{2+}-doped quantum dots (CdSe, CdS, CdSe/CdS): physical property dependence on dopant locale. Journal of the American Chemical Society 129: 9808–9818
[104] Erwin SC, Zu LJ, Haftel MI, Efros AL, Kennedy TA, Norris DJ (2005) Doping semiconductor nanocrystals. Nature 436: 91–94
[105] Pradhan N, Peng XG (2007) Efficient and color-tunable Mn-doped ZnSe nanocrystal emitters: control of optical performance via greener synthetic chemistry. Journal of the American Chemical Society 129: 3339–3347
[106] Pradhan N, Goorskey D, Thessing J, Peng XG (2005) An alternative of CdSe nanocrystal emitters: Pure and tunable impurity emissions in ZnSe nanocrystals. Journal of the American Chemical Society 127: 17586–17587
[107] Pradhan N, Battaglia DM, Liu YC, Peng XG (2007) Efficient, stable, small, and water-soluble doped ZnSe nanocrystal emitters as non-cadmium biomedical labels. Nano Letters 7: 312–317
[108] Green M (2002) Solution routes to III–V semiconductor quantum dots. Current Opinion in Solid State & Materials Science 6: 355–363
[109] Wells RL, Gladfelter WL (1997) Pathways to nanocrystalline III–V (13–15) Compound Semiconductors. Journal of Cluster Science 8(2): 217–238
[110] Mićić OI, Curtis CJ, Jones KM, Sprague JR, Nozik AJ (1994) Synthesis and characterization of InP quantum dots. Journal of Physical Chemistry 98: 4966–4969
[111] Guzelian AA, Katari JEB, Kadavanich AV, Banin U, Hamad K, Juban E, Alivisatos AP, Wolters RH, Arnold CC, Heath JR (1996) Synthesis of size-selected, surface-passivated InP nanocrystals. Journal of Physical Chemistry 100: 7212–7219
[112] Battaglia D, Peng XG (2002) Formation of high quality InP and InAs nanocrystals in a noncoordinating solvent. Nano Letters 2: 1027–1030
[113] Xu S, Kumar S, Nann T (2006) Rapid synthesis of high-quality InP nanocrystals. Journal of the American Chemical Society 128: 1054–1055
[114] Guzelian AA, Banin U, Kadavanich AV, Peng X, Alivisatos AP (1996) Colloidal chemical synthesis and characterization of InAs nanocrystal quantum dots. Applied Physics Letters 69: 1432–1434
[115] Mićić OI, Sprague JR, Curtis CJ, Jones KM, Machol JL, Nozik AJ et al (1995) Synthesis and characterization of InP, GaP, and $GaInP_2$ quantum dots. Journal of Physical Chemistry 99: 7754–7759
[116] Kim YH, Jun YW, Jun BH, Lee SM, Cheon JW (2002) Sterically induced shape and crystalline phase control of GaP nanocrystals. Journal of the American Chemical Society 124: 13656–13657
[117] Furis M, Sahoo Y, MacRae DJ, Manciu FS, Cartwright AN, Prasad PN (2003) Surfactant-imposed interference in the optical characterization of GaP nanocrystals. Journal of Physical Chemistry B 107: 11622–11625
[118] Sardar K, Dan M, Schwenzer B, Rao CNR (2005) A simple single-source precursor route to the nanostructures of AlN, GaN and InN. Journal of Materials Chemistry 15: 2175–2177
[119] Green M, O'Brien P (2004) The synthesis of III–V semiconductor nanoparticles using indium and gallium diorganophosphides as single-molecular precursors. Journal of Materials Chemistry 14: 629–636
[120] Sargent EH (2005) Infrared quantum dots. Advanced Materials 17: 515–522

[121] Rogach AL, Eychmüller A, Hickey SG, Kershaw SV (2007) Infrared-emitting colloidal nanocrystals: synthesis, assembly, spectroscopy, and applications. Small 3: 536–557
[122] Murray CB, Sun SH, Gaschler W, Doyle H, Betley TA, Kagan CR (2001) Colloidal synthesis of nanocrystals and nanocrystal superlattices. IBM Journal of Research and Development 45: 47–56
[123] Sun SH, Murray CB, Weller D, Folks L, Moser A (2000) Monodisperse FePt nanoparticles and ferromagnetic FePt nanocrystal superlattices. Science 287: 1989–1992
[124] Yu WW, Falkner JC, Shih BS, Colvin VL (2004) Preparation and characterization of monodisperse PbSe semiconductor nanocrystals in a noncoordinating solvent. Chemistry of Materials 16: 3318–3322
[125] Hines MA, Scholes GD (2003) Colloidal PbS nanocrystals with size-tunable near-infrared emission: observation of post-synthesis self-narrowing of the particle size distribution. Advanced Materials 15: 1844–1849
[126] Lu W, Fang J, Stokes KL, Lin J (2004) Shape evolution and assembly of monodisperse PbTe nanocrystals. Journal of the American Chemical Society 126: 11798–11799
[127] Murphy JE, Beard MC, Norman AG, Ahrenkiel SP, Johnson JC, Yu PR et al (2006) PbTe colloidal nanocrystals: synthesis, characterization, and multiple exciton generation. Journal of the American Chemical Society 128: 3241–3247
[128] Houtepen AJ, Koole R, Vanmaekelbergh DL, Meeldijk J, Hickey SG (2006) The hidden role of acetate in the PbSe nanocrystal synthesis. Journal of the American Chemical Society 128: 6792–6793
[129] Wehrenberg BL, Wang CJ, Guyot-Sionnest P (2002) Interband and intraband optical studies of PbSe colloidal quantum dots. Journal of Physical Chemistry B 106: 10634–10640
[130] Steckel JS, Coe-Sullivan S, Bulovic V, Bawendi MG (2003) 1.3 µm to 1.55 µm tunable electroluminescence from PbSe quantum dots embedded within an organic device. Advanced Materials 15 (21): 1862–1866
[131] Pietryga JM, Schaller RD, Werder D, Stewart MH, Klimov VI, Hollingsworth JA (2004) Pushing the band gap envelope: mid-infrared emitting colloidal PbSe quantum dots. Journal of the American Chemical Society 126: 11752–11753
[132] Baek IC, Il Seok S, Pramanik NC, Jana S, Lim MA, Ahn BY et al (2007) Ligand-dependent particle size control of PbSe quantum dots. Journal of Colloid and Interface Science 310: 163–166
[133] Sashchiuk A, Amirav L, Bashouti M, Krueger M, Sivan U, Lifshitz E (2004) PbSe nanocrystal assemblies: synthesis and structural, optical, and electrical characterization. Nano Letters 4: 159–165
[134] Nairn JJ, Shapiro PJ, Twamley B, Pounds T, von Wandruszka R, Fletcher TR et al (2006) Preparation of ultrafine chalcopyrite nanoparticles via the photochemical decomposition of molecular single-source precursors. Nano Letters 6: 1218–1223
[135] Czekelius C, Hilgendorff M, Spanhel L, Bedja I, Lerch M, Müller G et al (1999) A simple colloidal route to nanocrystalline ZnO/CuInS$_2$ bilayers. Advanced Materials 11: 643–646
[136] Banger KK, Jin MHC, Harris JD, Fanwick PE, Hepp AF (2003) A new facile route for the preparation of single-source precursors for bulk, thin-film, and nanocrystallite I–III–VI semiconductors. Inorganic Chemistry 42: 7713–7715
[137] Gurinowich LI, Gurin VS, Ivanov VA, Molochko AP, Solovei NP (1998) Optical properties of CuInS$_2$ nanoparticles in the region of the fundamental absorption edge. Journal of Applied Spectroscopy 63(3): 401–407
[138] Wakita K, Fujita F, Yamamoto N (2001) Photoluminescence excitation spectra of CuInS$_2$ crystals. Journal of Applied Physics 90: 1292–1296
[139] Castro SL, Bailey SG, Raffaelle RP, Banger KK, Hepp AF (2004) Synthesis and characterization of colloidal CuInS$_2$ nanoparticles from a molecular single-source precursor. Journal of Physical Chemistry B 108: 12429–12435
[140] Castro SL, Bailey SG, Raffaelle RP, Banger KK, Hepp AF (2003) Nanocrystalline chalcopyrite materials (CuInS$_2$ and CuInSe$_2$) via low-temperature pyrolysis of molecular single-source precursors. Chemistry of Materials 15: 3142–3147
[141] Choi SH, Kim EG, Hyeon T (2006) One-pot synthesis of copper-indium sulfide nanocrystal heterostructures with acorn, bottle, and larva shapes. Journal of the American Chemical Society 128: 2520–2521
[142] Nakamura H, Kato W, Uehara M, Nose K, Omata T, Otsuka-Yao-Matsuo S et al (2006) Tunable photoluminescence wavelength of chalcopyrite CuInS$_2$-based semiconductor nanocrystals synthesized in a colloidal system. Chemistry of Materials 18: 3330–3335
[143] Torimoto T, Adachi T, Okazaki K, Sakuraoka M, Shibayama T, Ohtani B et al (2007) Facile synthesis of ZnS-AgInS$_2$ solid solution nanoparticles for a color-adjustable luminophore. Journal of the American Chemical Society 129: 12388–12389

[144] Malik MA, O'Brien P, Revaprasadu N (1999) A novel route for the preparation of CuSe and CuInSe$_2$ nanoparticles. Advanced Materials 11: 1441–1444
[145] Zhong HZ, Li YC, Ye MF, Zhu ZZ, Zhou Y, Yang CH et al (2007) A facile route to synthesize chalcopyrite CuInSe$_2$ nanocrystals in non-coordinating solvent. Nanotechnology 18: 025602
[146] English DS, Pell LE, Yu ZH, Barbara PF, Korgel BA (2002) Size tunable visible luminescence from individual organic monolayer stabilized silicon nanocrystal quantum dots. Nano Letters 2: 681–685
[147] Holmes JD, Ziegler KJ, Doty RC, Pell LE, Johnston KP, Korgel BA (2001) Highly luminescent silicon nanocrystals with discrete optical transitions. Journal of the American Chemical Society 123: 3743–3748
[148] Hanrath T, Korgel BA (2003) Supercritical fluid–liquid–solid (SFLS) synthesis of Si and Ge nanowires seeded by colloidal metal nanocrystals. Advanced Materials 15: 437–440
[149] Pell LE, Schricker AD, Mikulec FV, Korgel BA (2004) Synthesis of amorphous silicon colloids by trisilane thermolysis in high temperature supercritical solvents. Langmuir 20: 6546–6548
[150] Tilley RD, Warner JH, Yamamoto K, Matsui I, Fujimori H (2005) Micro-emulsion synthesis of monodisperse surface stabilized silicon nanocrystals. Chemical Communications 1833–1835
[151] Warner JH, Hoshino A, Yamamoto K, Tilley RD (2005) Water-soluble photoluminescent silicon quantum dots. Angewandte Chemie-International Edition 44: 4550–4554
[152] Taylor BR, Kauzlarich SM, Delgado GR, Lee HWH (1999) Solution synthesis and characterization of quantum confined Ge nanoparticles. Chemistry of Materials 11: 2493–2500
[153] Yang CS, Bley RA, Kauzlarich SM, Lee HWH, Delgado GR (1999) Synthesis of alkyl-terminated silicon nanoclusters by a solution route. Journal of the American Chemical Society 121: 5191–5195
[154] Heath JR, Shiang JJ, Alivisatos AP (1994) Germanium quantum dots – optical properties and synthesis. Journal of Chemical Physics 101: 1607–1615
[155] Mokari T, Banin U (2003) Synthesis and properties of CdSe/ZnS core/shell nanorods. Chemistry of Materials 15: 3955–3960
[156] Kudera S, Zanella M, Giannini C, Rizzo A, Li YQ, Gigli G, Cingolani R, Ciccarella G, Spahl W, Parak WJ, Manna L (2007) Sequential growth of magic-size CdSe nanocrystals. Advanced Materials 19: 548–552
[157] Jun S, Jang E (2005) Interfused semiconductor nanocrystals: brilliant blue photoluminescence and electroluminescence. Chemical Communications 36: 4616–4618
[158] Wang S, Jarrett BR, Kauzlarich SM, Louie AY (2007) Core/shell quantum dots with high relaxivity and photoluminescence for multimodality imaging. Journal of the American Chemical Society 129: 3848–3856
[159] Revaprasadu N, Malik MA, O'Brien P, Wakefield G (1999) A simple route to synthesise nanodimensional CdSe–CdS core-shell structures from single molecule precursors. Chemical Communications 1573–1574
[160] Malik MA, O'Brien P, Revaprasadu N (2002) A simple route to the synthesis of core/shell nanoparticles of chalcogenides. Chemistry of Materials 14: 2004–2010
[161] Mekis I, Talapin DV, Kornowski A, Haase M, Weller H (2003) One-pot synthesis of highly luminescent CdSe/CdS core-shell nanocrystals via organometallic and "greener" chemical approaches. Journal of Physical Chemistry B 107: 7454–7462
[162] Pan DC, Wang Q, Jiang SC, Ji XL, An LJ (2005) Synthesis of extremely small CdSe and highly luminescent CdSe/CdS core-shell nanocrystals via a novel two-phase thermal approach. Advanced Materials 17: 176–179
[163] Danek M, Jensen KF, Murray CB, Bawendi MG (1996) Synthesis of luminescent thin-film CdSe/ZnSe quantum dot composites using CdSe quantum dots passivated with an overlayer of ZnSe. Chemistry of Materials 8: 173–180
[164] Reiss P, Bleuse J, Pron A (2002) Highly luminescent CdSe/ZnSe core/shell nanocrystals of low size dispersion. Nano Letters 2: 781–784
[165] Lee YJ, Kim TG, Sung YM (2006) Lattice distortion and luminescence of CdSe/ZnSe nanocrystals. Nanotechnology 17: 3539–3542
[166] Sung YM, Park KS, Lee YJ, Kim TG (2007) Ripening kinetics of CdSe/ZnSe core/shell nanocrystals. Journal of Physical Chemistry C 111: 1239–1242
[167] Steckel JS, Zimmer JP, Coe-Sullivan S, Stott NE, Bulovic V, Bawendi MG (2004) Blue luminescence from (CdS)ZnS core-shell nanocrystals. Angewandte Chemie-International Edition 43: 2154–2158
[168] Protière M, Reiss P (2006) Facile synthesis of monodisperse ZnS capped CdS nanocrystals exhibiting efficient blue emission. Nanoscale Research Letters 1: 62–67

[169] Yang YA, Chen O, Angerhofer A, Cao YC (2006) Radial-position-controlled doping in CdS/ZnS core/shell nanocrystals. Journal of the American Chemical Society 128: 12428–12429
[170] Lomascolo M, Creti A, Leo G, Vasanelli L, Manna L (2003) Exciton relaxation processes in colloidal core/shell ZnSe/ZnS nanocrystals. Applied Physics Letters 82: 418–420
[171] Chen HS, Lo B, Hwang JY, Chang GY, Chen CM, Tasi SJ et al (2004) Colloidal ZnSe, ZnSe/ZnS, and ZnSe/ZnSeS quantum dots synthesized from ZnO. Journal of Physical Chemistry B 108: 17119–17123
[172] Tsay JM, Pflughoefft M, Bentolila LA, Weiss S (2004) Hybrid approach to the synthesis of highly luminescent CdTe/ZnS and CdHgTe/ZnS nanocrystals. Journal of the American Chemical Society 126: 1926–1927
[173] Haubold S, Haase M, Kornowski A, Weller H (2001) Strongly luminescent InP/ZnS core-shell nanoparticles. Chem Phys Chem 2: 331–334
[174] Mićić OI, Smith BB, Nozik AJ (2000) Core-shell quantum dots of lattice-matched ZnCdSe$_2$ shells on InP cores: experiment and theory. Journal of Physical Chemistry B 104: 12149–12156
[175] Cao YW, Banin U (1999) Synthesis and characterization of InAs/InP and InAs/CdSe core/shell nanocrystals. Angewandte Chemie-International Edition 38: 3692–3694
[176] Cao YW, Banin U (2000) Growth and properties of semiconductor core/shell nanocrystals with InAs cores. Journal of the American Chemical Society 122: 9692–9702
[177] Kim SW, Zimmer JP, Ohnishi S, Tracy JB, Frangioni JV, Bawendi MG (2005) Engineering InAs$_x$P$_{1-x}$/InP/ZnSe III–V alloyed core/shell quantum dots for the near-infrared. Journal of the American Chemical Society 127: 10526–10532
[178] Sashchiuk A, Langof L, Chaim R, Lifshitz E (2002) Synthesis and characterization of PbSe and PbSe/PbS core-shell colloidal nanocrystals. Journal of Crystal Growth 240: 431–438
[179] Brumer M, Kigel A, Amirav L, Sashchiuk A, Solomesch O, Tessler N et al (2005) PbSe/PbS and PbSe/PbSe$_x$S$_{1-x}$ core/shell nanoparticles. Advanced Functional Materials 15: 1111–1116
[180] Lifshitz E, Brumer M, Kigel A, Sashchiuk A, Bashouti M, Sirota M et al (2006) Stable PbSe/PbS and PbSe/PbSe$_x$S$_{1-x}$ core-shell nanocrystal quantum dots and their applications. Journal of Physical Chemistry B 110: 25356–25365
[181] Xu J, Cui DH, Zhu T, Paradee G, Liang ZQ, Wang Q et al (2006) Synthesis and surface modification of PbSe/PbS core-shell nanocrystals for potential device applications. Nanotechnology 17: 5428–5434
[182] Kim S, Fisher B, Eisler HJ, Bawendi M (2003) Type-II quantum dots: CdTe/CdSe(core/shell) and CdSe/ZnTe(core/shell) heterostructures. Journal of the American Chemical Society 125: 11466–11467
[183] Yu K, Zaman B, Romanova S, Wang DS, Ripmeester JA (2005) Sequential synthesis of type II colloidal CdTe/CdSe core-shell nanocrystals. Small 1: 332–338
[184] Chen CY, Cheng CT, Yu JK, Pu SC, Cheng YM, Chou PT et al (2004) Spectroscopy and femtosecond dynamics of type-II CdSe/ZnTe core-shell semiconductor synthesized via the CdO precursor. Journal of Physical Chemistry B 108: 10687–10691
[185] Balet LP, Ivanov SA, Piryatinski A, Achermann M, Klimov VI (2004) Inverted core/shell nanocrystals continuously tunable between type-I and type-II localization regimes. Nano Letters 4: 1485–1488
[186] Zhong XH, Xie RG, Zhang Y, Basché T, Knoll W (2005) High-quality violet- to red-emitting ZnSe/CdSe core/shell nanocrystals. Chemistry of Materials 17: 4038–4042
[187] Xie RG, Zhong XH, Basché T (2005) Synthesis, characterization, and spectroscopy of type-II core/shell semiconductor nanocrystals with ZnTe cores. Advanced Materials 17: 2741–2744
[188] Milliron DJ, Hughes SM, Cui Y, Manna L, Li JB, Wang LW, Alivisatos AP (2004) Colloidal nanocrystal heterostructures with linear and branched topology. Nature 430: 190–195
[189] Xie RG, Kolb U, Basché T (2006) Design and synthesis of colloidal nanocrystal heterostructures with tetrapod morphology. Small 2: 1454–1457
[190] Burda C, Chen XB, Narayanan R, El-Sayed MA (2005) Chemistry and properties of nanocrystals of different shapes. Chemical Reviews 105: 1025–1102
[191] Kumar S, Nann T (2006) Shape control of II–VI semiconductor nanomateriats. Small 2: 316–329
[192] Carbone L, Nobile C, De Giorg M, Sala FD, Morello G, Pompa P et al (2007) Synthesis and micrometer-scale assembly of colloidal CdSe/CdS nanorods prepared by a seeded growth approach. Nano Letters 7: 2942–2950
[193] Talapin DV, Nelson JH, Shevchenko EV, Aloni S, Sadtler B, Alivisatos AP (2007) Seeded growth of highly luminescent CdSe/CdS nanoheterostructures with rod and tetrapod morphologies. Nano Letters 7: 2951–2959

Aqueous synthesis of semiconductor nanocrystals

By

Nikolai Gaponik[1], Andrey L. Rogach[2]

[1] Physical Chemistry, TU Dresden, Dresden, Germany
[2] Photonics and Optoelectronics Group, Physics Department and Center for NanoScience (CeNS), Ludwig-Maximilians-Universität München, Munich, Germany

1. Historical overview

Water is a natural medium for all forms of life. This is one reason why any solution-based techniques and processes proceeding in aqueous media are considered to be environmentally friendly and safe in comparison to others which demand, e.g. organic solvents or melts. In the field of the synthesis of colloidal semiconductor nanocrystals (NCs), which consist mainly of water insoluble II–VI and IV–VI compounds, the use of chemical precipitation reaction in aqueous media was historically the number one choice [1]. In the earlier work a commercially available colloidal silica sol (13 nm in diameter) was used as a carrier and stabilizer of the CdS [1] and ZnS [2] colloidal solutions as well as CdS–ZnS co-colloids [3]. The reported mean size of the NCs was relatively big, e.g. 37 nm diameter for CdS, as determined by fractional filtration through micropore filter, thus the optical spectra reported showed no serious deviations in comparison to the corresponding bulk materials [1]. Very soon the possibility of use of styrene/maleic acid anhydride copolymer (Brus group) or phosphates and polyphosphates (Henglein, Grätzel and Nozik groups) as stabilizers was recognised and colloidal (free of SiO_2) solutions of CdS [4–6], ZnS [2, 7], PbS [5], Cd_3P_2 [8], Zn_3P_2 [8], Cd_3As_2 [9], CdTe [10] and ZnTe [10] NCs were synthesized. Blue shifts of the absorption edge and the emission bands in comparison to the corresponding bulk materials were observed and reported. The theoretical background for this shift to higher energies/shorter wavelengths, the quantum confinement effect was introduced by brothers Efros [11] and Brus [12]. Further development of the stabilizing techniques included the use of short chelating peptides of general structure (g-Glu-Cys)n-Gly to control the nucleation and growth of CdS crystallites [13], micelles and vesicles [14] and finally the use of various short-chain thiols [15, 16]. The attempts to reach as small as possible sizes and as high as possible monodispersity of the NCs led to the establishment of the exciting field of ultra small molecular-like semiconductor clusters with definite size, structure and characteristic optical properties [16, 17], e.g. $[Cd_{17}S_4(SCH_2CH_2OH)_{26}]$ [16, 17] and $[Cd_{32}S_{14}(SCH_2CH(OH)CH_3)_{36}] \cdot (H_2O)_4$ [18]. These clusters correspond to minima of the free energy vs. particle size dependence owing to their closed

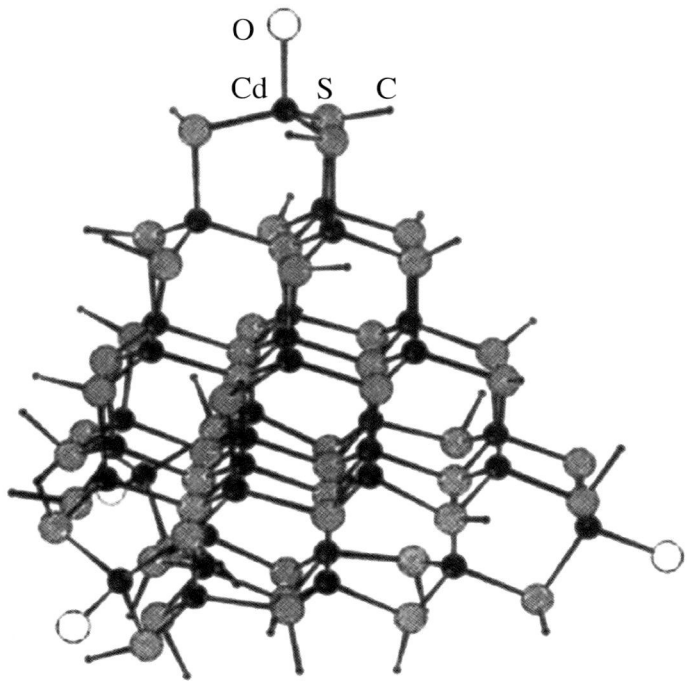

Fig. 1. Structure of a [$Cd_{32}S_{14}(SCH_2CH(OH)CH_3)_{36}$] · ($H_2O$)$_4$ cluster as derived from the single-crystal X-ray data. Reproduced from [18], © 1995, with permission from American Chemical Society

structural shells (the concept of so-called "clusters of magic size" in the earlier literature [19]) and are naturally "100% monodisperse". They can be crystallized in macroscopically large single crystals allowing their investigation by single-crystal X-ray analysis, including the exact determination of the atomic coordinates. Figure 1 shows the inner structure of a [$Cd_{32}S_{14}(SCH_2CH(OH)CH_3)_{36}$] · ($H_2O$)$_4$ cluster representing a piece of the zinc blende lattice in the shape of a tetrahedron.

Starting from the very early publications, the photoluminescence (PL) from semiconductor NCs was paid a lot of attention. It was shown, for example, that weak fluorescence of CdS NCs with maximum at 620–660 nm can be efficiently quenched in presence of nitrobenzene, various anions and cations [1, 2]. It was also shown that doping of the CdS NCs with Ag^+ or Cu^{2+} resulted in the increase of the fluorescence in comparison to the undoped samples [1]. One of the very interesting examples of fluorescence enhancement ("activation") in phosphate-stabilized CdS NCs was their treatment with NaOH and subsequent addition of excess of Cd^{2+} ions [20]. The precise control of the pH was found to be necessary as well. This treatment resulted in the many fold increase of the band gap photoluminescence, which was attributed to the formation of broad-band gap material, $Cd(OH)_2$, as a shell on the luminescent CdS core nanoparticles. This approach was essentially the first successful realization of the type I core-shell NCs, well-known and widely used nowadays (see the Chapter of P. Reiss of this book).

As a rule (with the exception of molecular-like, or magic-sized [19] clusters) the size distribution of the colloidal NCs prepared in water is relatively broad, and several post-preparative approaches where introduced to optimize it. One of the most important and widely used nowadays techniques, not only for water-soluble NCs, is the size-selective precipitation from solvent–non-solvent mixtures which was firstly introduced for CdS NCs [21]. The method is very simple and exploits the difference in solubility of smaller and larger particles [21, 22]. A typical example of carrying out the size-selective precipitation on a nanoparticle colloid is as follows: a sample of as-prepared nanoparticles with a broad size distribution is dispersed in a solvent and a non-solvent is added dropwise under stirring until the initially optically clear solution becomes slightly turbid. The largest nanoparticles in the sample exhibit the greatest attractive van der Waals forces and tend to aggregate before the smaller particles. The aggregates consisting of the largest nanoparticles can be isolated by centrifugation or filtration and re-dissolved in any appropriate solvent. The next portion of non-solvent is added to the supernatant to isolate the second size-selected fraction, and so on. The procedure can be repeated several times and allows for obtaining up to 10 or more size-selected fractions from one portion of the crude solution. Moreover, each size-selected fraction can be subjected again to size selection to further narrow the size distribution. Figure 2 shows an example of post-preparative size fractionation for CdTe NCs. All size-selected fractions possess sharp excitonic transitions in the absorption spectra which is a direct evidence of their narrow particle size distributions. TEM and HRTEM investigations show that a carefully performed size-selective precipitation allows for achieving size distributions of 5–10%. In some specific cases also exclusion chromatography [23, 24] and gel electrophoresis [25] were utilized for the size fractionation.

The introduction of the hot-injection synthesis in high-temperature boiling coordinating organic solvents (often referred to as TOP–TOPO synthesis, from initially used coordinating solvents trioctylphosphine, TOP and trioctylphosphine

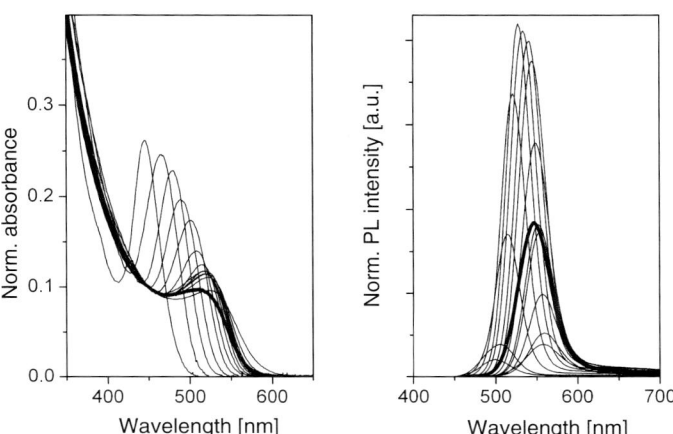

Fig. 2. Absorbance and photoluminescence of the size-selected fractions of the thioglycolic acid-capped CdTe nanocrystals. The spectra of the initial crude solution are highlighted in bold

oxide, TOPO) in 1993 made a revolution in colloidal synthesis of the NCs [22]. The CdSe NCs of different sizes synthesized by this approach became an object of numerous studies in the last 15 years. The methods of core passivation by wide-band gap inorganic shells were further developed [26], and synthetic approaches allowing shape control of NCs were introduced [27]. Resulting materials allowed first applications of colloidal NCs, among others in light-emitting diodes (LEDs) [28], photovoltaics [29] and bio-labelling [30]. A recent overview on semiconductor NCs synthesized in organic solvents can be found in [31] as well as in the Chapters of P. Reiss and Kudera/Carbone/Manna/Parak of this book.

First steps toward successful aqueous alternative to the organometallically synthesized CdSe-based NCs were done by Nozik's [15] and later by Weller's [32] groups in their developments of CdTe NCs. In the former work CdTe NCs were synthesized using the mixture of 3-mercapto-1,2-propane-diol and hexametaphosphate as stabilizers, and their size-dependent properties including absorption, band gap emission, energy level diagrams and extinction coefficient were reported. The latter work introduced CdTe NC synthesis in the presence of various thiols (2-mercaptoethanol or 1-thioglycerol (TG)) solely and included, in addition to the size-tuneable optical spectra, the TEM and X-ray diffraction (XRD) data of NCs. The following efforts in the field resulted in as high as 20% photoluminescence quantum efficiency (PL QE) of thioglycolic acid-capped CdTe NCs achieved by their proper post-preparative surface modification [33], in the demonstration of LEDs operating with these NCs [34, 35], and in their bio-conjugation with albumin [36]. Nowadays, aqueous syntheses of II–VI semiconductor NCs like CdS [16], CdSe [37], CdTe [38, 39], $Cd_xHg_{1-x}Te$ [40], HgTe [41] and ZnSe [42] by employing different short-chain thiols as stabilizing agents represent a useful alternative to synthetic routes in highly boiling organic solvents [22, 43–45], providing brightly emitting NCs with PL QE

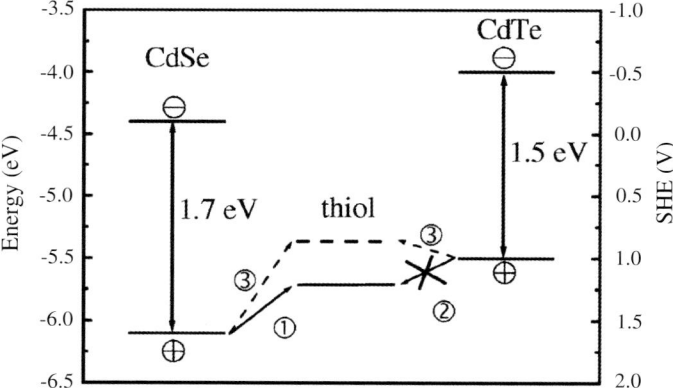

Fig. 3. Positions of bulk CdSe and bulk CdTe band edges shown both on a vacuum scale and with respect to a standard hydrogen electrode reference. The standard potential expected for a thiol that does not quench the CdTe exciton luminescence is given by a solid line between CdSe and CdTe. Hole trapping can occur from CdSe (process 1) but not from CdTe (process 2). The dashed line indicates the assumed position for the standard potential of a thiol that quenches the luminescence of both CdSe and CdTe (process 3). Reproduced from [46] © 2004, with permission from American Chemical Society

of 40–60%. The use of different thiols allows for the kinetic control of the NCs synthesis, to passivate surface dangling bonds, and to provide stability, solubility and surface functionality of the nanoparticles. It is of interest that thiol capping, although providing stable colloidal solutions of CdS and CdSe NCs, does not result in strong band gap emission of these nanoparticles. The reason for this, for the case of CdSe NCs, was addressed in [46] and is schematically explained in Fig. 3. The energy difference between the tops of the valence bands of bulk CdSe and CdTe is approx. 0.5 eV and this difference is even more pronounced in the case of CdS and CdTe (ca. 1.0 eV) [47]. Hole trapping from the semiconductor NCs on a thiol and subsequent PL quenching is energetically favorable only if the thiol redox energy level is situated at higher energies than the valence band top. As seen from Fig. 3, this probability is high for the CdSe (and should be even higher for CdS). This problem may be overcome if the appropriate inorganic passivation of the emitting NC core is done. One of the examples is the ZnSe(S) NCs in which sulfur-enriched shell may provide efficient wide-band gap associated screening of the hole trapping effect and by this the dramatic enhancement of the band gap PL of ZnSe NC core is observed, as will be discussed below.

In spite of the typically low PL QE of the core-only thiol-capped CdS and CdSe NCs, the interest to these nanoparticle systems synthesized by aqueous approaches remains to be high. For CdS NCs, capping with thiols has been proved to be successful in the above-mentioned formation of the molecular-like clusters [16, 17, 48], in the synthesis of NCs as functionalized building blocks for the sophisticated self-assembled structures with gold nanoparticles [49] or bio-conjugates [50, 51], and as model NCs for studies of the stability and size distributions in the nanoparticle assemblies [52]. Alternative synthetic approaches to water-soluble CdS NCs mainly addressed their biological applications, including syntheses in presence of D- and L-penicillamine [53], specially engineered peptides [54] or DNA [55]. Thiol-based synthesis of CdSe NCs also provided extremely small, molecular-like clusters [37]; the interest to larger CdSe NCs synthesized by this method has been fueled very recently as to the light-absorbing and photosensitizing species in solar cells [56]. Citrate-stabilized CdSe NCs have been developed [57], with a reasonably efficient band-edge emission being strongly increased upon photoactivation [58, 59].

Much of the literature based on the aqueous synthesis of the lead chalcogenide materials has concentrated on synthesizing PbS particles employing different surface ligands and demonstrating a degree of control over the final shape and morphology of the material. Using a wide variety of techniques such as electrochemistry, various polymer and mixed polymer systems and chemical synthesis in restricted spaces of various geometries, different shapes of PbS NCs such as hollow spheres [60], cubes [61–63], rods [64], belts [65], wires [65] and stars [66] have been reported. However, with few exceptions [67], water-based synthesis routes have not resulted in efficiently emissive PbS NC materials.

In what follows, we will focus on the description of the most successful examples of aqueous thiol-capped colloidal NCs, possessing strong band gap emission and by this being promising for the broad nano(bio)technological applications: the most widely studied CdTe NCs emitting in visible and near-infrared, UV-blue-emitting ZnSe NCs, and finally near-infrared-emitting HgTe and alloyed $Cd_xHg_{1-x}Te$ NCs.

2. CdTe nanocrystals

The basics of the aqueous synthesis of thiol-capped CdTe NCs has been described in details in [32, 38, 39]. In a typical standard synthesis [38], $Cd(ClO_4)_2 \cdot 6H_2O$ (or other soluble Cd salt) is dissolved in water in the range of concentrations of 0.02 M or less, and an appropriate amount of the thiol stabilizer is added under stirring, followed by adjusting the pH by dropwise addition of a 1 M solution of NaOH. The solution is placed in a flask fitted with a septum and valves and is deaerated by N_2 bubbling for 30 min. Under stirring, H_2Te gas is passed through the solution together with a slow nitrogen flow. Alternatively, an excess of H_2Te may be passed through the deaerated solution of the NaOH with known concentration forming equimolar NaHTe solution. After this the estimated amount of NaHTe solution can be taken out and injected in the reaction flask. It should be noted that NaHTe solutions are inherently very unstable and become pink due to the oxidation. The use of only freshly prepared NaHTe solution is recommended.

CdTe NC precursors are formed at the stage of Te precursor addition (Reaction 1 or 1a); formation and growth of NCs (Reaction 2) proceed upon refluxing at 100°C under open-air conditions with a condenser attached.

$$Cd^{2+} + H_2Te \xrightarrow{HS-R} Cd-(SR)_xTe_y + 2H^+ \qquad (1)$$

$$Cd^{2+} + NaHTe \xrightarrow{HS-R} Cd-(SR)_xTe_y + H^+ + Na^+ \qquad (1a)$$

$$Cd-(SR)_xTe_y \xrightarrow{100°C} CdTe\ NCs \qquad (2)$$

A schematic drawing of the typically used experimental setup is shown in Fig. 4, although the specific design of this setup may vary from one group to another.

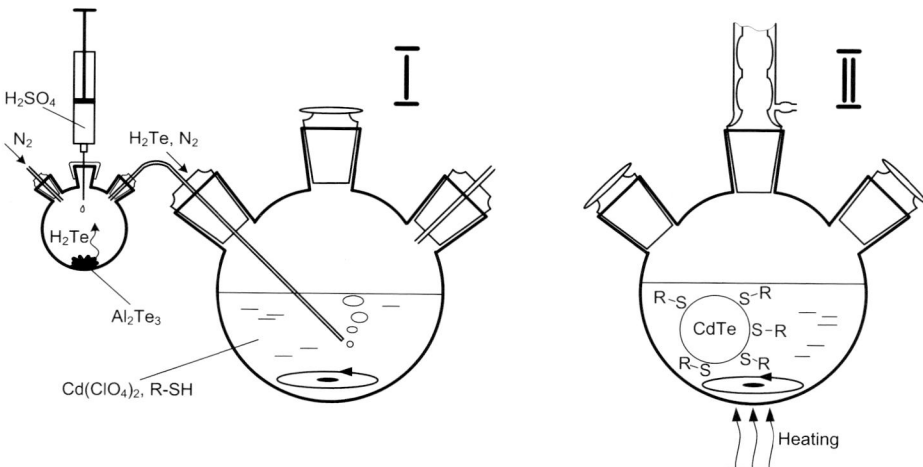

Fig. 4. Schematic presentation of the synthesis of thiol-capped CdTe NCs. First stage: formation of CdTe precursors by introducing H_2Te gas into the aqueous solution of Cd precursors complexed by thiols. Second stage: formation and growth of CdTe NCs promoted by reflux. Reproduced from [38], © 2002, with permission from American Chemical Society

Important part of this setup is the connecting tube for introducing the H$_2$Te gas, which should be as short as possible and made of glass or another inert material. The use of glass joints and connections is strongly recommended due to the high reactivity of H$_2$Te gas with rubber and common polymer tubes. The use of relatively small and well-deaerated flask for the generation of H$_2$Te may also help to reduce undesirable losses of this gas. Special precautions should be taken against the possible leakage of the non-reacted H$_2$Te. Exhaust traps filled with NaOH solution are proven to be efficient for this purpose. We note that the synthetic procedure described above is easily up-scalable. Indeed, even in laboratory conditions the Reactions (1) and (2) may be routinely performed in a few litres volume and by this, yield up to several grams or even tens of grams of CdTe NCs (estimated for dry weight).

H$_2$Te gas can be generated for the synthesis of CdTe NCs as well as other tellurides, like HgTe [41, 68] or ZnTe [10] by at least two different methods: chemical decomposition of Al$_2$Te$_3$ powder or lumps according to Reaction (3), and electrochemical reduction of Te electrode in acid media according to Reaction (4).

$$Al_2Te_3 + 3H_2SO_4 \rightarrow 3H_2Te \uparrow + Al_2(SO_4)_3 \qquad (3)$$

$$Te^0 + 2H^+ + 2e^- \rightarrow H_2Te \uparrow \qquad (4)$$

The use of Reaction (3) is the simplest way, if Al$_2$Te$_3$ is available. However, the limited amount of suppliers and the continuously increasing price for this reagent limit its availability for a lot of groups dealing with the synthesis of corresponding NCs, stimulating a search for alternative sources, e.g. NaHTe solution obtained by reduction of Te powder with NaBH$_4$ [69, 70]. Although this method provides an alternative for Te source in the synthesis of CdTe NCs, the direct injection of H$_2$Te gas is an easier, more controllable, cleaner and more reproducible way to produce high-quality CdTe NCs. To avoid the use of Al$_2$Te$_3$, electrochemical method (Reaction (4)) can be used to produce H$_2$Te gas, which has been known since the beginning of last century [71] and was recently generalized and reported in detail by Hodes et al. [72, 73]. The use of the electrochemically generated H$_2$Te gas for successful synthesis of both CdTe and HgTe NCs has been reported recently [39, 68, 74]. The tellurium cathode may be easily formed by melting of tellurium powder inside a glass tube and subsequent cutting the tube end. The reaction proceeds in a glass cell (or flask) and in the simplest case can be controlled by common power supplier instead of potentiostat, which makes the approach easily available for a broad scientific community. The electrochemical method also allows for control of H$_2$Te amount by measuring the charge passed through the cell, is applicable for the continuous generation of this gas, and can be easily scaled up.

Originally introduced [32] molar ratio of the main reaction components, Cd:Te:R–SH (R–SH stays for thiol stabilizer) being 1:0.5:2.45 was widely used with only slight deviations as "standard" synthetic approach [38]. As stabilizers, the wide family of short-chain (up to 4 carbon units) thiols possessing at least one or several functional groups (amino, carboxylic, hydroxylic, etc.) were employed [38]. Most popular are thioglycolic (or mercaptoacetic) acid (TGA), mercaptopropionic acid (MPA) and 2-mercaptoethylamine (or cysteamine). Both TGA and MPA allows the synthesis of the most stable (typically, for years) aqueous solutions of CdTe NCs possessing negative charge due to the presence of surface carboxylic groups. Cysteamine-stabilized NCs possess moderate photostability (although they may be stable for years as well being kept in darkness) and attract an interest due to surface amino-functionality and positive surface charge in neutral and slightly acidic media. Other thiol stabilizers are mainly used when some specific functionalities are envisaged, the overview of them may be found in [38]. The improvement of the standard synthesis of aqueous CdTe NCs focused on the optimization of the thiol:Cd ratio and pH. Upon testing different TGA:Cd ratios between 2.45 (as it was used in the standard synthesis of [38]) and 1.1, it was realized by several groups [75–77] that decreasing the TGA:Cd ratio leads to a drastic increase of the PL QE of the CdTe NCs. In an attempt to understand this influence on the properties of the NCs, the experimental data were compared with the results of a numerical simulation of the solution composition [76]. The results of this simulation show that the increase of the PL QE of CdTe NCs with the decrease in the TGA concentration can be attributed to the increase in the relative concentration of the Cd–SR complex (i.e. uncharged 1:1 complex of cadmium with TGA). This tendency has a natural limit at very low values of the TGA:Cd ratio (approaching 1) when the amount of stabilizer in the system becomes insufficient to stabilize NCs from aggregation. There is a competition of at least two different factors during the synthesis: (i) upon decrease in the concentration of TGA, the surface quality of the NCs improves as a result of the increase in the relative concentration of 1:1 Cd–SR complex in comparison with other possible complexes in solution and (ii) a sufficient amount of TGA as a stabilizer has to be present in solution to provide stability and surface passivation of the growing NCs. As a result, the optimum TGA:Cd ratio allowing to produce CdTe NCs emitting with PL QE 40–60% at room temperature as-prepared is slightly higher than 1 and the experimentally obtained optimal values are 1.30 [76], 1.32 [75] and 1.20 [77]. This tendency is of general use, for example, the synthesis of CdTe NCs employing 1-thioglycerol as a stabilizer at R–SH:Cd ratio of 1.3:1 provides as-prepared samples with PL QE in the range of 25%, which exceeds the initialy reported (3%) values [32, 38] by almost an order of magnitude.

We note that the solution of Cd precursors at low TGA:Cd ratios may look slightly turbid. This fact is an additional indirect evidence of the domination of the uncharged, less-soluble Cd–SR complex. The turbidity of the solution does not disappear during refluxing, but this does not influence the ongoing Reaction (2); filtration of the final solution of CdTe NCs easily removes the insoluble precipitate. The precipitate of Cd–SR may play an additional role as a source of cadmium. Gradual dissolution of the Cd–SR complex during the NC growth provides a constant rate of transport of Cd ions to the particles. A slow flux of the cadmium precursor in turn provides the possibility of growing the NCs under diffusion control which, as has been predicted theoretically [78, 79] is preferable for narrowing the size distribution and may be a key factor for the dynamic improvement of the surface quality of the growing NCs.

The pH value is another important factor which strongly influences the PL QE of thiol-capped NCs post-preparatively [33, 70, 80, 81]. Thus, it is reasonable to assume that the pH will influence the quality of the NCs during the synthesis as well.

According to [38] the optimum pH value for the synthesis employing different capping ligands strongly depends on the nature of the stabilizer. For example, the recommended value in case of cysteamine is 5.6–6.0 (which is a natural value of cadmium perchlorate mixture with cysteamine, no adjustment of the pH by addition of NaOH solution is necessary in this case). In case of TGA the recommended value [38] was 11.2–11.8. For this stabilizer, it was found later [39, 76] that an increase of pH of the initial solution is followed by a considerable acceleration of the NC growth. Moreover, as a result of this acceleration one can choose the synthetic conditions allowing the "focusing" of the NCs growth in term of narrowing their size distribution. For example, NCs synthesized at pH 12.0 and TGA:Cd ratio 1.3 possess full width at half maximum (FWHM) of the PL band of 39 nm, PL QE 45% and a Stokes shift as small as 100 meV when CdTe NCs are approximately 3 nm in diameter and show a PL maximum at ca. 600 nm. Further growth proceeds more slowly and is accompanied by a slight broadening of the size distribution [76]. A relatively fast NC growth leads to their low quality reflected in low crystallinity of the resulting particles and a large amount of defects and surface states. On the other hand, a comparatively slow growth rate leads to a high content of sulfur (as a product of the TGA decomposition [52, 82]) in the NCs.

CdTe NCs synthesized in water can be transferred to non-polar organic solvents like toluene through a partial ligand exchange with a long-chain thiol (1-dodecanethiol) in the presence of acetone [83]. The transfer efficiency reaches 90% and depends on the component ratio of the 1-dodecanethiol/toluene/acetone mixture (typically 1:1:4 volume ratio), which in turn depends on the concentration of NCs and their average size. For any particular batch of NCs to be transferred, the correct ratio has to be found experimentally, with a typical variation of acetone content in the above-mentioned three-component system from 1:1:3 to 1:1:8. NCs can be transferred more efficiently when they are washed (e.g. by size selective precipitation) from the reaction by-products and the excess of short-chain thiol ligands. Thiol-capped CdTe NCs transferred into organics were used as photosensitizers of fullerenes [84, 85], as building blocks for NCs/polymer composites [83] and as core material for the synthesis of stable and brightly emitting core-shell CdTe/ZnS nanoparticles [86]. Their absorption and emission wavelengths have been shown to be tuned by surface modification in the presence of dodecanethiol [87]. Alternative methods of transfer of CdTe NCs from water to organic media include the use of polymerizable surfactants [88–90], tetra-n-octylammonium bromide [87] and ionic liquids [91].

Furthermore, thiol-capped CdTe NCs are also available in polar organic solvents. Mercaptoethylamine-capped CdTe NCs synthesized in water are readily re-soluble in dimethylformamide (DMF) after being precipitated by 2-propanol and dried. A direct synthesis in DMF is possible by taking cadmium lactate as a precursor and 1-thioglycerol as a ligand [82]. The synthesis proceeds at higher temperature, the growth of the NCs is faster and it takes only a few hours to obtain red-emitting samples. CdTe NCs precipitated from the crude solution immediately after synthesis by addition of excess of non-solvent (e.g. diethylether) are not only readily re-dissolvable in DMF, but also in methanol. To the best of our knowledge, this is the only example of II–VI semiconductor colloidal NCs being soluble in methanol as

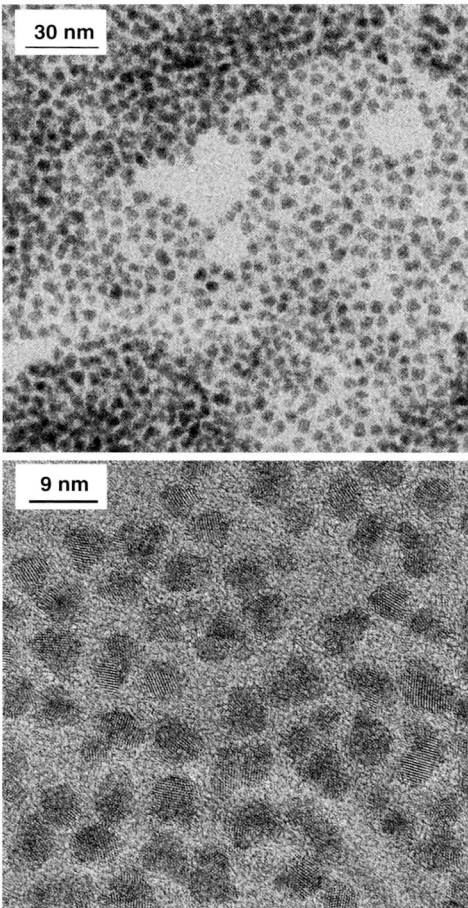

Fig. 5. TEM (top) and HRTEM (bottom) images of MPA-capped CdTe NCs, 5.5 nm average size, with PL maximum at 780 nm. Reproduced from [39], © 2007, with permission from American Chemical Society

synthesized. Indeed, methanol, among other short-chain alcohols is a commonly used non-solvent for size-selective precipitation of many types of organically and aqueously prepared NCs.

Typical TEM and HRTEM images of thiol-capped CdTe NCs with sizes from 2 to 6 nm can be found in [32, 38, 82]. Figure 5 shows TEM and HRTEM images of relatively large, 5.5 ± 0.5 nm CdTe NCs capped with MPA. In order to avoid aggregation on the TEM grid, which is common for aqueous solutions of thiol-capped NCs, the size-selected sample was transferred to toluene using partial ligand exchange with 1-dodecanethiol by applying the procedure of [83]. The images confirm the monodispersity of NCs; their non-spherical shape can be described within a recently proposed truncated tetrahedral model [92].

The sizes of thiol-capped CdTe NCs synthesized in aqueous solution can be determined from the so-called "sizing curve": the function of the size of thiol-capped

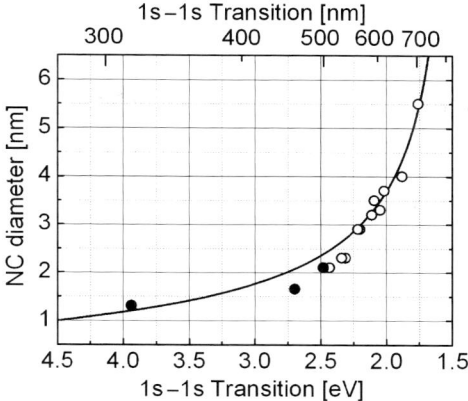

Fig. 6. Sizing curve for thiol-capped CdTe NCs synthesized in water. Filled circles represent sizes of NCs experimentally determined from powder XRD spectra; open circles represent sizes of NCs experimentally determined from TEM images. Solid line is a calculated dependence of the 1s–1s transition energy on CdTe NC size. Reproduced from [39], © 2007, with permission from American Chemical Society

CdTe NCs on the energy of the 1s–1s electronic transition estimated from the position of the first absorption peak (Fig. 6). The points (open circles) in Fig. 6 were derived from statistical analyses of NC sizes obtained by TEM measurements. For the smallest NCs (filled circles in Fig. 6) for which the precise TEM analyses are difficult we used the sizes derived from the powder XRD spectra as described in detail in [32]. Calculation of the 1s–1s transition energy of CdTe NCs (treated as spheres) as a function of their size has been done using an extended theoretical approach described in detail in [93], and is presented in Fig. 6 as solid line. The extension over the common effective mass approximation includes the implementation of the Coulomb interaction and finite potential wells at the particle boundaries in water as the surrounding dielectric medium. The physical parameters of bulk CdTe put into the model can also be found in [32]. The agreement between experiment and theory is quite well. We note that the sizing curve from [94] which is widely used in scientific community has been derived for CdTe NCs prepared by high-temperature organic syntheses and does not include data for small NCs (first absorption maximum at wavelengths shorter than 570 nm, Fig. 6) which are very easy to synthesize in water.

Thiol molecules can release S^{2-} in the course of the prolonged refluxing in the basic medium, which build into the lattice of the growing NCs. Thus, the positions of the powder XRD reflexes of CdTe NCs synthesized under prolonged refluxing in the presence of thioglycolic acid are intermediate between the values of the cubic CdTe and the cubic CdS phases [82]. Mixed $CdTe_xS_{1-x}$ NCs with some gradient of sulfur distribution from inside the nanoparticles to the surface are formed under these conditions. They represent a kind of core-shell system with a naturally CdS-capped surface created by mercapto-groups covalently attached to the surface cadmium atoms. Importantly, the synthesis of such kind of core-shell NCs occurs in one step, as the sulfur originates from the stabilizing thiol molecules and releases during the particle growth. At the bulk CdTe/CdS interface, the conduction band step, i.e. the

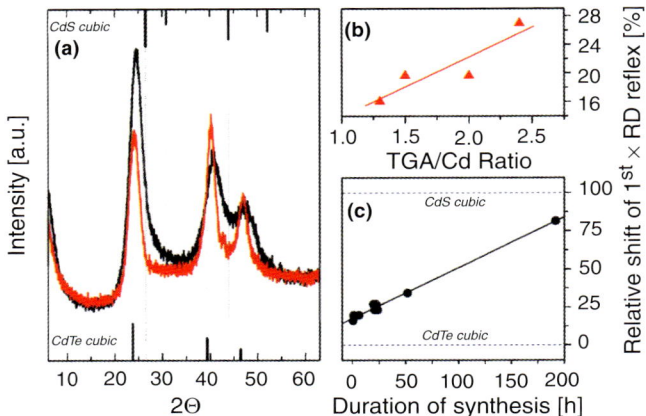

Fig. 7. XRD patterns of the CdTe NCs synthesized with a TGA:Cd ratio of 1.3 (red) and 2.45 (black) after 20 h of synthesis (**a**). **b** Shows the influence of the TGA:Cd ratio on the relative position of the (1 1 1) XRD reflex of the CdTe NCs. **c** Presents the evolution of the relative shift of the (1 1 1) XRD reflex of the CdTe NCs during the synthesis. Reproduced from [39], © 2007, with permission from American Chemical Society

offset of the absolute band position, is close to zero, whereas the valence band step is ~1 eV [47]. The wave functions calculated for a CdTe/CdS system with the particle-in-a-box model [93] show a delocalization of the electron through the entire structure and the confinement of the hole to the CdTe core – the same phenomenon as reported for the organically synthesized core-shell CdSe/CdS NCs [43] providing their photostability and electronic accessibility.

The reduction of the amount of TGA at the synthesis stage leads to a reduction of the sulfur content in the NCs. XRD patterns of the CdTe NCs (Fig. 7) show that at a TGA:Cd ratio of 1.3 and a pH 12, a smaller shift of the reflexes toward the position corresponding to cubic CdS is observed. It can be explained by the fact that decreasing the amount of the stabilizer in solution and the acceleration of the NCs growth leads to a decreasing probability of TGA hydrolysis and, as a result, to a lower sulfur content. As discussed in [38, 95] the sulfur-enriched shell itself may be important for the improvement of the stability and luminescence properties of CdTe NCs. At the same time, a few monolayers of this shell is enough for the efficient protection of the NCs and further growth of CdS only reduce the NCs quality similar to the effect of the ZnS shell on the properties of CdSe NCs [96].

Typical absorption and PL spectra of size-selected [21, 38] fractions of TGA- and MPA-capped CdTe NCs are shown in Fig. 8. PL spectra of TGA-capped CdTe NCs are tunable in the range of 500–700 nm, while those of MPA-capped NCs are tunable between 530 and 800 nm. The MPA capping allows for a relatively quick and controllable growth of CdTe NCs up to 6 nm in diameter. The energy gap of bulk CdTe estimated from the absorption measurements at 300 K is 1.43 eV or ca. 867 nm [97]. The superior tunability of the absorption over the very broad spectral range is important for the use of thiol-capped CdTe NCs as absorbers in solar cells [98], for choosing optimal donor–acceptor pairs for FRET-based structures [99–101], as well

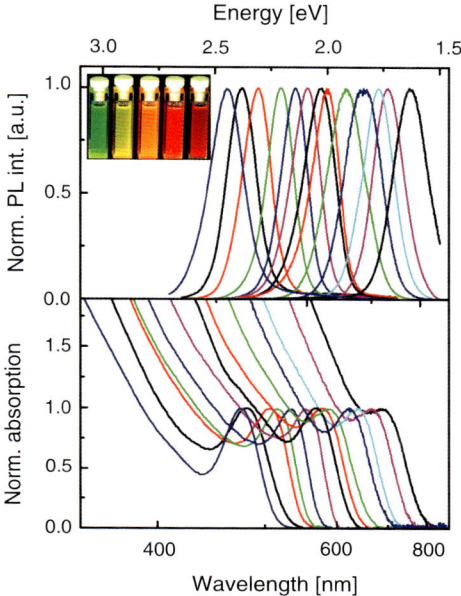

Fig. 8. A set of typical PL (top) and absorption (bottom) spectra of TGA-capped and MPA-capped CdTe NCs demonstrates their tuneability over a broad spectral range in the visible and near-infrared. Excitation wavelength is 450 nm. The inset shows a photograph of brightly emitting CdTe NCs of different sizes taken under UV lamp excitation. Reproduced from [39], © 2007, with permission from American Chemical Society

as for tuning an optimal resonance condition in nanoplasmonics systems [102, 103]. Narrow PL spectra in combination with their tuneability and high PL QE are of a special interest for bio-labelling applications [104], imaging [105] and LEDs [34] based on thiol-capped CdTe NCs.

The broad interest to thiol-capped CdTe NCs has triggered a search for improvements and specific adaptations of their conventional synthetic procedures described above [32, 33, 38]. This includes the hydrothermal synthesis [69, 77], illumination [95], ultrasonic [106] or microwave irradiation [107, 108] treatment, the use of an inert atmosphere [109], variation of reagent concentrations [75–77, 80] and pH [76, 80], and the use of alternative capping agents like glutathione [110].

3. ZnSe nanocrystals

Currently, a lot of attention is paid to the safe handling of nanometer-sized materials [111], which demands among others the development of syntheses of colloidal semiconductor NCs from low-toxic materials and the use of environmentally friendly technologies [112–116]. ZnSe NCs [117] synthesized in water is one of the prominent examples of such kind of nanomaterials.

The synthetic procedure for ZnSe NCs is very similar to the synthesis of CdTe NCs [42, 118, 119]. In a typical synthesis $Zn(ClO_4)_2 \cdot 6H_2O$ is dissolved in water in the range of concentrations of 0.02 M or less, and an appropriate amount of the thiol stabilizer (1-thioglycerol, TGA or MPA) is added under stirring, followed by adjusting the pH by dropwise addition of 1 M solution of NaOH to 6.5 in the case of TGA or MPA capping, or to 11.2–11.8 in the case of TG. The mentioned pH values were experimentally found to be optimal for the synthesis of stable colloids. The solution is deaerated by N_2 bubbling for 1 h. Under stirring, H_2Se gas (generated by the reaction of Al_2Se_3 lumps with an excess amount of 1 N H_2SO_4 under N_2 atmosphere [42] or electrochemically [73]) is passed through the solution together with a slow nitrogen flow. ZnSe precursors are formed at this stage. The further nucleation and growth of the NCs proceed upon refluxing at 100°C under open-air conditions with a condenser attached.

A typical temporal evolution of the absorption of the ZnSe NCs is shown in Fig. 9. A growth of the NCs during reflux is indicated by a low-energy shift of the absorption. The PL efficiency of these solutions is negligible and shows mainly a broad trap-emission band (400–600 nm). An additional very weak band-edge emission appears only after long times of reflux. Among the capping agents used a relatively stronger trap-emission is found to be characteristic for TG-capped ZnSe NCs. The synthesis and characterization of this type of white-blue-emitting NCs was reported in details recently [118, 119]. Widely used ratio of the precursors Zn:Se: R–SH is near 1:0.5:2.5, which is similar to the traditional one for the CdTe NCs

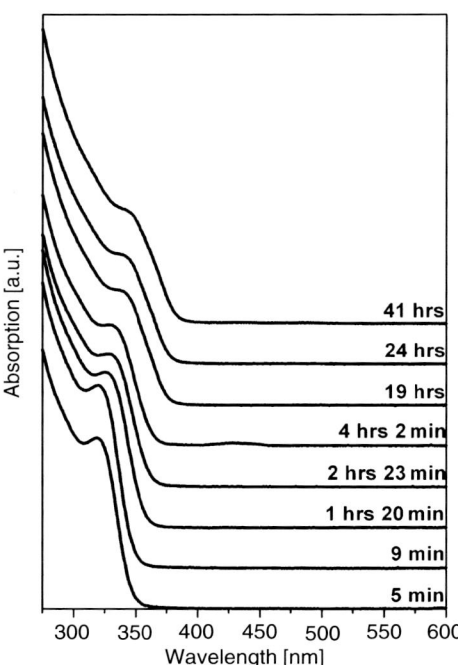

Fig. 9. Evolution of the absorption spectra of a crude solution of ZnSe NCs during the synthesis. Reproduced from [42], © 2004, with permission from American Chemical Society

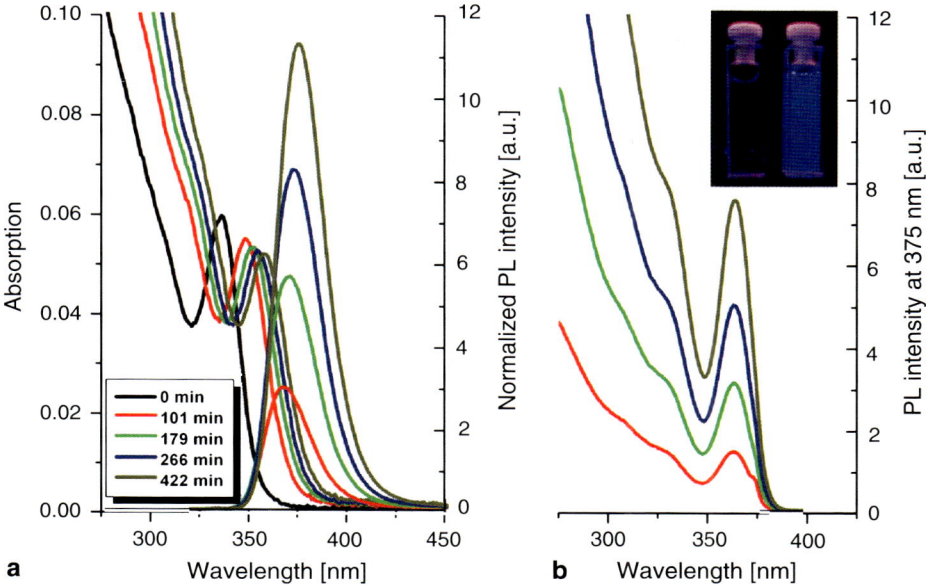

Fig. 10. Evolution of the absorption and PL (**a**) and of the PL excitation spectra ($\lambda_{observation}$ 375 nm) (**b**) of ZnSe NCs during illumination. Inset shows a true color fluorescence image (λ_{ex} 366 nm) of the ZnSe NCs before (left) and after (right) the photo-treatment. Reproduced from [121], © 2006, with permission from American Chemical Society

synthesis. The varying of this ratio towards lower amounts of the stabilizer allowed improvement of the photoluminescence in the case of 1-thioglycerol stabilizer resulting in strong whitish-blue trap-emission [119], but did not result in any considerable improvement in the cases of MPA and TGA.

In order to improve the PL properties of the ZnSe NCs (enhancement of the band-edge and suppression of the trap-emission), the colloidal solutions can be irradiated with a xenon lamp [42] or alternatively with a UV-lamp [120]. The presence of excess of Zn^{2+} ions and TGA molecules was found to be necessary. The dependence of the PL spectra on the duration of the irradiation is shown in Fig. 10. Under illumination, the PL QE increases from ca. 0.1% being characteristic for the as-prepared solutions up to 10–30% [42] and even 50% [120] if the pH during the treatment maintained to be 6.5 and 11, respectively.

The position of the PL maximum and the absorption edge shifted to the lower energy region during the irradiation. The PL emissions with maxima in the region 330–400 nm were achieved under the white-light irradiation and pH 6.5 [42], while larger NCs with PL maxima up to 435 nm may be prepared under UV light at pH 11 [120]. Since the ZnSe NCs studied are in the regime of size confinement, i.e. a low-energy band-edge corresponds to larger particles, we can assume, that the colloidal ZnSe particles grow under irradiation. Additionally, powder XRD and a HRTEM analysis show that the improvement of the PL QE is accompanied by the growth of the NCs. The XRD peaks shift to values which are characteristic for ZnSe/ZnS alloys (the sulfur appears as a product of photodecomposition of the TGA in solution [82]). Electrochemical studies were performed to show that surface Se-related states in the photochemically treated ZnSe NCs are efficiently exchanged with S-related states [121]. The formation of such a shell from a larger band gap material (ZnS) should provide an additional stabilization of the core particles and hence leads to better PL properties. Solutions treated by this way show narrow PL bands being almost free from trap-associated emission (Fig. 10). Moreover, the observed evolution of the PL properties of the ZnSe(S) NCs is generally followed by a decrease of the Stokes shift. ZnSe NCs capped with TGA, MPA or TG showed a similar increase in PL efficiency after irradiation. However, for both MPA- and TG-capped ZnSe NCs a pronounced increase of the trap-emission band was observed as well. The resulting colloids show a reasonable stability: several months of storage in the dark under air result neither in coagulation nor in recognizable changes in the optical properties. In a very recent report [122] the possibility of direct (without photochemical treatment) synthesis of strongly emitting ZnSe NCs capped with glutathione has been demonstrated.

4. HgTe and $Cd_xHg_{1-x}Te$ nanocrystals

In a typical aqueous synthesis of HgTe NCs [41, 122] metal Hg^{2+} ions react under N_2 atmosphere in aqueous solution with H_2Te gas in the presence of a thiol stabilizer. Adjustment of the pH to an appropriate value (11–12), as well as the judicious choice of absolute and relative concentrations of the reaction components allows the reaction rate, the quality of the product and its PL QE to be efficiently controlled [41, 122]. 1-Thioglycerol was found to be the best for controlling the synthesis at the precursors ratio of Hg:Te:R–SH being 1:0.25:2.5. The HgTe NCs grow upon reaction of the precursors at room temperature; the reaction may be stopped by cooling the reaction solutions down in ice pad and keeping them later in a fridge. Reaction lasting for approx. 2 h produces highly concentrated solutions of HgTe NCs of approx. 4 nm in size, with a strong (40–50% PL QE) emission. The drawback of this reaction is the broad size distribution of the resulting samples, leading to a broad luminescence peak of the as-prepared NCs covering the spectral region from 800 to 1400 nm with a maximum located at 1080 nm. However, this broad spectral coverage is advantageous for telecommunication applications, as it coincides with the 1.3-µm telecommunications window. To obtain HgTe NC fractions with various mean sizes and narrower size distributions, a standard size-selective precipitation technique can be applied. Recently it was reported that size and emission maxima of the HgTe NCs may be varied by the controllable growth in presence of 2-mercaptoethanol or 1-thioglycerol in the region from 1200 to 3700 nm [68]. HgTe nanoparticles so prepared belong to the cubic (coloradoite) HgTe phase. Upon gentle (70°C) heating of the as-prepared HgTe NC solutions for progressively longer times, the NCs grow through the Ostwald ripening mechanism, which results in a shift of the PL band towards longer wavelengths with the PL intensity gradually declining to a value of ~10% quantum efficiency. A similar gradual red shift was observed during the storage of as-synthesized HgTe NC solutions at room temperature on the time scale of weeks, which was accompanied by a gradual decline of the emission intensity. To

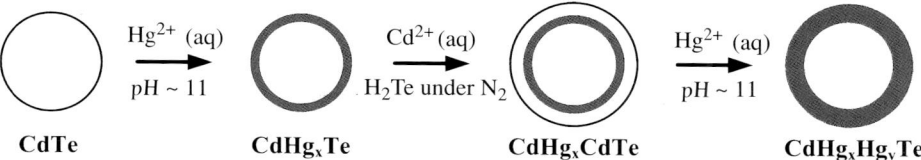

Fig. 11. Schematic diagram of the idealized synthetic routes to achieve mixed-phase $Cd_xHg_{1-x}Te$ nanoparticles

prevent this undesirable effect, capping of HgTe NCs with a shell of a wider-band gap material CdS can be undertaken [123], resulting in positive improvements both for the long-term storage and for the stability of the samples during high-temperature treatments involved in device fabrication.

HgTe NCs synthesized in water can easily be transferred into non-polar organic solvents, such as toluene, styrene, chloroform, chlorobenzene etc., by exchange of the stabilizer with long-chain thiols [68, 83]. Once the NCs have been transferred to non-polar solvents they can be easily processed by traditional spin-coating or casting techniques and are also suitable for the preparation of blends with optically transparent and/or conjugated polymers [124].

Alloyed $Cd_xHg_{1-x}Te$ NCs emitting in the spectral region between 800 and 1100 nm have been synthesized in water through the chemical modification of pre-synthesized thiol-capped CdTe NCs by Hg^{2+} ions [40, 125]. Due to the lower solubility of HgTe in comparison to CdTe in water, the Hg^{2+} ions substitute Cd^{2+} ions at the surface of the nanoparticles forming a $Cd_xHg_{1-x}Te$ alloy in the near-surface region, as schematically depicted in Fig. 11. A layer of CdTe can then be grown on the surface of $Cd_xHg_{1-x}Te$ NCs by addition of more Cd^{2+} ions reacting with H_2Te gas, and the substitution process can be repeated further leading to an increase in both the NC size and the relative Hg content in the alloyed particles (Fig. 11). The substitution reaction occurs with a finite rate and starts at some thermodynamically preferred site. Alloying in bulk $Cd_xHg_{1-x}Te$ systems (MCT) is well-known and is in fact the basis of long-wavelength infrared photodetectors, such that with time Hg^{2+} ions first incorporate on a surface and then into the volume of the particles. The near-surface region of the NCs can therefore be described as a solid solution, alloy or mixed crystal of $Cd_xHg_{1-x}Te$, possibly with a concentration gradient decreasing towards the particle interior. The greater the amount of Hg^{2+} ions added and the longer the time period allowed, the greater the concentration of mercury in the interior of the particles until a real $Cd_xHg_{1-x}Te$ alloy results. The band gap of bulk $Cd_xHg_{1-x}Te$ alloy varies approximately linearly with the composition from $+1.6\,eV$ at $x=1$ (pure CdTe) to $-0.3\,eV$ for $x=0$ (pure HgTe) [126]. This has been observed experimentally for the $Cd_xHg_{1-x}Te$ NCs as a red shift in both the absorption and luminescence spectra with increasing Hg content resulting in emission wavelengths ranging from 700 to 1350 nm depending on the composition. PL quantum efficiencies as high as 50% have been measured for the alloy $Cd_xHg_{1-x}Te$ NCs which is similar to that of HgTe NCs.

Figure 12 shows a series of normalized PL spectra of $Cd_xHg_{1-x}Te$ and HgTe NC fractions (the latter were measured in D_2O) which cover the spectral region between

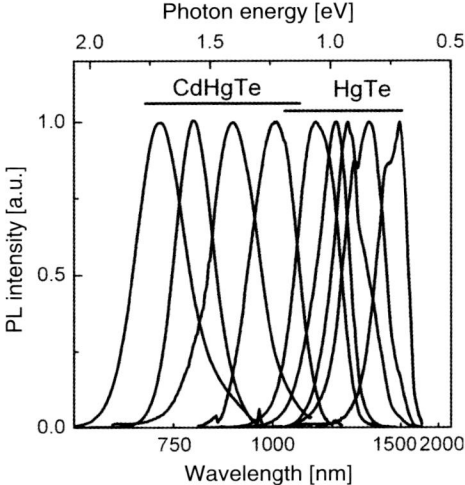

Fig. 12. Typical photoluminescence spectra of the $Cd_xHg_{1-x}Te$ and HgTe NCs. Reprinted with permission from [130], © 2007 Wiley VCH

700 and 1900 nm. Both $Cd_xHg_{1-x}Te$ and HgTe NCs are near-infrared-emitting materials whose recognition has grown rapidly all over the world in the last few years. A potentially very significant application of these NCs is their deployment as optical amplifier media for telecommunications systems based on silica fibre technology which has optimal transmission windows around 1.3 and 1.55 μm, as discussed in several topical reviews [127, 128]. Another evolving fields with a great potential are applications of near-infared-emitting NCs as fluorescent contrast agents for biomedical imaging in living tissue, electroluminescent devices, photodetectors and photovoltaics for solar energy conversion [129, 130]. It was found that $Cd_xHg_{1-x}Te$ NCs possess exceptionally high stability against photooxidation and degradation in typical biological buffer media like phosphate buffered saline [131]. This fact, together with the strong emission in the biological spectral window (both haemoglobin and water show minimum absorption of light between 800 and 1100 nm) makes this kind of NCs especially attractive for bio-imaging and bio-labelling. Recently, $Cd_xHg_{1-x}Te$ NCs additionally protected by CdS shell were successfully used for the in vivo imaging [132].

5. Summary and outlook

Sufficient progress has been made during last decade in the preparation and the design of the surface properties of thiol-capped water-soluble NCs whose luminescence covers broad spectral range depending on the material and the particle size. Among the advantages of the aqueous synthesis its simplicity and high reproducibility should be mentioned. Keeping in mind the potential importance of highly luminescent NCs for large-scale applications, it should be mentioned that the aqueous synthesis of thiol-capped NCs can be carried out equally effective on a vast scale, yielding up to 10 g of NCs per synthesis. These NCs can be precipitated, washed and kept in the dry state under ambient conditions for years, being stable and re-soluble in water.

NCs synthesized by the aqueous approach do not possess the degree of crystallinity of the NCs prepared in organic solvents by the hot-injection technique, where high annealing temperatures (200–360°C) are used during the synthesis [22, 133, 134]. A very effective separation of nucleation and growth stages achieved in the hot-injection technique allows to reach narrower size distributions of the NCs in comparison with those prepared in aqueous solutions. However, the aqueous approach generally allows the synthesis of smaller NCs, both CdS, CdSe and CdTe, and is the only existing reliable method nowadays allowing to produce mercury chalcogenide NCs. The post-preparative size-selective precipitation procedure works more reliable in case of aqueous colloids in terms of retaining the luminescence properties, than for their organically synthesized counterparts. The possibility to control the surface charge and other surface properties of thiol-capped NCs simply by choice of stabilizing thiol molecules with appropriate free functional groups used in the synthesis is extremely important, especially when water-soluble NCs are to be used for fluorescence tagging applications. The possibility to vary surface functional groups is also important when specific binding of NCs to other nanoparticles or to the surfaces is foreseen.

Semiconductor NCs prepared in aqueous solution have found several applications in material science and nanotechnology. Among others is the fabrication of polymer-NC [35, 42, 90, 124, 135–138] and glass-NC [120, 139] light-emitting composites which are robust and easily processable. Applications in optoelectronics cover the LEDs [34, 35, 124, 140], microarrays of multicolored light-emitting pixels [141] and photosensitive films [142, 143]. This is closely connected to photonic applications in which NCs play a role of subwavelength emitters [144], tunable light-sources coupled to optical modes of photonic crystals [145–147] and heterocrystals [148], spherical microresonators [149, 150] photonic molecules [151], and waveguides [152]. The ability of NCs to interact with neighboring nanoparticles or molecules gives rise to the fabrication of FRET-based [101, 153] and nanoplasmonic [103] devices as well as sensors [154–156]. The demands in light emitters which are compatible with water and the most common biological buffers [157] open for thiol-capped NCs such fields as bio-labelling [36, 131, 158–160] and bio-imaging [105, 110, 161]. They have been used as building blocks for self-organizing superstructures like luminescent nanowires [162–168], nanotubes [169] or nanosheets [92], for chemiluminescence generation [170], for fabrication of temperature-sensitive nanoassemblies [171], as luminescent components of multifunctional microbeads [144, 172–177] and polymer microcapsules [104, 131, 177–180]. Brightly emitting water-soluble NCs with a flexible surface chemistry determined by easy choice of capping ligands have already secured and will secure in future a wide field of applications, ranging from life sciences to photonics and optoelectronics. In the field of biological imaging of cellular processes, the ability to fabricate NCs with well-defined surface passivation is important in studying transport processes within living cells [181].

References

[1] Henglein A (1982) Photodegradation and fluorescence of colloidal cadmium sulfide in aqueous solution. Ber Bunsen-Ges Phys Chem 86: 301–305
[2] Weller H, Koch U, Gutierrez M, Henglein A (1984) Photochemistry of colloidal metal sulfides. 7. Absorption and fluorescence of extremely small zinc sulfide particles (the world of the neglected dimensions). Ber Bunsen-Ges Phys Chem 88: 649–656
[3] Alfassi Z, Bahnemann D, Henglein A (1982) Photochemistry of colloidal metal sulfides. 3. Photoelectron emission from cadmium sulfide and cadmium sulfide–zinc sulfide cocolloids. J Phys Chem 86: 4656–4657
[4] Rossetti R, Nakahara S, Brus LE (1983) Quantum size effects in the redox potentials, resonance Raman spectra, and electronic spectra of cadmium sulfide crystallites in aqueous solution. J Chem Phys 79: 1086–1088
[5] Nozik AJ, Williams F, Nenadovic MT, Rajh T, Micic OI (1985) Size quantization in small semiconductor particles. J Phys Chem 89: 397–399
[6] Ramsden JJ, Graetzel M (1984) Photoluminescence of small cadmium sulfide particles. J Chem Soc, Faraday Trans 1 80: 919–933
[7] Rossetti R, Hull R, Gibson JM, Brus LE (1985) Excited electronic states and optical spectra of zinc sulfide and cadmium sulfide crystallites in the ~15 to 50 Å size range: evolution from molecular to bulk semiconducting properties. J Chem Phys 82: 552–559
[8] Weller H, Fojtik A, Henglein A (1985) Photochemistry of semiconductor colloids: properties of extremely small particles of cadmium phosphide (Cd_3P_2) and zinc phosphide (Zn_3P_2). Chem Phys Lett 117: 485–488
[9] Fojtik A, Weller H, Henglein A (1985) Photochemistry of semiconductor colloids. Size quantification effects in Q-cadmium arsenide. Chem Phys Lett 120: 552–554
[10] Resch U, Weller H, Henglein A (1989) Photochemistry and radiation chemistry of colloidal semiconductors. 33. Chemical changes and fluorescence in CdTe and ZnTe. Langmuir 5: 1015–1020
[11] Efros AL, Efros AL (1982) Interband absorption of light in a semiconductor sphere. Sov Phys Semiconduct 16: 772–775
[12] Brus LE (1983) A simple model for the ionization potential, electron affinity, and aqueous redox potentials of small semiconductor crystallites. J Chem Phys 79: 5566–5571
[13] Dameron CT, Reese RN, Mehra RK, Kortan AR, Carroll PJ, Steigerwald ML, Brus LE, Winge DR (1989) Biosynthesis of cadmium sulfide quantum semiconductor crystallites. Nature 338: 596–597
[14] Fendler JH (1987) Atomic and molecular clusters in membrane mimetic chemistry. Chem Rev 87: 877–899
[15] Rajh T, Micic OI, Nozik AJ (1993) Synthesis and characterization of surface-modified colloidal cadmium telluride quantum dots. J Phys Chem 97: 11999–12003
[16] Vossmeyer T, Katsikas L, Giersig M, Popovic IG, Diesner K, Chemseddine A, Eychmüller A, Weller H (1994) CdS nanoclusters: synthesis, characterization, size dependent oscillator strength, temperature shift of the excitonic transition energy, and reversible absorbance shift. J Phys Chem 98: 7665–7673
[17] Vossmeyer T, Reck G, Katsikas L, Haupt ETK, Schulz B, Weller H (1995) A "double-diamond superlattice" built up of $Cd_{17}S_4(SCH_2CH_2OH)_{26}$ clusters. Science 267: 1476–1479
[18] Vossmeyer T, Reck G, Schulz B, Katsikas L, Weller H (1995) Double-layer superlattice structure built up of $Cd_{32}S_{14}(SCH_2CH(OH)CH_3)_{36} \cdot 4H_2O$ clusters. J Am Chem Soc 117: 12881–12882
[19] Henglein A (1988) Mechanism of reactions of colloidal microelectrodes and size quantization effects. Topics Curr Chem 143: 113–180
[20] Spanhel L, Haase M, Weller H, Henglein A (1987) Photochemistry of colloidal semiconductors. 20. Surface modification and stability of strong luminescing CdS particles. J Am Chem Soc 109: 5649–5655
[21] Chemseddine A, Weller H (1993) Highly monodisperse quantum sized cadmium sulfide particles by size selective precipitation. Ber Bunsen-Ges Phys Chem 97: 636–637
[22] Murray CB, Norris DJ, Bawendi MG (1993) Synthesis and characterization of nearly monodisperse CdE (E = sulfur, selenium, tellurium) semiconductor nanocrystallites. J Am Chem Soc 115: 8706–8715
[23] Fischer CH, Weller H, Fojtik A, Lume-Pereira C, Janata E, Henglein A (1986) Photochemistry of colloidal semiconductors. 10. Exclusion chromatography and stop flow experiments on the formation of extremely small cadmium sulfide particles. Ber Bunsen-Ges Phys Chem 90: 46–49

[24] Fischer CH, Weller H, Katsikas L, Henglein A (1989) Photochemistry of colloidal semiconductors. 30. HPLC investigation of small CdS particles. Langmuir 5: 429–432
[25] Eychmüller A, Katsikas L, Weller H (1990) Photochemistry of semiconductor colloids. 35. Size separation of colloidal cadmium sulfide by gel electrophoresis. Langmuir 6: 1605–1608
[26] Hines MA, Guyot-Sionnest P (1996) Synthesis and characterization of strongly luminescing ZnS-capped CdSe nanocrystals. J Phys Chem 100: 468–471
[27] Peng ZA, Peng XG (2001) Mechanisms of the shape evolution of CdSe nanocrystals. J Am Chem Soc 123: 1389–1395
[28] Colvin VL, Schlamp MC, Alivisatos AP (1994) Light-emitting diodes made from cadmium selenide nanocrystals and a semiconducting polymer. Nature 370: 354–357
[29] Huynh WU, Dittmer JJ, Alivisatos AP (2002) Hybrid nanorod-polymer solar cells. Science 295: 2425–2427
[30] Bruchez M Jr, Moronne M, Gin P, Weiss S, Alivisatos AP (1998) Semiconductor nanocrystals as fluorescent biological labels. Science 281: 2013–2016
[31] Donega CdM, Liljeroth P, Vanmaekelbergh D (2005) Physicochemical evaluation of the hot-injection method, a synthesis route for monodisperse nanocrystals. Small 1: 1152–1162
[32] Rogach AL, Katsikas L, Kornowski A, Su D, Eychmüller A, Weller H (1996) Synthesis and characterization of thiol-stabilized CdTe nanocrystals. Ber Bunsen-Ges Phys Chem 100: 1772–1778
[33] Gao M, Kirstein S, Möhwald H, Rogach AL, Kornowski A, Eychmüller A, Weller H (1998) Strongly photoluminescent CdTe nanocrystals by proper surface modification. J Phys Chem B 102: 8360–8363
[34] Gao M, Lesser C, Kirstein S, Möhwald H, Rogach AL, Weller H (2000) Electroluminescence of different colors from polycation/CdTe nanocrystal self-assembled films. J Appl Phys 87: 2297–2302
[35] Gaponik NP, Talapin DV, Rogach AL, Eychmüller A (2000) Electrochemical synthesis of CdTe nanocrystal/polypyrrole composites for optoelectronic applications. J Mater Chem 10: 2163–2166
[36] Mamedova NN, Kotov NA, Rogach AL, Studer J (2001) Albumin-CdTe nanoparticle bioconjugates: preparation, structure, and interunit energy transfer with antenna effect. Nano Lett 1: 281–286
[37] Rogach AL, Kornowski A, Gao M, Eychmüller A, Weller H (1999) Synthesis and characterization of a size series of extremely small thiol-stabilized CdSe nanocrystals. J Phys Chem B 103: 3065–3069
[38] Gaponik N, Talapin DV, Rogach AL, Hoppe K, Shevchenko EV, Kornowski A, Eychmüller A, Weller H (2002) Thiol-capping of CdTe nanocrystals: an alternative to organometallic synthetic routes. J Phys Chem B 106: 7177–7185
[39] Rogach AL, Franzl T, Klar TA, Feldmann J, Gaponik N, Lesnyak V, Shavel A, Eychmüller A, Rakovich YP, Donegan JF (2007) Aqueous synthesis of thiol-capped CdTe nanocrystals: state-of-the-art. J Phys Chem C 111: 14628–14637
[40] Harrison MT, Kershaw SV, Burt MG, Eychmüller A, Weller H, Rogach AL (2000) Wet chemical synthesis and spectroscopic study of CdHgTe nanocrystals with strong near-infrared luminescence. Mater Sci Eng B B69–70: 355–360
[41] Rogach A, Kershaw S, Burt M, Harrison M, Kornowski A, Eychmüller A, Weller H (1999) Colloidally prepared HgTe nanocrystals with strong room-temperature infrared luminescence. Adv Mater 11: 552–555
[42] Shavel A, Gaponik N, Eychmüller A (2004) Efficient UV-blue photoluminescing thiol-stabilized water-soluble alloyed ZnSe(S) nanocrystals. J Phys Chem B 108: 5905–5908
[43] Peng X, Schlamp MC, Kadavanich AV, Alivisatos AP (1997) Epitaxial growth of highly luminescent CdSe/CdS core/shell nanocrystals with photostability and electronic accessibility. J Am Chem Soc 119: 7019–7029
[44] Reiss P, Bleuse J, Pron A (2002) Highly luminescent CdSe/ZnSe core/shell nanocrystals of low size dispersion. Nano Lett 2: 781–784
[45] Talapin DV, Rogach AL, Kornowski A, Haase M, Weller H (2001) Highly luminescent monodisperse CdSe and CdSe/ZnS nanocrystals synthesized in a hexadecylamine-trioctylphosphine oxide-trioctylphosphine mixture. Nano Lett 1: 207–211
[46] Wuister SF, de MelloDonega C, Meijerink A (2004) Influence of thiol capping on the exciton luminescence and decay kinetics of CdTe and CdSe quantum dots. J Phys Chem B 108: 17393–17397
[47] Nethercot AH Jr (1974) Prediction of Fermi energies and photoelectric thresholds based on electronegativity concepts. Phys Rev Lett 33: 1088–1091

[48] Döllefeld H, Weller H, Eychmüller A (2002) Semiconductor nanocrystal assemblies: experimental pitfalls and a simple model of particle–particle interaction. J Phys Chem B 106: 5604–5608
[49] Kolny J, Kornowski A, Weller H (2002) Self-organization of cadmium sulfide and gold nanoparticles by electrostatic interaction. Nano Lett 2: 361–364
[50] Sondi I, Siiman O, Koester S, Matijevic E (2000) Preparation of aminodextran-CdS nanoparticle complexes and biologically active antibody-aminodextran-CdS nanoparticle conjugates. Langmuir 16: 3107–3118
[51] Narayanan SS, Sarkar R, Pal SK (2007) Structural and functional characterization of enzyme-quantum dot conjugates: covalent attachment of CdS nanocrystal to α-chymotrypsin. J Phys Chem C 111: 11539–11543
[52] Döllefeld H, Hoppe K, Kolny J, Schilling K, Weller H, Eychmüller A (2002) Investigations on the stability of thiol stabilized semiconductor nanoparticles. Phys Chem Chem Phys 4: 4747–4753
[53] Moloney MP, Gun'ko YK, Kelly JM (2007) Chiral highly luminescent CdS quantum dots. Chem Commun 3900–3902
[54] Spoerke ED, Voigt JA (2007) Influence of engineered peptides on the formation and properties of cadmium sulfide nanocrystals. Adv Funct Mater 17: 2031–2037
[55] Ma N, Yang J, Stewart KM, Kelley SO (2007) DNA-passivated CdS nanocrystals: luminescence, bioimaging, and toxicity profiles. Langmuir 23: 12783–12787
[56] Mora-Sero I, Bisquert J, Dittrich T, Belaidi A, Susha AS, Rogach AL (2007) Photosensitization of TiO_2 layers with CdSe quantum dots: correlation between light absorption and photoinjection. J Phys Chem C 111: 14889–14892
[57] Rogach AL, Nagesha D, Ostrander JW, Giersig M, Kotov NA (2000) "Raisin bun"-type composite spheres of silica and semiconductor nanocrystals. Chem Mater 12: 2676–2685
[58] Wang Y, Tang Z, Correa-Duarte MA, Liz-Marzan LM, Kotov NA (2003) Multicolor luminescence patterning by photoactivation of semiconductor nanoparticle films. J Am Chem Soc 125: 2830–2831
[59] Wang Y, Tang Z, Correa-Duarte MA, Pastoriza-Santos I, Giersig M, Kotov NA, Liz-Marzan LM (2004) Mechanism of strong luminescence photoactivation of citrate-stabilized water-soluble nanoparticles with CdSe cores. J Phys Chem B 108: 15461–15469
[60] Wang SF, Gu F, Lu MK (2006) Sonochemical synthesis of hollow PbS nanospheres. Langmuir 22: 398–401
[61] Jiang Y, Wu Y, Xie B, Yuan S, Liu X, Qian Y (2001) Hydrothermal preparation of uniform cubic-shaped PbS nanocrystals. J Cryst Growth 231: 248–251
[62] Flores-Acosta M, Sotelo-Lerma M, Arizpe-Chavez H, Castillon-Barraza FF, Ramirez-Bon R (2003) Excitonic absorption of spherical PbS nanoparticles in zeolite A. Solid State Commun 128: 407–411
[63] Hao E, Yang B, Yu S, Gao M, Shen J (1997) Formation of orderly organized cubic PbS nanoparticles domain in the presence of TiO_2. Chem Mater 9: 1598–1600
[64] Huang NM, Shahidan R, Khiew PS, Peter L, Kan CS (2004) In situ templating of PbS nanorods in reverse hexagonal liquid crystal. Colloid. Surf A 247: 55–60
[65] Zhou SM, Zhang XH, Meng XM, Fan X, Lee ST, Wu SK (2005) Sonochemical synthesis of mass single-crystal PbS nanobelts. J Solid State Chem 178: 399–403
[66] Ma YR, Qi LM, Ma JM, Cheng HM (2004) Hierarchical, star-shaped PbS crystals formed by a simple solution route. Cryst Growth Des 4: 351–354
[67] Lifshitz E, Sirota M, Porteanu H (1999) Continuous and time-resolved photoluminescence study of lead sulfide nanocrystals, embedded in polymer film. J Cryst Growth 196: 126–134
[68] Kovalenko MV, Kaufmann E, Pachinger D, Roither J, Huber M, Stangl J, Hesser G, Schaffler F, Heiss W (2006) Colloidal HgTe nanocrystals with widely tunable narrow band gap energies: from telecommunications to molecular vibrations. J Am Chem Soc 128: 3516–3517
[69] Zhang H, Wang L, Xiong H, Hu L, Yang B, Li W (2003) Hydrothermal synthesis for high-quality CdTe nanocrystals. Adv Mater 15: 1712–1715
[70] Zhang H, Zhou Z, Yang B, Gao M (2003) The influence of carboxyl groups on the photoluminescence of mercaptocarboxylic acid-stabilized CdTe nanoparticles. J Phys Chem B 107: 8–13
[71] Dennis LM, Anderson RP (1914) Hydrogen tellurid and the atomic weight of tellurium. J Am Chem Soc 36: 882–909
[72] Engelhard T, Jones ED, Viney I, Mastai Y, Hodes G (2000) Deposition of tellurium films by decomposition of electrochemically-generated H_2Te: application to radiative cooling devices. Thin Solid Films 370: 101–105
[73] Bastide S, Huegel P, Levy-Clement C, Hodes G (2005) Electrochemical preparation of H_2S and H_2Se. J Electrochem Soc 152: D35–D41

[74] Kovalenko MV, Bodnarchuk MI, Stroyuk AL, Kuchmii SY (2004) Spectral, optical, and photocatalytic characteristics of quantum-sized particles of CdTe. Theor Exp Chem 40: 220–225
[75] Li C, Murase N (2005) Surfactant-dependent photoluminescence of CdTe nanocrystals in aqueous solution. Chem Lett 34: 92–93
[76] Shavel A, Gaponik N, Eychmüller A (2006) Factors governing the quality of aqueous CdTe nanocrystals: calculations and experiment. J Phys Chem B 110: 19280–19284
[77] Guo J, Yang W, Wang C (2005) Systematic study of the photoluminescence dependence of thiol-capped CdTe nanocrystals on the reaction conditions. J Phys Chem B 109: 17467–17473
[78] Talapin DV, Rogach AL, Haase M, Weller H (2001) Evolution of an ensemble of nanoparticles in a colloidal solution: theoretical study. J Phys Chem B 105: 12278–12285
[79] Talapin DV, Rogach AL, Shevchenko EV, Kornowski A, Haase M, Weller H (2002) Dynamic distribution of growth rates within the ensembles of colloidal II–VI and III–V semiconductor nanocrystals as a factor governing their photoluminescence efficiency. J Am Chem Soc 124: 5782–5790
[80] Li L, Qian H, Fang N, Ren J (2005) Significant enhancement of the quantum yield of CdTe nanocrystals synthesized in aqueous phase by controlling the pH and concentrations of precursor solutions. J Lumin 116: 59–66
[81] Jeong S, Achermann M, Nanda J, Ivanov S, Klimov VI, Hollingsworth JA (2005) Effect of the thiol–thiolate equilibrium on the photophysical properties of aqueous CdSe/ZnS nanocrystal quantum dots. J Am Chem Soc 127: 10126–10127
[82] Rogach AL (2000) Nanocrystalline CdTe and CdTe(S) particles: wet chemical preparation, size-dependent optical properties and perspectives of optoelectronic applications. Mater Sci Eng B B69–70: 435–440
[83] Gaponik N, Talapin DV, Rogach AL, Eychmüller A, Weller H (2002) Efficient phase transfer of luminescent thiol-capped nanocrystals: from water to nonpolar organic solvents. Nano Lett 2: 803–806
[84] Biebersdorf A, Dietmüller R, Susha AS, Rogach AL, Poznyak SK, Talapin DV, Weller H, Klar TA, Feldmann J (2006) Semiconductor nanocrystals photosensitize C60 crystals. Nano Lett 6: 1559–1563
[85] Guldi DM, Zilbermann I, Anderson G, Kotov NA, Tagmatarchis N, Prato M (2004) Versatile organic (fullerene)–inorganic (CdTe nanoparticle) nanoensembles. J Am Chem Soc 126: 14340–14341
[86] Tsay JM, Pflughoefft M, Bentolila LA, Weiss S (2004) Hybrid approach to the synthesis of highly luminescent CdTe/ZnS and CdHgTe/ZnS nanocrystals. J Am Chem Soc 126: 1926–1927
[87] Akamatsu K, Tsuruoka T, Nawafune H (2005) Band gap engineering of CdTe nanocrystals through chemical surface modification. J Am Chem Soc 127: 1634–1635
[88] Zhang H, Cui Z, Wang Y, Zhang K, Ji X, Lu C, Yang B, Gao M (2003) From water-soluble CdTe nanocrystals to fluorescent nanocrystal-polymer transparent composites using polymerizable surfactants. Adv Mater 15: 777–780
[89] Zhang H, Wang C, Li M, Ji X, Zhang J, Yang B (2005) Fluorescent nanocrystal-polymer composites from aqueous nanocrystals: methods without ligand exchange. Chem Mater 17: 4783–4788
[90] Zhang H, Wang C, Li M, Zhang J, Lu G, Yang B (2005) Fluorescent nanocrystal-polymer complexes with flexible processability. Adv Mater 17: 853–857
[91] Nakashima T, Kawai T (2005) Quantum dots–ionic liquid hybrids: efficient extraction of cationic CdTe nanocrystals into an ionic liquid. Chem Commun: 1643–1645
[92] Tang Z, Zhang Z, Wang Y, Glotzer SC, Kotov NA (2006) Self-assembly of CdTe nanocrystals into free-floating sheets. Science 314: 274–278
[93] Schooss D, Mews A, Eychmüller A, Weller H (1994) Quantum-dot quantum well CdS/HgS/CdS: theory and experiment. Phys Rev B 49: 17072–17078
[94] Yu WW, Qu L, Guo W, Peng X (2003) Experimental determination of the extinction coefficient of CdTe, CdSe, and CdS nanocrystals. Chem Mater 15: 2854–2860
[95] Bao H, Gong Y, Li Z, Gao M (2004) Enhancement effect of illumination on the photoluminescence of water-soluble CdTe nanocrystals: toward highly fluorescent CdTe/CdS core-shell structure. Chem Mater 16: 3853–3859
[96] Baranov AV, Rakovich YP, Donegan JF, Perova TS, Moore RA, Talapin DV, Rogach AL, Masumoto Y, Nabiev I (2003) Effect of ZnS shell thickness on the phonon spectra in CdSe quantum dots. Phys Rev B 68: 165306
[97] Landolt-Börnstein (1982) Numerical data and functional relationship in science and technology. vol 17b, New Series, Group III. Springer-Verlag, Berlin

[98] Barnham K, Marques JL, Hassard J, O'Brien P (2000) Quantum-dot concentrator and thermodynamic model for the global redshift. Appl Phys Lett 76: 1197–1199
[99] Franzl T, Koktysh DS, Klar TA, Rogach AL, Feldmann J, Gaponik N (2004) Fast energy transfer in layer-by-layer assembled CdTe nanocrystal bilayers. Appl Phys Lett 84: 2904–2906
[100] Franzl T, Shavel A, Rogach AL, Gaponik N, Klar TA, Eychmüller A, Feldmann J (2005) High-rate unidirectional energy transfer in directly assembled CdTe nanocrystal bilayers. Small 1: 392–395
[101] Franzl T, Klar TA, Schietinger S, Rogach AL, Feldmann J (2004) Exciton recycling in graded gap nanocrystal structures. Nano Lett 4: 1599–1603
[102] Ray K, Badugu R, Lakowicz JR (2006) Metal-enhanced fluorescence from CdTe nanocrystals: a single-molecule fluorescence study. J Am Chem Soc 128: 8998–8999
[103] Komarala VK, Rakovich YP, Bradley AL, Byrne SJ, Gun'ko YK, Gaponik N, Eychmüller A (2006) Off-resonance surface plasmon enhanced spontaneous emission from CdTe quantum dots. Appl Phys Lett 89: 253118/1–253118/3
[104] Gaponik N, Radtchenko IL, Sukhorukov GB, Weller H, Rogach AL (2002) Toward encoding combinatorial libraries: charge-driven microencapsulation of semiconductor nanocrystals luminescing in the visible and near IR. Adv Mater 14: 879–882
[105] Lovric J, Bazzi HS, Cuie Y, Fortin GRA, Winnik FM, Maysinger D (2005) Differences in subcellular distribution and toxicity of green and red emitting CdTe quantum dots. J Mol Med 83: 377–385
[106] Wang C, Zhang H, Zhang J, Li M, Sun H, Yang B (2007) Application of ultrasonic irradiation in aqueous synthesis of highly fluorescent CdTe/CdS core-shell nanocrystals. J Phys Chem C 111: 2465–2469
[107] Li L, Qian H, Ren J (2005) Rapid synthesis of highly luminescent CdTe nanocrystals in the aqueous phase by microwave irradiation with controllable temperature. Chem Commun 528–530
[108] He Y, Sai L-M, Lu H-T, Hu M, Lai W-Y, Fan Q-L, Wang L-H, Huang W (2007) Microwave-assisted synthesis of water-dispersed CdTe nanocrystals with high luminescent efficiency and narrow size distribution. Chem Mater 19: 359–365
[109] Liu Y, Chen W, Joly AG, Wang Y, Pope C, Zhang Y, Bovin J-O, Sherwood P (2006) Comparison of water-soluble CdTe nanoparticles synthesized in air and in nitrogen. J Phys Chem B 110: 16992–17000
[110] Zheng Y, Gao S, Ying JY (2007) Synthesis and cell-imaging applications of glutathione-capped CdTe quantum dots. Adv Mater 19: 376–380
[111] Maynard AD, Aitken RJ, Butz T, Colvin V, Donaldson K, Oberdoerster G, Philbert MA, Ryan J, Seaton A, Stone V, Tinkle SS, Tran L, Walker NJ, Warheit DB (2006) Safe handling of nanotechnology. Nature 444: 267–269
[112] Peng ZA, Peng XG (2001) Formation of high-quality CdTe, CdSe, and CdS nanocrystals using CdO as precursor. J Am Chem Soc 123: 183–184
[113] Deng D-W, Yu J-S, Pan Y (2006) Water-soluble CdSe and CdSe/CdS nanocrystals: a greener synthetic route. J Colloid Interf Sci 299: 225–232
[114] Sapra S, Rogach AL, Feldmann J (2006) Phosphine-free synthesis of monodisperse CdSe nanocrystals in olive oil. J Mater Chem 16: 3391–3395
[115] Mekis I, Talapin DV, Kornowski A, Haase M, Weller H (2003) One-pot synthesis of highly luminescent CdSe/CdS core-shell nanocrystals via organometallic and "greener" chemical approaches. J Phys Chem B 107: 7454–7462
[116] Pradhan N, Goorskey D, Thessing J, Peng X (2005) An alternative of CdSe nanocrystal emitters: pure and tunable impurity emissions in ZnSe nanocrystals. J Am Chem Soc 127: 17586–17587
[117] Reiss P (2007) ZnSe based colloidal nanocrystals: synthesis, shape control, core/shell, alloy and doped systems. New J Chem 31: 1843–1852
[118] Murase N, Gao MY, Gaponik N, Yazawa T, Feldmann J (2001) Synthesis and optical properties of water soluble ZnSe nanocrystals. Int J Modern Phys B 15: 3881–3884
[119] Murase N, Gao M (2004) Preparation and photoluminescence of water-dispersible ZnSe nanocrystals. Mater Lett 58: 3898–3902
[120] Li CL, Nishikawa K, Ando M, Enomoto H, Murase N (2007) Highly luminescent water-soluble ZnSe nanocrystals and their incorporation in a glass matrix. Colloid Surf A 294: 33–39
[121] Osipovich NP, Shavel A, Poznyak SK, Gaponik N, Eychmüller A (2006) Electrochemical observation of the photoinduced formation of alloyed ZnSe(S) nanocrystals. J Phys Chem B 110: 19233–19237
[122] Harrison MT, Kershaw SV, Burt MG, Rogach A, Eychmüller A, Weller H (1999) Investigation of factors affecting the photoluminescence of colloidally-prepared HgTe nanocrystals. J Mater Chem 9: 2721–2722

[123] Harrison MT, Kershaw SV, Rogach AL, Kornowski A, Eychmüller A, Weller H (2000) Wet chemical synthesis of highly luminescent HgTe/CdS core/shell nanocrystals. Adv Mater 12: 123–125
[124] Koktysh DS, Gaponik N, Reufer M, Crewett J, Scherf U, Eychmüller A, Lupton JM, Rogach AL, Feldmann J (2004) Near–infrared electroluminescence from HgTe nanocrystals. Chem Phys Chem 5: 1435–1438
[125] Rogach AL, Harrison MT, Kershaw SV, Kornowski A, Burt MG, Eychmüller A, Weller H (2001) Colloidally prepared CdHgTe and HgTe quantum dots with strong near-infrared luminescence. Phys Stat Sol B 224: 153–158
[126] Balcerak R, Gibson JF, Gutierrez WA, Pollard JH (1987) Evolution of a new semiconductor product: mercury cadmium telluride focal plane arrays. Opt Eng 26: 191–200
[127] Harrison MT, Kershaw SV, Burt MG, Rogach AL, Kornowski A, Eychmuller A, Weller H (2000) Colloidal nanocrystals for telecommunications. Complete coverage of the low-loss fiber windows by mercury telluride quantum dots. Pure Appl Chem 72: 295–307
[128] Kershaw SV, Harrison M, Rogach AL, Kornowski A (2000) Development of IR-emitting colloidal II–VI quantum-dot materials. IEEE J. Sel. Top. Quant Electron 6: 534–543
[129] Sargent EH (2005) Infrared quantum dots. Adv Mater 17: 515–522
[130] Rogach AL, Eychmüller A, Hickey SG, Kershaw SV (2007) Infrared-emitting colloidal nanocrystals: synthesis, assembly, spectroscopy, and applications. Small 3: 536–557
[131] Gaponik N, Radtchenko IL, Gerstenberger MR, Fedutik YA, Sukhorukov GB, Rogach AL (2003) Labeling of biocompatible polymer microcapsules with near-infrared emitting nanocrystals. Nano Lett 3: 369–372
[132] Qian H, Dong C, Peng J, Qiu X, Xu Y, Ren J (2007) High-quality and water-soluble near-infrared photoluminescent CdHgTe/CdS quantum dots prepared by adjusting size and composition. J Phys Chem C 111: 16852–16857
[133] Talapin DV, Haubold S, Rogach AL, Kornowski A, Haase M, Weller H (2001) A novel organometallic synthesis of highly luminescent CdTe nanocrystals. J Phys Chem B 105: 2260–2263
[134] Guzelian AA, Banin U, Kadavanich AV, Peng X, Alivisatos AP (1996) Colloidal chemical synthesis and characterization of InAs nanocrystal quantum dots. Appl Phys Lett 69: 1432–1434
[135] Tekin E, Smith PJ, Hoeppener S, van den Berg AMJ, Susha AS, Rogach AL, Feldmann J, Schubert US (2007) Inkjet printing of luminescent CdTe nanocrystal-polymer composites. Adv Funct Mater 17: 23–28
[136] Li J, Hong X, Liu Y, Li D, Wang Y, Li J, Bai Y, Li T (2005) Highly photoluminescent CdTe/poly(N-isopropylacrylamide) temperature-sensitive gels. Adv Mater 17: 163–166
[137] Gaponik NP, Talapin DV, Rogach AL (1999) A light-emitting device based on a CdTe nanocrystal/polyaniline composite. Phys Chem Chem Phys 1: 1787–1789
[138] Mamedov AA, Belov A, Giersig M, Mamedova NN, Kotov NA (2001) Nanorainbows: graded semiconductor films from quantum dots. J Am Chem Soc 123: 7738–7739
[139] Li C, Murase N (2004) Synthesis of highly luminescent glasses incorporating CdTe nanocrystals through sol-gel processing. Langmuir 20: 1–4
[140] Bertoni C, Gallardo D, Dunn S, Gaponik N, Eychmüller A (2007) Fabrication and characterization of red-emitting electroluminescent devices based on thiol-stabilized semiconductor nanocrystals. Appl Phys Lett 90: 034107
[141] Gao M, Sun J, Dulkeith E, Gaponik N, Lemmer U, Feldmann J (2002) Lateral patterning of CdTe nanocrystal films by the electric field directed layer-by-layer assembly method. Langmuir 18: 4098–4102
[142] Talapin DV, Poznyak SK, Gaponik NP, Rogach AL, Eychmüller A (2002) Synthesis of surface-modified colloidal semiconductor nanocrystals and study of photoinduced charge separation and transport in nanocrystal-polymer composites. Physica E 14: 237–241
[143] Guldi DM, Zilbermann I, Anderson G, Kotov NA, Tagmatarchis N, Prato M (2005) Nanosized inorganic/organic composites for solar energy conversion. J Mater Chem 15: 114–118
[144] Olk P, Buchler BC, Sandoghdar V, Gaponik N, Eychmüller A, Rogach AL (2004) Subwavelength emitters in the near-infrared based on mercury telluride nanocrystals. Appl Phys Lett 84: 4732–4734
[145] Solovyev VG, Romanov SG, Sotomayor Torres CM, Müller M, Zentel R, Gaponik N, Eychmüller A, Rogach AL (2003) Modification of the spontaneous emission of CdTe nanocrystals in TiO_2 inverted opals. J Appl Phys 94: 1205–1210
[146] Romanov SG, Chigrin DN, Sotomayor Torres CM, Gaponik N, Eychmüller A, Rogach AL (2004) Emission stimulation in a directional band gap of a CdTe-loaded opal photonic crystal. Phys Rev E 69: 046606/1–046606/4

[147] Richter S, Steinhart M, Hofmeister H, Zacharias M, Goesele U, Gaponik N, Eychmüller A, Rogach AL, Wendorff JH, Schweizer SL, von Rhein A, Wehrspohn RB (2005) Quantum dot emitters in two-dimensional photonic crystals of macroporous silicon. Appl Phys Lett 87: 142107/1–142107/3
[148] Gaponik N, Eychmüller A, Rogach AL, Solovyev VG, Sotomayor Torres CM, Romanov SG (2004) Structure-related optical properties of luminescent hetero-opals. J Appl Phys 95: 1029–1035
[149] Rakovich YP, Donegan JF, Gaponik N, Rogach AL (2003) Raman scattering and anti-stokes emission from a single spherical microcavity with a CdTe quantum dot monolayer. Appl Phys Lett 83: 2539–2541
[150] Rakovich YP, Yang L, McCabe EM, Donegan JF, Perova T, Moore A, Gaponik N, Rogach A (2003) Whispering gallery mode emission from a composite system of CdTe nanocrystals and a spherical microcavity. Semiconduct Sci Technol 18: 914–918
[151] Rakovich YP, Donegan JF, Gerlach M, Bradley AL, Connolly TM, Boland JJ, Gaponik N, Rogach A (2004) Fine structure of coupled optical modes in photonic molecules. Phys Rev A 70: 051801
[152] Roither J, Pichler S, Kovalenko MV, Heiss W, Feychuk P, Panchuk O, Allam J, Murdin BN (2006) Two- and one-dimensional light propagations and gain in layer-by-layer-deposited colloidal nanocrystal waveguides. Appl Phys Lett 89: 111120/1–111120/3
[153] Müller F, Götzinger S, Gaponik N, Weller H, Mlynek J, Benson O (2004) Investigation of energy transfer between CdTe nanocrystals on polystyrene beads and dye molecules for FRET-SNOM applications. J Phys Chem B 108: 14527–14534
[154] Susha AS, Javier AM, Parak WJ, Rogach AL (2006) Luminescent CdTe nanocrystals as ion probes and pH sensors in aqueous solutions. Colloid Surf A 281: 40–43
[155] Chen B, Ying Y, Zhou ZT, Zhong P (2004) Synthesis of novel nanocrystals as fluorescent sensors for Hg^{2+} ions. Chem Lett 33: 1608–1609
[156] Li J, Bao D, Hong X, Li D, Li J, Bai Y, Li T (2005) Luminescent CdTe quantum dots and nanorods as metal ion probes. Colloid Surf A 257–258: 267–271
[157] Boldt K, Bruns OT, Gaponik N, Eychmüller A (2006) Comparative examination of the stability of semiconductor quantum dots in various biochemical buffers. J Phys Chem B 110: 1959–1963
[158] Wang S, Mamedova N, Kotov NA, Chen W, Studer J (2002) Antigen/antibody immunocomplex from CdTe nanoparticle bioconjugates. Nano Lett 2: 817–822
[159] Wolcott A, Gerion D, Visconte M, Sun J, Schwartzberg A, Chen S, Zhang JZ (2006) Silica-coated CdTe quantum dots functionalized with thiols for bioconjugation to IgG proteins. J Phys Chem B 110: 5779–5789
[160] Li J, Zhao K, Hong X, Yuan H, Ma L, Li J, Bai Y, Li T (2005) Prototype of immunochromatographic assay strips using colloidal CdTe nanocrystals as biological luminescent label. Colloid Surf B 40: 179–182
[161] Byrne SJ, Corr SA, Rakovich TY, Gun'ko YK, Rakovich YP, Donegan JF, Mitchell S, Volkov Y (2006) Optimization of the synthesis and modification of CdTe quantum dots for enhanced live cell imaging. J Mater Chem 16: 2896–2902
[162] Tang Z, Kotov NA, Giersig M (2002) Spontaneous organization of single CdTe nanoparticles into luminescent nanowires. Science 297: 237–240
[163] Tang Z, Ozturk B, Wang Y, Kotov NA (2004) Simple preparation strategy and one-dimensional energy transfer in CdTe nanoparticle chains. J Phys Chem B 108: 6927–6931
[164] Volkov Y, Mitchell S, Gaponik N, Rakovich YP, Donegan JF, Kelleher D, Rogach AL (2004) In-situ observation of nanowire growth from luminescent CdTe nanocrystals in a phosphate buffer solution. Chem Phys Chem 5: 1600–1602
[165] Zhang H, Wang D, Möhwald H (2006) Ligand-selective aqueous synthesis of one-dimensional CdTe nanostructures. Angew Chem, Int Ed 45: 748–751
[166] Zhang L, Gaponik N, Müller J, Plate U, Weller H, Erker G, Fuchs H, Rogach AL, Chi L (2005) Branched wires of CdTe nanocrystals using amphiphilic molecules as templates. Small 1: 524–527
[167] Zhang H, Wang D, Yang B, Möhwald H (2006) Manipulation of aqueous growth of CdTe nanocrystals to fabricate colloidally stable one-dimensional nanostructures. J Am Chem Soc 128: 10171–10180
[168] Rakovich YP, Volkov Y, Sapra S, Susha AS, Doeblinger M, Donegan JF, Rogach AL (2007) CdTe nanowire networks: fast self-assembly in solution, internal structure, and optical properties. J Phys Chem C 111: 18927–18931
[169] Niu H, Gao M (2006) Diameter-tunable CdTe nanotubes templated by 1D nanowires of cadmium thiolate polymer. Angew Chem, Int Ed 45: 6462–6466

[170] Wang Z, Li J, Liu B, Hu J, Yao X, Li J (2005) Chemiluminescence of CdTe nanocrystals induced by direct chemical oxidation and its size-dependent and surfactant-sensitized effect. J Phys Chem B 109: 23304–23311
[171] Lee J, Govorov AO, Kotov NA (2005) Nanoparticle assemblies with molecular springs: A nanoscale thermometer. Angew Chem, Int Ed 44: 7439–7442
[172] Salgueirino-Maceira V, Correa-Duarte MA, Spasova M, Liz-Marzan LM, Farle M (2006) Composite silica spheres with magnetic and luminescent functionalities. Adv Funct Mater 16: 509–514
[173] Susha AS, Caruso F, Rogach AL, Sukhorukov GB, Kornowski A, Möhwald H, Giersig M, Eychmüller A, Weller H (2000) Formation of luminescent spherical core-shell particles by the consecutive adsorption of polyelectrolyte and CdTe(S) nanocrystals on latex colloids. Colloid Surf A 163: 39–44
[174] Rogach A, Susha A, Caruso F, Sukhorukov G, Kornowski A, Kershaw S, Möhwald H, Eychmüller A, Weller H (2000) Nano- and microengineering. Three-dimensional colloidal photonic crystals prepared from submicrometer-sized polystyrene latex spheres pre-coated with luminescent polyelectrolyte/nanocrystal shells. Adv Mater 12: 333–337
[175] Radtchenko IL, Sukhorukov GB, Gaponik N, Kornowski A, Rogach AL, Möhwald H (2001) Core-shell structures formed by the solvent-controlled precipitation of luminescent CdTe nanocrystals on latex spheres. Adv Mater 13: 1684–1687
[176] Wang D, Rogach AL, Caruso F (2002) Semiconductor quantum dot-labeled microsphere bioconjugates prepared by stepwise self-assembly. Nano Lett 2: 857–861
[177] Shavel A, Gaponik N, Eychmüller A (2005) The assembling of semiconductor nanocrystals. Eur J Inorg Chem: 3613–3623
[178] Gaponik N, Radtchenko IL, Sukhorukov GB, Rogach AL (2004) Luminescent polymer microcapsules addressable by a magnetic field. Langmuir 20: 1449–1452
[179] Sukhorukov GB, Rogach AL, Zebli B, Liedl T, Skirtach AG, Koehler K, Antipov AA, Gaponik N, Susha AS, Winterhalter M, Parak WJ (2005) Nanoengineered polymer capsules: tools for detection, controlled delivery, and site-specific manipulation. Small 1: 194–200
[180] Zebli B, Susha AS, Sukhorukov GB, Rogach AL, Parak WJ (2005) Magnetic targeting and cellular uptake of polymer microcapsules simultaneously functionalized with magnetic and luminescent nanocrystals. Langmuir 21: 4262–4265
[181] Nabiev I, Mitchell S, Davies A, Williams Y, Kelleher D, Moore R, Gun'ko YK, Byrne S, Rakovich YP, Donegan JF, Sukhanova A, Conroy J, Cottell D, Gaponik N, Rogach A, Volkov Y (2007) Nonfunctionalized nanocrystals can exploit a cell's active transport machinery delivering them to specific nuclear and cytoplasmic compartments. Nano Lett 7: 3452–3461

Multishell semiconductor nanocrystals

By

Dirk Dorfs, Alexander Eychmüller

Physical Chemistry, TU Dresden, Dresden, Germany

1. Introduction

Nanocrystals consisting of more than one material are a topic of special interest almost since the beginning of the wet chemical synthesis of these materials. So-called core shell nanocrystals were synthesized, e.g. from water-based CdS nanocrystals to improve their emission quantum yield significantly [1]. Nowadays, quite a number of different and even better defined core shell particles can be synthesized also in high-boiling organic solvents. Well known examples are, e.g. CdSe/ZnS [2] and CdSe/CdS [3].

A more complex structure, namely the quantum dot quantum well (QDQW) system consisting of CdS particles with an embedded layer of HgS was first synthesized in 1993 [4]. These particles were examined with different characterization techniques like, e.g. static and time resolved photoluminescence, transient photobleaching and high-resolution electron microscopy (HRTEM) and the results were compared with theory. The findings on this model system will be summarised in the first part of this chapter (2.1).

In the latter parts of this chapter the research progress on different multi-shell nanocrystals will be surveyed. Three types of nanostructures will be covered: Ternary core shell shell (CSS) systems with an intermediate layer as a "lattice adapter", double quantum dot quantum well (double-QDQW) systems and an "inversed QDQW" system. Sorting multilayered nanocrystals into these three categories is justified by the different potential steppings of the semiconductor materials involved. Figure 1 shows the principle potential steppings of the valence and conduction bands in these three kinds of multilayered systems.

The potential stepping on the left in Fig. 1 causes both charge carriers (electron and hole) to be confined in the core of the nanocrystal. In the case of QDQW systems (middle), electron and hole are confined in the potential well which is embedded in the quantum dot. In the case of "inversed" quantum dot quantum well systems (right), the charge carriers are located in the core of the nanocrystal and in the outer shell. Depending on the different potential steppings, each of these systems shows unique properties, as will be outlined in this review.

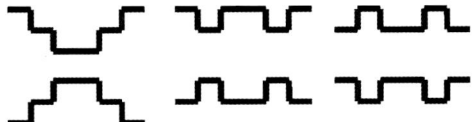

Fig. 1. Potential stepping in CSS nanocrystals (left), QDQW nanocrystals (middle) and "inverse" QDQW nanocrystals (right)

2. Water-based "quantum dot quantum well" systems

2.1 CdS/HgS – QDQW. Based on polyphosphate stabilized water-soluble CdS nanocrystals the quantum dot quantum well system consisting of CdS nanocrystals with an embedded layer of HgS (a quantum well within a quantum dot) became a model system for a number of fundamental studies.

The synthetic concept was developed by Mews et al. [4, 5]. To a solution of CdS quantum dots, a calculated amount of Hg^{2+} -ions is added resulting in a substitution reaction on the particle surface where the outermost layer of Cd-ions is replaced by Hg-ions and the Cd-ions are released into the solution. By analyzing the concentration of the free Cd- and Hg-ions in solution (see Fig. 3) it could be shown that for excessive addition of Hg-ions, no further substitution reaction takes place, since no further increase in Cd-ion concentration occurs while the Hg-ion concentration suddenly starts to rise. This indicates that only one ionic monolayer is substituted. Subsequently, the Cd-ions released into solution can be precipitated onto the particles by the addition of H_2S. The emerging colloidal particles consisted of a

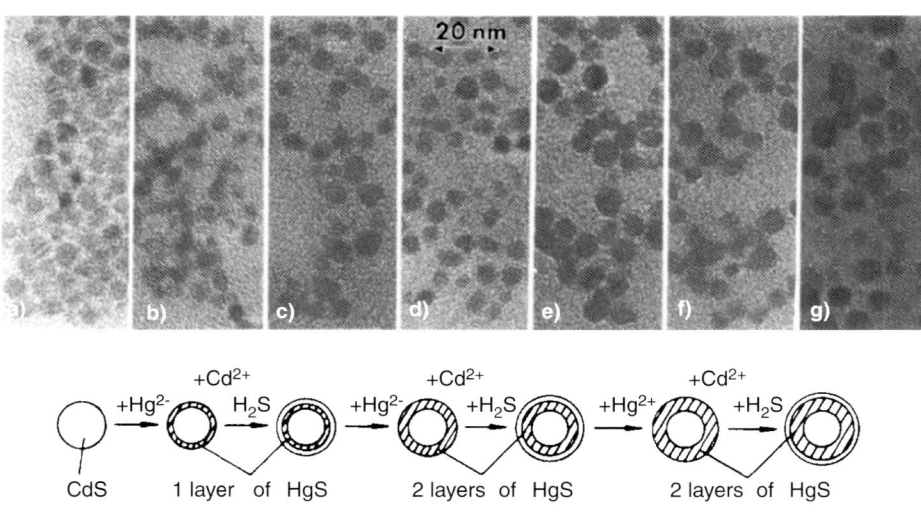

Fig. 2. Scheme of the synthesis of CdS/HgS/CdS QDQWs and TEM pictures at various stages of the synthesis. Reproduced with permission from Journal of Physical Chemistry 1994, 98, 934. © 1994 Am. Chem. Soc.

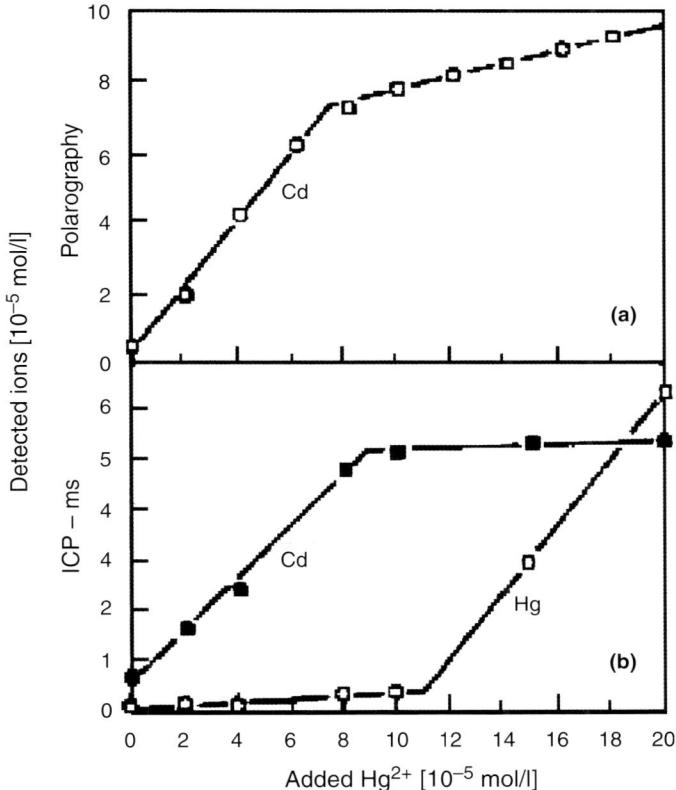

Fig. 3. Concentration of free Cd-ions and Hg-ions in a solution of CdS colloids upon the stepwise addition of Hg-ions, determined by polarography (upper panel) and inductively coupled plasma mass spectrometry (bottom). Reproduced with permission from Journal of Physical Chemistry 1994, 98, 934. © 1994 Am. Chem. Soc.

CdS core surrounded by a monolayer of HgS and almost one monolayer of CdS as the outermost shell (see below). From this point the preparation divides into two branches: either increasing the HgS layer thickness or increasing the thickness of the outermost CdS layer. The thickening of the HgS layer was simply achieved by repeating the substitution and reprecipitation steps described above (see Fig. 2). An increase in the CdS layer thickness was performed independently of the thickness of the formerly prepared HgS layer by additional addition of Cd-ions and precipitation of these onto the nanocrystals with H_2S.

In Fig. 4 the spectral evolution in the course of the further preparation is depicted. The major finding is a strong shift of the absorption onset towards lower energies with an increase in the thickness of the HgS well. This behavior can be explained with the small band gap of HgS (0.5 eV) with respect to CdS (2.5 eV) and a localization of the charge carriers in the HgS wells (see chapter 2.2 for details). Remarkable is the finding that also the CdS capping of the particles leads to a significant shift of the

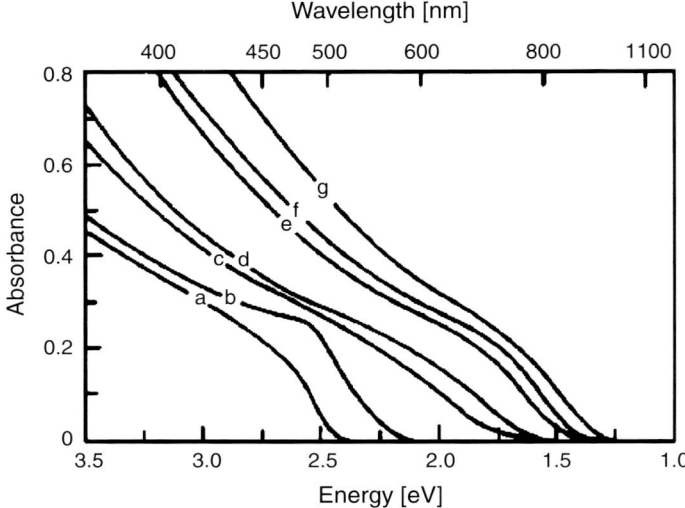

Fig. 4. Absorption spectra of the colloidal solutions of (a) CdS, (b) a + 8×10^{-5} M Hg^{2+}, (c) b + H_2S, (d) c + 8×10^{-5} M Hg^{2+}, (e) d + H_2S, (f) e + 8×10^{-5} M Hg^{2+}, and (g) f + H_2S (see Fig. 2). Reproduced with permission from Journal of Physical Chemistry 1994, 98, 934. © 1994 Am. Chem. Soc.

absorption onset towards lower energies even so CdS itself cannot absorb in this spectral region.

Theoretical calculations for these structures have been performed by Schoos et al. in 1994 [6] applying the particle in the box model with the effective approximation resulting in calculated values for the first electronic transition and the corresponding wave functions for the charge carriers. The presented results were in good agreement with the measured optical data. A further theoretical treatment was performed by Bryant et al. from 1995 onwards [7, 8].

The CdS/HgS/CdS quantum dot quantum well system acted as a prototype system for numerous spectroscopic investigations:

Transient photobleaching experiments have been performed with these structures in 1994 [9]. The most interesting result from these experiments was that the photobleaching follows spectrally the newly evolving 1s–1s electronic transition of the composite particles. Some considerations concerning the charge carrier dynamics in the novel QDQWs were outlined in this article.

Subpicosecond photoexcitation of CdS/HgS/CdS QDQW nanoparticles at wavelengths shorter than their interband absorption (390 nm) leads to a photobleach spectrum at longer wavelengths (440–740 nm) [10]. The photobleach spectrum changes and its maximum red shifts with delay time. These results are explained by the rapid quenching of the initially formed laser-excited excitons by two types of energy acceptors (traps); one is proposed to be due to CdS molecules at the CdS/HgS interface, and the other trap is that present in the CdS/HgS/CdS well. The results of the excitation at longer wavelengths as well as the formation and decay of the bleach spectrum at different wavelengths support this description.

The homogeneous absorption and fluorescence spectra of the CdS/HgS quantum dot quantum well system were investigated by transient hole burning and fluorescence line-narrowing spectroscopy. Also these photophysical measurements provide evidence for a charge carrier localization within the HgS well [11].

High-resolution transmission microscopy studies on the CdS/HgS/CdS QDQW system have been performed to study details of the crystallography of the system [12] (Fig. 5). A typical HRTEM micrograph of a CdS nanocrystal (a2), shows a triangular feature against the speckled background arising from the amorphous carbon substrate. The spacings and angles between lattice planes show that the nanocrystal is aligned along the (1 1 0) axis of the zinc-blende crystal structure of CdS. The decrease of contrast in going from the apex to the base implies a decrease in thickness. This suggests that the nanocrystal is a tetrahedron terminated in (1 1 1) surfaces (a1). This shape represents a polyhedron in which only (1 1 1) surfaces, either cadmium or sulfur terminated, are present. Since anionic polyphosphate ligands are at the surfaces of the nanocrystals, the best explanation for the observed morphology is to assume that the exposed surfaces are cadmium rich, as illustrated. The corresponding HRTEM simulation [Fig. 5(a3)] agrees with this interpretation of the experimental image, but only a small fraction of the crystallites are aligned along the proper crystallographic axis to allow the shape to be discerned. Seventy-five percent of that fraction showed the triangular projection with (1 1 1) surfaces and a

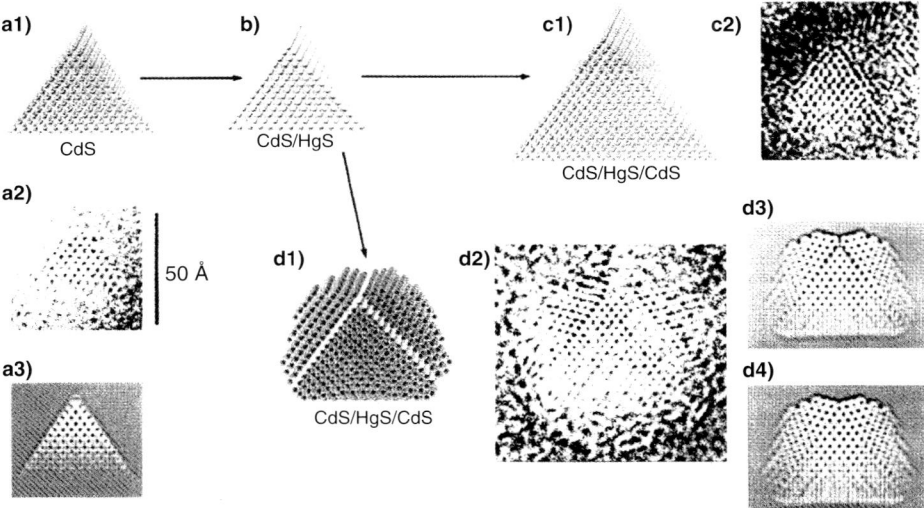

Fig. 5. HRTEM study of the structural evolution of the CdS/HgS/CdS nanostructure. The micrograph of a CdS core cluster (**a2**), exhibits tetrahedral morphology, which is in agreement with the TEM simulation (**a3**). The corresponding molecular model **a1** shows that all surfaces are cadmium-terminated (1 1 1). **b** Shows a model of the CdS particle after surface modification with Hg. A typical micrograph of a tetrahedral CdS/HgS/CdS nanocrystal is shown in **c2** along with a corresponding model **c1**. Model **d1** and micrograph **d2** represent a CdS/HgS/CdS nanocrystal after twinned epitaxial growth. The arrow marks the interfacial layer exhibiting increased contrast due to the presence of HgS, in agreement with the simulation (**d3**). No contrast change is seen in a simulation of a model with all Hg replaced by Cd (**d4**). Reproduced with permission from Physical Review B 1996, 53, 13242. © 1996 American Physical Society

mean edge length of 50 Å (±17%). The remaining structures can be assigned to truncated octahedral and heavily faulted crystals. The basic morphology is preserved in the next step of the synthesis, in which the surface cadmium-ions of the CdS crystallites are exchanged with mercury. Figure 5(b) shows a model of the structure resulting from single monolayer substitution. Within the statistical error, the mean size and size distribution is unchanged. However, defect structures become more prevalent and 64% of the oriented particles have tetrahedral shape. The final coating of these particles is carried out by adding excess cadmium-ions to the solution and growing CdS on top of the HgS layer via slow H_2S injection. Sixty-three percent of the resulting nanocrystals shows tetrahedral shape [Fig. 5(c)]. The average size increases to 62 Å (±24%) edge length. Furthermore, no "amorphous region" can be detected in any of the micrographs, suggesting epitaxial growth of the layers. The close match of the CdS and HgS lattice parameters ($a_{HgS} = 5.852$ Å, $a_{CdS} = 5.818$ Å) and the presence of faceted crystallites with only one exposed plane favor this growth mode. Half of the nontetrahedral nanocrystals show a different structure similar to that in Fig. 5(d). Here twin faults on the tetrahedral surfaces have resulted in the final CdS layers growing out of phase with the core. This arises by introducing one layer of hexagonal (wurtzite) stacking into an otherwise cubic (zinc-blende) structure and does not lead to a loss of passivation, as no bonds are broken. The cap layers on adjacent faces are crystallographically mismatched and cannot grow into each other. A model for this structure is shown in Fig. 5(d1). When viewed along the (1 1 0) crystallographic axis, two of the HgS planes are viewed edge-on and the initial CdS core can clearly be seen as a triangle. Close inspection of the micrograph shows a line of enhanced contrast along the twin fault [Fig. 5(d2), arrow] corresponding to the higher contrast of mercury relative to cadmium. The contrast change is reproduced in the simulation of the HRTEM image [Fig. 5(d3)], while the same model with cadmium in place of the mercury shows no such contrast change [Fig. 5(d4)]. The simulation for the tetrahedral model [Fig. 5(c)] shows a mild contrast change, but it is undetectable above background noise in the experiment.

The influence of the crystallography of the interface between the CdS and the HgS well on the optical properties of the CdS/HgS/CdS system was subject of further characterization utilizing optically detected magnetic resonance spectroscopy (ODMR) [13].

In the field of quantum dot quantum well structures, also new material combinations have been used to prepare these structures. The El-Sayed group first reported on a ZnS/CdS QDQW system [14]. These structures were characterized optically and compared with theoretical calculations. A quantum dot quantum well system consisting of CdS nanocrystals with an embedded monolayer of CdSe was presented by Battaglia et al. [15]. These QDQW structures were prepared in high-boiling organic solvents by the SILAR technique which was introduced by Peng et al. [16], yielding very monodisperse QDQW systems with high emission quantum efficiencies.

2.2 Double-QDQW. In 2001 first results on an extended CdS/HgS QDQW with two embedded HgS wells were reported by us [17] and Braun et al. [18]. Further characterization of these structures was done using X-ray photoelectron spectroscopy [19] as depth profiling technique. The obtained results were in good agreement with the predicted structure.

These structures are of interest because they allow studying the distance dependent interaction between two quantum wells within one quantum dot. In a later report the spectroscopic properties of these double well quantum dots were also compared with theoretical calculations in the framework of the effective mass approximation [20]. In principle, the synthetic procedure for the double well quantum dot nanocrystals is the same as for the normal quantum dot quantum well systems. Briefly, the outermost ion monolayer of the CdS nanocrystals is substituted with HgS by the addition of Hg$(ClO_4)_2$. The Cd^{2+}-ions released into solution are then grown onto the particles by the addition of H_2S. Different double well quantum dot systems are obtained by a sequence of growing of CdS shells and substituting them by HgS. For ease of discussion, all CdS/HgS/CdS/HgS/CdS samples are named according to Fig. 6 as CdS/HgS-ABCD, where each letter stands for the thickness of the corresponding layer in monolayers. Thus, for example, CdS/HgS-1213 relates to a nanocrystal

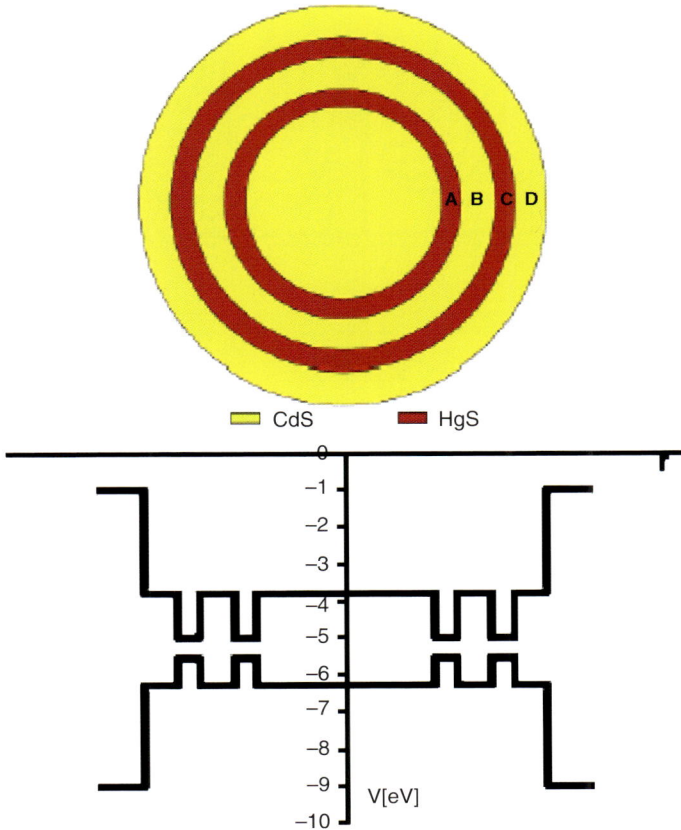

Fig. 6. Idealized picture of the double well quantum dot structure with introduction of the CdS/HgS-ABCD nomenclature and the corresponding radial potential for electron and hole [20]. Reproduced with permission from Journal of Physical Chemistry B 2004, 108, 1578. © 2004 Am. Chem. Soc.

consisting of a CdS core, followed by 1 monolayer of HgS, 2 monolayers of CdS, 1 monolayer of HgS, and again 3 monolayers of CdS. CdS/HgS/CdS QDQWs are analogously named as CdS/HgS-AB. Of course, this nomenclature refers to an idealized situation. In reality, the particles have variations in the thickness of all layers and are expected to have inhomogeneities within the different layers.

Figure 7 shows calculated radial probabilities of presence for the electron and the hole for the CdS/HgS-1x13 series of nanocrystals according to the nomenclature described above, where x is varied from $x = 0$ to $x = 7$. All these nanocrystals

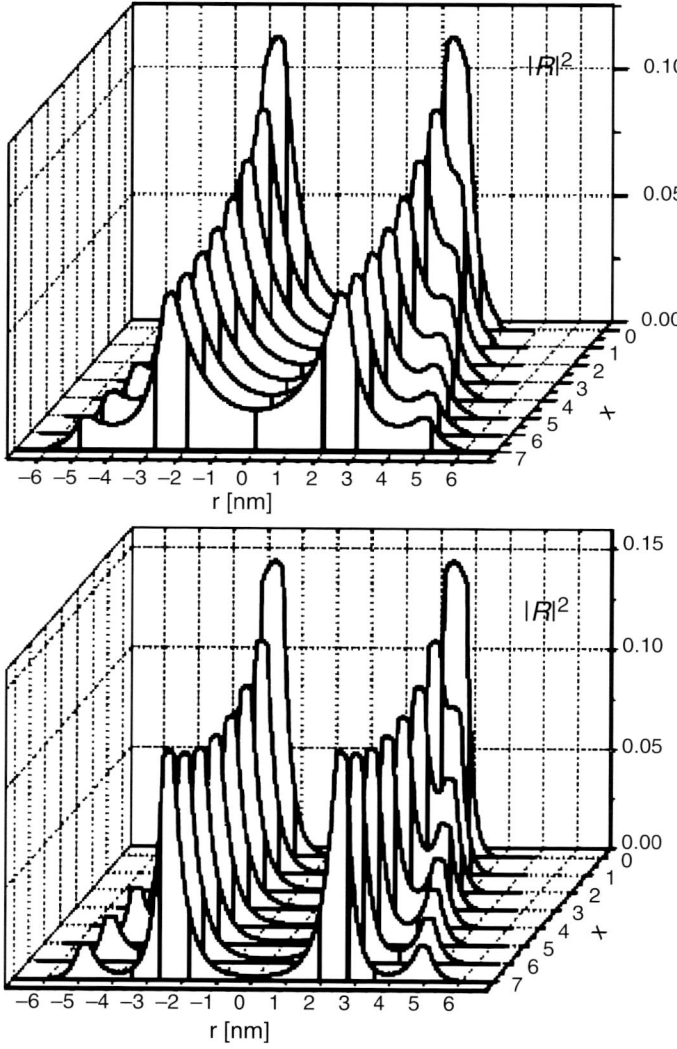

Fig. 7. Radial probability of presence in the CdS/HgS-1x13 systems ($x = 0$–7) for the electron (top) and the hole (bottom), r is the radial distance from the particle center [20]. Reproduced with permission from Journal of Physical Chemistry B 2004, 108, 1578. © 2004 Am. Chem. Soc.

contain the same CdS core, two wells each consisting of one monolayer of HgS and three outer cladding layers of CdS, the only difference being the distance between the two HgS wells which is varied from 0 to 7 monolayers of CdS. As expected, the probability of presence has a maximum within the HgS wells for both the electron and the hole, thus giving rise to a spatial overlap of the two wave functions within the HgS wells. Because of the higher effective mass of the hole compared to that of the electron in both materials, the localization is much stronger for the hole than for the electron in the same systems. Increasing the distance between the two HgS monolayers results in a stronger separation of the two maxima of the probability of presence. Thus, in the CdS/HgS-1713 system the two maxima of the probability of presence for the hole are almost totally separated from each other. This separation becomes more and more smeared out with a decrease in distance of the two HgS layers. A similar behavior is observed for the electron but because of the lower effective mass the maxima are not separated to the same degree even for the system with the largest distance between two HgS layers (i.e. CdS/HgS-1713).

Figure 8 (left) shows the UV/vis-absorption spectra of the CdS/HgS-1x13 series of nanocrystals with $x = 0$ to $x = 4$. The vertical bars represent the calculated first electronic transition (E_{gap}) for the corresponding ideal system. In general, the point of maximum curvature is believed to represent the first electronic transition of a sample. For the samples CdS/HgS-23, CdS/HgS-1113, and CdS/HgS-1213, this point matches quite well with the calculated values. For the samples CdS/HgS-1313 and CdS/HgS-1413, such a point is difficult to discern but still the absorption onset of those samples is shifted towards higher energies which is in good agreement with the calculated values (cf. vertical bars). This may be explained by a decreasing interaction between the two HgS layers with increasing distance between those layers. The high-energy absorption (above 2.5 eV) appears to depend mainly on the total amount of absorbing material. This is concluded from the experimental procedure: all samples were taken out of the crude reaction mixture and thus the particle concentration is assumed to be constant in all measurements. Consequently we observe the systems with the thicker CdS layer between the HgS layers having a

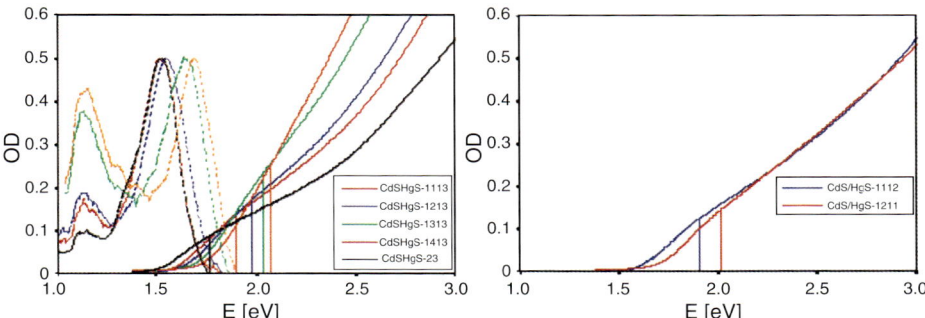

Fig. 8. UV/vis absorption and emission spectra (lines and dotted lines, respectively) of the CdS/HgS-1x13 systems together with the calculated eigenvalues of the first electronic transition (left) and the comparison of two systems with the same amount of CdS and HgS but different layer structure (right). Lines of the same color refer to the same sample in the left figure [20]. Reproduced with permission from Journal of Physical Chemistry B 2004, 108, 1578. © 2004 Am. Chem. Soc.

higher oscillator strength in this region. The absorption behavior between 1.8 and 2.2 eV is difficult to explain since the presented simple theoretical treatment is unable to explain the oscillator strengths in different regions of the spectra. The dotted lines in Fig. 8 (left) are the corresponding normalised emission spectra of these samples. Each sample shows an emission close to the band gap absorption. The relative position of the emission maxima is in good agreement with the absorption onsets of the samples. Furthermore, each sample shows a second emission at 1.15 eV which is likely to be "trap emission" because of stacking faults at the interfaces of the layers as previously shown by ODMR measurements on the CdS/HgS/CdS systems [13]. To show that the absorption onsets of those systems are not only affected by the molar ratio of CdS to HgS but also indeed depend on the layer structure, Fig. 8 (right) shows a comparison of two structures with the same molar ratio but a different layer structure (CdS/HgS-1112 and CdS/HgS-1211). The structure in which the HgS wells are separated by two monolayers of CdS and with a capping layer of one monolayer shows an absorption onset at higher energy compared to the system with a separation of one layer of CdS and two capping layers of CdS. In both cases, the agreement with the calculated transition energy (again given by the vertical bars) is satisfactory.

3. Core shell shell nanocrystals synthesized in high-boiling organic solvents

3.1 Lattice adapting spacer layers. The first report of a ternary core shell shell structure was given by Reiss et al. A synthetic procedure yielding CdSe/ZnSe/ZnS core shell shell particles was presented [21].

Talapin et al. synthesized similar particles in 2004 [22]. CdSe nanocrystals stabilized with TOP/TOPO (n-trioctylphospine/n-trioctylphosphinoxide) were coated with a shell of CdS or ZnSe. The second shell is composed of ZnS in both cases. According to the authors, the main purpose of the outermost ZnS shell is to avoid charge carrier penetration towards the surface of the particles. ZnS is a good candidate because of its large band gap (3.7 eV). A problem is the large lattice mismatch between CdSe and ZnS. This problem shall be overcome by using the intermediate ZnSe or CdS shell as a "lattice adapter".

Early reports on simple CdSe/ZnS core shell particles have shown that the quantum yield of the nanocrystals as a function of the ZnS layer thickness passes through a maximum at a ZnS layer thickness of approximately two monolayers [2]. One reason given for this behavior is the strain induced in the system by the mismatching lattices of ZnS and CdSe.

Figure 9 shows the principle structure of the CSS structures (**a**), as well as the potential stepping of valence and conduction band edges (**b**) and the absolute band gaps of the materials used as a function of the lattice spacing (**c** and **d**). It shows, that not only the band gaps are important but also the lattice spacing of CdS and ZnSe, which are the materials used as "lattice adapter" layers between those of CdSe and ZnS.

Figure 10 shows the development of the absorption and emission spectra during the coating procedure as well as the development of the quantum efficiency as a function of the shell thickness for different shell compositions. The red shift of the

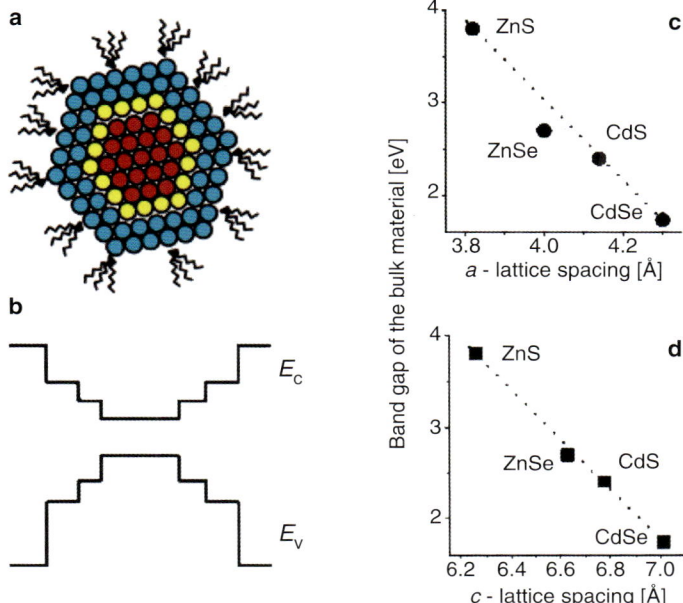

Fig. 9. Core shell shell nanocrystal: **a** schematic outline and **b** the schematic energy level diagram; **c, d** relationship between band gap energy and lattice parameter of bulk wurzite phase CdSe, ZnSe, CdS, and ZnS [22]. Reproduced with permission from Journal of Physical Chemistry B 2004, 108, 18826. © 2004 Am. Chem. Soc.

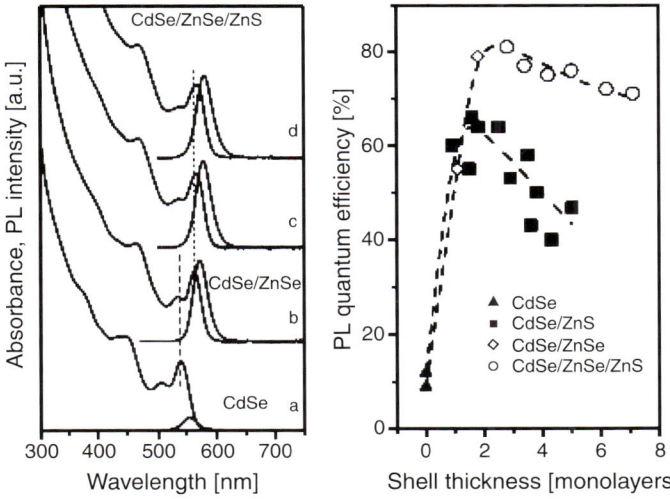

Fig. 10. Left: absorption and PL spectra of (a) CdSe cores, (b) CdSe/ZnSe core shell nanocrystals (thickness of ZnSe shell ~2 monolayers), (c, d) CdSe/ZnSe/ZnS nanocrystals with the thickness of the ZnS shell ~2 monolayers (c) and ~4 monolayers (d). Right: room-temperature PL quantum yields of CdSe, CdSe/ZnSe, and CdSe/ZnSe/ZnS nanocrystals dissolved in chloroform. For comparison, the dependence of the PL quantum yield on the shell thickness for various samples of CdSe/ZnS nanocrystals is shown [22]. Reproduced with permission from Journal of Physical Chemistry B 2004, 108, 18826. © 2004 Am. Chem. Soc.

first absorption signal is remarkably stronger for the ZnSe coating than for the ZnS coating which is interpreted as a stronger "leakage" of the exciton into the ZnSe shell compared to the ZnS shell. According to the authors another significant finding is the fact that the quantum efficiencies of the CSS structures do not drop as strongly for an increased ZnS layer thickness as for "normal" CdSe/ZnS core shell structures. This is

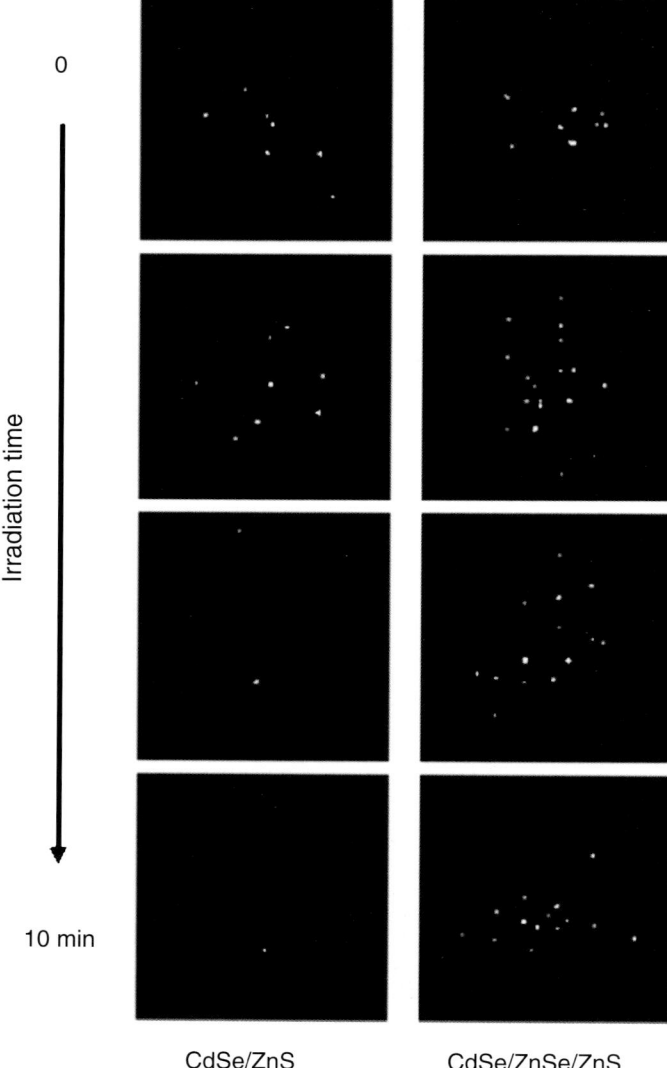

Fig. 11. Single particle luminescence images of CdSe/ZnS (left column) and CdSe/ZnSe/ZnS (right column) nanocrystals excited by an intense laser beam (514 nm) in air. The excitation intensity was chosen high enough to irreversibly bleach both nanocrystal samples on a short time scale. Each image is averaged over 10 s [22]. Reproduced with permission from Journal of Physical Chemistry B 2004, 108, 18826. © 2004 Am. Chem. Soc.

attributed to a higher crystallinity as a consequence of stress release of the CSS structures compared to the CdSe/ZnS structure.

According to the authors another benefit of the CSS structures is their greatly increased photostability. Figure 11 shows single particle luminescence images of CdSe/ZnS core shell structures and CdSe/ZnSe/ZnS CSS structures both deposited on a glass substrate and illuminated with a laser beam under ambient conditions. After 10 min, almost all core shell shell structures are still luminescent, while most of the core shell particles have already extinguished. This is assigned to a higher stability against photo-oxidation of the CSS structures in comparison with the core shell particles. Recently, Jun et al. presented a simplified one-step synthesis for CdSe/CdS/ZnS CSS nanocrystals [23].

Another example for core shell shell structures with a lattice adapting layer was reported by Xie et al. in 2005 [24]. In principle, the presented structure was a CdSe/CdS/ZnS CSS structure as presented above. Additionally the authors show the possibility to include an alloyed layer of $Zn_{0.5}Cd_{0.5}S$ into the structure resulting in CdSe/CdS/$Zn_{0.5}Cd_{0.5}$S/ZnS multishell particles. The coating steps were done using the SILAR technique introduced by Peng et al. [16]. Xie et al. also report very high quantum efficiencies for the multishell particles. Additionally their results concerning the stability versus photo oxidation show the same trend as in [22].

The TEM pictures presented in [24] (Fig. 12) demonstrate the growth of the particles and show very nicely that the particles retain a very narrow size distribution throughout the coating procedure.

Recently, the Banin group gave an example for a core shell shell structure with InAs and thus a III–V core material [25]. These InAs cores were covered with an intermediate layer of CdSe and an outermost layer of ZnSe. Similar as in the literature for the pure II–VI CSS particles, very high emission quantum efficiencies

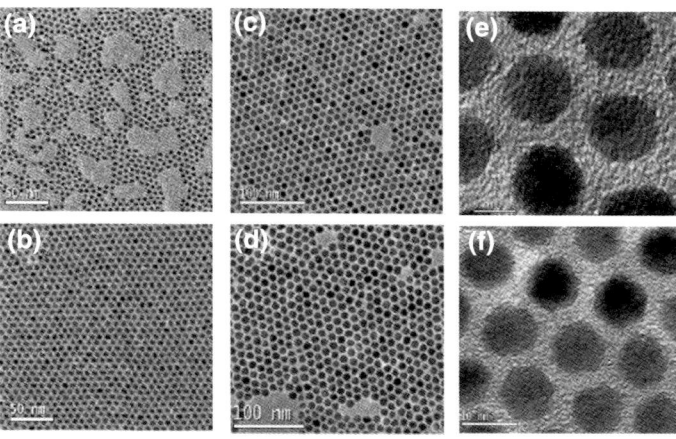

Fig. 12. TEM images of the plain CdSe cores and core/shell nanocrystals obtained under typical reaction conditions. **a** TEM images of CdSe cores (before injection of Cd^{2+} solution); **b** (a) plus 2 monolayers of CdS; **c/e** (b) plus 3.5 monolayers of $Zn_{0.5}Cd_{0.5}S$; **d/f** (c) plus 2 monolayers of ZnS [24]. Reproduced with permission from Journal of the American Chemical Society 2005, 127, 7480. © 2005 Am. Chem. Soc.

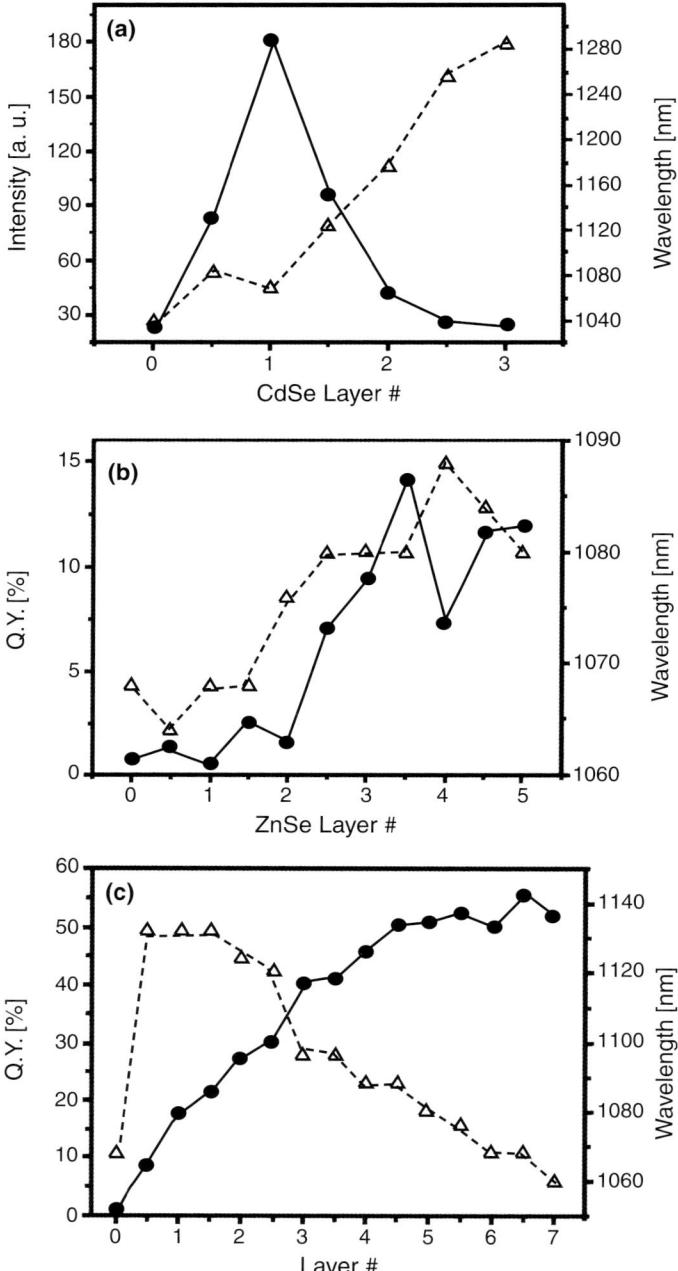

Fig. 13. Quantum yields (black dots with full lines) and emission wavelengths (white triangles with dotted lines) as a function of the shell thickness in monolayers for InAs/CdSe core shell nanocrystals (top), InAs/ZnSe core shell nanocrystals (middle) and InAs/CdSe/ZnSe CSS nanocrystals (bottom, first layer is CdSe here, all other layers are ZnSe) [25]. Reproduced with permission from Journal of the American Chemical Society 2006, 128, 257. © 2006 Am. Chem. Soc.

(up to 70%) are reported. However, due to the use of InAs as the core material, the emission wavelength is tuneable throughout the near infrared region of the spectrum in this case (800–1600 nm). This report shows nicely that the properties of the CSS nanocrystals can neither be obtained with simple InAs/CdSe nor with InAs/ZnSe core shell nanocrystals. Figure 13 shows the evolution of the quantum yield as a function of the coating thickness for InAs/CdSe core shell nanocrystals, InAs/ZnSe nanocrystals and the CSS system (where the first added layer is a CdSe layer and the following ones are composed of ZnSe). In the case of the InAs/CdSe system there is a clear maximum of the quantum yield for one monolayer of CdSe. For the InAs/ZnSe the quantum efficiencies increase continuously but reach only about 15%. In the CSS system the quantum efficiencies increase continuously to about 50%.

In summary, the CSS structures with a lattice adapting intermediate layer have been proven to be excellent systems concerning high luminescence quantum yields and a high stability versus photo-oxidation.

3.2 Decoupling spacer layers. Another interesting structure developed by the Peng group is the CdSe/ZnS/CdSe core shell shell system [26]. Since here the ZnS is the wide band gap material, these are not quantum dot quantum well systems in the common sense but nevertheless of great fundamental interest. It is the first report of two distinct emission signals from one nanocrystal where none of the signals is trap induced. Thus, the embedded ZnS layer seems to be thick enough to completely

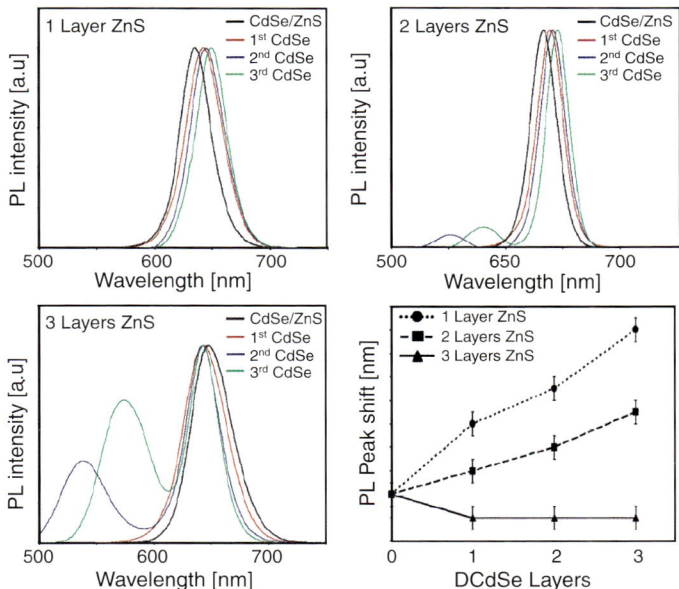

Fig. 14. Emission spectra of CdSe/ZnS/CdSe nanocrystals with a ZnS barrier layer of one (top left), two (top right) and three (bottom left) monolayers. All spectra are shown for the simple CdSe/ZnS core shell structure and for an outer shell of one, two and three monolayers of CdSe. The shift of the emission signals is summarized in the lower right figure [26]. Reproduced with permission from Journal of the American Chemical Society 2005, 127, 10889. © 2005 Am. Chem. Soc.

decouple the outer shell of CdSe from the inner CdSe core of the particle. Figure 14 shows the evolution of the emission signal in this kind of inversed QDQW. For particles with a ZnS layer of one monolayer thickness an increase in the outermost CdSe quantum well leads to a shift of the emission signal to lower energies but not to the occurrence of a second emission signal. In contrast, at a ZnS layer thickness of three monolayers, no shift of the emission signal can be observed. Instead, a second emission signal occurs at higher energies which shifts to lower energies with increasing thickness of the outer CdSe well. This behavior is explained as the phenomenon of "coupled and decoupled quantum systems in one semiconductor nanocrystal".

The shift towards lower energies for small ZnS buffer layers can be interpreted as an interaction between the core and the outer shell which is comparable to the interaction between the two quantum wells in the above presented "double well quantum dot structure". The two distinct emissions occurring with thick intermediate ZnS layers are interpreted as a complete electronic decoupling of the outer CdSe quantum well and the CdSe core. These structures were subject to further spectroscopic investigations [27] and are also investigated as a possible sources of white light emission [28]. However, since the presented measurements are ensemble measurements it can not doubtlessly be excluded that in some or all cases the single nanocrystals have only an emitting core or an emitting shell and not both. Only single particle spectroscopy on these structures could finally answer this question.

4. Conclusions

In this chapter we summarized the progress of the past years in the field of multishell nanocrystals. Three categories of particles were discussed. The core shell shell nanocrystals with an intermediate layer which can relax crystal strain. These structures show extremely high photoluminescence quantum yields and high stability versus photo-oxidation and therefore are of great interest for many applications like LEDs and biolabelling. The quantum dot quantum well structures as well as the "inverse" QDQW structures presented here are of great fundamental interest for understanding charge carrier dynamics and interactions across extremely thin semiconductor layers. All three categories will most likely remain a hot research topic in the next years.

References

[1] Spanhel L, Haase M, Weller H, Henglein A (1987) Photochemistry of colloidal semiconductors: surface modification and stability of strong luminescing CdS particles. J Am Chem Soc 109: 5649–5655

[2] Dabbousi BO, Rodriguez Viejo J, Mikulec FV, Heine JR, Mattoussi H, Ober R, Jensen KF, Bawendi MG (1997) (CdSe)ZnS core-shell quantum dots: synthesis and characterization of a size series of highly luminescent nanocrystallites. J Phys Chem B 101: 9463–9475

[3] Peng XG, Schlamp MC, Kadavanich AV, Alivisatos AP (1997) Epitaxial growth of highly luminescent CdSe/CdS core/shell nanocrystals with photostability and electronic accessibility. J Am Chem Soc 119: 7019–7029

[4] Eychmüller A, Mews A, Weller H (1993) A quantum-dot quantum-well – CdS/HgS/CdS. Chem Phys Let 208: 59–62

[5] Mews A, Eychmüller A, Giersig M, Schooss D, Weller H (1994) Preparation, characterization, and photophysics of the quantum-dot quantum-well system CdS/HgS/CdS. J Phys Chem 98: 934–941

[6] Schooss D, Mews A, Eychmüller A, Weller H (1994) Quantum-dot quantum-well CdS/HgS/CdS – theory and experiment. Phys Rev B 49: 17072–17078
[7] Bryant GW (1995) Theory for quantum-dot quantum wells: pair correlation and internal quantum confinement in nanoheterostructures. Phys Rev B 52: 16997–17000
[8] Jaskolski W, Bryant GW (1998) Multiband theory of quantum-dot quantum wells: dim excitons, bright excitons, and charge separation in heteronanostructures. Phys Rev B 57: R4237–R4240
[9] Eychmüller A, Vossmeyer T, Mews A, Weller H (1994) Transient photobleaching in the quantum-dot quantum-well CdS/HgS/CdS. J Luminescence 58: 223–226
[10] Kamalov VF, Little R, Logunov SL, El-Sayed MA (1996) Picosecond electronic relaxation in CdS/HgS/CdS quantum dot quantum well semiconductor nanoparticles. J Phys Chem 100: 6381–6384
[11] Banin U, Mews A, Kadavanich AV, Guzelian AA, Alivisatos AP (1996) Homogeneous optical properties of semiconductor nanocrystals. Molecular Crystals and Liquid Crystals Science and Technology Section a – Molecular Crystals and Liquid Crystals 283: 1–10
[12] Mews A, Kadavanich AV, Banin U, Alivisatos AP (1996) Structural and spectroscopic investigations of CdS/HgS/CdS quantum-dot quantum wells. Phys Rev B 53: 13242–13245
[13] Lifshitz E, Porteanu H, Glozman A, Weller H, Pflughoefft M, Eychmüller A (1999) Optically detected magnetic resonance study of CdS/HgS/CdS quantum dot quantum wells. J Phys Chem B 103: 6870–6875
[14] Little RB, El-Sayed MA, Bryant GW, Burke S (2001) Formation of quantum-dot quantum-well heteronanostructures with large lattice mismatch: ZnS/CdS/ZnS. J Chem Phys 114: 1813–1822
[15] Battaglia D, Li JJ, Wang YJ, Peng XG (2003) Colloidal two-dimensional systems: CdSe quantum shells and wells. Angew Chem Int Ed 42: 5035–5039
[16] Li JJ, Wang YA, Guo WZ, Keay JC, Mishima TD, Johnson MB, Peng XG (2003) Large-scale synthesis of nearly monodisperse CdSe/CdS core/shell nanocrystals using air-stable reagents via successive ion layer adsorption and reaction. J Am Chem Soc 125: 12567–12575
[17] Dorfs D, Eychmüller A (2001) A series of double well semiconductor quantum dots. Nano Lett 1: 663–665
[18] Braun M, Burda C, El-Sayed MA (2001) Variation of the thickness and number of wells in the CdS/HgS/CdS quantum dot quantum well system. J Phys Chem A 105: 5548–5551
[19] Borchert H, Dorfs D, McGinley C, Adam S, Moller T, Weller H, Eychmüller A (2003) Photoemission study of onion like quantum dot quantum well and double quantum well nanocrystals of CdS and HgS. J Phys Chem B 107: 7486–7491
[20] Dorfs D, Henschel H, Kolny J, Eychmüller A (2004) Multilayered nanoheterostructures: Theory and experiment. J Phys Chem B 108: 1578–1583
[21] Reiss P, Carayon S, Bleuse J, Pron A (2003) Low polydispersity core/shell nanocrystals of CdSe/ZnSe and CdSe/ZnSe/ZnS type: preparation and optical studies. Synthetic Metals 139: 649–652
[22] Talapin DV, Mekis I, Gotzinger S, Kornowski A, Benson O, Weller H (2004) CdSe/CdS and CdSe/ZnSe/ZnS core-shell-shell nanocrystals. J Phys Chem B 108: 18826–18831
[23] Jun S, Jang E, Lim JE (2006) Synthesis of multi-shell nanocrystals by a single step coating process. Nanotechnology 17: 3892–3896
[24] Xie RG, Kolb U, Li JX, Basche T, Mews A (2005) Synthesis and characterization of highly luminescent CdSe-core CdS/Zn0.5Cd0.5S/ZnS multishell nanocrystals. J Am Chem Soc 127: 7480–7488
[25] Aharoni A, Mokari T, Popov I, Banin U (2006) Synthesis of InAs/CdSe/ZnSe core/shell1/shell2 structures with bright and stable near-infrared fluorescence. J Am Chem Soc 128: 257–264
[26] Battaglia D, Blackman B, Peng XG (2005) Coupled and decoupled dual quantum systems in one semiconductor nanocrystal. J Am Chem Soc 127: 10889–10897
[27] Dias EA, Sewall SL, Kambhampati P (2007) Light harvesting and carrier transport in core/barrier/shell semiconductor nanocrystals. J Phys Chem C 111: 708–713
[28] Sapra S, Mayilo S, Klar TA, Rogach AL, Feldmann J (2007) Bright white-light emission from semiconductor nanocrystals: by chance and by design. Adv Mater 19: 569–572

Self-assembly of semiconductor nanocrystals into ordered superstructures

By

Elena V. Shevchenko[1], Dmitri V. Talapin[2]

[1] Center for Nanoscale Materials, Argonne National Lab, Argonne, IL, USA
[2] Department of Chemistry, University of Chicago, Chicago, IL, USA

1. Introduction

Organization of uniform objects into periodic structures can be found in many natural systems, such as atomic and molecular solids, opals, sponges and bacterial colonies – self-assembly is the fundamental phenomenon that generates structural organization on all scales [1]. In this chapter we discuss the structures spontaneously formed by nanoparticles which attracted significant interest from different branches of science and technology. The progress in colloidal synthesis of inorganic nanomaterials enabled preparation of different materials (metals, semiconductors, magnetic and ferroelectric materials) in the form of uniform nanometer-size crystals with amazing levels of size and shape control [2]. Nowadays colloidal synthesis allows for creation of nanostructures where composition, size, shape and connectivity of multiple parts of a multicomponent structure can be tailored in an independent and predictable manner (Fig. 1). In many nanoscale materials size and shape control provides additional degrees of freedom for designing physical and chemical properties. Thus, the effect of quantum confinement allows fine-tuning of the optical and electronic properties of semiconductor nanoparticles through varying particle size [3]. Exchange-biased ferromagnetism [4], size-dependent magnetic and catalytic properties of sub-20 nm particles are all examples of how material properties can be tailored by size and shape engineering at the nanoscale [5, 6].

Unique size tunable physical and electronic properties of the individual nanoparticles have prompted some researchers to refer to them as "artificial atoms". As the number of nanoparticle systems under strict synthetic control has expanded, the parallels to the development of a "new periodic table" have been discussed. Following this analogy, the ability to assemble "artificial atoms" into ordered one-, two- and three-dimensional structures (superlattices) should lead to creation of novel class of "artificial solids" with tunable properties determined both by the properties of individual nanoparticle constituents and the collective physical properties of the superlattice. In contrast to random mixtures of nanoparticles, the ordered nanoparticle arrays provide precise uniformity of packing and rigorous control of the

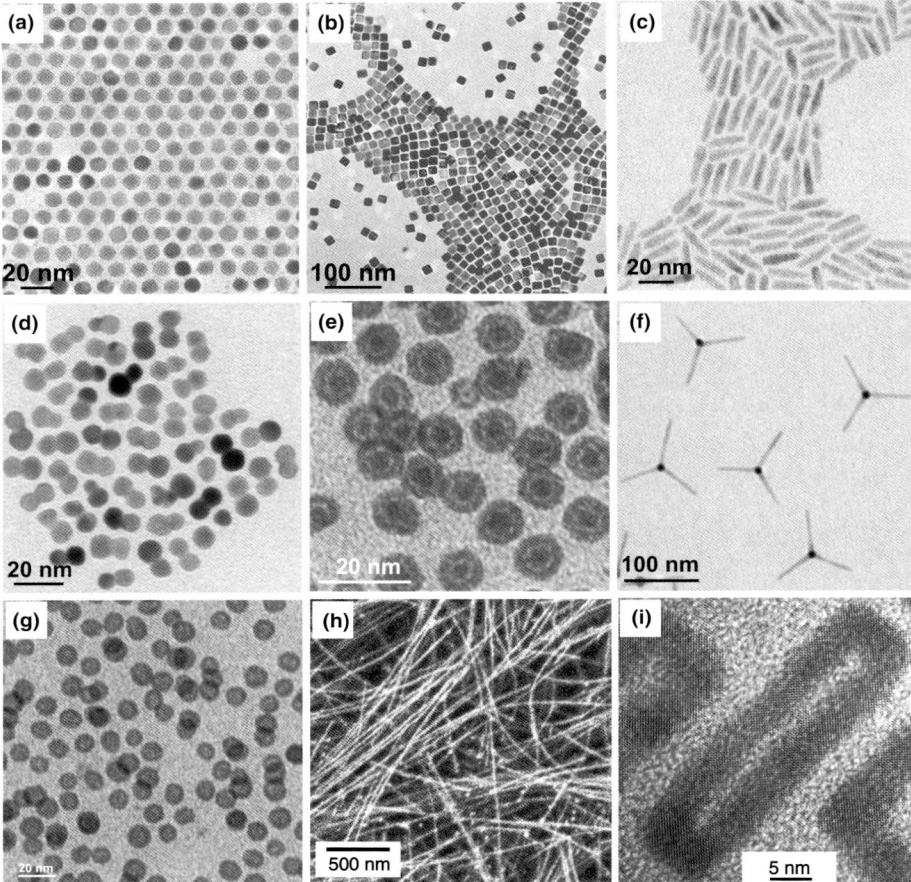

Fig. 1. Examples of semiconductor and magnetic nanomaterials of different sizes, shapes and compositions synthesized by colloidal chemistry techniques: **a** nearly spherical PbS nanocrystals; **b** PbSe nanocubes; **c** heterostructured CdSe/CdS nanorods; **d** $CoPt_3$/Au nanodumbbells; **e** nested iron–iron oxide nanostructures; **f** CdSe/CdS nanotetrapods; **g** hollow iron oxide nanoparticles; **h** PbSe nanowires; **i** PbSe nanorings

interparticle distances. Such structures, too small to be produced by existing lithographic "top down" approaches, can be generated by self-assembly which takes advantage of natural forces and organizes different pieces into desired architectures. In other words, novel materials or devices can be built by themselves and scientists merely have to find means to direct the self-assembly process [1].

2. Self-assembly of monodisperse quasi-spherical nanoparticles

Among the variety of objects that can self-assemble into ordered structures, uniform spheres are the simplest model system. Ordering of spherical particles has been extensively studied both theoretically and experimentally [7–9]. In the case of nano- and microparticles, the most popular approach to form ordered superstructures is the dispersing particles in an appropriate carrier solvent followed by gradual increase of particle concentration by solvent evaporation. Depending on nanoparticle concentration, solvent, temperature, substrate, etc., different types of nanoparticle aggregates can be obtained. These can be characterized using optical, transmission electron (TEM) and scanning electron (SEM) microscopies, small-angle X-ray scattering (SAXS), X-ray diffraction (XRD) and electron diffraction (ED) techniques. Optical microscopy allows characterization of the shape of nanoparticle superlattices only, whereas electron microscopy and X-ray diffraction allow for insight into details of the local structure of nanoparticle assemblies. Both TEM and SEM provide real-space two-dimensional projections of superlattice structure. TEM generally has significantly higher resolution than SEM, however, it can be applied only to thin (typically sub-100 nm) samples transparent to the electron beam. The details of three-dimensional arrangements of particles can be reconstructed from a series of TEM images taken by tilting the sample by different angles. We would like to point the reader to the excellent review by Wang on TEM characterization of nanoparticle assemblies [10]. SAXS, small-angle XRD and ED techniques are frequently used to obtain quantitative information about symmetry and lattice parameters of nanoparticle assemblies [11].

2.1 Particle interactions. The assembly of nanoscale building blocks into macroscopic structures is driven by the interactions of particles with other particles, substrate, solvent, etc. Sometimes the driving force can be rather counterintuitive. Thus, *entropy* can drive *ordering* of hard spheres in three dimensions because of the increased local free space available for each sphere (i.e., its translational entropy) in the ordered lattice compared to the disordered state [7]. The entropy driven crystallization well describes self-assembly of micron-size polymer beads which can be modeled by non-interacting hard spheres. At the same time, metal and semiconductor nanoparticles usually exhibit strong interparticle electrostatic interactions [12, 13]. In addition to static electric charge, nanocrystals can exhibit large dipole moments and polarizabilities [14–16]; large Hamaker constants result in strong van der Waals attraction [17]. The pair potential for two charged nanocrystals should take into account multiple interactions including sterical and osmotic repulsion, van der Waals attraction, and electrostatic terms for zeroth and higher order moments of charge distribution (charge–charge, charge–dipole, dipole–dipole, charge–induced dipole, etc.):

$$V = V_{steric} + V_{osmotic} + V_{vdW} + V_{Coulomb} + V_{charge-dipole} \\ + V_{dipole-dipole} + V_{charge-ind.dipole} \quad (1)$$

The Coulomb potential between two nanoparticles with charges Z_A and Z_B is

$$V_{Coulomb} \approx \frac{Z_A Z_B e^2}{4\pi\varepsilon\varepsilon_0 R} \quad (2)$$

where R is the center-to-center interparticle distance and ε is the static dielectric constant of the surrounding medium.

Fig. 2. Charge–dipole and dipole–dipole nanoparticle interactions. Interaction energy depends on the orientation of nanoparticle dipoles described by angles θ_1, θ_2, φ_1 and φ_2 in Eqs. (3) and (4)

The potential energy for charge–dipole, dipole–dipole and charge–induced dipole interactions between the particles A and B can be calculated as follows:

$$\text{charge–dipole}: V_{\text{charge–dipole}} = -\frac{e}{4\pi\varepsilon\varepsilon_0 R^2}(Z_A\mu_B\cos\theta_1 - Z_B\mu_A\cos\theta_2) \quad (3)$$

$$\text{dipole–dipole}: V_{\text{dipole–dipole}} = -\frac{\mu_A\mu_B(2\cos\theta_2\cos\theta_2 - \sin\theta_1\sin\theta_2\cos(\phi_2-\phi_1))}{4\pi\varepsilon\varepsilon_0 R^3} \quad (4)$$

$$\text{charge–induced dipole}: V_{\text{charge–ind.dipole}} = -\frac{e^2}{32\pi^2\varepsilon^2\varepsilon_0^2 R^4}(\alpha_A Z_B^2 + \alpha_B Z_A^2) \quad (5)$$

where μ_A and μ_B are dipole moments of particles A and B, respectively; θ_1, θ_2, φ_1 and φ_2 are the angles describing dipole orientations (Fig. 2); α_A and α_B are polarizabilities of particles A and B, respectively.

Dispersive (van der Waals) potential between two nanoparticles of radii R_A and R_B at the distance of closest approach D is

$$V_{\text{vdW}} = -\frac{A}{12}\left\{\frac{S}{D[1+D/2(R_A+R_B)]} + \frac{1}{1+D/S+D^2/4R_AR_B} + 2\ln\left(\frac{D[1+D/2(R_A+R_B)]}{R[1+D/S+D^2/4R_AR_B]}\right)\right\} \quad (6)$$

where A is the solvent-retarded Hamaker constant and $S = 2R_AR_B/(R_A+R_B)$ is the reduced radius.

The short-range repulsion can be estimated as follows [18]:

$$V_{\text{repulsion}} \approx \frac{100S\delta^2}{(R-2S)\pi\sigma^2}kT\exp\left(-\frac{\pi(R-2S)}{\delta}\right) \quad (7)$$

where δ is the apparent thickness of the ligand shell around nanoparticle core, and σ is the diameter of the area occupied by the ligand anchor group on the particle

Table 1. Magnitudes of interactions between two 3 nm/5 nm semiconductor or metal nanoparticles with charges $Z = \pm e$ and dipoles oriented along the particle–particle axis

Interaction	$V_{\text{semiconductor–semiconductor}}$ (eV)		$V_{\text{metal–metal}}$ (eV)	
	$R = 6$ nm	$R = 10$ nm	$R = 6$ nm	$R = 10$ nm
Coulomb ($\sim 1/R$)	0.12	0.072	0.12	0.072
charge–dipole ($\sim 1/R^2$)	0.082	0.029	–	–
dipole–dipole ($\sim 1/R^3$)	0.028	5.8×10^{-3}	–	–
charge–induced dipole ($\sim 1/R^4$)	8.4×10^{-3}	1.1×10^{-3}	0.018	2.3×10^{-3}
van der Waals ($\sim 1/R^6$)	0.016	2.1×10^{-4}	0.097	1.3×10^{-3}

R is the particle center-to-center distance

surface. More precise descriptions for the short-range nanoparticle repulsion can be found in [19, 20].

Substitution of the expressions (2)–(6) into (1) allows estimation of the interparticle potentials for typical dielectric (semiconducting) and metallic nanoparticles. Table 1 provides estimated magnitudes of the interactions between two 3 nm/5 nm semiconductor nanocrystals with charge $Z = \pm e$, dipole moments $\mu = 100$ D and polarizabilities $\alpha/4\pi\varepsilon_0 = 30$ nm³ (these values are typical for CdSe nanocrystals [14, 15]) and two 3 nm/5 nm metal (e.g., Au) nanoparticles with charge $Z = \pm e$ and polarizabilities $\alpha/4\pi\varepsilon_0 = 65$ nm³ ($\sim$nanoparticle volume). One can see that for semiconductors the energy of charge–dipole, dipole–dipole and charge–induced dipole interactions can approach $\sim 68\%$, 23% and 7% of Coulomb energy, respectively, when the particle separation is ~ 1 nm (the typical thickness of interdigitated capping ligands). Depending on the orientation of the nanoparticle dipole moments,

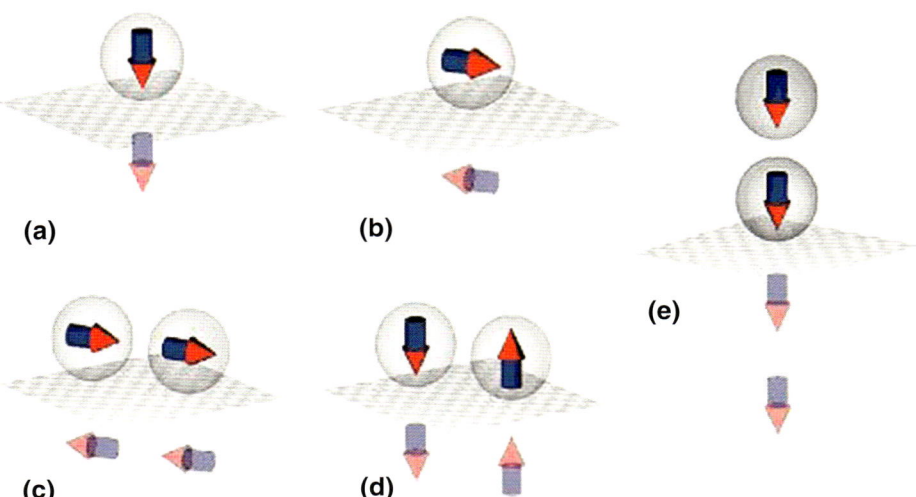

Fig. 3. Interaction of charged and dipolar nanoparticles with substrate can occur through the image charges. Reproduced from [23], © 2007, with permission from American Chemical Society

these additional terms can double (or cancel) electrostatic interactions between adjacent nanoparticles. This situation is quite unique. For example, for charged small molecules the Coulomb interactions dominate over dipolar and dispersive interactions by orders of magnitude [21].

When charged or dipolar nanoparticles are in close proximity with a conducting wall (e.g., a conducting carbon substrate), free electrons in the substrate screen the electric field of the charges and dipoles [22], which leads to an additional energy term due to the Coulombic coupling of all charges and dipoles with their mirror images (Fig. 3) [23].

In addition to electrostatic interactions, the hard inorganic core of a nanoparticle is surrounded with a "soft" layer of surfactant molecules. It was shown that surfactants can influence the assembly of NPs [24]. Thus, $Au_{140}(C_{12}H_{25}S)_{62}$ clusters capped with long-chain thiols are predicted to pack into 3D tetragonal superlattice whereas $Au_{140}(C_4H_9S)_{62}$ should form bcc superlattice [24]. Hydrophobic interactions between the ligands at the surface of CdTe nanocrystals stabilized with 2-(dimethylamino)ethanethiol can lead to spontaneous formation of 2D, free-floating sheets consisting of individual CdTe nanoparticles, as will be discussed in Sect. 2.3 [25].

The intrinsic complexity of the nanoparticle interactions introduces significant flexibility and the ability to manipulate the thermodynamics of their self-assembly by external parameters such as solvent, surfactants, temperature, external electric and magnetic fields. Proper tailoring of experimental conditions can lead to the formation of various amazing structures discussed in the following sections of this chapter.

2.2 Self-assembly of nanoparticles into one-dimensional structures.

Semiconductor nanocrystals with large dipole moments can form chains in a colloidal solution due to strong dipole–dipole interactions. Nanoparticle chaining in solution has been recently documented by direct cryo-TEM studies of colloidal solutions of PbSe nanocrystals [26]. In the chains, the nanoparticles can fuse together by the mechanism of oriented attachment along identical crystal faces forming single-crystalline nanowires [27, 28]. The formation of nanowires via self-assembly of nanoparticles has been studied for several materials [27–29]. Weller and coworkers reported the formation of ZnO nanorods by oriented attachment of ZnO nanoparticles along the unique axis of its wurtzite crystal lattice [28]. The suggested aggregation mechanism is based on the difference in surface structure and reactivity of the (0 0 2) and (0 0 $\bar{2}$) faces of the wurtzite lattice. Tang et al. demonstrated that wurtzite, CdTe nanowires can be formed from zinc blend, CdTe nanocrystals after partial removal of the stabilizing agents from the nanocrystals' surfaces [27]. The cubic, zinc blend, CdTe nanocrystals formed chain like aggregates due to dipole–dipole interactions at the early stage of wire formation and then slowly re-crystallized into hexagonal, wurtzite, CdTe nanowires.

The inherent anisotropy of crystal structure or crystal surface reactivity was identified in previous studies as the driving force for the one-dimensional growth. In nanoparticles with wurtzite structure (e.g., CdSe), an electric dipole moment originates from the noncentrosymmetric atomic lattice and scales with the nanoparticle volume [14, 30]. However, in some nanoparticle systems the chaining of nanoparticles into one-dimensional structure may look very counterintuitive. For example, PbSe nanocrystals have rocksalt type highly symmetric cubic crystal lattice. To form a 10-μm long nanowire, more than 10^3 individual nanocrystals must assemble and attach along one ⟨1 0 0⟩ axis while each PbSe nanocrystal has six equivalent ⟨1 0 0⟩ facets. Here, the dipolar interactions should be the driving force directing PbSe nanocrystals to assemble into chains. The origin of the dipoles in nanoparticles with centrosymmetric atomic unit cell could be explained by asymmetric lattice truncations [14] or noncentrosymmetric arrangement of Pb- and Se-terminated {1 1 1} facets [29]. Indeed, the reconstruction of the high-resolution TEM images revealed that PbSe nanocrystals are terminated by six {1 0 0} and eight {1 1 1} facets. The {1 0 0} facets are formed by both Pb and Se atoms while the {1 1 1} facets must be either Se- or Pb-terminated (Fig. 4a–c). Due to the difference in electronegativities between Pb and Se, {1 1 1} facets are polar and their arrangement will determine the distribution of electric charge within the PbSe nanocrystal.

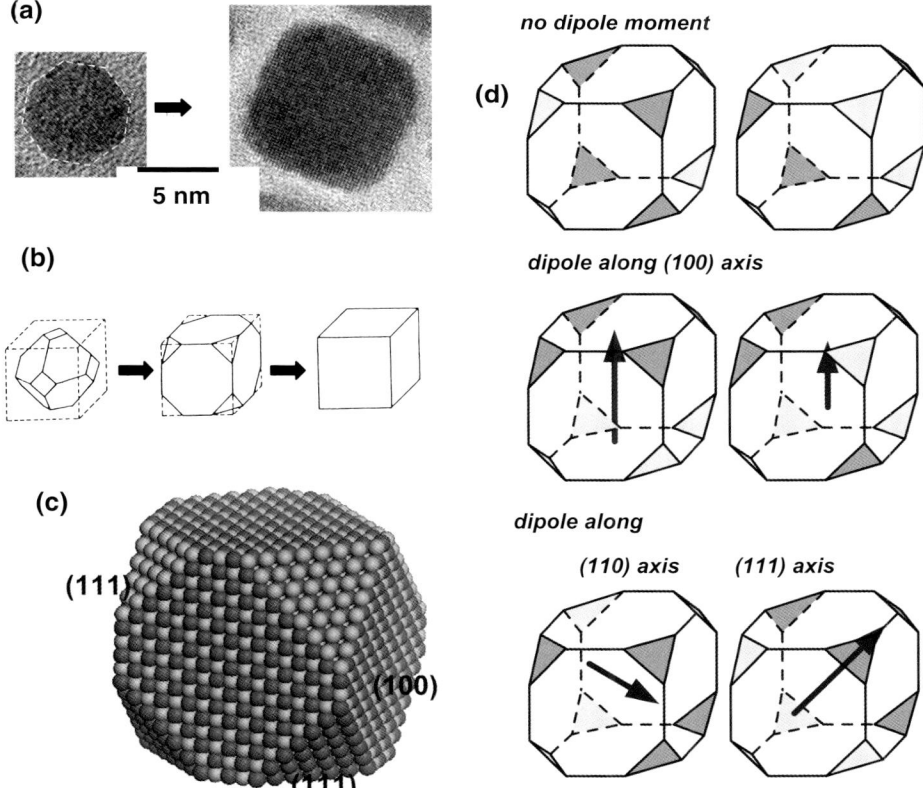

Fig. 4. a HRTEM images showing the typical evolution of the PbSe nanocrystal shape upon growth, as schematically depicted in b. c Atomic reconstruction of a rocksalt PbSe nanocrystal showing the structural difference between the ⟨1 0 0⟩ and ⟨1 1 1⟩ facets. d Different arrangements of polar ⟨1 1 1⟩ facets result in various orientations and magnitudes of the nanocrystal dipole moment. Reproduced from [29], © 2005, with permission from American Chemical Society

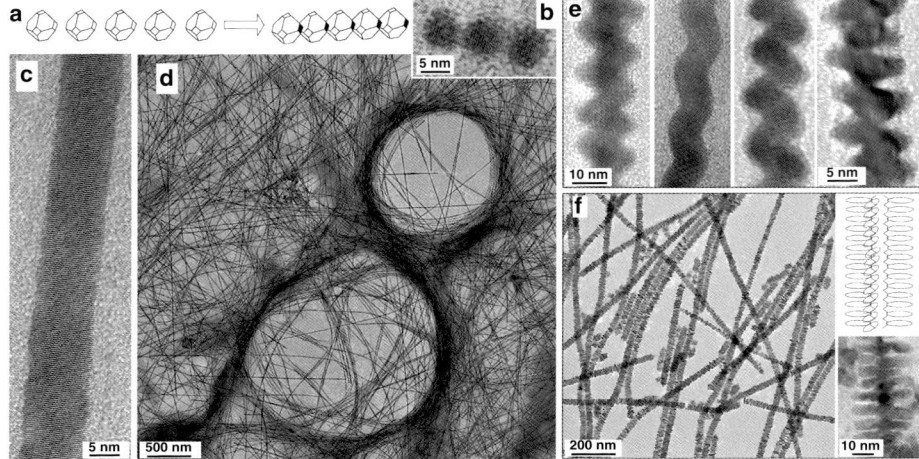

Fig. 5. PbSe nanowires synthesized by one-dimensional assembly and oriented attachment of nanoparticle building blocks. **a** Schematic representation of oriented attachment process. **b** High-resolution TEM image of "trimer" formed at early stage of the reaction. **c** High-resolution and **d** overview TEM images of straight PbSe nanowires. **e** TEM images of zig-zag and helical PbSe nanowires. **f** TEM images and a corresponding scheme of branched PbSe nanowires

Depending on the mutual arrangement of the ⟨1 1 1⟩ facets, the whole nanocrystal can either have central symmetry and thus a zero net dipole moment or it can lack central symmetry and possess a dipole moment along the ⟨1 0 0⟩, ⟨1 1 0⟩, or ⟨1 1 1⟩ axes, respectively (Fig. 4d). Assuming a random distribution of polar ⟨1 1 1⟩ facets and cuboctahedral shape PbSe, the majority of PbSe nanocrystals (∼ 89%) should have nonzero dipole moments, that drives their oriented attachment.

One-dimensional assembly and oriented attachment of nanoparticle building blocks allow us to obtain large quantities of high quality, catalyst-free semiconductor nanowires (Fig. 5). These nanowires can form stable colloidal dispersions and, therefore, are easy for solution-processing and device integration. Moreover, the nanowire synthesis through oriented attachment produces nanowires with control of wire dimensions and morphology. In addition to straight nanowires, zig-zag, helical, branched, and tapered nanowires as well as single-crystal nanorings could all be prepared by the adjustment of the reaction conditions (Fig. 5). Different nanowire morphologies may be advantageous depending on the potential applications. Straight nanowires with minimal surface roughness may provide the high carrier mobilities necessary for high-performance FETs [31]. On the other hand, the performance of nanowire-based sensors [32, 33] and photovoltaic devices [34] should improve by increasing the wire surface area, i.e., highly branched nanowires may best match these applications. The performance of nanowire-based thermoelectric devices [35, 36] may substantially benefit from the multiple scattering of acoustic phonons in zig-zag and helical nanowires.

2.3 Self-assembly of nanoparticles into two-dimensional structures.

Inorganic nanoparticles can form various structures upon evaporation of carrier

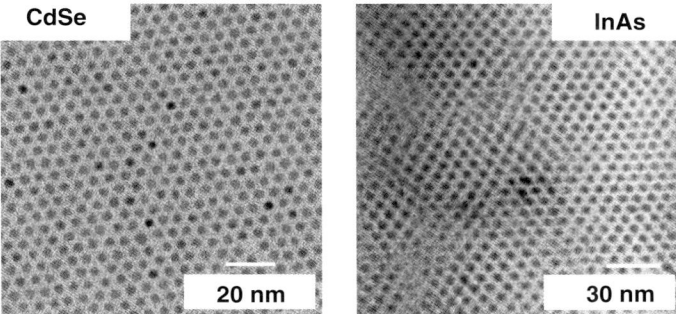

Fig. 6. Examples of two-dimensional ordered arrays of semiconductor nanocrystals viewed in TEM

solvent from a colloidal solution. Depending on nanoparticle concentration, solvent, temperature, substrate, etc., different types of nanoparticle aggregates can be obtained. If solvent and substrate show good wetting properties, dilute solutions of monodisperse nanoparticles can produce long-range ordered two-dimensional superlattices shown in Fig. 6. Such arrays typically show hexagonal packing corresponding to maximal packing density of nanoparticles in 2D and the strongest van der Waals interactions. The spacing between crystallites in the monolayer is determined by the length of the surfactant molecules and can be adjusted by tailoring the capping ligands. The interparticle spacing determines degree of electronic or magnetic coupling between the nanoparticles and plays an important role for design of nanoparticle-based thin-film materials and devices [5, 37].

Formation of nanoparticle films on a substrate induced by solvent evaporation was modeled by Rabani et al. using coarse-grain model of nanoparticle self-assembly [38]. Irreversible evaporation of solvent was considered as a non-equilibrium process. Because of non-equilibrium nature of this process, the nanoparticle assemblies can exhibit complex transitory structures, even when equilibrium fluctuations are mundane [39]. Depending on coverage and mobility of nanoparticles different patterns of nanoparticle aggregates were obtained. Disk-like and ribbon-like morphologies were found to represent different stages of the same growth mechanism [38] as a result of spatially homogeneous solvent evaporation (Fig. 7). When evaporation is heterogeneous in space, the dynamic of self-assembly can be dramatically different and determined by the nucleation and growth of vapor bubbles. If nanoparticles are sufficiently mobile to track the fronts of growing vapor nuclei, the aggregate patterns will be shaped by the structural history of the evaporation. In other words, the distribution of nanoparticle aggregates traces the intersection lines of the colliding vapor nuclei, leading to the network morphologies [38, 39]. Even though coarse-grain model does not take into account hydrodynamic convection flows, film thickness, substrate roughness and non-local interactions, it successfully explains a variety of experimentally observed patterns [38].

Self-assembly of nanoparticles can occur both on solid substrate and liquid surface. Reasonably ordered multilayer films of nanoparticles can be deposited by the different modifications of Langmuir–Blodgett (LB) technique. The LB approach is based on the formation of a monolayer of charged long-chain molecules at the

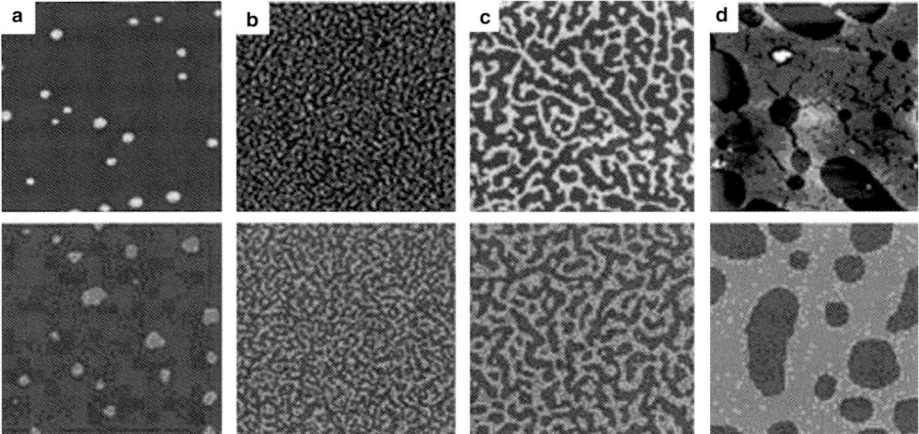

Fig. 7. Self-assembled morphologies resulting from homogeneous evaporation and wetting of nanoparticle domains. The bottom panels demonstrate the coarse-grain model simulations assuming homogeneous evaporation for coverages of 5% **a**, 30% **b**, 40% **c** and 60% **d** and top panels show the experimental results for CdSe nanoparticles with the same coverage. The details of modeling can be found in [38]. Reproduced from [38], © 2003, with permission from Nature Publishing Group

air–liquid interface. These monolayers can be further used for organization of charged particles [1]. It was shown that LB technique can be directly utilized to prepare various phases of organically functionalized nanoparticles by spreading the colloidal solution across the liquid surface of Langmuir trough. Self-assembly of nanoparticles into close-packed 2D or 3D arrays was achieved upon different regimes of film compression [40, 41]. The quality of the Langmuir films is sensitive to particle size distribution, compression pressure, temperature, etc. The great advantage of this method is that the film formed at the air–liquid interface can be further transferred to any substrate. The LB technique was used to deposit monolayers of nearly monodisperse nanometer-size CdSe crystallites onto various substrates. Size-selected CdSe nanocrystals capped with tri-octylphosphine oxide (TOPO) were directly applied onto the water surface of a LB trough and served as the LB-active species [42]. Absorption and luminescence studies of monolayers transferred onto glass slides indicate that the monolayers retain the general optical properties of isolated nanocrystals. Transmission electron micrographs of monolayers on amorphous carbon show the formation of two-dimensional hexagonal close-packed domains. This technique allows us to obtain 2D monolayers of quasi-spherical particles in a controllable and reproducible way. However, the concentration of stacking faults and vacancies in films obtained by LB technique is relatively high [42].

Dynamics of the solvent evaporation is very important and can be varied by changing the temperature and/or solvent. The efficiently screened particle–particle attractions in solution become significant as solvent evaporates. As a result large nanoparticle layers tend to crack upon drying. The appearance of cracks is associated with increasing of van der Waals attractions that shrink the interparticle distances [43]. The compositional inhomogeneity of deposited structures negatively affects the

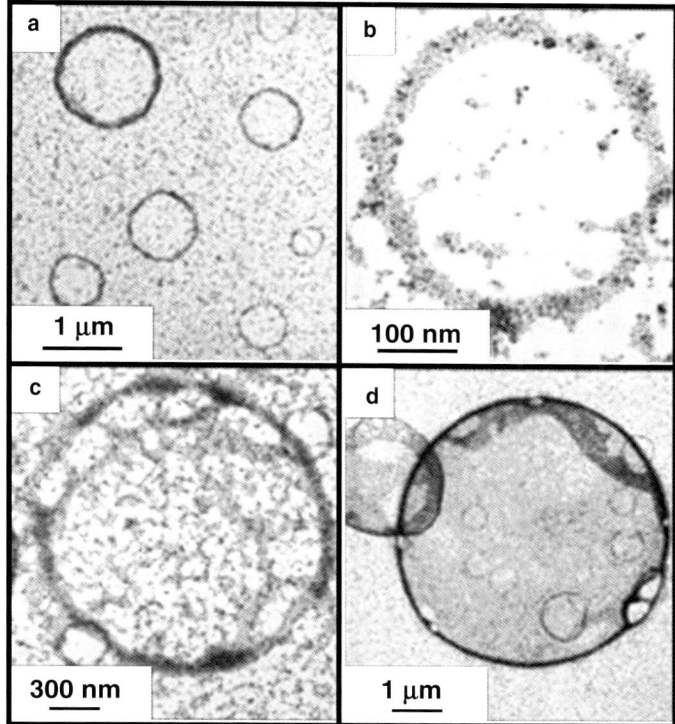

Fig. 8. Various rings obtained from diluted suspensions of CdS nanoparticles. Coverage ratio is 0.5. Reproduced from [44], © 2000, with permission from the American Chemical Society

performance of the self-assembled device. Experimentally, it is hard to avoid the cracking of the drying films. The formed cracks can be filled by infiltration of the deposited nanoparticle structures with extra portion of colloidal solution. However, the deposited layer has to be protected against the dissolution. Striping of the capping ligands from the surface of nanoparticles helps to fix nanoparticles at the substrates. Plasma-, thermo- and chemo-treatments can be used to remove or modify the capping ligands [37].

Controllable solvent evaporation can lead to formation of more sophisticated patterns such as nanoparticle rings, dendrites and islands with radial particle size gradients. Several mechanisms can be responsible for the formation of rings (Fig. 8) [17, 44]. One of the possible explanations of the appearance of micron wide, uniform in size rings consisting of nanoparticles is nucleation and growth of holes in the solvent films completely wetting the substrate. Hole nucleation in an evaporating film can be modeled as substrate-mediated boiling and hole formation can be considered as bubble nucleation in a bulk of superheated liquid with the only difference that holes in thin films are "open" to the vapor above the film [40]. The growth of holes pushes the particles and solvent out toward the "bulk" wet film of colloidal solution. If the nucleation and growth of holes are separated in time, rings uniform in width will be formed. The force acting on the hole rim increases linearly with the size of the hole, and the number of particles being swept out by the hole area increases quadratically [45]. The concentration of nanoparticles inside the ring is significantly lower as compared to particle concentration outside. In this model the particle–particle attractions are considered not to exceed thermal energies. As a result, particles can be moved until the friction arising from their attraction to the substrate pin the contact line. The resulting hole size depends on particle size and their concentration [46]. In the case of low concentrations, the contact lines of the holes will not be pinned before the "percolation" of the growing holes and compact domains of nanoparticles will be formed. The smaller size of nanoparticles also favors the growth of compact nanoparticle domains: small particles can more easily overcome their static friction [45] and as a result, the pinning size of hole becomes larger than the average distance between holes.

The Marangoni effect can also lead to ring-like nanoparticle patterns. Evaporation of a thin film of volatile solvent can create significant temperature variation between the substrate and the free upper surface leading to the temperature gradient across the film. The interfacial tension depends on the temperature and, as a result, local "hot" and "cold" spots generate low and high interfacial areas pulling the solvent from the warmer to cooler areas. At the same time solvent moves upward to the warmer places in order to compensate the concentration perturbation. This generates a convective flow through the liquid film and leads to the appearance of a network of cells. Inside those the fluid goes up by the center of the cells and goes down by the walls of the cells.

Dimensionless parameter determining the threshold of this dynamic solvent instability is called the Marangoni number (M_a):

$$M_a = -\frac{d\sigma}{dT}\frac{1}{\eta\alpha}L\Delta T \tag{8}$$

where σ is the surface tension (N/m), η is the dynamic viscosity (kg/(s m)), α is the thermal diffusivity (m^2/s), L is the film thickness and ΔT is the temperature difference (°C).

The critical value for the Marangoni number is $M_a = 80$. Below this value interfacial instabilities do not appear. Above 80 different patterns can be generated, depending on the system [47]. Increase in temperature and interfacial tension fluctuations, as well as in the thickness of the liquid film promote the appearance of interfacial instabilities, while fluctuation enhancement is suppressed by thermal diffusivity and dynamic viscosity of the liquid. Marangoni effect was found to be responsible for the formation of ring-like structures and hexagonal networks for a variety of nanoparticles of different shapes [44] (Fig. 8). Systematic study performed by Pileni and co-workers [44] demonstrated that ring formation was related on the evaporation rate of the solvent used to disperse nanoparticles. Replacement of highly volatile hexane with decane, as well as evaporation of hexane solution under the "saturated" vapor of hexane, led to the precipitation of random dispersions of nanoparticles [44].

Dendrites are another interesting example of self-assembled nanoparticle structures. Dendrite-like structures are generated as a result of the interplay between the ordering effect of crystallization, the disordering effect of local orientation fluctu-

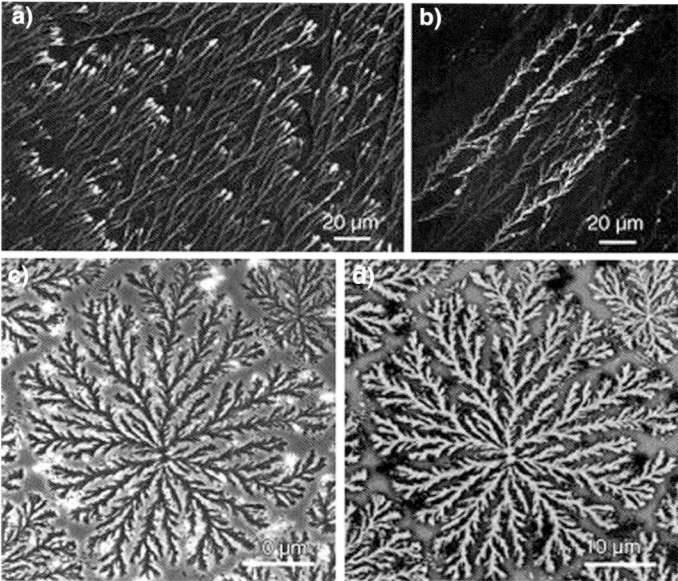

Fig. 9. Phase contrast (**a**, **c**) and fluorescent (**b**, **d**) images of dendrites formed by water-solubilized CdSe/ZnS quantum rods (**a**, **b**) and by quantum dots (**c**, **d**) upon their drying-mediated self-assembly at room temperature at 0.3 mg/mL concentration. Reproduced from [48], © 2006, with permission from Wiley

ations of the nanoparticles and the freezing of the domain edges after complete solvent evaporation [48]. Recently, dendrites consisting of CdSe/ZnS nanodots and nanorods were reported (Fig. 9) [48]. Typically, dendrites formed from nearly spherical nanoparticles consisted of highly luminescent core with six-to-ten branches. Self-assembly of nanorods under the same conditions led to the formation of long, highly asymmetric dendrites with 2–3 branches growing towards the droplet center. The difference in the morphology of nanodot and nanorod dendrite structures can be attributed to stronger dipole–dipole interactions between rods and their shape-related anisotropy [48]. Increase of the particle concentration as well as evaporation temperature was found to cause gradual decrease of the lateral sizes of dendrites and an increase of their thickness.

Some size distribution is always present in nanoparticle samples and this could have a substantial influence on both thermodynamics and kinetics of formation of nanoparticle superlattices. As an example, pair wise van der Waals attraction in a sample of polydisperse nanoparticles is strongly dependent on particle size (Eq. (6)). Attractions between the largest particles are the strongest. As a result they energetically prefer to be clustered together forming characteristic structures where the largest particles at the center are surrounded by smaller particles (Fig. 10a) [17]. This experimental observation has been supported by Monte Carlo calculations performed for a system of particles with mutual attraction described by Eq. (6) (Fig. 10b).

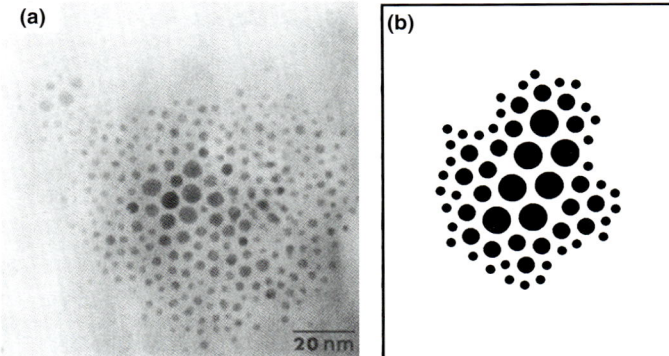

Fig. 10. a TEM micrograph revealing the size-dependent phase separation in polydisperse Au nanocrystals capped with dodecanethiol. **b** Simulated pattern for a system containing 7A, 21B and 36C Au spheres with radii of 1.0, 0.6 and 0.3, respectively. The energy of the system was minimized via Monte Carlo simulated annealing routine. The minimum allowable distance between particle cores is 0.5. Reproduced from [17], © 1995, with permission from American Institute of Physics

Formation of two-dimensional nanoparticle assemblies often occurs at the interfaces, liquid–solid, liquid–air or liquid–liquid. The interface typically plays a role of template directing assembly of nanoparticles in two dimensions. However, Kotov and co-workers recently reported that CdTe nanoparticles stabilized with 2-(dimethylamino)ethanethiol can spontaneously form monolayers of two-dimensional free-floating sheets consisting of individual CdTe nanoparticles in the bulk of solution (Fig. 11) [25]. This very counterintuitive observation has been explained by simultaneous effects of electrostatic interactions arising from both dipole moment

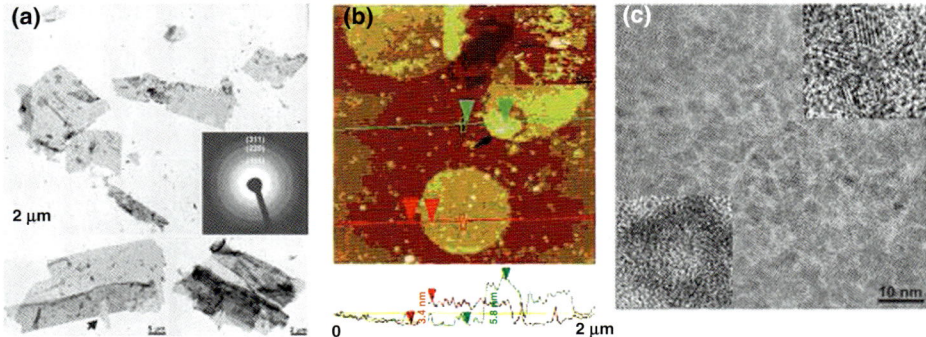

Fig. 11. a TEM images of free-floating films of CdTe NPs. (Inset) Electron diffraction pattern obtained from the films. **b** Tapping-mode AFM and corresponding topography cross sections of the monolayer films on a silica surface. Inset: large-scale AFM image of the films. The morphology of the films is the same for hydrophilic and hydrophobic TEM grids, as well as for AFM images obtained on hydrophilic and hydrophobic Si substrates. **c** High-resolution TEM images of the monolayer films of CdTe NPs. The inset at top right represents a detailed arrangement of single NPs obtained by HRTEM. Each particle has a distinct cubic crystal structure identified from the spacing between adjacent crystal planes inside the NPs equal to 0.38 nm, which corresponds to (111) surfaces of zinc blend CdTe. Inset at bottom left shows characteristic assembled rings of cubic CdTe NPs with zinc blend cubic crystal structure. Reproduced from [25], © 2006, with permission from AAAS

and a small positive particle charge combined with hydrophobic particle–particle interactions. The positive charge of CdTe particles stabilized with 2-(dimethylamino)ethanethiol was estimated to be $1e$ to $3e$. Monte Carlo simulations confirmed the importance of the "net" charge for the formation of 2D nanoparticle aggregates [25].

2.4 Self-assembly of nanoparticles into three-dimensional superlattices.

Nanocrystals can form three-dimensional structures with different degree of short- and long-range ordering. Individual nanoparticles can be brought together as freestanding structures or thin films in two forms: (i) amorphous (glassy) and (ii) crystalline solids. Glassy solids are isotropic materials with only short-range order and randomly oriented nanoparticles (Fig. 12). Close-packed glassy films usually form if the nanoparticles are polydisperse or if the rate of destabilization of a colloidal solution is high (e.g., very fast solvent evaporation) [11]. If colloidal nanoparticles are poorly soluble in a given solvent due to insufficient sterical repulsion or the lack of the capping ligands on nanocrystal surface, no long-range ordered arrays is expected. Also, if repulsive forces dominate particle–particle

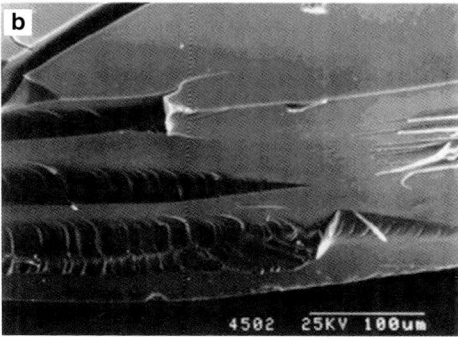

Fig. 12. a Low and high (inset) magnification SEM micrographs of a glassy solid prepared from 5.6 nm CdSe nanoparticles. **b** SEM micrograph of the fracture surface from a $\sim 150\,\mu m$ thick glassy solid of 3.8 nm CdSe nanoparticles. The NC solid was imaged by tilting it 45° with respect to the incident electron beam. Reproduced from [11], © 2002, with permission from Annual Reviews

interactions and these interactions are weak, there is no sufficient energy driving the formation of an ordered lattice [11]. As the concentration rises with solvent evaporation, the viscosity of the dispersion increases until it freezes the local structure of nanoparticles in the form of glassy-like solid. High-resolution SEM images (Fig. 12a) demonstrate examples of glassy-like film formed from CdSe nanoparticles. The nanoparticle glass is a brittle material and can fracture under the stress (Fig. 12b) [11]. Even though glassy-like structures allow realizing dense packing of nanoparticles, they are characterized by the absence of compositional uniformity because of the lack of "global" periodicity in particle distribution. In the case of periodic structures the positions of particles and, as a result, chemical composition, are known. This makes periodic nanoparticle structures (e.g., films, colloidal crystals) more attractive for practical applications and fundamental studies of their properties.

Generally, both entropy and isotropic attractive van der Waals forces should favor assembly of spherical particles into the structures with highest packing density such as face centered cubic (*fcc*) and hexagonally close-packed (*hcp*) lattices. In *fcc* and *hcp* lattices hexagonally close-packed layers are shifted with respect to each other in a way that centers of mass in one layer are above the hole in the layer below. This gives three different layer positions marked as A, B and C. As schematically shown in Fig. 13, *hcp* and *fcc* lattices have characteristic ABAB- and ABCABC-type stacking of hexagonally packed nanocrystal layers.

In an ensemble of non-interacting hard spheres entropy can be a sole driving force for the transition from disordered into long-range ordered state. This transition spontaneously occurs when the volume fraction of monodisperse hard spheres exceeds a threshold value of 0.494 [7]. Theoretical calculations and simulations of hard-sphere colloids predict that *fcc* structure should be slightly more stable compared to *hcp* [8, 9]. The free energy difference between these two close-packed structures with identical packing density (~ 0.7405) is very small, $\sim 10^{-3}$ kT per particle. In agreement with this prediction, monodisperse micron-size latex and silica

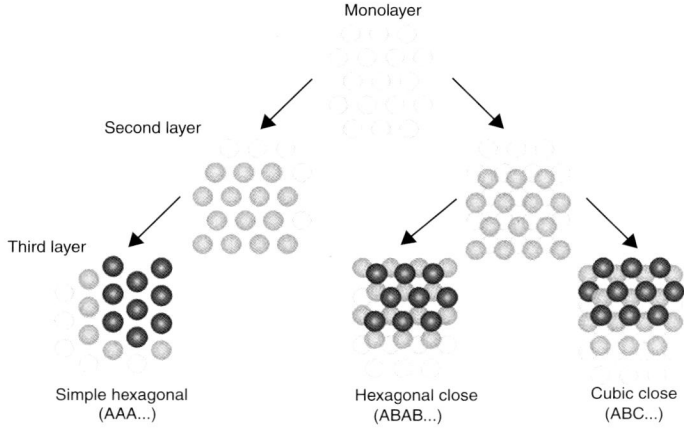

Fig. 13. Schematic depiction of nanoparticle packing in different periodic structures

spheres, whose behavior is similar to hard spheres, exhibit predominantly *fcc* superlattices, also known as synthetic opals, whereas the *hcp* phase does not form [49, 50]. Kinetic factors associated with solvent flow can also play an important role

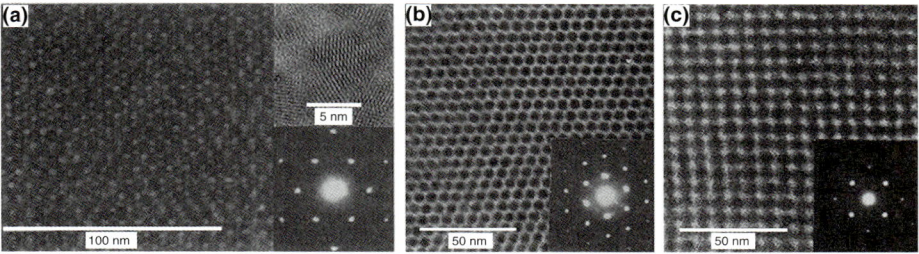

Fig. 14. TEM images and ED patterns of different projections of *fcc* nanoparticle superlattices: **a** ⟨1 1 1⟩-, **b** ⟨1 0 1⟩- and ⟨1 0 0⟩-oriented arrays of 6.4 nm (**a**) and 4.8 nm (**b**, **c**) CdSe nanoparticles, respectively. The lower insets in **a**–**c** show the small-angle ED patterns. The upper inset in **a** is high-resolution image of a single 6.4 nm nanoparticle sitting on the top surface of the array with its (1 1 0) axis parallel to the electron beam and its (0 0 2) axis in the plane of the superlattice. Reproduced from [51], © 1995, with permission from AAAS

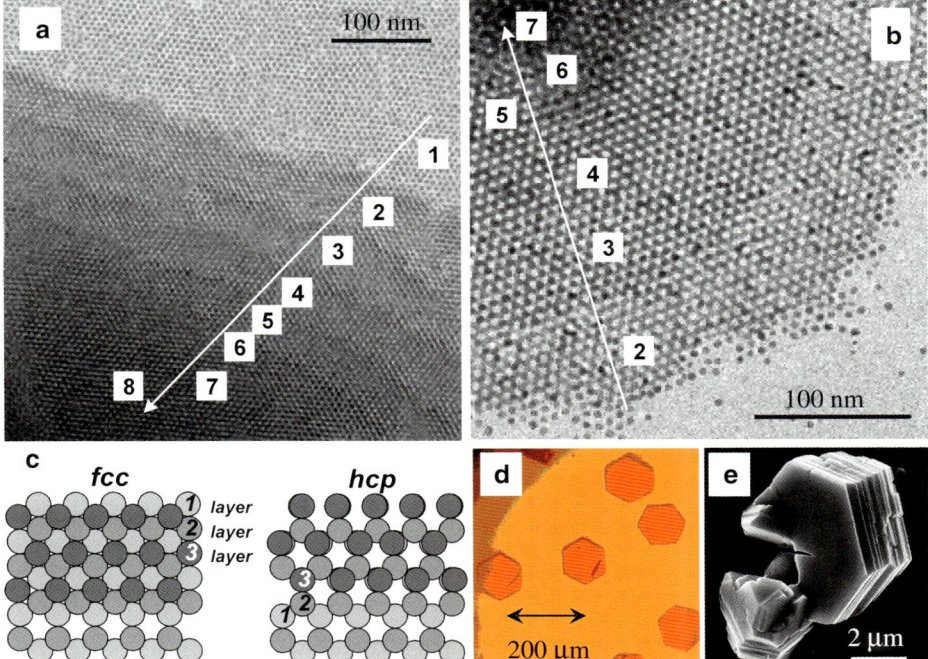

Fig. 15. Nanoparticle superlattices with *hcp* packing symmetry. **a** TEM image of [111] projection of a superlattice self-assembled from CdSe nanocrystals. The numbers highlight nanocrystal layers with ABAB... layer stacking, characteristic to the *hcp* lattice. **b** TEM image of an [111] projection of *hcp* superlattice self-assembled from CoPt$_3$ nanocrystals. **c** Scheme showing the difference between [111] projections of *fcc* and *hcp* packing. **d** Optical micrograph of macroscopic 3D superlattices of CdSe nanocrystals faceted as hexagonal platelets, typical for *hcp* lattice. **e** SEM image of 3D superlattice of CoPt$_3$ nanocrystals. Reproduced from [23], © 2007, with permission from American Chemical Society

in determining structure of nanoparticle superlattices, usually favoring the formation of *fcc* phase.

Both *fcc* and *hcp* structures were found in CdSe nanocrystal superlattices (Figs. 14 and 15) [51]. In real nanoparticle superlattices, the effects associated with the slight deviation of nanoparticle shape from spherical, the presence of dipole moments or coupling of higher order multipole moments can influence the relative stabilities of *fcc* and *hcp* phases. As an example, dipolar coupling of nanoparticles in a superlattice can provide higher stability of *hcp* structure compared to *fcc*, due to more favorable dipolar coupling in the *hcp* packing, with odd layers sitting on the top of each other. This explains the experimentally observed formation and stability of *hcp* nanocrystal superlattices (Fig. 15) [23]. The calculations for 5.8 nm CdSe nanocrystals with

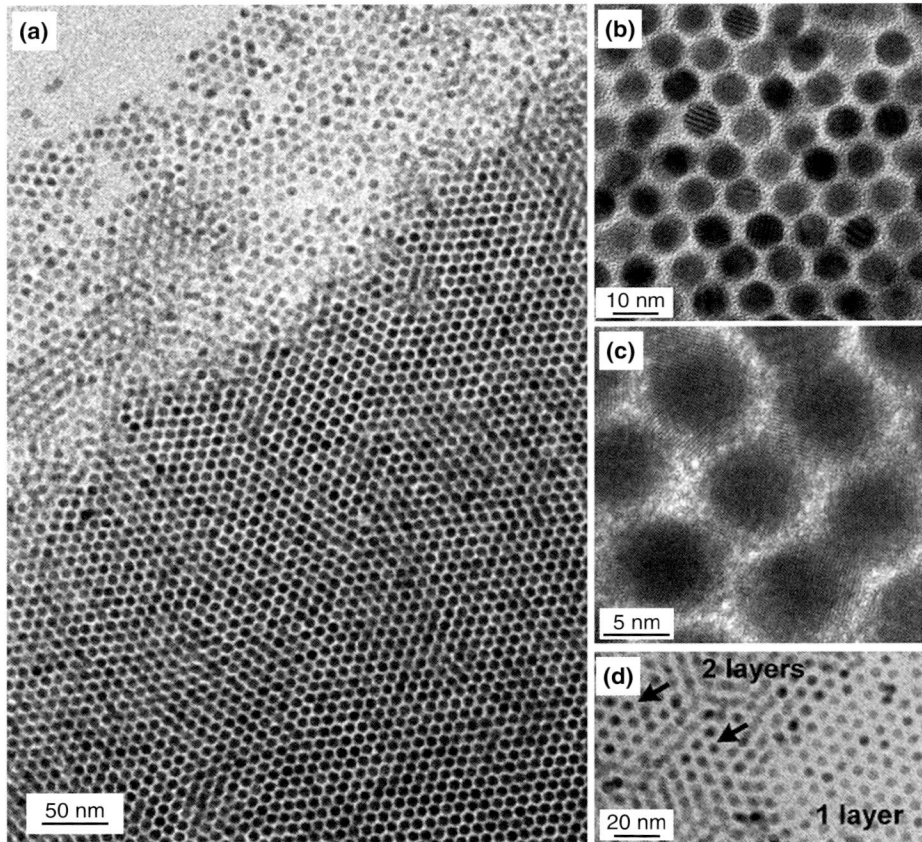

Fig. 16. *shp*-Type superlattices of 7.2 nm PbSe nanocrystals. **a** Overview TEM image showing interface between superlattice and unstructured phase. **b** TEM image showing moiré fringes originating from the diffraction of electrons on nanocrystals with mutually rotated atomic planes and **c** corresponding high-resolution TEM image. **d** TEM image showing formation of the second layer of PbSe nanoparticles with AA layer stacking (i.e., directly above the particles from the first layer). Arrows show hexagonally ordered nanocrystal columns. Reproduced from [23], © 2007, with permission from American Chemical Society

dipole moments of 100 D predict a small ($\sim 1\%$) difference between the binding energies of *fcc* and *hcp* lattices of nanoparticles due to dipolar coupling [23]. These results correlate well with the calculations of the Madelung energies for bulk *fcc* and *hcp* lattices, which predict higher stability of the *hcp* phase in dipolar spheres with ferroelectric ordering [23]. The Madelung energy difference between *hcp* and *fcc*, proportional to μ^2, can overcome the entropic contribution to the free energy, for 5.8 nm nanoparticles the transition from *fcc* to *hcp* superlattice is predicted to occur at $\mu \sim 50$ D at room temperature [23].

Both *fcc* and *hcp* types of packing are very common for nanoparticles of semiconducting and magnetic materials, however, ordering of nanoparticle dipole moments can stabilize superlattices with more open structures. Thus, non-close-packed structures with simple hexagonal packing (*shp*) have been observed for lead chalcogenide nanocrystals [23]. In *shp* lattice, nanocrystals form layers with hexagonal ordering, and these layers assemble one-on-one in the vertical direction (i.e., AAA-type layer stacking) (Fig. 13). It was found that nearly spherical PbSe and PbS nanocrystals with diameters above ~ 7 nm often self-assemble into *shp*-type superlattices shown in Fig. 16. Instead of growing layer-by-layer, these structures form abrupt boundaries with the unstructured phase. Diffraction of electrons in a column of nanocrystals with mutually rotated atomic lattice planes gives rise to rotational moiré fringes, seen in Fig. 16b. The PbSe nanocrystal superlattice exhibits an undistorted hexagonal arrangement of nanocrystal columns, with the angles between the lattice planes very close to 60° what makes them distinctively different from [0 1 1] projections of *fcc* lattice which is always stretched by $\sim 15\%$ in one direction (Fig. 17). The observation of non-close-packed structures self-assembled from spherical particles provides additional evidence that real semiconductor nanocrystals cannot be approximated by hard spheres and possess strong particle–particle interactions. The formation of *shp* structure was explained by considering the interactions between non-local dipoles of individual nanoparticles [23]. In addition to dipoles, higher order moments of charge distribution can also affect the binding energy in nanoparticle superlattices. For example, cuboctahedral PbSe nanocrystals with eight polar $\langle 1\,1\,1 \rangle$ facets can possess high quadrupole and octapole moments [29] that may stabilize other "open" superlattices (e.g., simple cubic) [23].

Perfection of self-assembled structure strongly depends on the superlattice growth conditions. Self-assembly requires that the components either equilibrate between aggregated and non-aggregated states, or adjust their positions relative to each other in an aggregate [1]. The arriving particles must have enough energy and time to find equilibrium superlattice sites on the growing structure. This might be achieved by diffusing on a terrace and along the superlattice step edge. Generally, slow growth is important to obtain highly ordered structures. If new nanoparticles are arriving too fast or stick together irreversibly, they will land on the top of each other and block the diffusion pathways for each other that will lead to the formation of amorphous solid. Tailoring the evaporation conditions by changing the temperature and/or composition of the dispersing media allows us to adjust the destabilization of colloidal solution. For instance, hexane (95%)/octane (5%) solvent mixture leads to the formation of glassy-like structures, while more highly boiling octane (95%)/octanol (5%) media gives long-ranged periodic structures [11]. This approach allows the

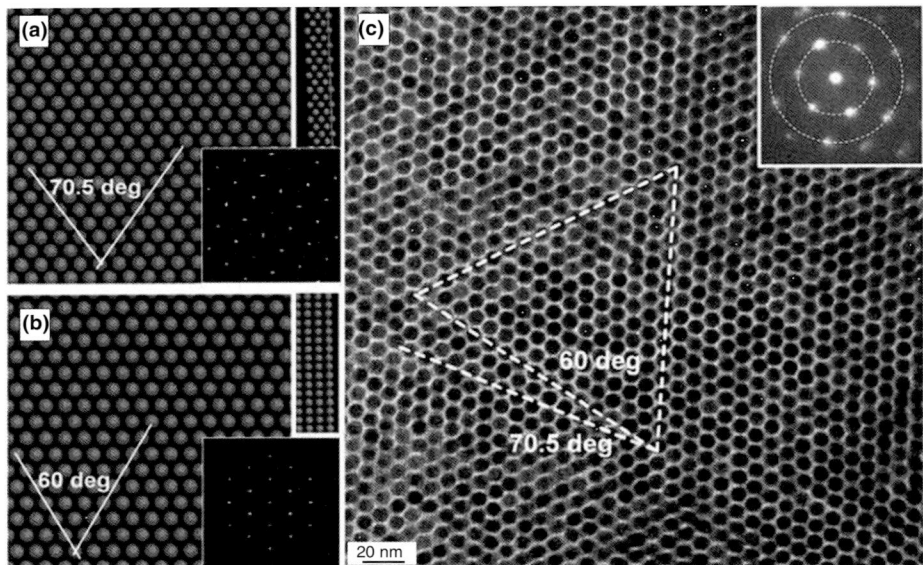

Fig. 17. Modeled **a** [0 1 1] projection of *fcc* lattice and **b** [0 0 1] projection of *shp* lattice. Right bottom insets show FFT of these lattice projections with **a** 2-fold and **b** 6-fold rotational symmetry. Right top insets show side view of the superlattices. **c** TEM image of 7.2 nm PbSe nanocrystal superlattice. The angle between superlattice planes is 60°, typical for *shp* lattice. The inset shows small-angle electron diffraction pattern collected from $6\mu^2$ superlattice area. Reproduced from [23], © 2007, with permission from American Chemical Society

production of three-dimensional NC superlattices coherent over hundreds of microns [51]. Faceted colloidal crystals 5–50 μm in size made from CdSe nanoparticles are shown in the optical micrograph (Fig. 18). The red color of the triangles in the optical micrograph is arising from the optical spectrum for the 5.7 nm CdSe nanoparticle building blocks.

Another approach to create three-dimensional nanoparticle superlattices is controllable oversaturation of a colloidal solution without solvent evaporation. The nanoparticle self-assembly is induced by slow diffusion of a non-solvent into a concentrated solution of nanoparticles (Fig. 19) [52]. The difference in the density of colloidal solution of nanoparticles (e.g., in toluene) and non-solvent (e.g., isopropanol or methanol) provides initial sharp interface between two liquids which slowly interdiffuse. The rate of mixing of the colloidal solution of nanoparticles with non-solvent determines the rate of the destabilization of the former, and hence the rate of the crystallization of nanoparticles. For example, slow diffusion of methanol into toluene solution of monodisperse CdSe nanoparticles induces the spontaneous nucleation and growth of 3D supercrystals (Fig. 20). Both irregular-shaped colloidal crystals with sizes up to 200 μm and perfectly faceted hexagonal colloidal crystals with sizes of about 100 μm have been built from monodisperse CdSe nanocrystals, depending on the conditions of the crystal growth. Irregular shape of colloidal crystals can be a result of faster destabilization of toluene solution of CdSe nanoparticles by methanol that leads to faster nucleation of multiple nuclei which

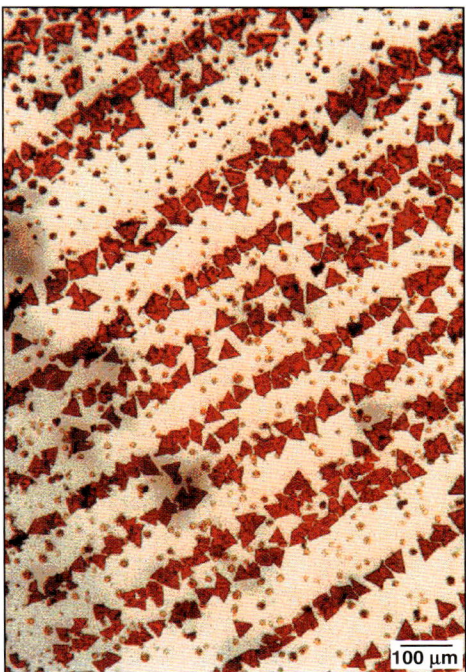

Fig. 18. Optical micrograph of three-dimensional colloidal crystals of 5.7 nm CdSe nanoparticles. The red color of the triangles is characteristic of the size-dependent absorption for the 5.7 nm CdSe nanoparticle building blocks. Reproduced from [11], © 2002, with permission from Annual Reviews

grow and fuse forming large "polycrystalline" crystals consisting of several domains. Isopropanol is a weaker non-solvent compared to methanol and intermixing of isopropanol and toluene results in gentler destabilization of nanoparticle solutions leading to the formation of relatively uniform highly periodic structures [52]. The method of controllable oversaturation of a colloidal solution has been successfully used for growing faceted three-dimensional superlattices of different materials: CdSe [52], FePt [53], CoPt$_3$ [54], PbS [55], and CdSe/ZnS core-shell nanocrystals. The crystals assembled from semiconductor nanocrystals can have various shapes corresponding to terminations by different crystallographic planes (Fig. 21) [55].

The ordering of nanocrystals in superstructures is also observed in SAXS patterns [11] through appearance of sharp reflections in the region of 2Θ angles of ~ 1–$15°$. Figure 22a shows SAXS patterns measured on superlattices of CdSe nanocrystals of different size. SAXS patterns show a set of reflections that are characteristic for a given symmetry of nanocrystal packing. These reflections can be modeled and assigned to certain crystallographic directions (Fig. 22). The 2Θ positions of small-angle reflections provide information about lattice constant of superlattice unit cell that can be used to calculate center-to-center distance between nearest neighbors. If the size of nanocrystal core is known from independent measurements (e.g., TEM or absorption data), the difference between center-to-center distance and nanocrystal diameter gives information about separation of nanocrystals in the superlattice. The

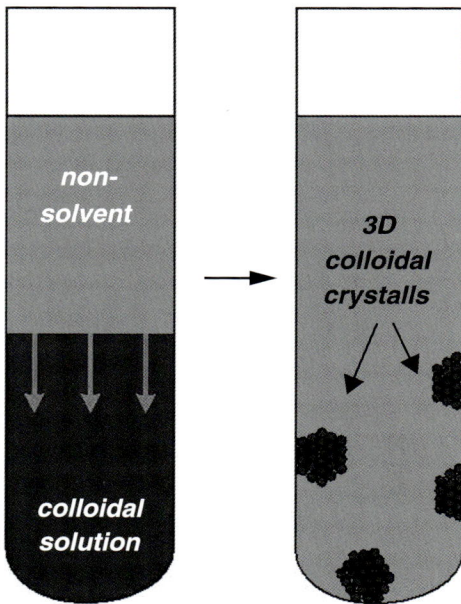

Fig. 19. Controllable oversaturation of a colloidal solution of nanocrystals. Left tube: the non-solvent (methanol) diffuses into a colloidal solution of CdSe nanocrystals in toluene. Right tube: nanocrystal superlattices nucleate and grow on walls of the tube

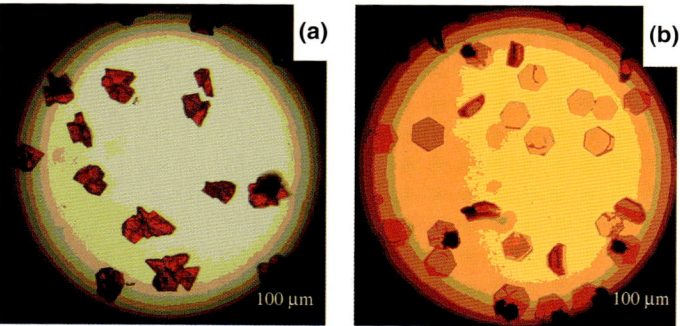

Fig. 20. Optical micrographs of colloidal crystals of CdSe nanocrystals taken by a digital camera through the objective of an optical microscope. **a** Faster nucleation, irregular-shaped crystals. **b** Slower nucleation results in perfectly faceted hexagonal platelets. Reproduced from [52], © 2001, with permission from Wiley

spacing between nanocrystals is determined by the length of the capping ligands and can be adjusted by changing the capping groups (Fig. 22b). The important advantage of SAXS over TEM or SEM is that SAXS data are averaged over large number of nanocrystals, whereas the TEM and SEM data always represent a tiny fraction of nanocrystals in the sample [56].

The wide-angle part of X-ray diffraction patterns corresponds to the diffraction of X-rays on atoms the nanocrystals consist of, and allows the estimation of average size

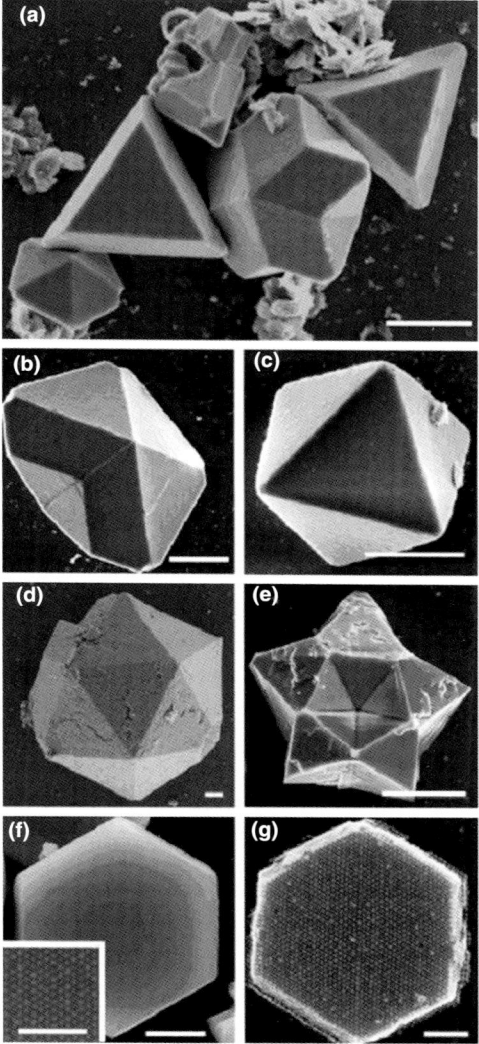

Fig. 21. SEM and HRSEM images of three-dimensional crystals self-assembled from PbS nanoparticles. The scale bars correspond to 1 μm in parts **a–d**, and **f**, 20 μm for part **e**, and 100 nm for parts **g** and the inset of **f**. Reproduced from [55], © 2007, with permission from Oldenbourg Wissenschaftsverlag

of crystalline domains within each nanocrystal. Wide-angle XRD of nanoparticles reveals the internal structure of the average nanocrystal core and permits measurement of nanoparticle size using Debye–Sherrer approximation [56]. Experimentally, WAXS patterns not just demonstrate broadening associated with finite particle size but also provide information about the lattice alignment of individual nanoparticles within superlattice. The ordering and orientation of the individual nanoparticles in the superlattice can be evidenced by the dramatic enhancement in the intensity obtained from certain crystallographic reflections. Thus ∼ 90% of CdSe nanopar-

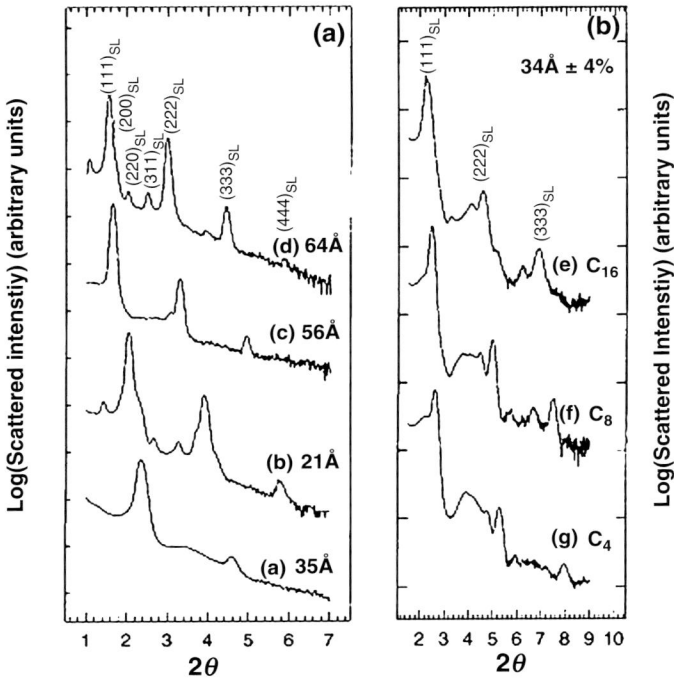

Fig. 22. a SAXS patterns for superlattices prepared from CdSe nanocrystals ranging from 3.5 to 6.4 nm in diameter. The *fcc* superlattice reflections are indicated at the top of **a**. **b** SAXS patterns of highly ordered thin films prepared from 3.4 nm CdSe nanocrystals with (e) tricetylphosphate, giving a 1.7 nm inter particle spacing; with (f) the native tri-octylphosphine chalcogenide, giving a 1.1 nm spacing; and with (g) tri-butylphosphine oxide, giving a 0.7 nm spacing. Reproduced from [11], © 2002, with permission from Annual Reviews

ticles were found to be aligned with (1 1 0) planes of the wurtzite *c*-axis in the plane of the substrate (Fig. 23a). The same information can be obtained from wide-angle electron diffraction (Fig. 23b). Preferential alignment of lattices of individual nanoparticles leads to the strong modulation of the diffraction pattern. Preferential alignment of nanoparticles within superlattice can be associated with the fact that during the superlattice growth, the individual nanoparticles may align themselves in order to maximize their attractive interaction (e.g., van der Waals and dipolar) with the substrate and neighboring nanoparticles [57]. In the case of glassy thin films wide-angle pattern is represented by continuous rings and consisted with a randomly oriented ensemble of nanoparticles.

Some size distribution is always present in nanoparticle samples and this could have a substantial influence on both thermodynamics and kinetics of formation of nanoparticle superlattices. Experiments on crystallization of latex particles that can be modeled as hard spheres showed the suppression of crystal formation in suspensions with polydispersity exceeding 12% [58]. From our experiments on self-assembly of semiconductor nanoparticles from non-polar solvents we found that particle size distribution is one of the key parameters determining if crystallization

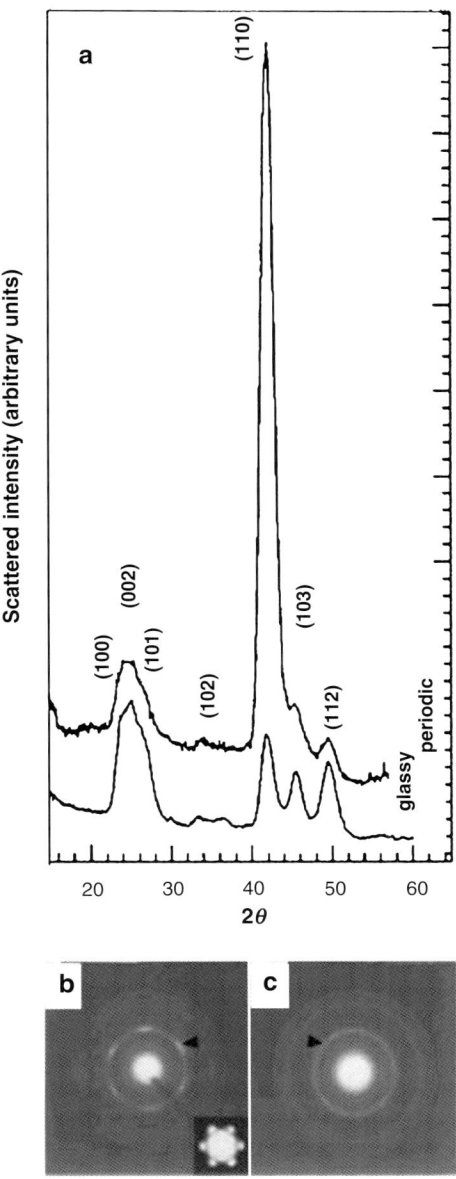

Fig. 23. a WAXS for an oriented periodic and a glassy thin film prepared from 6.4 nm CdSe nanoparticles assembled in *fcc* superlattice. **b, c** Wide-angle electron diffraction patterns from a ∼2 μm area with *fcc* and randomly packed nanoparticles, respectively. Reproduced from [11], © 2002, with permission from Annual Reviews

occurs or not. The best particle size distribution achieved for sub-10 nm semiconductor nanoparticles is ∼4–6% which is sufficient to obtain periodic superlattices over tens of microns [11, 51, 52]. Increase in polydispersity up to ∼10–12% results in the significant decrease in the size of ordered domains. Monte Carlo simulations performed for hard spheres revealed that the nucleation barrier is independent of the polydispersities below 5% [59]. Increase in particle size distribution above 5% leads to the rapid increase of the nucleation barrier and formation of critical nuclei is suppressed. Calculations also demonstrated that crystallization is not observed in hard-sphere suspensions with a polydispersity greater than 12% and under these conditions the colloidal glass is truly amorphous [59].

Any crystal is characterized by certain concentration of defects, such as vacancies, interstitials, dislocations, etc. Numerical calculations showed that equilibrium concentration of interstitials in hard-sphere colloidal crystals is almost undetectable [60]. It can be associated with high mobility of interstitials. Being trapped during the crystal growth, interstitials diffuse rapidly to the crystal surface. On the other hand, the concentration of vacancies was found to be three orders of magnitude higher than the concentration of interstitials [60]. Size distribution can also affect the concentration of stacking faults and point defects. Concentration of interstitials in crystals grown from polydisperse colloids can be up to six orders of magnitude higher than in crystals formed from monodisperse colloids [61]. Such a dramatic increase in interstitial concentration was attributed to the presence of small particles that have an increasing probability of fitting in a hole of the underlying crystalline lattice. Calculations demonstrated that the concentration of interstitial depends sensitively on the tail of the size distribution in the liquid phase [61]. Control over defects in crystalline nanoparticle superlattice is important for many device applications. However, to date there is no established techniques for quantitative analysis of defect concentrations in nanoparticle superstructures.

3. Self-assembly of anisotropic nanoparticles

The recent progress in colloidal synthesis of inorganic nanomaterials established robust techniques for tailoring not only the size but also the shape of semiconductor nanocrystals. For example, the shape of CdSe nanocrystals can be tuned from nearly spherical to rod-, disk-, arrow-, rise-, teardrop- and tetrapod-like [62, 63]. PbSe nanocrystals can be synthesized in form of spherical particles as well as cubes and octahedrons [29]. These shape-controlled nanocrystals attract great interest because they cannot be treated as purely "zero-dimensional" quantum dots [64]. Thus, CdSe nanorods can have the aspect ratio up to ~30 and exhibit the properties being intermediate to that of 0D and 1D objects [65]. Highly anisotropic inorganic semiconductor nanostructures are perspective candidates for photovoltaic applications because they can exhibit benefits of quantum-confined systems without compromising facile charge separation and transport. The photogenerated charges can immediately be harvested and transported within the 1D unit. Semiconductor nanorods have tunable linearly polarized photoluminescence and anisotropic non-linear optical properties that opens up the possibility of using them in electro optical devices such as flat panel displays, polarized light-emitting devices, etc. Tuned aspect ratios (length/diameter), replaceable surface capping molecules and diversity in composition of semiconductor nanorods make them an ideal system to study the fundamental aspects of self-assembly of anisotropic particles.

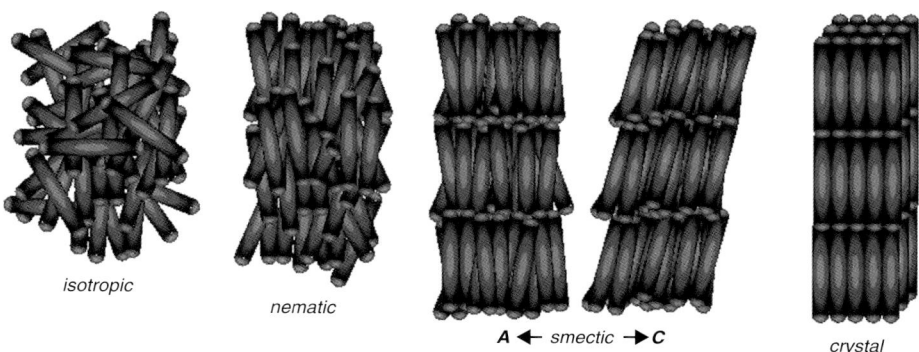

Fig. 24. Typical assemblies of rod-like particles with different degrees of ordering

This section is primarily focused on self-assembly of nanorods into periodic structures which are of interest for many technologically important applications [66]. The phase behavior of rods was extensively studied both experimentally and theoretically. Onsager's theory as well as extensive computer simulations [67–69] predict formation of a series of orientationally and/or positionally ordered phases upon assembly of anisotropic particles. The most of theoretical models and numerical calculations consider rods as non-interacting hard spherocylinders. It is known that a collection of rods can exist in at least four stable phases: isotropic fluid, nematic liquid crystals, crystalline solid and a smectic (mainly A) phase [70] (Fig. 24). Isotropic crystals are represented by randomly oriented rods with no orientational and positional order. There is a large variety of different types and degrees of positional and orientational ordering. The most common are nematic, smectic-A, and smectic-C phases. Nematic liquid crystal is the simplest liquid crystalline phase characterized by lower degree of orientational order than smectic phase and it also has no positional order. In smectic phases the rods form defined layers being positionally ordered along one direction. The rods are typically oriented either normally to the layers (smectic-A phase) or tilted to the same angle (smectic-C phase, Fig. 24). In crystalline structures the rods show both positional and orientational ordering. Typically, in crystalline phase the rods have the same arrangement as spherical particles packed into the simple hexagonal lattice discussed in the previous section: the rods in each hexagonally close-packed layer are located directly above and below the rods from adjacent top and bottom layers (Fig. 24).

Self-assembly of rods is less entropically favorable compared to spherical particles because it requires both positional and orientational ordering of individual rods. Recent computer simulations predict stunted crystal growth in ensembles of hard spherocylinders because of kinetic "self-poisoning" of lamellar crystal nuclei [71]. Simulations predict that single-layered lamellar crystallite forms at the early nucleation stage. However, the top and bottom surfaces of this crystallite prefer to be covered with rods that align parallel to the surface. The preference of rods to lie flat on the lamella surface facilitates lateral growth of the layer, however, it makes growth of the top/bottom surface difficult. Numerical study demonstrates that the rods have to overcome a barrier of $\sim 1.5\,\mathrm{kT}$ per particle to stand up on the surface and align with the director while the lateral growth of nanorod layer is easy [71]. Real semiconductor nanorods have strong longitudinal dipole moments [15]. Pairing of the dipole moments can significantly contribute to thermodynamics and kinetics of self-assembly process. Monte Carlo simulations for hard spherocylinders demonstrate that longitudinal dipoles can stabilize the layered smectic phases due to antiparallel side-by-side dipole pairing within a layer and between the adjacent layers [72].

At low concentrations, colloidal solution of CdSe nanorods is isotropic and shows polarization independent extinction, no birefringent patterns are observed. Evaporation of solvent (e.g., cyclohexane) causes gradual increase of the nanorod concentration and at certain point birefringent liquid crystalline phase appears. Figure 25 demonstrates two stages following the evaporation of cyclohexane from a solution of CdSe nanorods with a length of 40 nm and a width of 4 nm. First, the isotropic phase undergoes a phase transition breaking into multiple droplets. The intense red color of these droplets, called tactoids, indicates the high concentration of CdSe nanorods in the droplets. The droplets are birefringent – the anisotropic refractive index of the liquid crystalline phase alters the polarization of transmitted light, leading to light and dark patterns on the micron scale viewed in optical microscope at crossed polarization [73]. The variations in the intensity of the transmitted light through the

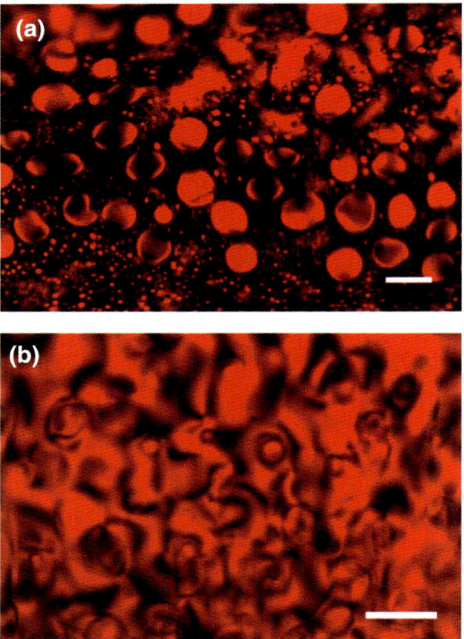

Fig. 25. Optical micrographs obtained for CdSe nanorods with the length of 40 nm and the width of 4 nm dispersed in cyclohexane at different stages of solvent evaporation: **a** phase-separated droplets of concentrated CdSe nanorods in cyclohexane and **b** Schlieren texture. The images were obtained by illumination with linearly polarized white light and detected at crossed polarization. The red color is due to strong band edge absorption of the CdSe nanorods. The scale bars are 20 μm. Reproduced from [73], © 2002, with permission from American Chemical Society

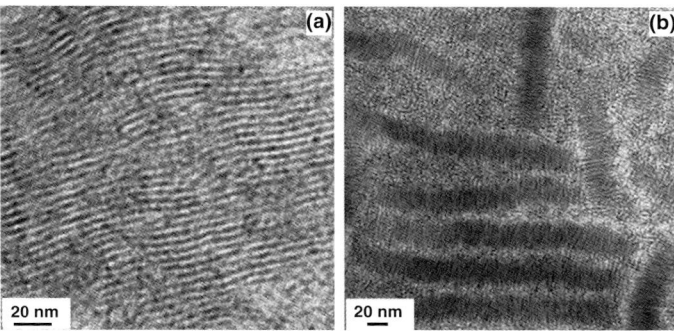

Fig. 26. Assemblies with (**a**) nematic and (**b**) smectic ordering of 22 nm × 3.5 nm CdSe nanorods

droplets indicate alignment of the rods within each droplet. The patterns are generally dipolar (Fig. 25a). At later stage of solvent evaporation so-called Schlieren structures characteristic to nematic phase form. Dark thread-like brushes are observed sticking out from some singular points. The singular points (disclinations) have been identified as the positions where the alignment direction of the nanorods is not well defined. The dark brushes are due to the particles with alignment parallel to either of the polarizers. When the sample is rotated, these singular points remain fixed but the dark brushes move around them. When one end of the capillary tube is lifted, the patterns flow and deform easily, suggesting the fluidity of the birefringent phase [73].

Fast evaporation of concentrated toluene solutions of monodisperse CdSe nanorods leads to the formation of solids with nematic order (Fig. 26a) [74]. However, if a high boiling solvent such as butyl ether is added to the concentrated nanorod solution followed by drying at 40–60°C under reduced pressure, long tracks of CdSe nanorods would be stacked side by side. At higher concentrations, these tracks align parallel to each other, resulting in smectic-A ordered superstructures (Fig. 26b). Experimentally, nanorods with high aspect ratios were found to be more difficult to self-assemble into structures with orientational and/or positional order [75].

Slow destabilization of toluene solution of CdSe nanorods by the diffusion of a non-solvent (e.g., ethanol, isopropanol, etc.) allows us to obtain ~ 10–50 μm large droplet-like aggregates (Fig. 27a, b). Sonication in ethanol breaks the droplet-like aggregates into sheets (Fig. 27c). HRSEM investigation revealed that these sheets consist of CdSe nanorods packed into long-range ordered crystalline phase (Fig. 27d). The nanorod aggregates shown in Fig. 27a and b exhibit characteristic birefringence, called Maltese crosses (Fig. 28a–c). The simultaneous appearance of Maltese crosses and concentric rings can be explained if the layer-by-layer growth results in a texture where smectically ordered layers are separated by ledges as shown in Fig. 28f. In vicinity of the ledges, nanorods are tilted with respect to layer planes as shown in Fig. 28f and rotate polarization of incident light [74]. The glassy films of CdSe nanorods prepared by fast deposition from a hexane solution exhibit neither Maltese crosses nor concentric rings (Fig. 28d, e). Similar polarized optical micrographs are observed for spherulite textures. Growth of a spherulite originates from a

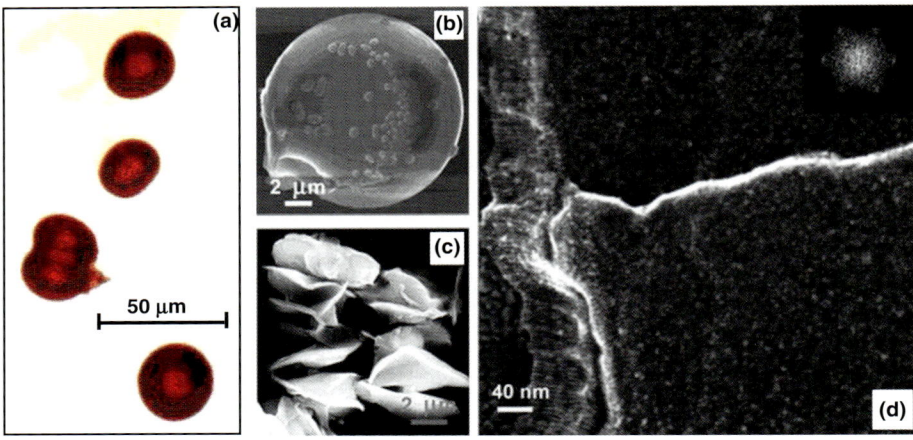

Fig. 27. a Optical micrograph of self-assembled aggregates consisting of 18 nm × 3 nm CdSe nanorods. **b** SEM image of an assembly of 22 nm × 3.5 nm CdSe nanorods. **c** Sonication of the nanorod assemblies results in their cleavage in several flat sheets. **d** High-resolution SEM image showing internal packing of CdSe nanorods inside the sheets. Vertical columns of nanorods at the edge of the superlattice show ordering in longitudinal direction. Inset: FFT from selected area of top surface of the nanorod sheet shown in frame **d** demonstrates hexagonal ordering of the nanorods within the layers. On the left, part of the image the ITO substrate can be seen. Reproduced from [74], © 2004, with permission from American Chemical Society

single nucleus (lamellae) followed by uniform growth in three dimensions to form ball-like structures [76]. In the case of a liquid crystal with spherulitic texture, the director rotates around the center of spherulite resulting in a cross-like extinction "Maltese cross", which occurs along the polarization axes of the polarizers. Spherulites of polyethylene and some other polymers exhibit similar concentric extinction rings (radial banding) usually assigned to lamellar twisting along the radial direction during crystal growth [77].

Another interesting example of nanorod assemblies is based on highly luminescent CdSe/CdS heterostructured nanorods [74, 75, 78]. Self-assembled CdSe/CdS nanorod solids (Fig. 29) exhibit strong, stable PL under UV excitation. CdSe/CdS heterostructures are known to be very efficient harvesters of short wavelength light because the CdS part provides large extinction below 500 nm [78]. The photogenerated carriers in the CdS rod are efficiently captured in the CdSe core within ~ 5 ps. The CdSe/CdS heterostructure nanorods emit in the spectral region where only the CdSe core can reabsorb the emitted light [79]. The extinction coefficients in this spectral region are relatively small. This means that reabsorption in CdSe/CdS heterostructure nanorod solids is considerably smaller than reabsorption in CdSe or any other single-phase nanorod solids. This makes CdSe/CdS nanorods particularly interesting for solid-state luminescence applications such as phosphors, light-emitting diodes and lasers, as well as for detectors and for light harvesting applications. CdSe/CdS nanorods deposited from concentrated toluene solutions preferentially self-assemble into smectic-A and highly periodic hexagonally packed layers (Fig. 29). These layers are oriented perpendicular to the substrate and typically grow laterally rather than in third dimension. CdSe/CdS nanorods can form aligned

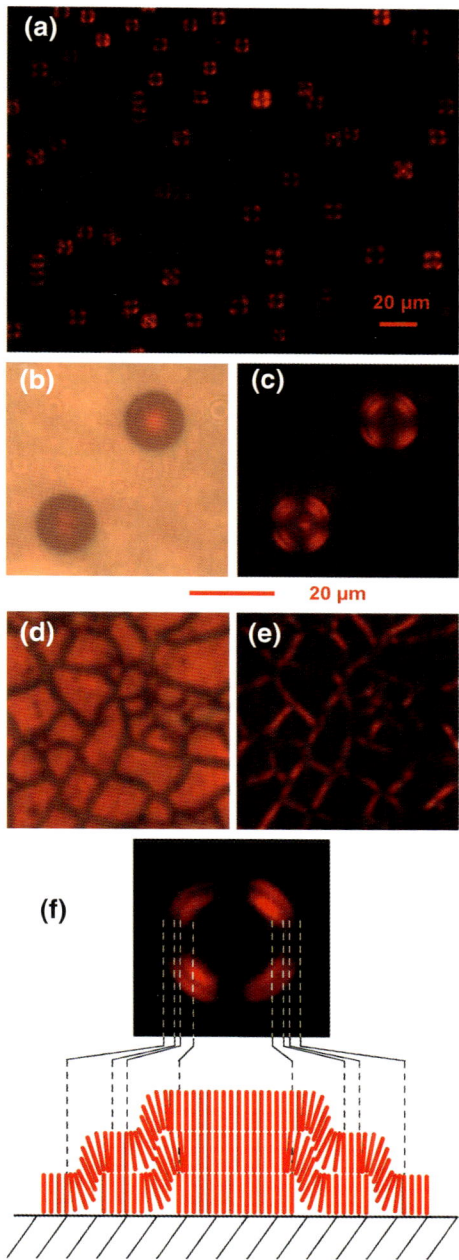

Fig. 28. a Optical micrograph of assemblies of 22 nm × 3.5 nm CdSe nanorods viewed between crossed polarizers. Optical micrographs of self-assembled nanorods observed without (**b**) and with (**c**) crossed polarizers at higher magnification. Optical micrographs of a glassy film of CdSe nanorods observed without (**d**) and with (**e**) crossed polarizers. **f** Scheme showing a possible packing of nanorods in the assemblies that is consistent with simultaneous appearance of Maltese crosses and concentric rings. Reproduced from [74], © 2006, with permission from American Chemical Society

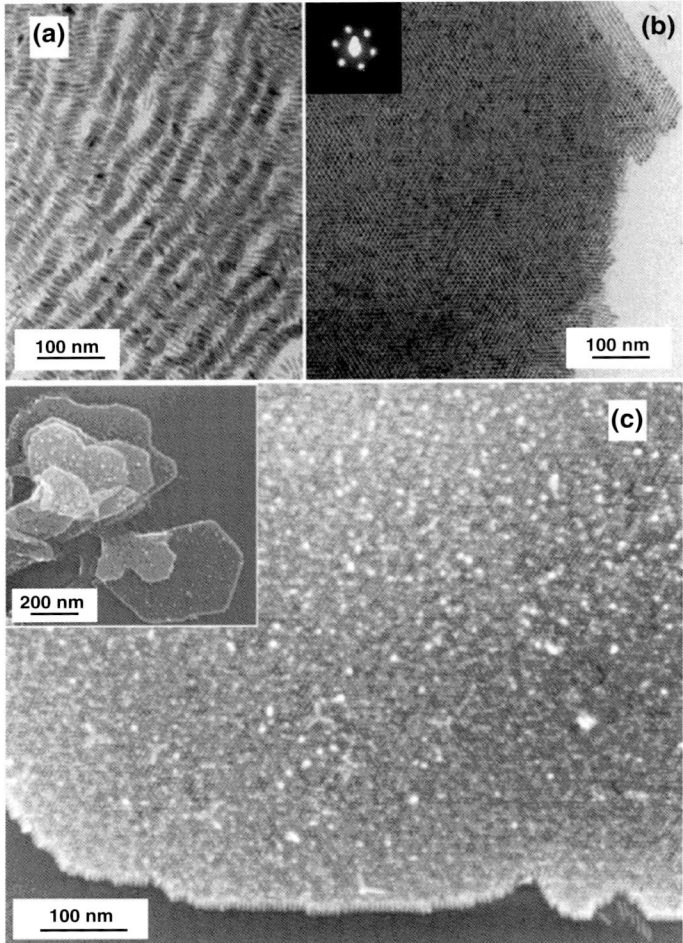

Fig. 29. Self-assembly of 24 nm × 5.2 nm CdSe/CdS nanorods. **a** TEM image of a superstructure with smectic-A ordering. **b** TEM image of a superlattice with simple hexagonal packing of nanorods assembled perpendicular to the substrate. Inset shows electron diffraction from a superlattice domain. **c** HRSEM image of CdSe/CdS nanorods assembled perpendicular to the substrate (silicon wafer). Inset shows layer-by-layer growth of a nanorods superlattice. Reproduced from [78], © 2007, with permission from American Chemical Society

structures floating at the surface of water. Regular patters of vertically stacked nanorods were obtained by evaporating concentrated toluene solution of CdSe/CdS nanorods at the surface of water [75]. The nanorods form hexagonally close-packed monolayers. The AAA stacking of monolayers results in the growth of crystalline phase with high positional and orientational order [75].

Evaporation of nanorod solutions in the presence of external electric field facilitates alignment of individual nanorods and is favorable for the formation of nanorod superlattices [75, 80]. The electric field interacts with nanorod dipole moments, providing high orientational order in the final structures. For example, evaporation of toluene solution of 30 nm × 5 nm CdS nanorods in electric field (1 V/ μm) leads to close-packed structures where all nanorods align along the field [80]. DC electric field was also used to align CdSe/CdS nanorods. The use of high concentrations of CdSe/CdS nanorods allowed us to grow large unidirectionally aligned superstructures with nematic and smectic ordering (Fig. 30).

The recent progress in synthesis of nanostructures with complex morphologies such as, e.g., CdTe and CdSe/CdS nanotetrapods [78, 81] with well-controlled nanoscale dimensions opens up the possibility of new kinds of self-assembled

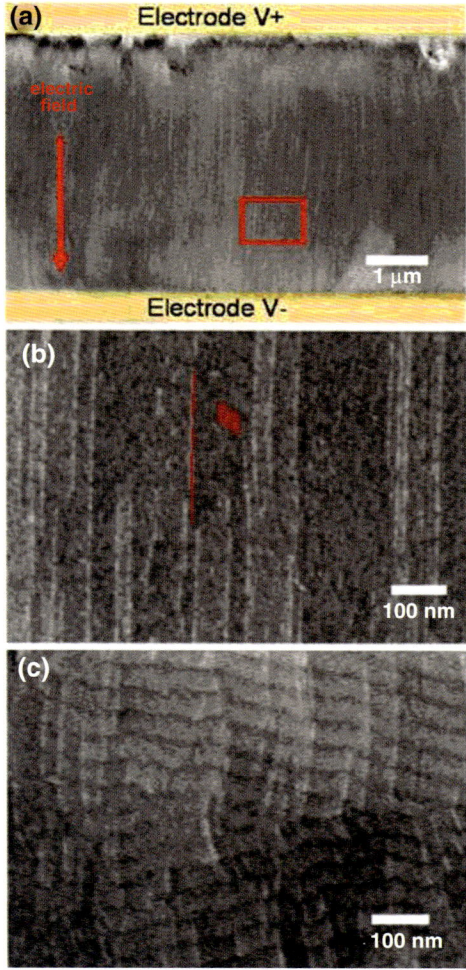

Fig. 30. Lateral alignment of nanorods with electric fields. **a–c** SEM images showing aligned arrays of nanorods with an aspect ratio of 10. The red arrow indicates the direction of the electric field that was applied during the evaporation of the nanorod solution, and the red square marks the region that is displayed in **a**. **b** Micrometer size area where the individual rods can be resolved. Some rods are highlighted in red as a guide to the eye. **c** Zoom that displays an area of nanorods assembled in ribbon-like structures. Reproduced from [75], © 2007, with permission from American Chemical Society

structures. For example, according to theoretical predictions tetrapods with identical arms can assemble into cubatic phase if no kinetic arrest occurs [82]. Cubatic phase is a special case of the biaxial nematic phase with two mutually perpendicular preferred axes.

4. Multicomponent nanoparticle assemblies

The ability to mix and match different nanoparticles and assemble them systematically into ordered binary superlattices, with precisely controlled stoichiometry and symmetry, extends the analogy to multifunctional nanocomposites constituting "nanoparticle compounds". Binary nanocrystal superlattices (BNSLs) raise the possibility of combining the properties of individual components with new properties that arise from the interactions between the nanocrystals. Binary mixtures naturally provide a much richer class of compositions and structures as compared to the monodisperse nanoparticles. When two types of particles cocrystallize, their individual assembly tendencies must adjust themselves to space constrains. They must meet, as well as possible, the requirement of certain geometrical principles. Many theoretical studies have been focused on predicting the probability of the formation of various ordered binary structures and comparison of their stability [83–86]. In the simplest approach, the formation of a binary assembly of hard spheres is expected only if its packing density exceeds the packing density of single-component crystals in *fcc* or in *hcp* structure (~ 0.7405) [86]. This mechanistic space-filling principle formulated by Murray and Sanders is still very useful and allows us to predict the behavior of hard non-interacting particles in binary mixtures. The particle size ratio ($\gamma = R_{small}/R_{large}$) and relative concentrations are considered as the factors determining structure of binary assemblies. In the case of hard spheres, a binary superlattice can form only if its entropy is higher than the sum of entropies of separated *fcc*-packed small and large spheres. Taking into account geometrical considerations, we can expect the formation of superlattices isostructural with NaCl, $NaZn_{13}$, and AlB_2. Detailed computer simulations show that the formation of NaCl-, AlB_2-, and $NaZn_{13}$-type structures of hard spheres can be driven by entropy alone without any specific energetic interactions between the particles [83, 84, 87]. Indeed, $NaZn_{13}$- and AlB_2-type assemblies of silica particles were found in natural Brazilian opals [87] and could be grown from latex spheres [88, 89]. At the same time, the lattices of hard spheres with CsCl, zinc blend, wurtzite structures, as well as any lattices with AB_3, AB_4, and AB_5 stoichiometries were predicted to be unstable for any particle size ratio. Table 2 summarizes the predicted ranges of stability for different hard-sphere binary superlattices. One can see that stable binary lattices usually have the packing density exceeding the packing density of *fcc* and *hcp* lattices (~ 0.7405). This is in agreement with a simple scheme: the higher packing density is, the larger is the excluded volume and, in turn, the higher entropy of the system can be achieved during crystallization of the superlattice [87]. However, the intrinsic entropy of binary superlattice also affects its stability. For example, the maximum packing density of $NaZn_{13}$-type superlattice is slightly below 0.74, whereas the detailed entropy calculations predict stability of this structure in a rather broad range of γ (Table 2) [90].

Table 2. Maximum packing density and range of stability calculated for binary lattices of non-interacting hard spheres. Reproduced from [12], © 2006, with permission from American Chemical Society

Stoichiometry	Type of structure	Maximum packing density (γ)	Range of stability	References
AB	NaCl	0.793 (0.414)	Below 0.458	[26]
			$0.2 \leqslant \gamma \leqslant 0.42$	[24]
	NiAs	0.793 (~ 0.4)	~ 0.4	[57]
	CsCl	0.729, unstable	0.732	[58]
AB$_2$	AlB$_2$	0.778 (0.58)	$0.482 \leqslant \gamma \leqslant 0.624$	[26]
			$0.42 \leqslant \gamma \leqslant 0.59$	[24]
	Laves phases – hexagonal: MgZn$_2$, MgNi$_2$: cubic MgCu$_2$	0.71 (0.813), unstable	$0.606 \leqslant \gamma \leqslant 0.952$	[26]
	CaF$_2$	0.757, unstable	0.225	[24]
A	NaZn$_{13}$ (*ico*-AB$_{13}$)	0.738 (0.58)	$0.54 \leqslant \gamma \leqslant 0.61$	[24]
			$0.474 \leqslant \gamma \leqslant 0.626$	[22, 23]
B	*ico*-AB$_{13}$, with some size distribution for B spheres	Above 0.755	$0.537 \leqslant \gamma \leqslant 0.583$	[26]
	cub-AB13	0.700 (0.565)	Unstable	[26]

In contrast to these predictions, an amazing variety of BNSLs can self-assemble from colloidal solutions of nearly spherical nanoparticles of different materials (Fig. 31). Coherently packed domains extend up to 10 µm in lateral dimensions and can display well defined facets. In many cases several BNSL structures form simultaneously on the same substrate and under identical experimental conditions [12, 91]. The same nanoparticle mixture can assemble into BNSLs with very different stoichiometry and packing symmetry. Figure 32 shows 11 different BNSL structures prepared from the same batches of 6.2 nm PbSe and 3.0 nm Pd nanoparticles. In general, BNSLs tolerate much broader γ ranges than hard spheres. For example, Fig. 33 shows AlB$_2$-type BNSLs assembled from different combinations of PbSe, PbS, Au, Ag, Pd, Fe$_2$O$_3$, CoPt$_3$ and Bi nanoparticles in a broad γ-range. We also observed BNSLs structures that we could not assign to certain intermetallic compounds. Two of these structures are shown in Fig. 32. The observed structural diversity of BNSLs defies traditional expectations, and shows the great potential of modular self-assembly at the nanoscale. The formation of binary structures with packing density significantly lower than the density of single-phase *fcc* packing (0.7405) rules out entropy as the main driving force for nanoparticle ordering. Experimentally, $\sim 5–8\%$ distribution in nanoparticle size of individual nanoparticles was sufficient to grow binary superlattices with dimensions of several microns. Notably smaller fragments (~ 400 nm^2) of binary superlattices can be obtained from solutions of nanoparticles with polydispersity of $\sim 10–12\%$ SD. However, suppression of crystallization from mixtures containing particles with broad size distribution ($\geqslant 14\%$ SD) was observed.

Three-dimensional descriptions of the superlattices can be developed by surveying large regions of the samples, to categorize all the crystal orientations, and recording a series of 2D projections down the major symmetry axes. Tilting of the samples allows observation of additional orientations not expressed in the plan view

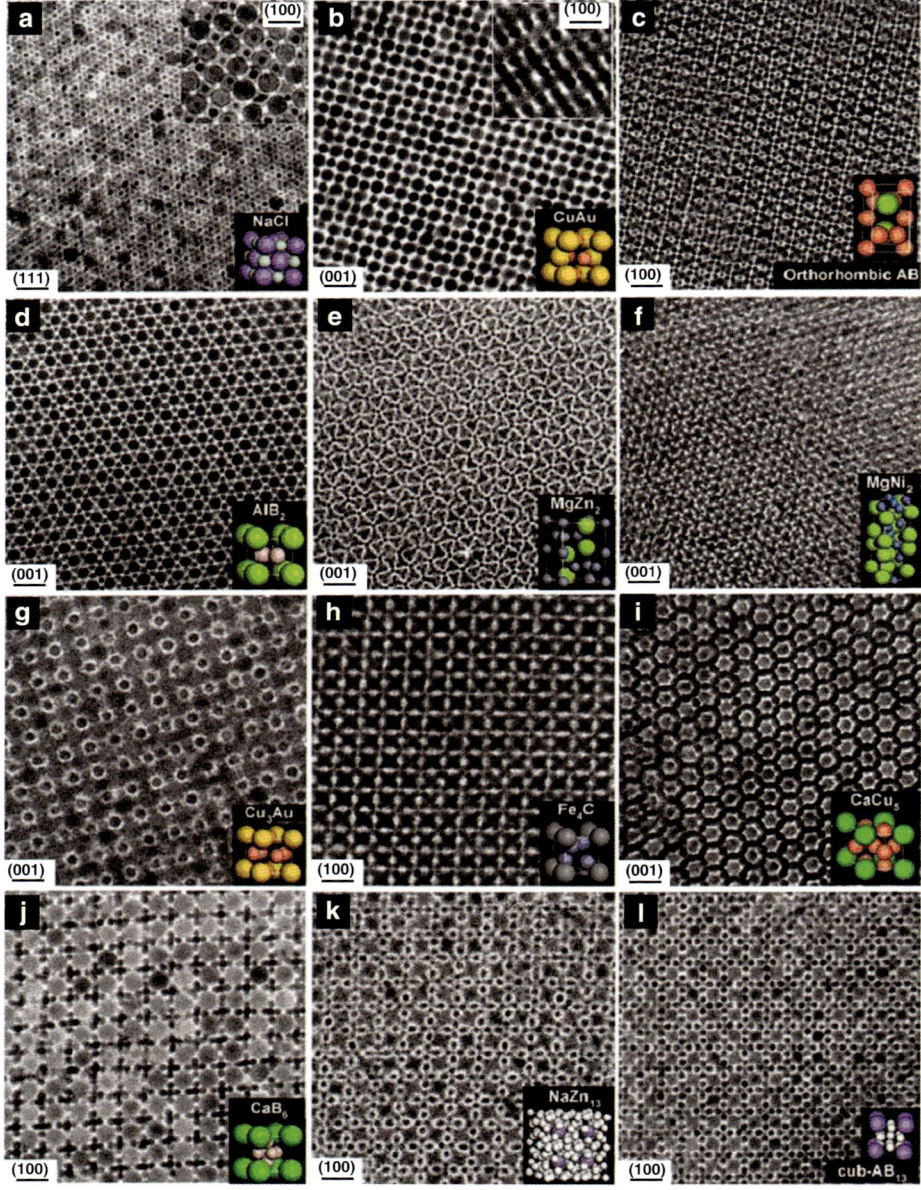

Fig. 31. TEM images of the characteristic projections of the binary superlattices, self-assembled from different nanoparticles, and modeled unit cells (insets) of the corresponding three-dimensional structures. The superlattices are assembled from **a** 13.4 nm γ-Fe$_2$O$_3$ and 5.0 nm Au; **b** 7.6 nm PbSe and 5.0 nm Au; **c** 6.2 nm PbSe and 3.0 nm Pd; **d** 6.7 nm PbS and 3.0 nm Pd; **e** 6.2 nm PbSe and 3.0 nm Pd; **f** 5.8 nm PbSe and 3.0 nm Pd; **g** 7.2 nm PbSe and 4.2 nm Ag; **h** 6.2 nm PbSe and 3.0 nm Pd; **i** 7.2 nm PbSe and 5.0 nm Au; **j** 13.4 nm γ-Fe$_2$O$_3$ and 5.0 nm Au; **k** 7.2 nm PbSe and 4.2 nm Ag; **l** 6.2 nm PbSe and 3.0 nm Pd nanoparticles. Scale bars: **a–c**, **e**, **f**, and **i–l**, 20 nm; **d**, **g**, and **h**, 10 nm. The lattice projection is labeled in each panel above the scale bar. Reproduced from [91], © 2006, with permission from Nature Publishing Group

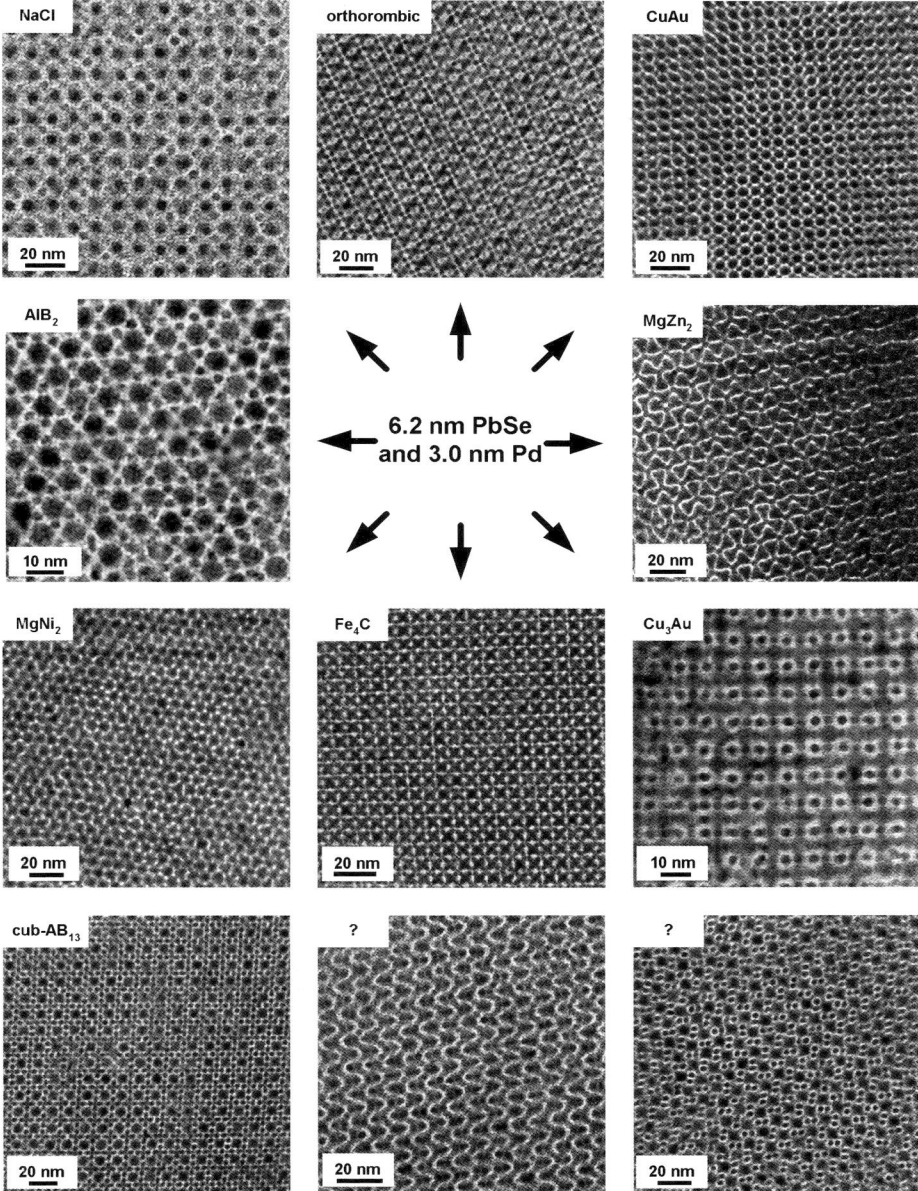

Fig. 32. TEM images of different binary superlattices self-assembled from the same batches of 6.2 nm PbSe and 3.0 nm Pd nanoparticles. Reproduced from [91], © 2006, with permission from Nature Publishing Group

images of the films. In many cases the analysis of series of these 2D lattice projections resulted in a single-crystal structure. To assign the observed structures to crystallographic space groups we built 3D lattice models for the 180 most common space groups using Accelrys MS Modeling 3.1 software. The TEM images were compared with simulated projections to match the symmetry of the superlattices. We also performed a comparison of experimental small-angle electron diffraction patterns taken over larger areas, and the 2D Fourier transformation power spectra of real-space TEM images and the FFT power spectra of the simulated projections to assure consistency.

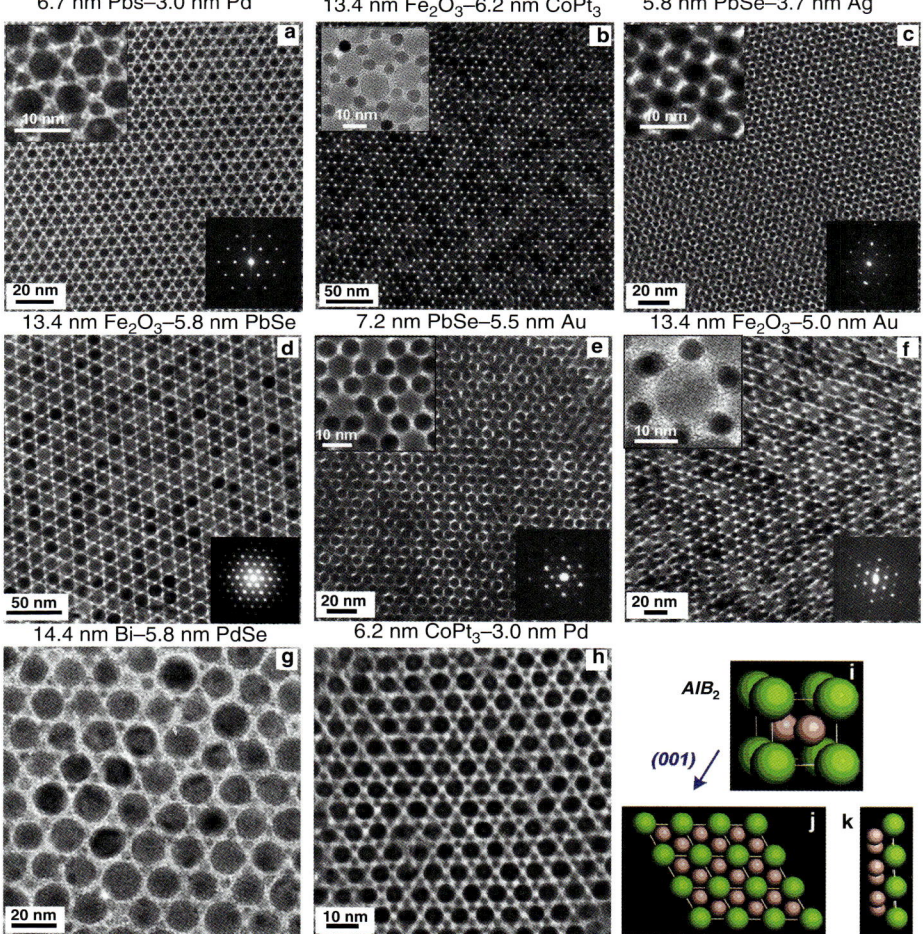

Fig. 33. a–h TEM micrographs of (0 0 1) planes of binary superlattices isostructural with AlB_2. Top left insets in **a–c**, **e** and **f** show the arrangement of NPs in AlB_2-type superlattices at higher magnification. Bottom right insets show small-angle ED patterns measured along the (0 0 1) projection. **i** A unit cell of AlB_2 lattice. **j** Depiction of the (0 0 1) plane. **k** Depiction of minimum number of nanoparticle layers along the (0 0 1) direction necessary to reproduce the observed experimental patterns. The ratio of particle radii (γ) for the reported AlB_2-type binary superlattices varied from 0.43 to 0.79. The γ-range of stability for AlB_2-type packing of hard spheres is shown in Table 2. Reproduced from [91], © 2006, with permission from the American Chemical Society

Theoretical predictions developed for hard spheres fail in the case of real nanoparticles because they cannot be approximated by non-interacting spheres. The combination of space-filling principle and soft (Coulomb) interparticle interactions can generate a rather complex phase diagram, as it has been recently demonstrated for binary superlattices of charged PMMA beads [92]. Crystalline nanoparticles cannot be considered as non-interacting because they undergo strong attractive van der Waals and repulsive sterical interactions [12]. However van der Waals, sterical or dipolar interparticle interactions are not sufficient to explain why these low-density BNSLs form, instead of their constituents separating into single-component superlattices. It was found that opposite electrical charges on nanoparticles can impart a specific affinity of one type of particle (e.g., dodecanethiol-capped Au, Ag, Pd) for another (e.g., PbSe, PbS, Fe_2O_3, $CoPt_3$ capped with long-chain carboxylic acids). If nanoparticles are oppositely charged, the Coulomb potential stabilizes the BNSL while destabilizing the single-component superlattices. The electrical charges can exist on sterically stabilized nanoparticles even in non-polar solvents [13]. Figure 34 shows the distribution of electrophoretic mobilities within a colloidal solution of PbSe and Au nanoparticles. The electrical charge (Z, in units of e) of a spherical

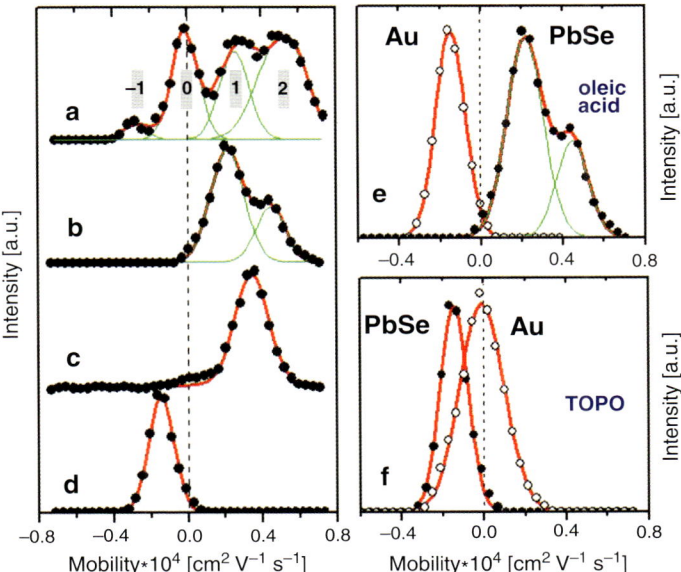

Fig. 34. Electrophoretic mobility of PbSe and Au nanocrystals in chloroform. **a–d** Distribution of electrophoretic mobility for 7.2 nm PbSe nanocrystals. **a** PbSe nanocrystals washed to remove excess of capping ligands. The grey bars show mobilities predicted for nanocrystals with charges of $-1, 0, 1$ and 2 (in units of e). **b–d** Electrophoretic mobility of PbSe nanocrystals in the presence of **b** 0.02 M oleic acid, **c** 0.06 M oleic acid and **d** 0.05 M tri-n-octylphosphine oxide. **e, f** Comparison of electrophoretic mobilities of 7.2 nm PbSe and 4.8 nm Au nanocrystals in the presence of **e** 0.02 M oleic acid and **f** 0.05 M tri-n-octylphosphine oxide. a.u., arbitrary units. Reproduced from [91], © 2006, with permission from Nature Publishing Group

particle in a low-dielectric solvent can be calculated from the electrophoretic mobility (μ_e) expressed as

$$\mu_e = \frac{Ze}{3\pi\eta a} \tag{9}$$

where η is the viscosity of the solvent and a is the hydrodynamic diameter of a particle [93]. With $a = 10$ nm, we obtain $\mu_e \sim 0.27 \times 10^{-4}\, Z\,\text{cm}^2\,\text{V}^{-1}\,\text{s}^{-1}$. These calculated values agree well with the peaks in the experimental mobility distribution for 7.2 nm diameter PbSe nanocrystals in chloroform shown in Fig. 34a. Due to the organic coating (oleic acid) the effective hydrodynamic radius of PbSe nanocrystals extends beyond the crystalline core by 1–2 nm, depending on the density of surface coverage. This observation indicates the presence of particles with charges $-e$, 0, e and $2e$ in a colloidal solution of monodisperse PbSe nanocrystals. The charges on PbSe nanocrystals can be altered by adding surfactant molecules like carboxylic acids and tri-alkylphosphine oxide. Addition of oleic acid increases the population of positively charged PbSe nanocrystals at the expense of the negatively charged and neutral nanocrystals. Depending on the amount of acid added, the majority of nanocrystals can be adjusted to have either one or two positive charges (Fig. 34b, c). Addition of oleic acid increases viscosity of solution, causing the peaks to shift towards lower mobility (compare Fig. 34a–c). The addition of TOPO increases the population of negatively charged PbSe nanocrystals and reduces the concentration of positively charged nanocrystals (Fig. 34d). Surveys of many samples revealed that the additives reliably shifted the distribution of charge states; however, the initial proportion of particles in each charge state was dependent on the sample processing [91]. After addition of oleic acid, the most Au nanoparticles become negatively charged (Fig. 34e), whereas the addition of TOPO neutralizes Au nanoparticles (Fig. 34f). The charges on PbSe and Au nanoparticles could originate from deviations in nanocrystal stoichiometry and adsorption/desorption of charged capping ligands. In the presence of oleic acid, PbSe and Au nanoparticles are oppositely charged (Fig. 34e). The Coulomb potential between two oppositely charged nanoparticles ($Z = \pm 1$) separated by 10 nm of a solvent like chloroform is comparable with kT at room temperature, and solutions of mixed PbSe and metal nanoparticles retain stability for several weeks. The relatively small interparticle potential favors annealing of the BNSLs as they grow. For a NaCl-type BNSL with $Z_+ = 1, Z_- = -1$ and the nearest-neighbor distance $R_0 = 11.5$ nm (Fig. 31c), the Coulomb binding energy per unit cell is estimated to be

$$U_{\text{Coul}} \approx -\frac{MZ_+Z_-e^2}{4\pi\varepsilon_0 R_0} \sim -0.2\,\text{eV} \tag{10}$$

(or $\sim -8kT$ at the superlattice growth temperature, 50°C), where $M = 1.7476$ is the Madelung constant. The contribution from the Coulomb binding energy is comparable to the van der Waals attractive energy expected in a NaCl-type BNSL [22]. In an AB_x BNSL where A and B hold opposite charges, the Coulomb potential per AB_x "molecule" is $U_{\text{Coul}} \sim -\alpha + \beta(xZ_- + Z_+)^2 N^{2/3}$ where α and β are positive constants and N is the number of assembled nanoparticles [91]. Coulomb energy determines the stoichiometry of the growing BNSL. An extended three-dimensional BNSL can form only if the positive and negative charges compensate each other. If during growth the BNSL accumulates non-compensated charge, eventually U_{Coul} changes sign from negative to positive and the growth is self-limiting. The superlattice nucleation stage should be less sensitive to the Coulomb interactions. Indeed, many small domains with different BNSL structure can simultaneously nucleate on the same substrate, but their size does not exceed $\sim 10^2$ nanoparticles [12]. Only one or two structures grow to larger length scales ($\sim 10^6$ to 10^8 particles). BNSLs with many particles per unit cell (e.g., AB_4, AB_5, AB_6, AB_{13}) might form when both negatively charged and neutral Au nanoparticles are incorporated into the structure. The presence of differently charged nanoparticles in the colloidal solutions (Fig. 34a) could also contribute to the simultaneous formation of different BNSLs.

Tuning the charge state of the nanoparticles allows us to direct the self-assembly process. Reproducible switching between different BNSL structures has been achieved by adding small amounts of carboxylic acids, TOPO or dodecylamine to colloidal solutions of PbSe (PbS, Fe_2O_3, etc.) and metal (Au, Ag, Pd) nanocrystals. Figure 35 demonstrates how these additives direct the formation of specific BNSL structures. Combining native solutions of 6.2 nm PbSe and 3.0 nm Pd nanoparticles (particle concentration ratio: $\sim 1:5$) results in the formation of several BNSL structures with $MgZn_2$ and cub-AB_{13} dominating. However, the same nanoparticles assemble into orthorhombic AB- and AlB_2-type superlattices after adding oleic acid (Fig. 35a) and into $NaZn_{13}$ or cub-AB_{13} BNSLs after the addition of dodecylamine or TOPO, respectively (Fig. 35b). In the AB_{13}-type BNSL metal particles assemble into icosahedral ($NaZn_{13}$) or cuboctahedral (cub-AB_{13}) clusters, with each large PbSe particle surrounded by 24 metal spheres at the vertices of a snub cube [9]. In the presence of TOPO the metal nanoparticles are neutral (Fig. 34f), favoring formation of the Pd_{13} (Au_{13}, Ag_{13}) clusters. The clusters of metal nanoparticles in turn provide screening of the charges on PbSe nanocrystals in the AB_{13}-type BNSL.

Surveys of many samples show that the addition of a carboxylic acid to solutions of PbSe–Pd, PbSe–Au, PbSe–Ag and PbSe–Fe_2O_3 nanocrystals results in either AB or AB_2 superlattices (Fig. 35c, e), while the addition of TOPO to mixtures of the same nanoparticles favors growth of AB_{13} (if $\gamma \leqslant 0.65$) or AB_5 (if $\gamma \geqslant 0.65$) BNSLs (Fig. 35d, f). Thus the space-filling principles and particle charging work in combination to determine the structure. Adjusting the relative concentration of A and B particles can be used as an additional tool with which to control the BNSL structure. For example, in presence of TOPO AB_4 or AB_5 BNSLs can form when the A:B ratio is $\sim 1:1$ while exclusively AB_{13} forms in the presence of excess B particles.

BNSLs can exhibit various effects of preferential nanoparticle orientation. Preferred orientation of the atomic lattices of individual nanocrystals packed into a superlattice can be studied by wide-angle selected area electron diffraction (SAED). Randomly oriented nanocrystals result in uniform diffraction rings in wide-angle SAED patterns as shown in Fig. 36a for a superlattice of 13.4 nm Fe_2O_3 nanoparticles. In contrast, the wide-angle X-ray scattering from Fe_2O_3 nanoparticles packed into the NaCl-type BNSL with 5.5 nm Au nanoparticles (Fig. 36d) appears as a series of arcs rather than continuous rings. These arcs originate from the preferential orientation of Fe_2O_3 nanoparticles in the superlattice. Preferential orientation can sometimes be observed in fcc single-component superlattices if nanoparticle shape

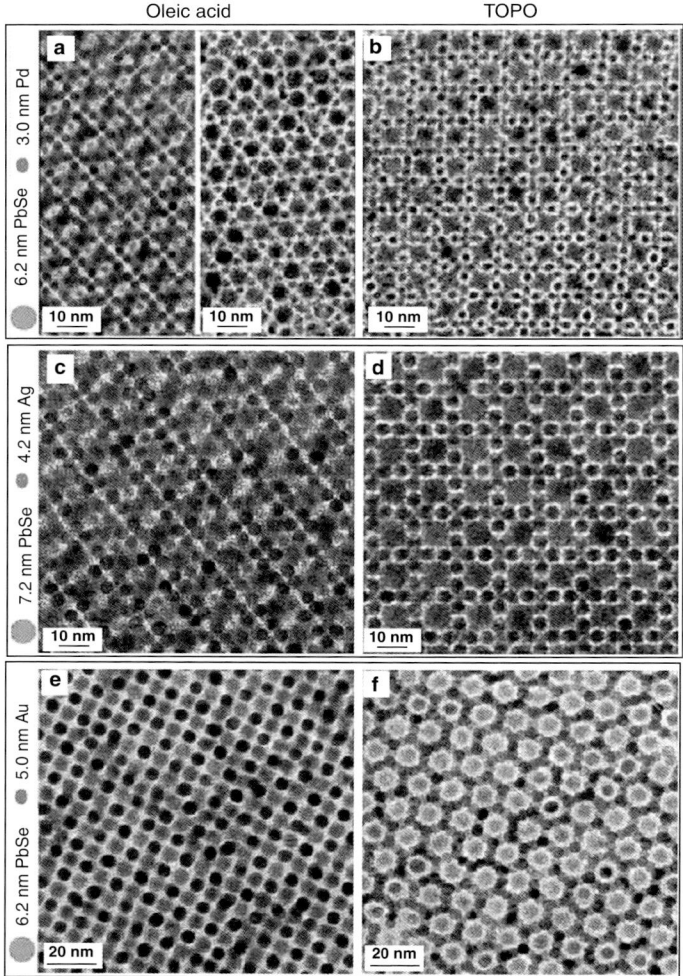

Fig. 35. TEM images of binary superlattices self-assembled in the presence of 4 mM oleic acid (left column) and 6 mM tri-n-octylphosphine oxide (right column). **a** 6.2 nm PbSe and 3.0 nm Pd nanoparticles self-assembled into orthorhombic AB- and AlB$_2$-type BNSLs, and (**b**) into NaZn$_{13}$-type BNSL. **c, d** 7.2 nm PbSe and 4.2 nm Ag nanoparticles self-assembled into orthorhombic AB and cuboctahedral AB$_{13}$ BNSLs, respectively. **e, f** 6.2 nm PbSe and 5.0 nm Au nanoparticles self-assembled into CuAu- and CaCu$_5$-type BNSLs, respectively. Reproduced from [91], © 2006, with permission from Nature Publishing Group

deviates from spherical due to faceting [11]. Figure 36b, c shows examples of preferential orientation in *fcc* arrays of PbSe nanocrystals.

Single-component *fcc* and *hcp* superlattices of nanocrystals usually show unidirectional orientation of atomic lattices while BNSLs can allow more complex preferential orientation effects. For example, wide-angle SAED pattern of (1 1 1) plane of NaCl-type Fe$_2$O$_3$–Au superlattice shows 12 equidistant arcs for the (2 2 0) reflections of Fe$_2$O$_3$ NPs (Fig. 36d). If all Fe$_2$O$_3$ nanoparticles orient in the same

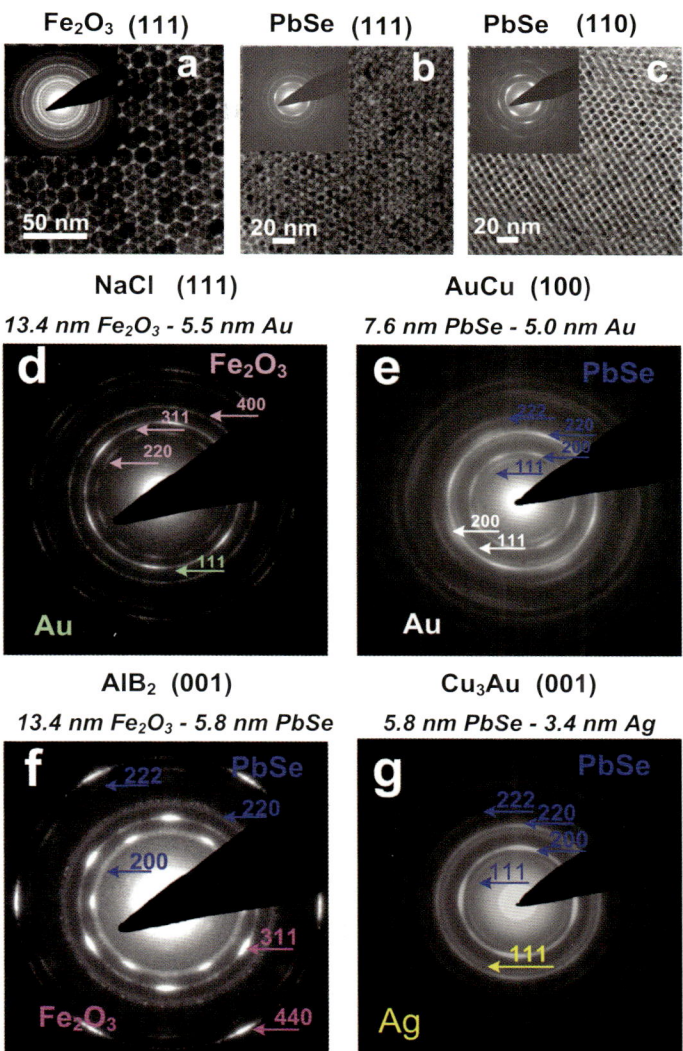

Fig. 36. TEM micrographs of **a** (111) planes of superlattices formed by 13.3 nm Fe_2O_3 nanoparticles; **b** (1 1 1) and **c** (1 1 0) planes of superlattices formed by 6.2 nm PbSe nanoparticles. Insets demonstrate the wide-angle selected area ED patterns measured from different planes. **d–g** Wide-angle selected area electron diffraction patterns measured from single-crystal domains of different BNSLs. Reproduced from [12], © 2006, with permission from the American Chemical Society

direction, we should observe four (2 2 0) diffraction spots. The 12 arcs for Fe_2O_3 (2 2 0) reflections in NaCl-type superlattice can result from 13 different orientations of Fe_2O_3 nanoparticles tilted by 60°.

Preferential orientation can be observed either for a single component or both components of binary BNSLs. Figure 36e and f shows preferential orientations for both components in PbSe–Au AuCu-type superlattice and Fe_2O_3–PbSe AlB_2-type superlattice, respectively. Figure 36g demonstrates an example of BNSL where the larger 5.8 nm PbSe NPs show preferential orientation while smaller 3.4 nm Ag NPs are disordered. Preferential orientation should be typical for superlattices where the arrangement of nearest neighbors matches the symmetry of nanoparticle facets "locking" each nanoparticle in a specific orientation.

Different types of nanoparticle superlattices can form simultaneously on the same substrate [12]. The domains with different symmetry are usually spatially separated. However, sometimes a smooth transition between the two phases is observed. For example, Fig. 37a shows an epitaxial "super-heterostructure" formed by the (0 1 1) planes of AlB$_2$-type and (1 0 0) plane of cub-AB$_{13}$-type superlattices. Figure 37b demonstrates the example of the defect heterostructure between (1 1 1) plane of

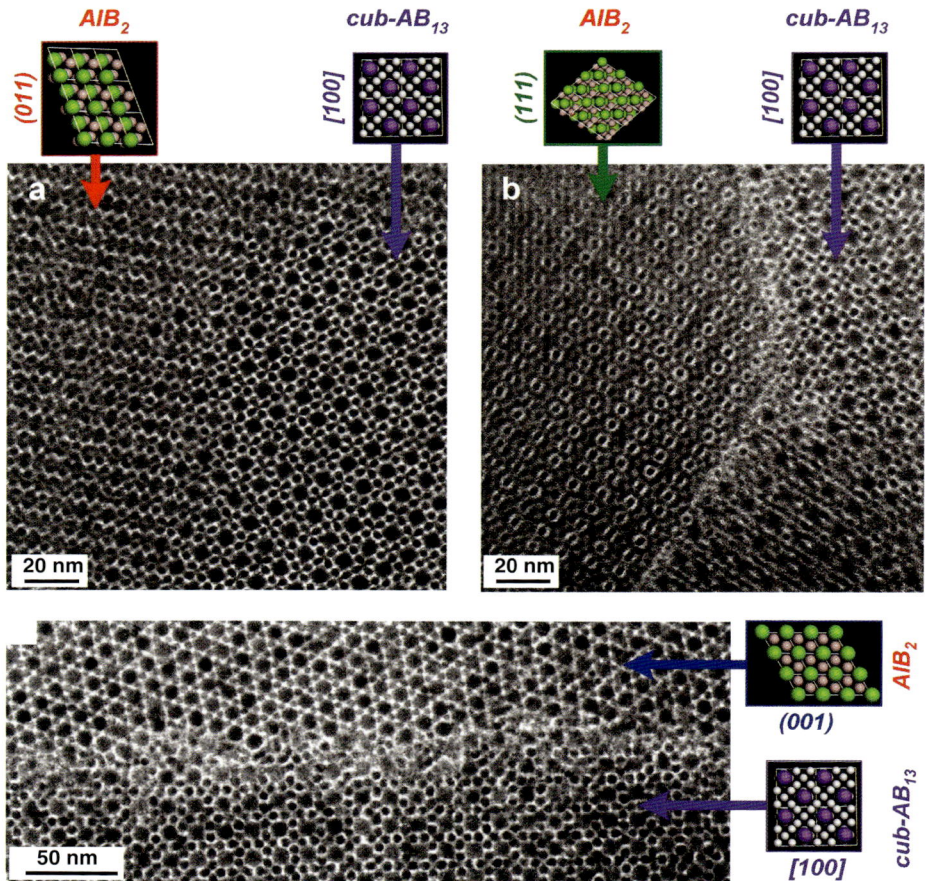

Fig. 37. TEM micrograph of **a** the epitaxial "super-heterostructure" formed by the (0 1 1) plane of AlB$_2$-type and (1 0 0) plane of cub-AB$_{13}$-type superlattices; **b** heterostructure formed by the (1 1 1) plane of AlB$_2$-type and (1 0 0) plane of cub-AB$_{13}$-type superlattices consisting of 5.8 nm PbSe and 3.0 nm Pd nanoparticles and **c** heterostructure formed by the (0 0 1) plane of AlB$_2$-type and (1 0 0) plane of cub-AB$_{13}$-type superlattices consisting of 7.2 nm PbSe and 4.2 nm Ag nanoparticles. Reproduced from [12], © 2006, with permission from the American Chemical Society

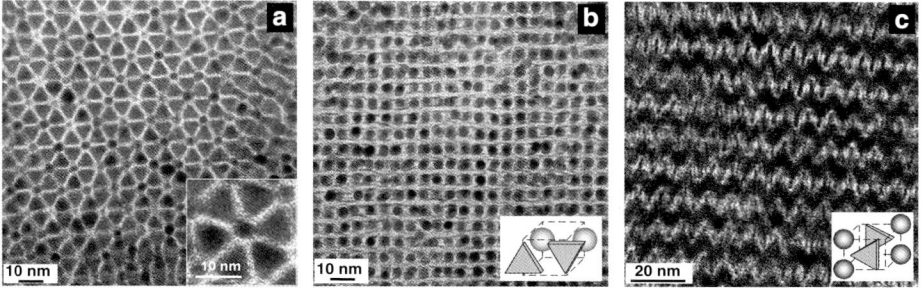

Fig. 38. TEM images of binary superlattices self-assembled (**a**, **b**) from LaF$_3$ triangular nanoplates (9.0 nm side) and 5.0 nm Au nanoparticles; **c** from LaF$_3$ triangular nanoplates and 6.2 nm PbSe nanocrystals. The insets show **a** a magnified image and **b**, **c** proposed unit cells of the corresponding superlattices. The structure shown in **a** forms on silicon nitride surface while structures shown in **b** and **c** form preferentially on amorphous carbon substrates. Reproduced from [91], © 2006, with permission from the American Chemical Society

AlB$_2$-type and (1 0 0) plane of cub-AB$_{13}$-type superlattice and Fig. 37c shows the example of heterostructure formed by (0 0 1) plane of AlB$_2$-type and (1 0 0) plane of cub-AB$_{13}$-type superlattice. This type of heterostructure can be formed as a collision of laterally growing islands of simultaneously nucleated AlB$_2$-type and cub-AB$_{13}$-type superlattices. Similar heterostructures were observed by Sanders in Brazilian opals [94].

In contrast to particles with amorphous or polycrystalline morphology, nanocrystals allow exploitation of the inherent crystal anisotropy to precisely engineer nanocrystal shape. The nanocrystal shape can in turn be used as a powerful tool to engineer the structure of the self-assembled BNSLs. For example, Fig. 38 shows several BNSLs self-assembled from LaF$_3$ triangular nanoplates and spherical Au or PbSe nanocrystals. In the LaF$_3$–Au system, LaF$_3$ nanoplates lay flat on silicon nitride membranes (Fig. 38a) and stand on edge when assembled on amorphous carbon (Fig. 38b, c) demonstrating how the choice of substrate can be used to control the structure of assembled BNSL.

It is specifically at nanoscale that the van der Waals, electrostatic, sterical repulsion, and the directional dipolar interactions can contribute to the interparticle potential with comparable weights [12]. These, together with the effects of particle substrate interactions, space-filling (entropic) factors, and specific ligand–ligand interactions determine enormous richness and diversity of structures assembled from nanoparticle building blocks. The non-equilibrium nature of evaporative self-assembly process adds additional complexity. Precise control of nanoparticle size, shape and composition allows us to engineer electronic, optical and magnetic properties of nanoparticle building blocks. Assembling these nanoscale building blocks into a wide range of BNSL systems provides powerful modular approach to the design of "metamaterials" with programmable physical and chemical properties.

5. Concluding remarks

During the last two decades semiconductor nanoparticles have developed into the new class of materials with their own sets of properties, some of them being unique and not observable in "molecular" and "macroscopic" worlds. Recent advances in synthesis and self-assembly introduced entirely new ways of controlling the size, shape, arrangement, connectivity, and even the topology of crystalline metallic, semiconducting and oxide nanomaterials. Further progress in this direction will enable complex multi-component nanostructures to be created by design in cost-effective ways. These structures will be further used as the building blocks for different electronic, optoelectronic, magnetic and catalytic materials.

Precise assembly of nanocrystals with desired optical, electronic, and magnetic properties into single-component and binary nanoparticle superlattices provides a route to 'metamaterials' combining metallic, semiconducting and magnetic components into ordered superlattices; a diverse collection of these structures have been recently reported and many novel structures will follow. The modular design of multicomponent solids using nanometer scale building blocks provides access to unique combinations of properties not available in single-component bulk solids. The controllable assembly of nanoparticle into macroscopic structures with desired arrangement of the building blocks requires fundamental understanding of self-organization phenomena at the nanoscale. Significant progress has been already achieved in this area; however, little is known about collective phenomena in the nanoparticle assemblies. The presence of long-range translational ordering in nanoparticle superstructures makes them different from common amorphous and polycrystalline solids. The coupling among ordered quantum dots can lead to a splitting of the quantized carrier energy levels of single particles and result in the formation of three-dimensional minibands (Fig. 39) [95]. By changing the size of the nanocrystals, interparticle distances, barrier height, and regimentation, one can control the electronic band structure of these artificial quantum dot crystals [96]. Model calculations show that the carrier density of states, effective mass tensor and other properties of quantum dot crystals are different from those of bulk solids and quantum well superlattices [95]. Generally speaking, nanocrystal assemblies can be

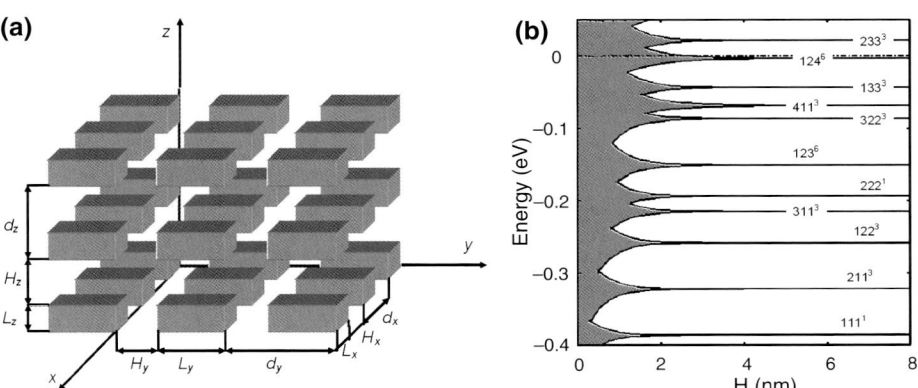

Fig. 39. a Schematic structure of a silicon quantum dot crystal and (**b**) its calculated electronic structure as a function of interparticle distance H. The size of the nanoparticles is $L = 6.5$ nm. At small H splitting of the quantized energy levels of single dots results in the formation of three-dimensional minibands.
Reproduced from [95], © 2001, with permission from American Institute of Physics

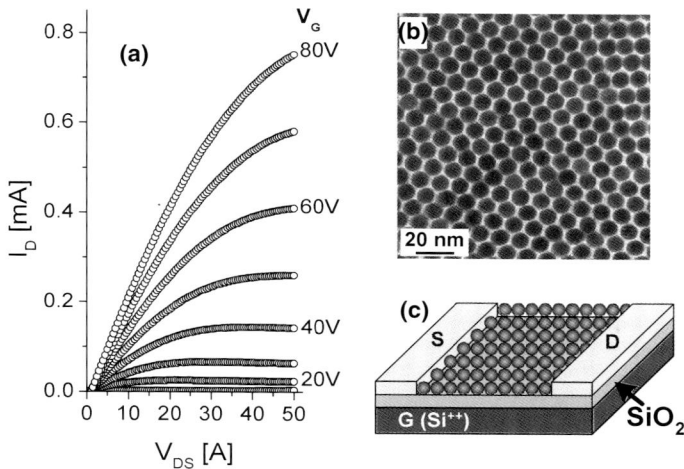

Fig. 40. a Electronic characteristics (drain current I_D vs. source-drain voltage V_{DS} measured for different gate voltages V_G) of a field-effect transistor with channel assembled from 9 nm PbSe nanocrystals shown in (**b**). **c** Schematics of nanocrystal-based field-effect device. Channel length 40 μm channel width 500 μm. Thickness of gate dielectric 100 nm

considered as a novel type of condensed matter, whose behavior depend both on the properties of the individual building blocks and the many body exchange interactions.

The recent studies of charge transport in superlattices of semiconductor nanocrystals demonstrated both n- and p-type transport with respectable charge carrier mobilities (>1 cm^2/V s for electrons and ~ 0.2 cm^2/V s for holes) [37] pointing to the possibility of miniband formation in properly engineered superlattices of semiconductor nanocrystals. There is a strong believe that nanocrystal assemblies can find use in inexpensive solution-processed field-effect transistors (Fig. 40), photovoltaic and thermoelectric devices [97]. At the same time, the electronic structure, phonon spectra and other characteristics of single- and multicomponent nanoparticle assemblies have never been a subject of thorough theoretical studies. Because of the absence of theoretical background, the superlattice approaches to materials design are still in the early stages of development. The generation of truly high-performance materials, requiring the optimization of many competing parameters, is still some way off. At the same time, first results represent the important steps in the rational design of nanocrystal superlattices for targeted applications.

Acknowledgements

We are deeply indebted to all our colleagues and collaborators, their names appear in the cited literature. We wish to specially thank Christopher B. Murray (University of Pennsylvania), Daan Frenkel (FOM-Institute for Atomic and Molecular Physics, Netherlands), Louis Brus, Stephen O'Brien and Irving Herman (Columbia University) for helpful discussions. E.V.S. acknowledges financial support by DOE, BES-Materials Sciences, under Contract DE-AC02-06CH11357. D.V.T. acknowledges financial support by MRSEC Program of the National Science Foundation under Award Number DMR-0213745 and FACCTS Program of the University of Chicago.

References

[1] Whitesides GM, Grzybowski B (2002) Self-assembly at all scales. Science 295: 2418–2421
[2] Yin Y, Alivisatos AP (2005) Colloidal nanocrystal synthesis and the organic–inorganic interface. Nature 437: 664–670
[3] Schmid G (ed) (2004) Nanoparticles: from theory to applications, Wiley–VCH
[4] Zeng H, Li J, Liu JP, Wang Z, Sun S (2002) Exchange-coupled nanocomposite magnets by nanoparticle self-assembly. Nature 420: 395–398
[5] Sun S, Murray CB, Weller D, Folks L, Moser A (2000) Monodisperse FePt nanoparticles and ferromagnetic nanocrystal superlattices. Science 287: 1989–1992
[6] Narayanan R, El-Sayed MA (2004) Shape-dependent catalytic activity of platinum nanoparticles in colloidal solution. Nano Letters 4: 1343–1348
[7] Alder BJ, Hoover WG, Young DA (1968) Studies in molecular dynamics. V. High-density equation of state and entropy for hard disks and spheres. Journal of Chemical Physics 49: 3688–3694
[8] Bolhuis PG, Frenkel D, Mau SC, Huse DA (1997) Entropy difference between crystal phases. Nature 388: 235–236
[9] Rudd RE, Broughton JQ (1998) Coarse-grained molecular dynamics and the atomic limit of finite elements. Physical Review B 58: 5893–5896
[10] Wang ZL (1998) Structural analysis of self-assembling nanocrystal superlattices. Advance of Materials 10: 13–30
[11] Murray CB, Kagan CR, Bawendi MG (2002) Synthesis and characterization of monodisperse nanocrystals and close-packed nanocrystal assemblies. Annual Review of Material Science 30: 545–610
[12] Shevchenko EV, Talapin DV, Murray CB, O'Brien S (2006) Structural characterization of self-assembled multifunctional binary nanoparticle superlattices. Journal of the American Chemical Society 128: 3620–3637
[13] Islam MA, Xia Y, Steigerwald ML, Yin M, Liu Z, O'Brien S, Levicky R, Herman IP (2003) Addition, suppression, and inhibition in the electrophoretic deposition of nanocrystal mixture films for CdSe nanocrystals with γ-Fe_2O_3 and Au nanocrystals. Nano Letters 3: 1603–1606
[14] Shim M, Guyot-Sionnest P (1999) Permanent dipole moment and charges in colloidal semiconductor quantum dots. Journal of Chemical Physics 111: 6955–6964
[15] Li L-S, Alivisatos AP (2003) Origin and scaling of the permanent dipole moment in CdSe nanorods. Physical Review Letters 90: 097402
[16] Rabani E, Hetenyi B, Berne DJ, Brus LE (1999) Electronic properties of CdSe nanocrystals in the absence and presence of a dielectric medium. Journal of Chemical Physics 110: 5355–5361
[17] Ohara PC, Leff DV, Heath JR, Gelbart WM (1995) Crystallization of opals from polydisperse nanoparticles. Physical Review Letters 75: 3466–3469
[18] Korgel BA, Fullam S, Connolly S, Fitzmaurice D (1998) Assembly and self-organization of silver nanocrystal superlattices: ordered "soft spheres". Journal of Physical Chemistry B 102: 8379–8388
[19] Saunders AE, Korgel BA (2004) Second virial coefficient measurements of dilute gold nanocrystal dispersions using small-angle X-ray scattering. Journal of Physical Chemistry B 108: 16732–16738
[20] Rabideau BD, Bonnecaze RT (2005) Computational predictions of stable 2D arrays of bidisperse particles. Langmuir 21: 10856–10861
[21] Berry RS, Rice SA, Ross J (2000) Physical Chemistry. Oxford University Press
[22] Chung DDL (2004) Electrical applications of carbon materials. Journal of Materials Sciences 39: 2645–2661
[23] Talapin DV, Shevchenko EV, Murray CB, Titov AV, Kral P (2007) Dipole–dipole interactions in nanoparticle superlattices. Nano Letters 7: 1213–1219
[24] Luedtke WD, Landman U (1996) Structure, dynamics, and thermodynamics of passivated gold nanocrystallites and their assemblies. Journal of Physical Chemistry 100: 13323–13329
[25] Tang ZY, Zhang ZL, Wang Y, Glotzer SC, Kotov NA (2006) Self-assembly of CdTe nanocrystals into free-floating sheets. Science 314: 274–278
[26] Klokkenburg M, Houtepen AJ, Koole R, de Folter JWJ, Erhe BH, van Faassen E, Vanmaekelbergh D (2007) Dipolar structures in colloidal dispersions of PbSe and CdSe quantum dots. Nano Letters 7: 2931–2936
[27] Tang Z, Kotov NA, Giersig M (2002) Spontaneous organization of single CdTe nanoparticles into luminescent nanowires. Science 297: 237–240
[28] Pacholski C, Kornowski A, Weller H (2002) Self-assembly of ZnO: from nanodots to nanorods. Angewandte Chemie International Edition 41: 1188–1191

[29] Cho K-S, Talapin DV, Gaschler W, Murray CB (2005) Designing PbSe nanowires and nanorings through oriented attachment of nanoparticles. Journal of the American Chemical Society 127: 7140–7147
[30] Shanbhag S, Kotov NA (2006) On the origin of a permanent dipole moment in nanocrystals with a cubic crystal lattice: effects of truncation, stabilizers, and medium for CdS tetrahedral homologues. Journal of Physical Chemistry B 110: 12211–12217
[31] Cui Y, Zhong Z, Wang D, Wang WU, Lieber CM (2003) High performance silicon nanowire field effect transistors. Nano Letters 3: 149–152
[32] Patolsky F, Zheng G, Hayden O, Lakadamyali M, Zhuang X, Lieber CM (2004) Electrical detection of single viruses. Proceedings of the National Academy of Sciences of the United States of America 101: 14017–14022
[33] Hahm J-I, Lieber CM (2004) Direct ultrasensitive electrical detection of DNA and DNA sequence variations using nanowire nanosensors. Nano Letters 4: 51–54
[34] Huynh WU, Dittmer JJ, Alivisatos AP (2002) Hybrid nanorod–polymer solar cells. Science 295: 2425–2427
[35] Hicks LD, Dresselhaus MS (1993) Thermoelectric figure of merit of a one-dimensional conductor. Physical Review B 47: 16631–16634
[36] Lin YM, Dresselhaus MS (2003) Thermoelectric properties of superlattice nanowires. Physical Review B 68: 075304
[37] Talapin DV, Murray CB (2005) PbSe nanocrystal solids for n- and p-channel thin film field-effect transistors. Science 310: 86–89
[38] Rabani E, Reichman PL, Brus LE (2003) Drying-mediated self-assembly of nanoparticles. Nature 426: 271–274
[39] Tanaka H (2000) Viscoelastic phase separation. Journal of Physics: Condensed Matter 12: R207–R264
[40] Ohara PC, Heath JR, Gelbart WM (1997) Self-assembly of submicrometer rings of particles from solutions of nanoparticles. Angewandte Chemie International Edition 15: 1078–1080
[41] Heath JR, Knobler CM, Leff DV (1997) Pressure/temperature phase diagrams and superlattices of organically functionalized metal nanocrystal monolayers: the influence of particle size, size distribution, and surface passivant. Journal of Physical Chemistry B 101: 189–197
[42] Dabbousi BO, Murray CB, Rubner MF, Bawendi MG (1994) Langmuir–Blodgett manipulation of size-selected CdSe nanocrystallites. Chemistry of Materials 6: 216–219
[43] Ge G, Brus L (2000) Evidence for spinodal phase separation in two-dimensional nanocrystal self-assembly. Journal of Physical Chemistry B 104: 9573–9575
[44] Maillard M, Motte A, Ngo AT, Pileni MP (2000) Rings and hexagons made of nanocrystals: a Marangoni effect. Journal of Physical Chemistry B 104: 11871–11877
[45] Gelbart WM, Sear RP, Heath JR, Chaney S (1999) Array formation in nano-colloids: theory and experiment in 2D. Faraday Discussions 112: 299–307
[46] Ohara PC, Gelbart WM (1998) Interplay between hole instability and nanoparticle array formation in ultrathin liquid films. Langmuir 14: 3418–3424
[47] Ondarcuhu T, Millan-Rodriguez J, Mancini HL, Garcimartin A, Perez-Garcia C (1993) Benard–Marangoni convective patterns in small cylindrical layers. Physical Review E 48: 1051–1057
[48] Sukhanova A, Baranov AV, Perova TS, Cohen JHM, Nabiev I (2006) Controlled self-assembly of nanocrystals into polycrystalline fluorescent dendrites with energy-transfer properties. Angewandte Chemie International Edition 45: 2048–2052
[49] Norris DJ, Arlinghaus EG, Meng L, Heiny R, Scriven LE (2004) Opaline photonic crystals: how does self-assembly work? Advanced Materials 16: 1393–1399
[50] Hynninen A-P, Dijkstra M (2005) Phase diagram of dipolar hard and soft spheres: manipulation of colloidal crystal structures by an external field. Physical Review Letters 94: 138303–138306
[51] Murray CB, Kagan CR, Bawendi MG (1995) Self-organization of CdSe nanocrystallites into three-dimensional quantum dot superlattices. Science 270: 1335–1338
[52] Talapin DV, Shevchenko EV, Kornowski A, Gaponik N, Haase M, Rogach AL, Weller H (2001) A new approach to crystallization of CdSe nanoparticles into ordered three-dimensional superlattices. Advanced Materials 13: 1868–1871
[53] Shevchenko E, Talapin D, Kornowski A, Wiekhorst F, Koetzler J, Haase M, Rogach A, Weller H (2002) Colloidal crystals of monodisperse FePt nanoparticles grown by a three-layer technique of controlled oversaturation. Advanced Materials 14: 287–290
[54] Shevchenko EV, Talapin DV, Rogach, AL, Kornowski A, Haase M, Weller H (2002) Colloidal synthesis and self-assembly of $CoPt_3$ nanocrystals. Journal of the American Chemical Society 124: 11480–11485

[55] Nagel M, Hickey SG, Frömsdorf A, Kornowski A, Weller H (2007) Synthesis of monodisperse PbS nanoparticles and their assembly into highly ordered 3D colloidal crystals. Zeitschrift für Physikalische Chemie 221: 427–437
[56] Borchert H, Shevchenko EV, Robert A, Mekis I, Kornowski A, Grubel G, Weller H (2005) Determination of nanocrystal sizes: a comparison of TEM, SAXS, and XRD studies of highly monodisperse CoPt$_3$ particles. Langmuir 21: 1931–1936
[57] Blanto SA, Leheny RL, Hines MA, Guyot-Sionnest P (1997) Dielectric dispersion measurements of CdSe nanocrystal colloids: observation of a permanent dipole moment. Physical Review Letters 79: 865–868
[58] Pusey PN (1991) Colloidal suspensions. In: Liquids, freezing and glass transition Hansen JP, Levesque D, Zinn-Justin J (eds.), North-Holland, Amsterdam, pp. 765–942
[59] Auer S, Frenkel D (2001) Suppression of crystal nucleation in polydisperse colloids due to increase of the surface free energy. Nature 413: 711–713
[60] Pronk S, Frenkel D (2001) Point defects in hard-sphere crystals. Journal of Physical Chemistry B 105: 6722–6727
[61] Pronk S, Frenkel D (2004) Large effect of polydispersity on defect concentrations in colloidal crystals. Journal of Chemical Physics 120: 6764–6772
[62] Manna L, Scher EC, Alivisatos AP (2000) Synthesis of soluble and processable rod-, arrow-, teardrop-, and tetrapod-shaped CdSe nanocrystals. Journal of the American Chemical Society 122: 12700–12706
[63] Peng X (2002) Green chemical approaches toward high quality semiconductor nanocrystals. Chemistry-A European Journal 8: 335–340
[64] Peng X, Manna L, Yang W, Wickham J, Scher E, Kadavanich A, Alivisatos AP (2000) Shape control of CdSe nanocrystals. Nature 404: 59–61
[65] Li L-S, Hu J, Yang W, Alivisatos AP (2001) Band gap variation of size- and shape-controlled colloidal CdSe quantum rods. Nano Letters 1: 349–351
[66] Hu J, Li L-S, Yang W, Manna L, Wang L-W, Alivisatos AP (2001) Linearly polarized emission from colloidal semiconductor quantum rods. Science 292: 2060–2063
[67] Frenkel D, Lekkerkerker HNW, Stroobants A (1988) Thermodynamic stability of a smectic phase in a system of hard rods. Nature 332: 822–823
[68] Schilling T, Frenkel D (2004) Self-poisoning of crystal nuclei in hard-rod liquids. Physical Review Letters 92: 085505–085508
[69] McGrother SC, Williamson DC, Jackson G (1996) A re-examination of the phase diagram of hard spherocylinders. Journal of Chemical Physics 104: 6755–6759
[70] Frenkel D (1988) Structure of hard-core models for liquid crystals. Journal of Physical Chemistry 92: 3280–3284
[71] Schilling T, Frenkel D (2004) Self-poisoning of crystal nuclei in hard-rod liquids. Journal of Physics: Condensed Matter 16: S2029–S2036
[72] McGrother C, Gil-Villegas A, Jackson G (1996) The liquid-crystalline phase behavior of hard spherocylinders with terminal point dipoles. Journal of Physics: Condensed Mater 8: 9649–9655
[73] Li L-S, Walda J, Manna L, Alivisatos AP (2002) Semiconductor nanorod liquid crystals. Nano Letters 2: 557–560
[74] Talapin DV, Shevchenko EV, Murray CB, Kornowski A, Forster S, Weller H (2004) CdSe and CdSe/CdS nanorod solids. Journal of the American Chemical Society 126: 12984–12988
[75] Carbone L, Nobile C, De Giorgi M, Sala FD, Morello G, Pompa P, Hytch M, Snoeck E, Fiore A, Franchini IR, Nadasan M, Silvestre AF, Chiodo L, Kudera S, Cingolani R, Krahne R, Manna, L (2007) Synthesis and micrometer-scale assembly of colloidal CdSe/CdS nanorods prepared by a seeded growth approach. Nano Letters 7: 2942–2950
[76] Phillips PJ (1994) Spherulitic crystallization in macromolecules. In: Handbook of crystal growth (Hurtle DTJ, ed.) North-Holland: Amsterdam, pp. 1168–1215
[77] Singfield KL, Hobbs JK, Keller A (1998) Correlation between main chain chirality and crystal twist direction in polymer spherulites. Journal of Crystal Growth 183: 683–689
[78] Talapin DV, Nelson JH, Shevchenko EV, Aloni S, Sadtler B, Alivisatos AP (2007) Seeded growth of highly luminescent CdSe/CdS nanoheterostructures with rod and tetrapod morphologies. Nano Letters 7: 2951–2959
[79] Talapin DV, Koeppe R, Gotzinger S, Kornowski A, Lupton JM, Rogach AL, Benson O, Feldmann J, Weller H (2003) Highly emissive colloidal CdSe/CdS heterostructures of mixed dimensionality. Nano Letters 3: 1677–1681
[80] Ryan KM, Mastroianni A, Stancil KA, Liu H, Alivisatos AP (2006) Electric-field-assisted assembly of perpendicularly oriented nanorod superlattices. Nano Letters 6: 1479–1482

[81] Manna L, Milliron DJ, Meisel A, Scher EC, Alivisatos AP (2003) Controlled growth of tetrapod-branched inorganic nanocrystals. Nature Materials 2: 382–385
[82] Blaak R, Mulder BM, Frenkel D (2004) Cubatic phase for tetrapods. Journal of Chemical Physics 120: 5486–5492
[83] Eldridge MD, Madden PA, Frenkel D (1993) The stability of the AB13 crystal in a binary hard sphere system. Molecular Physics 79: 105–120
[84] Trizac E, Eldridge MD, Madden PA (1997) Stability of the AB crystal for asymmetric binary hard sphere mixtures. Molecular Physics 90: 675–678
[85] Cottin X, Monson PA (1995) Substitutionally ordered solid solutions of hard spheres. Journal of Chemical Physics 102: 3354–3361
[86] Murray MJ, Sanders JV (1980) Close-packed structures of spheres of two different sizes. II. The packing densities of likely arrangements. Philosophical Magazine A 42: 721–740
[87] Sanders JV, Murray MJ (1978) Ordered arrangements of spheres of two different sizes in opal. Nature 275: 201–203
[88] Hachisu S, Yoshimura S (1980) Optical demonstration of crystalline superstructures in binary mixtures of latex globules. Nature 283: 188–189
[89] Bartlett P, Ottewill RH, Pusey PN (1992) Superlattice formation in binary mixtures of hard-sphere colloids. Physical Review Letters 68: 3801–3804
[90] Eldridge MD, Madden PA, Frenkel D (1993) Entropy-driven formation of a superlattice in a hard-sphere binary mixture. Nature 365: 35–37
[91] Shevchenko EV, Talapin DV, Kotov NA, O'Brien S, Murray CB (2006) Structural diversity in binary nanoparticle superlattices. Nature 439: 55–59
[92] Leunissen ME, Christova CG, Hynninen A-P, Royall CP, Campbell AI, Imhof A, Dijkstra M, van Roij R, van Blaaderen A (2005) Ionic colloidal crystals of oppositely charged particles. Nature 437: 235–240
[93] O'Brien RW, White LR (1978) Electrophoretic mobility of a spherical colloidal particle. Journal of the Chemical Society-Faraday Transactions II 74: 1607–1626
[94] Sanders JV (1980) Close-packed structures of spheres of two different sizes I. Observations on natural opal. Philosophical Magazine A 42: 705–720
[95] Lazarenkova OL, Balandin AA (2001) Miniband formation in a quantum dot crystal. Journal of Applied Physics 89: 5509–5513
[96] Jiang C-W, Green MA (2006) Silicon quantum dot superlattices: Modeling of energy bands, densities of states, and mobilities for silicon tandem solar cell applications. Journal of Applied Physics 99: 114902
[97] Urban JJ, Talapin DV, Shevchenko EV, Kagan CR, Murray CB (2007) Synergism in binary nanocrystal superlattices leads to enhanced p-type conductivity in self-assembled PbTe/Ag_2Te thin films. Nature Materials 6: 115–121

Semiconductor nanocrystal–polymer composites: using polymers for nanocrystal processing

By

Dayang Wang*

Max Planck Institute of Colloids and Interfaces, Potsdam, Germany

1. Introduction

Semiconductor nanocrystals (NCs) in the size range of 1–10 nm exhibit unique size-dependent photoluminescence properties, distinct from either the corresponding molecules or bulk materials, which are a result of quantum confinement effect and enormously high specific surface area [1–5]. Accordingly, there is much speculation about the potential use of semiconductor NCs in a vast spectrum of high-technology fields such as optics, electronics, and biomedicine. In this context, the past decade has seen a great progress in tailoring a diversity of semiconductors into nanometer-sized particles with defined but varied size, shape, and surface chemistry [6–9]. Once prepared, however, NCs in general have a strong tendency to agglomerate owing to the presence of a great deal of highly active surface atoms, which dramatically deteriorates their physicochemical properties. Stabilization of NCs is necessitated for both exploring their intrinsic size-related properties and exploiting their technical applicability. Up to now numerous approaches have been developed to stabilize semiconductor NCs by surface charges [6], functionalized alkanes [6–9], silica [10–13], and polymers [14]. The stability of a NC is determined by the thermodynamic balance between repulsive interactions – mainly electrostatic repulsion and steric repulsion – and attractive interactions – mainly van der Waals and hydrophobic interaction; the NC is stable when the repulsive interactions are dominant. Since the electrostatic repulsion is rather sensitive to the size of NCs and the variation of the surrounding media, steric repulsion is envisioned ideal for stabilization of NCs. Among all materials used to stabilize NCs, polymers, by far, usually provide an excellent steric hindrance effect on the NCs, thus imparting NCs with a robust stability against the environmental variation.

In addition of stabilization, hybridizing NCs with polymers can also bring out a number of merits. First of all, NCs are envisaged to inherit the good compatibility, the excellent processing capability, and the high-engineering performance from the polymer stabilizers, which are sought in most technical applications of NCs [15]. In case of biomedical applications, the proper choice of hydrophilic polymer stabilizers

---

*Present address: Laboratory of Advanced Materials, Fudan University, Shanghai, P.R. China.

such as polyethylene glycol (PEG) based (co-)polymers also render semiconductor NCs biocompatibility and reduce the cytotoxicity of the NCs [16, 17]. The merits of using polymers to stabilize NCs can be also foreseen from the intelligent response of polymers to a vast variety of external stimuli [18] and their rich phase separation properties [19]. Using polymers for directing the self-assembly of semiconductor NCs allows for the construction of tailor-made sophisticated arrays of the NCs with the desired properties [20]. The stimuli-response of polymer stabilization matrices allows a fine control of the interaction of neighboring semiconductor NCs and thus the collective properties of NCs [21].

As a result, enormous effort has been devoted to the development of different methods to stabilize semiconductor NCs with polymers, including synthesis of NCs within polymer stabilizers, polymerization of monomers in the presence of NCs, (co-)polymerization of NCs capped with monomers, coating NCs with polymers, exchange of the original ligands on NCs with polymers, growth of polymer brushes from NCs, layer-by-layer (LbL) consecutive deposition of NCs and polymers via non-covalent bonding, and blending of NCs with polymers. All these methods can be classified into two categories: chemical synthesis and physical blending. The present chapter will summarize the recent progress on fabrication of semiconductor NC–polymer composites. It is centered mainly on fabrication methodologies of polymer–semiconductor NC composites. Thus, using polymers to direct self-assembly of semiconductor NCs, the technical use of semiconductor NCs and their polymer composites are only slightly touched upon. Although the strategies described here have also been extensively used to form polymer composites with metal NCs, metal oxide NCs, and organic NCs, we focus exclusively on semiconductor NCs.

In this chapter, we commence with in situ synthesis of NCs within polymer matrices – the earliest developed strategy – with highlighting current progress. In the section on chemical synthesis of NC–polymer composites, we also discuss the strategy of synthesis of polymers in the presence of NCs, in which the possibility of scaling-up for industrial production and the possibility of processing NCs into different device-prototypes are underlined. The following section is to discuss the methods of grafting polymers on NCs, which is more related with enhancement of colloidal stability of NCs and the biological use of NCs. In this section, two strategies of growth of polymers on NCs, grafting-to and grafting-from, are treated in detail. Layer-by-layer electrostatic deposition of NCs and polyelestrolytes into composite thin films is in principle one of blending strategies. Due to the large activity in the area and especially success in generalizing this strategy by means of other non-electrostatic interactions such as hydrogen bonding, we give this strategy a subsection to highlight its great processability and high applicability to numerous technical applications. The topic of layer-by-layer deposition is treated in details in the Chapter of Srivastava/Kotov of this book. The last section is devoted to discussion of the vast spectrum of blending of NCs and polymers, which is envisioned as the most processable for practical use. Due to the explosion of reports in the field with respect to fabrication semiconductor NC–polymer composites, the present chapter is not able to include all of the published work, but will focus on the most recent developments in preparation methodologies. We apologize to any authors for the unintentional non-inclusion of their excellent work.

2. Chemical synthesis of semiconductor NC–polymer composites

2.1 Synthesis of semiconductor NCs in the presence of polymers.

Before nanometer-sized objects distinguished themselves as innovative materials distinct from both molecular and bulk materials, synthesis of inorganic nanoparticles within polymer matrices has already been developed, mainly driven by pigment industry demanding incorporation of inorganic pigments such as ZnO and Fe_2O_3 into polymers [22]. In a typical preparation procedure, organometallic precursors of semiconductor NCs are first introduced into polymer matrices either through simple mixing or polymerization of monomers in presence of NC precursors or copolymerization of NC precursors and monomers if the precursors are polymerizable. Afterwards, the mixtures of polymers and NC precursors are exposed, for example, to chalcogenide solution or gas. Due to the polymer matrices preventing the infinite crystal growth, only nanometer-sized particles of semiconductors are obtained, thus yielding semiconductor NC–polymer composites. Well-established organometallic chemistry allows for design of NC precursors with good solubility in polymers, eliminating the incompatibility between NCs and polymers that is always problematic for hybridizing as-prepared NCs with polymers and even their monomers. Prior to synthesis of NCs, one can easily process the mixtures of polymers and NC precursors into different device prototypes, bulk materials, thin films, microspheres, and so on, which prevents chemically deterioration of semiconductor NCs during the postprocessing, for instance oxidation.

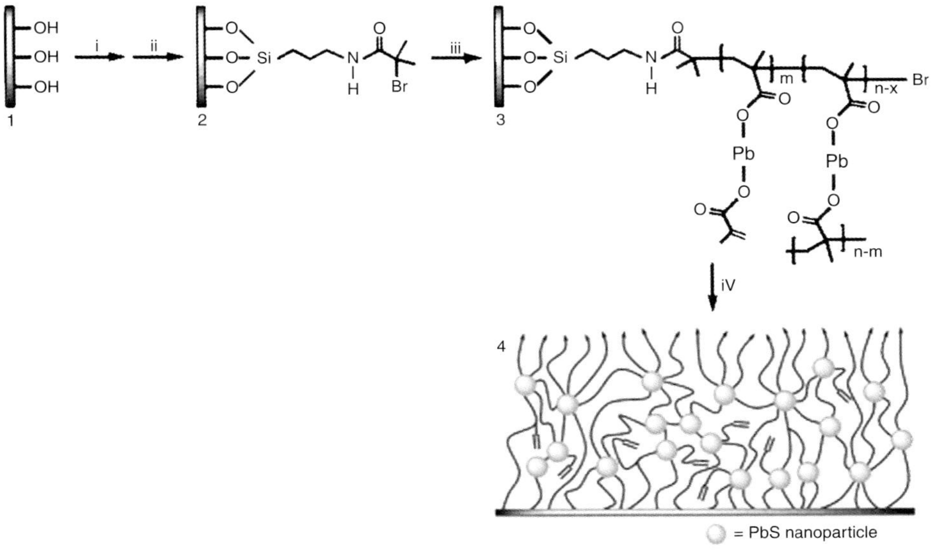

Fig. 1. Schematic illustration of the preparation procedure of PbS NC–polymer composites on a silicon wafer by surface-initiated ATRP of Pb-containing monomers, followed by reaction with H_2S. Conditions: (i) 3-aminopropyltriethoxysilane, toluene; (ii) 2-bromo-2-methylpropionic acid, DCC, DMAP, CH_2Cl_2; (iii) lead dimethacrylate, p-toluenesulfonyl chloride, Cu(I)Cl, 2,2′-bipyridine, DMF; (iv) H_2S gas. Reproduced with permission from [23], © 2002 ACS

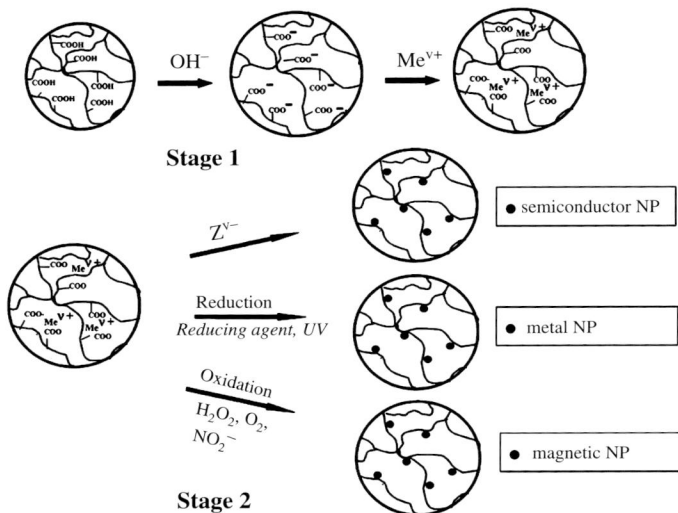

Fig. 2. Schematic of microgel-based synthesis of semiconductor, metal, and magnetic NCs. Reproduced with permission from [24], © 2004 ACS

Yang's group has recently synthesized new types of monomers and cross-linkers composed of methacrylate bearing Pb, Zn, or Cd [23]. After copolymerization of the Pb-containing monomers and cross-linkers via atom transfer radical polymerization (ATRP), as shown in Fig. 1, the resulting films were exposed to H_2S gas, yielding PbS NCs embedded within polymer thin films, which should hold great promise in optoelectronics.

Kumacheva and coworkers have used polymer microgels as in situ reactors for synthesis of semiconductor NCs, giving rise to micrometer-sized composite particles comprising polymer gels and semiconductor NCs (Fig. 2) [24]. They also found that the postheating treatment of the resulting composite particles may to some extent narrow the size distribution of the resulting semiconductor NCs and thus tune their photoluminescence properties.

Obviously, this in situ synthesis strategy cannot endorse a fine control of the size and shape of semiconductor NCs obtained in polymer matrices, so the photoluminescence properties of the NCs are generally rather poor. Providing the photoluminescence is not a targeted issue, on the other hand, this strategy is indeed the easy way for forming polymer–semiconductor NC composites and the resulting composites may be used as high-performance engineering plastic materials.

2.2 Polymerization in the presence of semiconductor NCs. In order to define the photoluminescence behavior of semiconductor NC–polymer composites, one has devoted enormous effort to the incorporation of preformed semiconductor NCs into polymers. According to the way polymer matrices are formed, the incorporation of preformed NCs into polymers can be implemented in two ways; directly blending preformed NCs with preformed polymers and polymerization of monomers in the presence of NCs. The former will be discussed in Sect. 3.1. The present section is centered on how to form polymer–semiconductor NC composites by polymerization of monomers in the presence of NCs.

According to the synthesis environment, NCs can be divided into two classes, organic and aqueous NCs. Organic semiconductor NCs are typically prepared via pyrolysis of organometallic precursors in hot organic coordinative solvents. Once prepared, they are capped by ligands with long alkyl chains such as trioctylphosphine oxide (TOPO) [7–9, 25–28]. In comparison, aqueous NCs are prepared via neutralization of metallic ions and chalcogenides in the presence of mercapto-ligands, such as 3-mercaptopropionic acid (MPA), thioglycolic acid (TGA), and mercaptoethanol (ME) [6, 29–32]. The surface wettability of NCs may be turned from hydrophilic to hydrophobic or vice versa via phase transfer with the aid of amphiphilic molecules [33]. Regardless if NCs are directly synthesized or obtained via phase transfer, in most cases organic NCs are stabilized by the long alkyl chain capping, while, aqueous NCs by the surface charges arising from ionization of their capping ligands. In the case of organic semiconductor NCs, their long alkyl capping may not usually guarantee them a good solubility in polymer matrices. Aggregation of semiconductor NCs in the polymer matrices, and the quenching of their photoluminescence, is usually observed [34, 35]. In this sense semiconductor NCs should first be dispersed in monomers, since a better compatibility between NCs and monomers is always easy to achieve. In order to construct optically transparent polymer–semiconductor NC composites, the appropriate choice of monomers is further required, because over the course of polymerization of the initially transparent NC–monomer solution, phase separation of the NCs from the polymer matrices may cause aggregation of the NCs, thereby lowering the transparency of the composites and especially the photoluminescence efficiency of the NCs in the composites. For instance polymerization of mixture solutions of semiconductor NCs with styrene usually causes phase separation of the NCs.

Bawendi and coworkers have succeeded in fabricating transparent polymer composites of organic ZnS-overcoated CdSe NCs with little deterioration of the photoluminescence properties of the NCs by polymerization of laurylmethacrylate

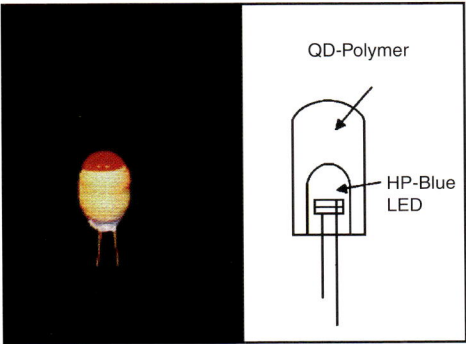

Fig. 3. A down-conversion light-emitting device derived from ZnS-overcoated CdSe NC–PLMA composites emitting at 590 nm. GaN light-emitting diode (LED) was used as the excitation source at 425 nm. Reproduced with permission from [36], © 2000 Wiley–VCH

in the presence of the NCs [36]. They found that the long alkyl side branches of polylaurylmethacrylate (PLMA) efficiently prevented the NCs from aggregation and phase separation from the polymer matrices during polymerization. A prototype of light-emitting device using the resulting composites was demonstrated (Fig. 3).

Capping semiconductor NCs with polymerizable ligands can create a strong coupling with polymer matrices. Yang's group capped octadecyl-p-vinyl-benzyl dimethylammonium chloride (OVDAC) on aqueous CdTe NCs via electrostatic interaction. OVDAC-coated CdTe NCs were readily to transfer into organic monomers, and followed by free radical bulk polymerization, CdTe NC–polymer bulk composites were obtained [37]. The strong coupling affinity between CdTe NCs and the polymer matrices to a large extent avoids the phase separation of the NCs from the polymer host during polymerization. As such their approach can be amenable for a wide range of monomers, such as styrene and methyl methacrylate to choose.

This strategy has been extended to fabricate composites of semiconductor NCs with conjugated polymers [38]. Most recently, Pyun and coworkers have capped 10-((3-methyl-3,4-dihydro-2H-thieno[3,4-b][1,4]dioxepin-3-yl)-methoxy)-10-oxo-decanoic acid (ProDOT-CA) on organic CdSe NCs and cast the mixtures of the resulting NCs and ProDOT-CA on ITO electrodes [39]. The following electrochemical polymerization turned ProDOT-CA into poly(3,4-ethylenedioxylthiophene). During polymerization of ProDOT-CA, ProDOT-CA coated NCs were cross-linked within the thin films. The potential of using the resulting mechanically robust and electronically active composite thin films composed for solar cells was demonstrated.

Suspension and emulsion polymerization in the presence of semiconductor NCs allow sculpture of semiconductor NC–polymer composites into microspheres, which should be very useful as fluorescence markers for biological detection. Nonetheless, the strategy of polymerization in the presence of semiconductor NCs has as a whole, the large problem that free radicals usually quench the photoluminescence of semiconductor NCs, lowering the optical properties of the resulting composites. Gao and coworkers have demonstrated that the appropriate choice of initiators is crucial to remain a better photoluminescence efficiency of the NCs when they were

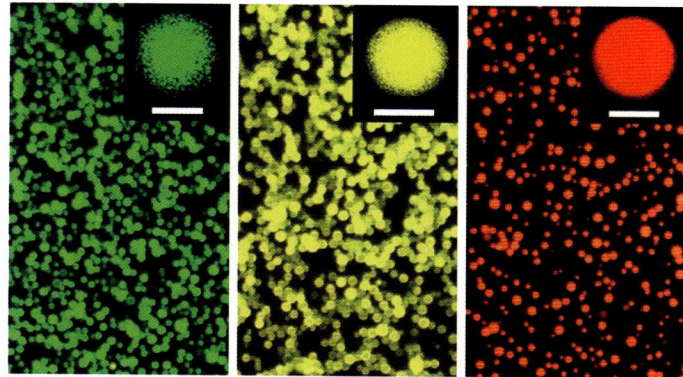

Fig. 4. Fluorescence images of PS beads loaded with green (left), yellow (middle), and red CdTe NCs (right). The insets are the corresponding confocal fluorescence images. The scale bars correspond to 2 μm. Reproduced with permission from [40], © 2006 Wiley–VCH

embedded in polymer matrices [40]. They used different initiators including benzoyl peroxide, $K_2S_2O_8$, H_2O_2 and 2,2'-azobisisobutyronitrile (AIBN) for miniemulsion polymerization of styrene in the presence of CdTe NCs, and found that only AIBN may preserve the photoluminescence of the NCs embedded in polystyrene (PS) microspheres, thus leading to chemically robust fluorescent beads (Fig. 4).

2.3 Grafting polymers on semiconductor NCs. Semiconductor NC–polymer composites obtained either via synthesis of NCs in the presence of polymers or via polymerization in the presence of NCs, are usually in the form of bulky blocks, thin films and fibers. Nevertheless, these two simple strategies are not capable of defining the separation distance between NCs, thus leading to a poor control of the interaction between the NCs and thus of the collective properties derived thereof. To overcome this problem, one has started to generate polymer shells with defined but

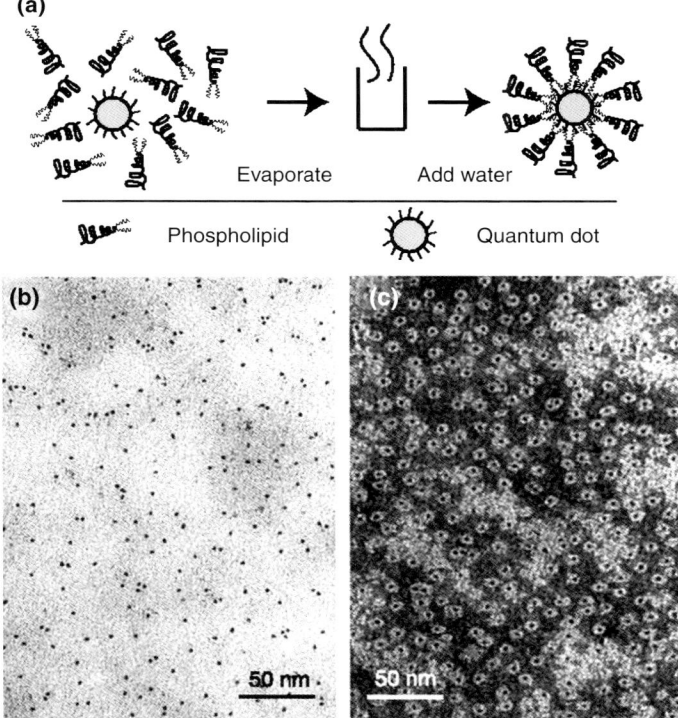

Fig. 5. NC–micelle formation and characterization. **a** Schematic of single-NC encapsulation in a phospholipid block-copolymer micelle. **b** TEM image of NC–micelles dried on a carbon-Formvar-coated 200-mesh nickel grid. Only the NCs inside the micelle core are visible and the particles appear evenly spread on the surface. Although some clusters of 2–4 NCs are visible, most of the NCs are isolated, suggesting that a majority of micelles contain a single NC. **c** TEM image of the phospholipid layer obtained by negative staining with 1% phosphotungstic acid at pH 7. With this technique, both the NC and the micelle can be visualized at the same time. The NC (dark spot) appears surrounded by a white disk of unstained phospholipids that stands out against the stained background. Reproduced with permission from [16], © 2002 Science

varied thickness either by coating NCs with polymers or growing polymer brushes from NCs. Developing a strategy of grafting polymers on semiconductor NCs was also promoted by the demand of enhancing the colloidal stability of NCs against the environmental variation because the polymer coating always give the NCs a better steric stabilization. The strategy of coating preformed NCs with polymer chains is referred to as "grafting-to", while that of growing polymer brushes from NCs as "grafting-from".

2.3.1 *Grafting-to strategy.* The simple grafting-to strategy should be encapsulation of semiconductor NCs by copolymers on the basis of the hydrophobic interaction between the polymer and the long alkyl chain capping of organic semiconductor NCs. Up to date, there exist a great deal of reports on encapsulating organic NCs into the hydrophobic cores of block-copolymer micelles, accompanied

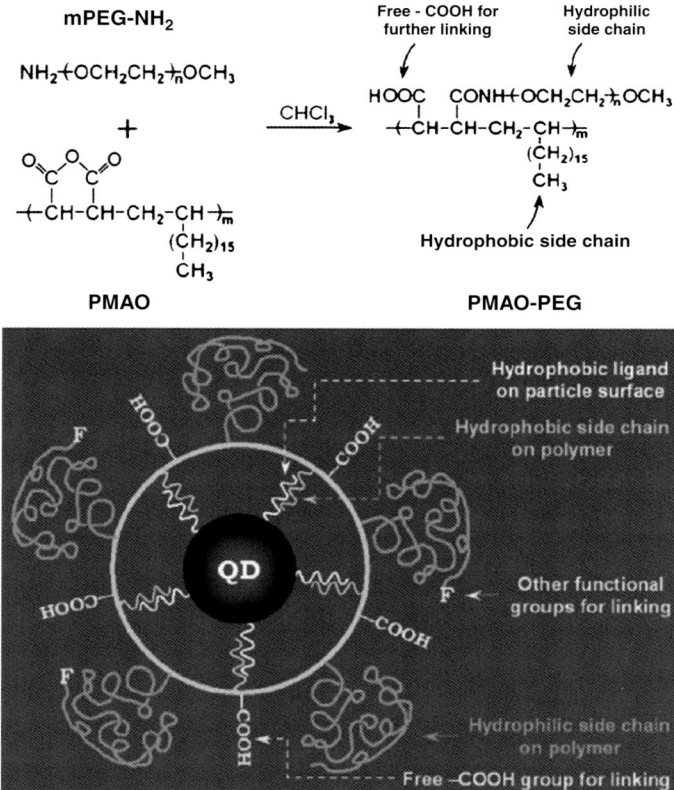

Fig. 6. Top: one-step formation of PEG-grafted amphiphilic polymers through reaction between maleic anhydride and amino groups. Bottom: schematic structure of water-soluble NCs (F stands for a functional group instead of $-OCH_3$, such as $-OH$, $-COOH$, $-NH_2$, etc.). NCs were encapsulated by PEG-grafted amphiphilic polymer hydrophobic interaction. Hydrophilic side chains of PEGs stayed exteriorly to make the whole structure soluble in water; carboxylic and F functional groups were used for bioconjugation. Reproduced with permission from [43], © 2007 ACS

by a phase transfer of organic NCs into aqueous media [16, 41, 42]. The main driving force behind this work is the promising potential of using semiconductor NCs for biomedical applications [5]. One of the simplest methods to incorporate organic NCs into block-copolymer micelles is to mix hydrophobic NCs with copolymers in a common solvent, followed by solvent evaporation.

Following this strategy, Dubertret et al. have succeeded in confining a ZnS-overcoated CdSe NC into the hydrophobic core of a micelle composed of PEG, phosphatidylethanolamine, and phosphatidylcholine (Fig. 5) [16]. Besides block copolymers, graft copolymers have also been used to coat semiconductor NCs. Colvin and coworkers have developed a series of PEG-based graft copolymers to render NCs a good biocompatibility (Fig. 6) [43]. Parak and coworkers have capped organic NCs with poly(maleic anhydride alt-1 tetradecene) and the subsequent cross-linking of the anhydride ring backbone of the polymer capping with bis(6-aminohexyl) amine has turned the NCs from hydrophobic to hydrophilic [44].

In the case of aqueous semiconductor NCs, their surfaces are usually negatively or positively charged arising from protonation or deprotonation of the capping ligands. Yang and coworkers have coated negatively charged CdTe NCs with positively charged polymers through electrostatic interactions [45, 46]. The resulting polymer-capped CdTe NCs have been used to construct various micrometer-sized patterned structures, microspheres, and bulk materials (Fig. 7) [45]. Recently negatively charged CdTe NCs have been capped with different carbazole-containing copolymers and thus a fine control of the photoluminescence behaviour of the NCs has been realized [47–49].

This encapsulation strategy has a number of advantages. The first advantage is alteration of NC solubility without damaging the original ligands capped on NC surfaces and therefore the preservation of NC photoluminescence. The second is that

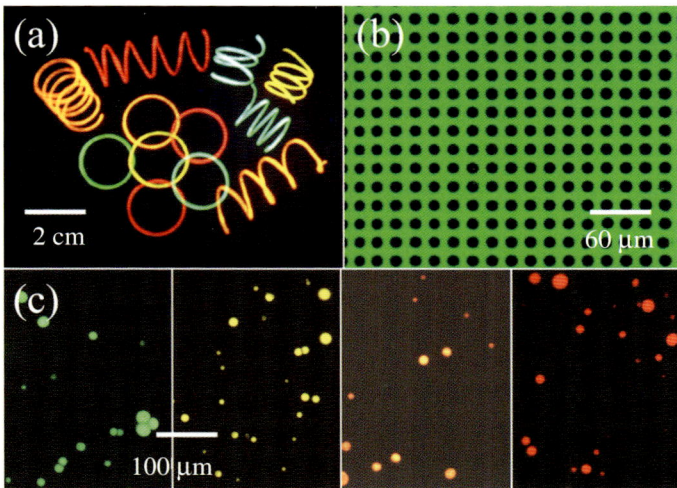

Fig. 7. Fluorescence images of CdTe NC–polymer composites. **a** In different macroscopic shapes, **b** patterned composites with green emission, and **c** microspheres with respective green, yellow, orange, and red emission. Reproduced with permission from [45], © 2005 ACS

the hydrophobic intermediate layer of block-copolymer coating provides a sufficiently strong hydrophobic barrier preventing NCs from attacks from the surrounding aqueous milieu, thus enhancing the chemical stability of NCs against, for instance, acid. The third is that one can add a number of functionalities into the block-copolymer coating without deterioration of NCs. The fourth is the flexibility to incorporate different NCs into polymer capsules, thus leading to multifunctional biological labels. The last but not least advantage is that one can use a diversity of phase-separation structures of amphiphilic copolymers to direct self-assembly of semiconductor NCs [20, 50].

The drawback of this encapsulation strategy is the limited capability of guaranteeing encapsulation of one NC within one polymer micelle. Accordingly, another grafting-to strategy – ligand exchange – has been also used to replace the original ligands capped on NCs with multifunctional polymers. The success of ligand exchange arises from the stronger coupling affinity of the polymer ligands than that of the original capping on NCs. Winnik and coworkers have used poly(dimethylaminoethyl methacrylate) (polyDMAEMA) or its functional copolymers as multidentate ligands to replace the original capping ligands of organic CdSe NCs, TOPO [51–53]. Recently, Weller's group has synthesized amino-functionalized PEG and copolymers of PEG and poly(ethylene imine) (PEI) and compared the stabilization derived from these two polymer ligands when capped on semiconductor NCs via ligand exchange [54]. The advantage of the use of PEG–PEI copolymers was observed, which arose not only from the excellent biocompatible PEG outer shell but also from the amine-rich inner shells of PEI capable for coupling with any molecules of interest, for instance cross-linking to further stabilize NCs.

Overall, the grafting-to strategy is the simplest way to cap semiconductor NCs with polymer. The well-developed macromolecular chemistry also allows easy diversification of the surface functionality of the NCs. But incubation of NCs with polymers always brings up substantial concerns about the colloidal stability of the NCs, especially during ligand exchange using multidentate polymer ligands. In this sense, grafting polymer ligands to NCs indeed requires a deliberate design of both polymer ligands and experimental condition to avoid agglomeration of NCs.

2.3.2 Grafting-from strategy. In order to minimize agglomeration of NCs during surface modification, the ideal way should be to grow polymers directly from NCs. To implement the grafting-from strategy the simple way is to cap NCs with polymerizable ligands with such as vinyl groups, following by polymerization triggered by initiators in the surrounding media. Kim and Bawendi have capped CdSe NCs with oligomeric phosphine ligands via ligand exchange and copolymerized the oligomeric phosphines with other hydrophobic monomers to form CdSe NC–polymer composites with a long-term stability without phase separation and a robust stability of photoluminescence [55]. Peng's group has succeeded in capping CdSe NCs by generation-3 dendron-thio ligands with vinyl groups at the termini. The subsequent polymerization of the vinyl group via ring-closing metathesis yielded cross-linked dendron shells on CdSe NCs, leading to a superior chemical, photochemical, and thermal stability [56]. By partially replacing the original ligands of CdSe NCs, TOPO with vinylbenzene functionalized TOPO or norbornene-thio,

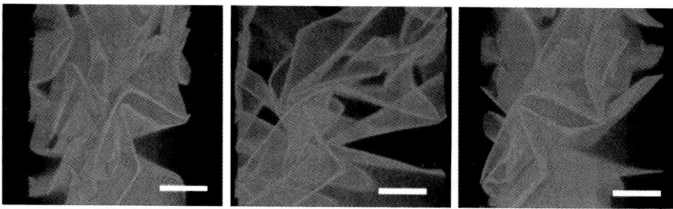

Fig. 8. Confocal microscopy image of a CdSe NC film prepared by cross-linking the associated organic ligands. Bar: 50 μm. Reproduced with permission from [57], © 2003 ACS

Emrick and coworkers have succeeded in cross-linking the NCs self-assembling at the water/oil interface, thus leaving behind freestanding films or capsules composed of close-packed NC monolayers (Fig. 8) [57, 58].

There exists another complementary strategy of grafting polymer brushes from the surfaces of NCs, that is, surface-initiated polymerization by capping initiators on NCs. An explosive growth in this field has been driven by development of living radical polymerization techniques including ring-opening metathesis polymerization (ROMP) [59], ATRP [60], nitroxide-mediated living radical polymerization (NMP) [61], reversible addition-fragmentaion chain transfer (RAFT) [62] and other new methods [64]. Before development of these new living polymerization techniques, implementation of surface-initiated polymerization on nanoparticulate inorganic materials has relied mainly on γ-ray irradiation [22]. Yang and coworkers have recently utilized γ-ray to create radicals on ZnS NCs to initiate copolymerization of DMAEDA, styrene, and divinylbenzene, forming transparent bulk composites with high loading of ZnS NCs and thus creating high-refractive-index materials (Fig. 9) [64].

Due to the absence of radicals during polymerization, ROMP, RAFT, and NMP have been successfully used to graft polymer brushes on semiconductor NCs. By partially replacing TOPO ligands on CdSe NCs by bis-(tricyclohexylphospine) benzulideneruthenium dichloride, Emrick and coworkers have successfully grafted polyolefin brushes on the NCs via ruthenium-based ROMP of for instance cyclooctane (Fig. 10) [65]. Following the similar strategy, Sill and Emrick have also grafted polystyrene brushes on CdSe NCs via surface initiated NMP [66]. In order to grow most conventional chain-growth polymers on CdSe NCs, Skaff and Emrick have

Fig. 9. Photograph of a bulk nanocomposite containing 20 wt% ME-capped ZnS (sample thickness: 4 mm). Reproduced with permission from [64], © 2006 Wiley–VCH

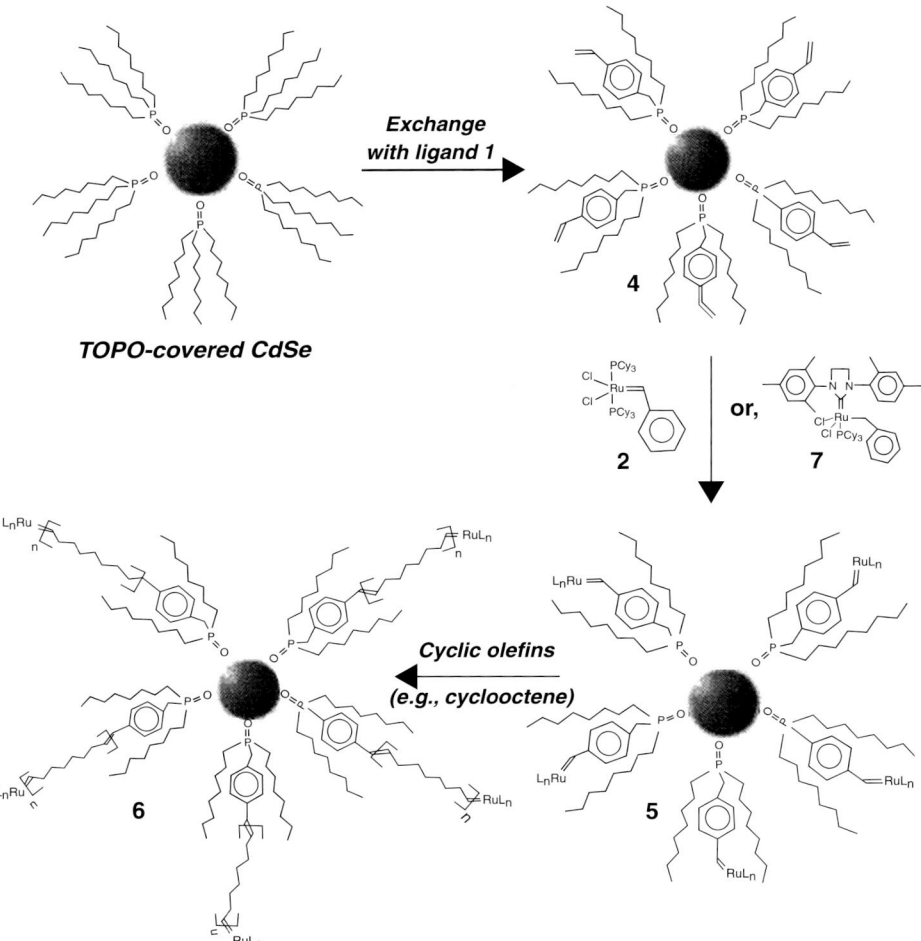

Fig. 10. Schematic depiction of grafting polyolefin brushes on CdSe NCs via surface-initiated ROMP. Reproduced with permission from [65], © 2002 ACS

synthesized new functional phosphine oxide ligands bearing trithiocarbonate moiety amenable to RAFT polymerization [67]. By capping the new ligands on CdSe NCs, surface-initiated RAFT allows grafting of various homopolymers, block copolymers, and graft copolymers on CdSe NCs such as PS, poly(butyl acrylate), poly(styrene-b-methyl acrylate), and poly(styrene-g-isoprene).

ROMP, NMP, and RAFT usually allow growth of hydrophobic polymers in organic media. In comparison, ATRP is able to synthesize both hydrophobic and hydrophilic polymers even in aqueous media; the latter bring up not only environmental values but also should be appealing for biological applications. Despite a great progress on metal and metal oxide NCs, however, surface-initiated ATRP has limited success on semiconductor NCs mainly because the ligand exchange with initiators and radicals

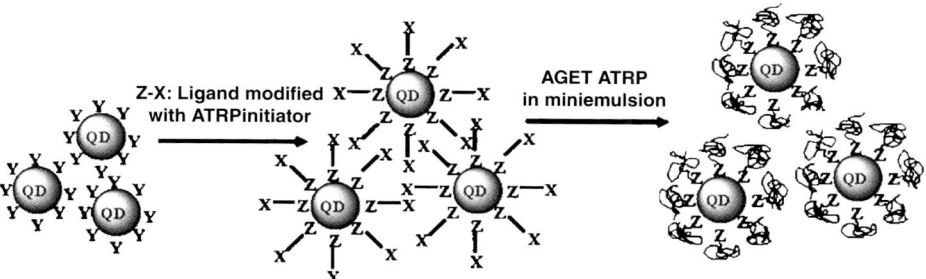

Fig. 11. Synthetic strategy for the preparation of CdS NC–polymer composites by AGET ATRP in miniemulsion. Reproduced with permission from [68], © 2007 Wiley–VCH

generated during polymerization may damage the surface structures of semiconductor NCs and thus lower the photoluminescence efficiency.

Most recently, Barros-Timmons and co-workers have employed a newly developed ATRP technique – activator generated by electron transfer (AGET) ATRP – to grow polymer brushes on semiconductor NCs [68]. They have synthesized tris (hydroxypropyl)phosphine and capped them on CdS NCs via ligand exchange. The following surface-initiated AGET ATRP on the NCs in miniemulsion led to aqueous suspension of CdS NCs coated with poly(n-butyl acrylate) shells (Fig. 11). The major advantage of surface-initiated AGET ATRP is that in this polymerization system, a reducing agent activates the Cu^{II} catalyst by electron transfer and thus avoids quenching of the photoluminescence of CdS NCs by free radicals.

The common drawback for all grafting-from strategies via surface-initiated living polymerization arises from the maximal grafting density of the polymer brushes obtained on NCs. This maximal grafting density forces the polymer brushes to adopt a stretching conformation, so the physicochemical properties of the polymer brushes grafted on semiconductor NCs should be envisioned rather different from those of their corresponding free polymer chains. For instance the stimuli-response behavior may be dramatically suppressed [69]. Thereby, optimization of the grafting density of polymer brushes on NCs should be paid more attention in future work on using grafting-from strategy to cap semiconductor NCs with polymer brushes.

3. Physical confinement of semiconductor NCs in polymers

Although the aforementioned chemical synthesis strategies allow elaboration of the structures of semiconductor NC–polymer composites and their chemical and physical functionality, however, incorporation of as-prepared NCs into polymer matrices should be the most straightforward way especially for the large-scale industrial use. In this case, obviously, the key requirement for blending semiconductor NCs with polymers is the compatibility between these two components. As mentioned above, semiconductor NCs derived from pyrolysis of organometallic precursors are capped by hydrophobic alkyl chains, which imparts them with a good compatibility with a certain number of conventional hydrophobic polymers. But enforcement of the structural stability of semiconductor NC–polymer composites may require strong interactions between NCs and the polymer host matrices, such as electrostatic interaction and even covalent bonding. This section is devoted to summarize the current progress in this field.

3.1 Blending of NCs and polymers.

The use of semiconductor NCs as a key component to high performance of LEDs was the original driving force of forming semiconductor NC–polymer composites. Bawendi and coworkers have succeeded in dispersing CdSe NCs into a chloroform solution of the mixture of polyvinylcarbozole, a hole conducting polymer, and an oxadiazole derivative, an electron transport species [70]. By spin-coating, they constructed thin films with tunable electroluminescence and photoluminescence from 530 to 650 nm by varying the NC size; the smaller NCs showed a higher threshold voltage of electroluminescence than the larger ones. Banin and coworkers have mixed core-shell InAs–ZnSe NCs and conjugated polymers, poly[2-methoxy-5-(2-ethylhexyloxy)-1,4-phenyl enevinylene] (MEH-PPV) or poly [(9,9-dihexylfluorenyl-2,7-diyl)-co-(1,4-{benzo-[2,19,3]thiadiazole})] and created thin films via spin–coating. LEDs derived from these composite thin films have a tunable emission depending on the sizes of the NCs and the molecular weight of the conjugated polymers [71]. The good compatibility of the alkyl capping of the NCs with the polymer matrices endorsed a homogeneous mixing and thus a good transferring of electron across the NC/polymer interface.

The removal of the surface-capping ligands of semiconductor NCs usually leads to a phase separation of the NCs from the polymer matrices. Greenham and coworkers have found that the removal of the TOPO capping caused the phase segregation of CdSe or CdS NCs from MEH-PPV with the length scale in the range of

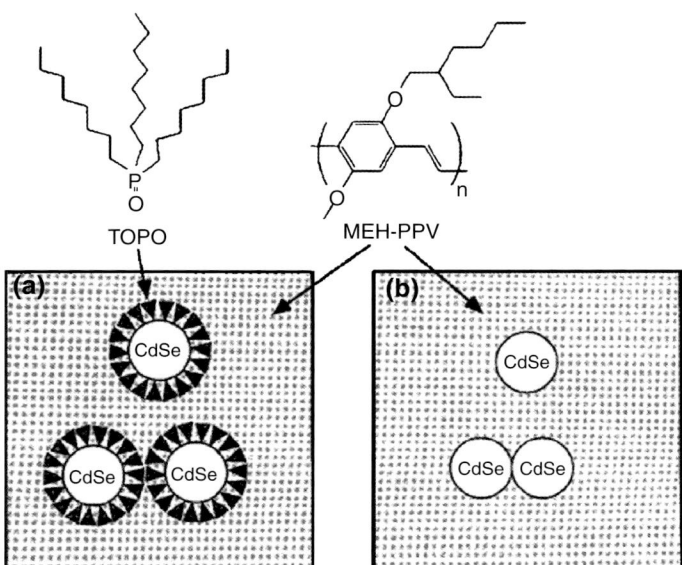

Fig. 12. Schematic diagram of MEH-PPV/NC composites, showing the chemical structure of MEH-PPV and TOPO. **a** CdSe NCs with surfaces coated by TOPO. **b** CdSe NCs with naked surfaces. Reproduced with permission from [72], © 1996 APS

20–200 nm, creating a large interface for charge separation and thus providing pathways for electron and holes travelling to the appropriate electrode without recombination (Fig. 12) [72]. Besides, the removal of the TOPO capping also enhanced the charge separation and charge transport between neighboring NCs and quenched the photoluminescence of the MEH-PPV. As a result, semiconductor NC/MEH-PPV composites showed significantly improved quantum efficiencies in comparison with those derived from pure polymers; the quantum efficiency was as high as 12%.

In most cases, however, the phase separation of semiconductor NCs from polymer matrices has to be minimized in order to form homogeneous and especially transparent composites for optical applications. To address this issue, two major ways have intensively used: one is to coat NCs with polymer shells either via grafting-to or via grafting-from and another to introduce a strong interaction between NCs and polymers.

3.2 Layer-by-layer assembly. Based on the strong interaction between NCs and polymers, alternatively depositing them on a substrate has paved a controlled way to tailor the chemical composition, thickness, and the vertical structure of NC–polymer composite films with a nanometer precision [73]. The interaction commonly used for LbL growing semiconductor NC–polymer composites is electrostatic interaction. The LbL strategy can work well on different substrates, such as silica, glass, and wood, with different curvatures.

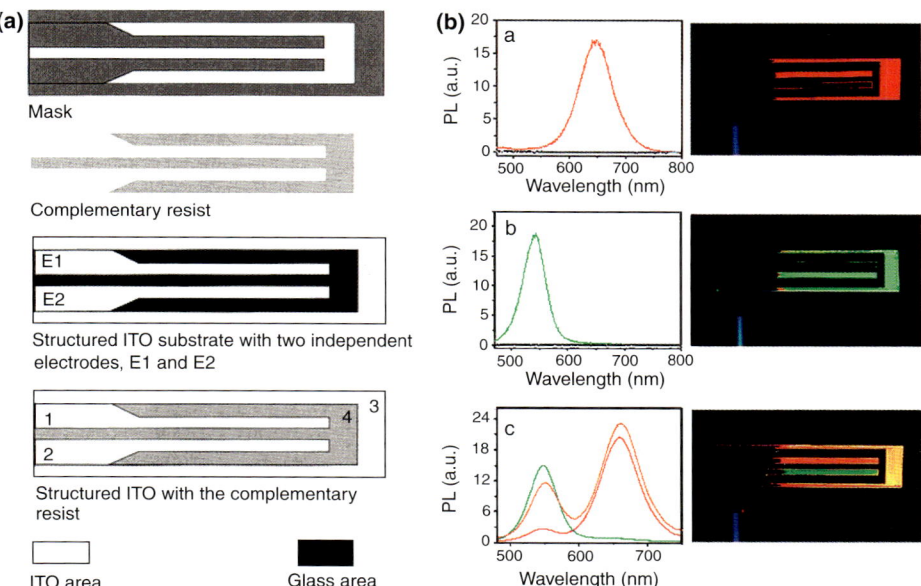

Fig. 13. a Lateral structures on patterned ITO substrates. **b** Photographs of the lateral structures formed by LbL deposition of green- and red-emission CdTe NCs and PDDA. Reproduced with permission from [77], © 2002 ACS

3.2.1 *LbL assembly on planar substrates.* As mentioned above, aqueous synthesized NCs, stabilized by mercapto-ligands, have negatively or positively charged surfaces, depending on the nature of the capping thio-ligands. Differently sized aqueous semiconductor NCs have been LbL-hybridized with polyelectrolytes to form NC–polymer composite films [74, 75]. Yang and coworkers have successfully implemented LbL assembly of aqueous CdTe NCs with poorly water-soluble carbazole-containing polymers by using poly(acrylic acid) as bridging layers. The photoluminescence properties of the resulting polymers even on curved substrates in particular and the function of films were tuned by the ratio of the NCs to carbazole units and the PAA concentration [76]. Gao and coworkers have demonstrated a flexible way of using the external electric field to manipulate the electrostatic interaction between CdTe NCs and poly(diallyldimethylammonium chloride) (PDDA) and thus the deposition rate and amount of the NCs [77]. Using patterned electrodes to precisely alter the intensity and location of the electric field, they also realized a selective deposition of NCs on tailor-designed areas, thus creating photoluminescence patterns on substrates (Fig. 13).

Of significance is that the LbL strategy allows one to stepwise deposit differently sized NCs in subsequent layers, thus creating a graded gap structure [78]. Differently sized CdTe NCs, stabilized with TGA, have been alternatively assembled with

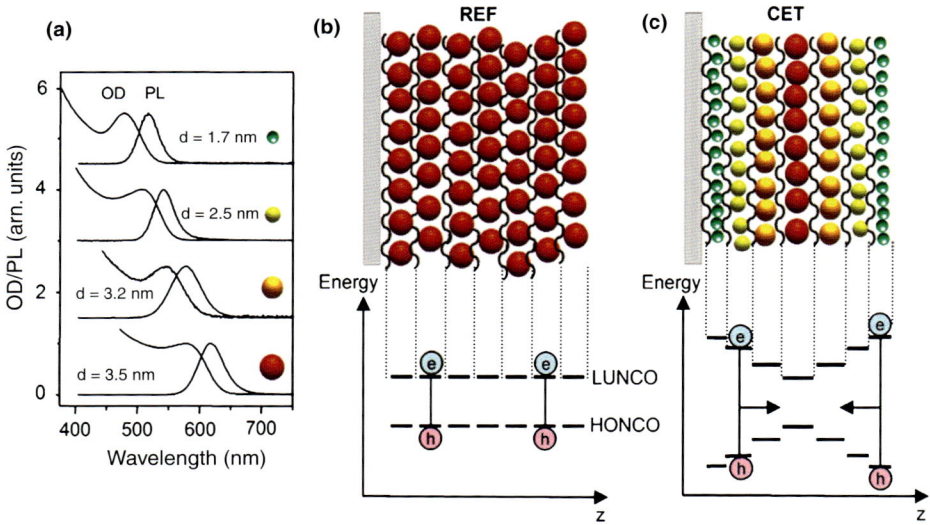

Fig. 14. a Absorbance (optical density, OD) and photoluminescence spectra of the CdTe NCs of four different sizes used to prepare the LbL samples. The sizes are chosen such that the emission spectrally matches the excitonic absorption of subsequently larger NCs. All spectra are taken in aqueous solution and vertically shifted for clarity. **b** Schematic sketch of the REF sample which consists of seven layers of red-emitting, 3.5-nm large CdTe NCs. Below the REF sample the energetic positions of the highest occupied nanocrystal orbital (HONCO) and the lowest unoccupied NC orbital (LUNCO) are sketched. Each excitation remains in the layer where it is created because of the absence of a gap gradient. **c** Sample CET consists of subsequent layers comprising green, yellow, orange, red, orange, yellow, and green emitting NCs. Below the CET (cascaded energy transfer) sample the HONCO and LUNCO are sketched, visualizing the cascaded band gaps used to facilitate cascaded energy transfer. Reproduced with permission from [79], © 2004 ACS

PDDA into composite multilayer films while altering the NC sizes in subsequent layers. Multilayer films comprising green, yellow, orange, red, orange, yellow, and green emitting CdTe NCs showed a cascaded band gaps and thus a cascade energy transfer (Fig. 14) [79]. The exciton density in the center layer composed of red-emitting NCs can be enhanced by a factor of 28.

In addition of electrostatic interaction, hydrogen bonding has also been used for LbL growth of semiconductor NC–polymer composite multilayer films. Hao and Lian have used hydrogen bonding between the capping ligands, 4-mercaptobenzoic acid of CdSe NCs and poly(vinylpyrindine) (PVP) for LbL assembly of the NCs and PVP into composite multilayer films in ethanol [80]. Semiconductor NC–polyelectrolyte composite thin films derived from LbL assembly based on non-covalent interactions are not robust against environment variation. In order to achieve a better performance for practical applications, further covalent cross-linking is needed. Yang et al. have LbL assembled TGA-stabilized CdTe NCs and nitro-containing diazoresin (NDR) into composite multilayer films via electrostatic interactions, and subsequently turned ionic bonds between the negatively charged NCs and the positively charged polymers into covalent bonds of ester via ultraviolet irradiation (Fig. 15) [81].

3.2.2 LbL assembly on curved substrates.

The most prominent development in LbL self-assembly is the extension from planar substrates to the surfaces of microparticles [82–85]. According to a specific requirement, the size, shape, and chemical composition of microparticle templates can be flexibly selected. Their size can range from a few nanometers to tens of micrometers and they can be latex spheres, silica particles, enzyme crystals, and so on. Similar to that on planar substrates, semiconductor NC/polyelectrolyte composite films can be easily LbL grown on microparticles. In order to form uniform NC layers, however, polyelectrolyte trilayers are required as primer layers sandwiched between the NC layers [84]. LbL

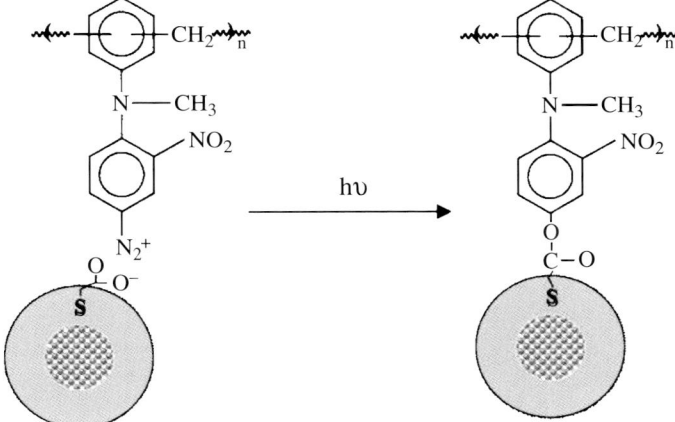

Fig. 15. Photoreaction of NDR- and TGA-stabilized CdTe NCs in a self-assembled film. Reproduced with permission from [81], © 2002 Elsevier

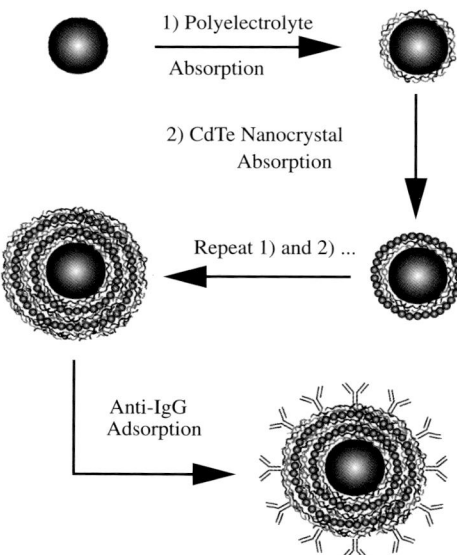

Fig. 16. Schematic illustration of the procedure used to prepare CdTe NCs–microsphere bioconjugations via LbL assembly. Reproduced with permission from [86], © 2002 ACS

growth of semiconductor NCs and polymer composite films on microparticles hold immense promise in biomedical applications. The composite films of CdTe NCs and polyelectrolytes capped on microparticles via LbL electrostatic deposition can endow the resulting composite microparticles with not only tunable photoluminescence functionality depending on the NC size, but also specific coupling affinity determined by the outmost layer coating, thus leading to new fluorescent markers for biological labeling (Fig. 16) [86]. LbL assembly also allows incorporation of different NCs on one microparticle, such as combination of luminescent CdTe NCs and magnetic Fe_3O_4 NCs using polyelectrolytes as glue, thus creating multifunctional biological markers.

On the other hand, micrometer-sized particles are comparable with the wavelength of light, their self-assemblies with highly ordered packing, so-called colloidal crystals, have demonstrated a possibility to manipulate the propagation of light [87, 88]. Caruso and coworkers have fabricated LbL composite films of CdTe NCs and polyelectrolytes on latex microspheres and studied the optical properties of the colloidal crystals derived from the coated microparticles [89, 90]. LbL growth of semiconductor NC/polyelectrolyte multilayers on microspheres has also been used to manipulate the refractive index and thus the optical properties of the colloidal crystals comprising the coated spheres and the inverse opals derived thereof [91, 92].

Kotov's group has recently grown LbL composite films of CdTe NCs and polyelectrolytes on highly curved surfaces, including ropes and tubes [93, 94]. In the case of tubes, both exterior and interior walls of the tubes can be easily and uniformly coated by CdTe NC–polyelectrolyte composite films [94]. Besides, two unanticipated effects were observed: (1) gradual red shift of the luminescence spectrum as the layer thickness increases; (2) two-stage luminescence transient including the significant enhancement of the luminescence during first 100 min of UV illumination.

3.3 Using microspheres to encapsulate semiconductor NCs.
Tailoring composites of semiconductor NCs and polymers into particles with sizes from submicrometers to micrometers is rather demanded by technical applications especially in biomedical labeling and assays. Creating microparticles of semiconductor NC–polymer composites usually relies on suspension, emulsion, miniemulsion polymerization [40], or emulsification in the presence of NCs [95]. But the radicals generated during polymerization are always problematic. Besides, control of polydispersity of the resulting composite microspheres is usually not easy. In order to circumvent these problems, a new strategy – using microspheres to encapsulate semiconductor NCs – has been developed [96–100]. Nie et al. have developed a simple way to fabricate semiconductor NC–polymer composite microspheres by swelling cross-linked PS beads in chloroform/isopropanol suspensions of organic ZnS–overcoated CdSe NCs (Fig. 17) [96]. Since the polarity of the mixture of chloroform and isopropanol (95/5, vol/vol) was stronger than that of the polymer, hydrophobic NCs were spontaneously imbibed by PS beads due to hydrophobic–hydrophobic interaction.

By mimicking the concept of using hydrogel microspheres to deliver proteins [101], the author's group has succeeded on loading aqueous semiconductor NCs into hydrogel microspheres in a controlled fashion [97–99]. The incubation of aqueous semiconductor NCs with stimuli-responsive hydrogel microspheres such as poly(N-isopropylacrylamide) (PNIPAM) at the swollen state, followed by altering the

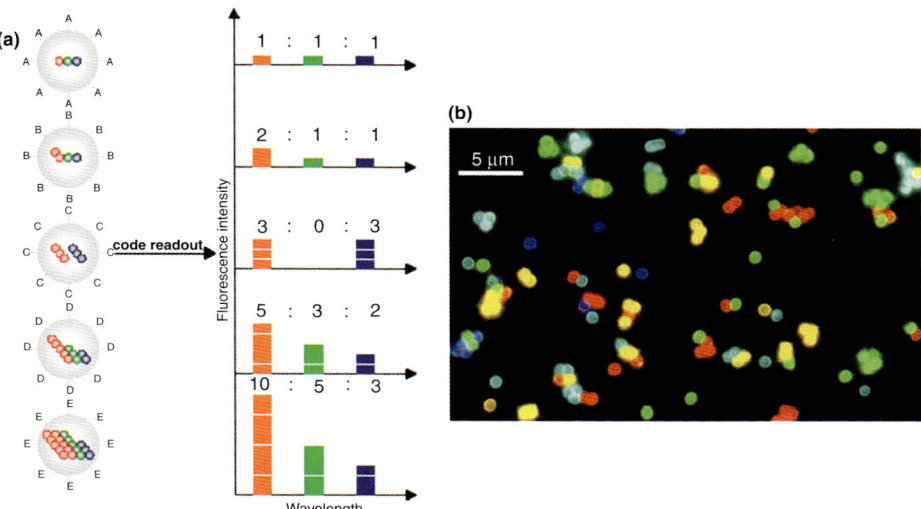

Fig. 17. **a** Schematic illustration of optical coding based on wavelength and intensity multiplexing. **b** Fluorescence micrograph of a mixture of CdSe/ZnS NC-tagged beads. Reproduced with permission from [96], © 2001 Nature

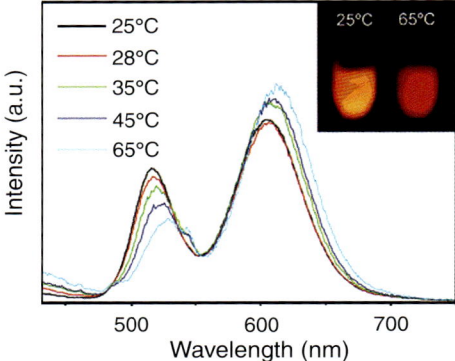

Fig. 18. Fluorescence spectra of PNIPAM spheres loaded with both 2.5 and 3.5 nm CdTe NCs against environmental temperature. All spectra presented were normalized by taking the temperature effects into account since both emissions from 2.5 to 3.5 nm CdTe NCs decrease with increasing temperature, but by a rather similar factor. The CdTe–PNIPAM spheres were obtained by incubating NCs and gel spheres at 45°C. The molar ratio of 2.5–3.5 nm NCs is 5:1. The loading amount of NCs is around 4.0×10^4 NCs (2.5 + 3.5 nm) per gel sphere. The inset shows the fluorescent pictures of these spheres under UV irradiation at 25 and 65°C, respectively. Reproduced with permission from [98], © 2005 ACS

environmental condition to shrink the gel, the NCs were firmly confined within the collapsed gel network of hydrogel microspheres. The driving force for loading aqueous NCs into hydrogel microspheres and confining them inside was mainly the electrostatic interaction between the NC surface and the gel network. In this case, hydrogel host microspheres not only provide semiconductor NCs an excellent biocompatibility but also allow the separation distance between the NCs and thus the interaction of neighboring NCs of being tunable in response to the external stimuli such as temperature (Fig. 18) [98]. Incorporation of semiconductor NCs into hydrogel allows not only formation of composite microspheres but also generation of other shaped nanostructures. Rogach and coworkers have incorporated TGA-stabilized CdTe NCs into polyvinyl alcohol and used inkjet printing to create photoluminescence patterns [102].

Besides the use of solid microspheres to encapsulate semiconductor NCs, hollow capsules have also been employed to do so. Rogach and coworkers have succeeded in loading aqueous CdTe NCs into polyelectrolyte capsules. The electrostatic attraction between CdTe NCs and the oppositely charged interiors of the capsules moved the NCs and trapped them inside the capsules [103]. Since the success of this approach was determined by the surface charge of semiconductor NCs, differently sized CdTe NCs can easily be encapsulated into one polyelectrolyte capsule (Fig. 19).

The major advantage of using preformed solid or hollow microparticles to encapsulate as-prepared semiconductor NCs is the capability of confining differently sized NCs into one microparticle. Since this encapsulation procedure has little influence on the photoluminescence behavior of semiconductor NCs and the optimization of the molar ratio between differently sized NCs can suppress the energy transfer between neighboring NCs, one can easily realize multiplexed optical coding. In principle, using a series of NCs with five emission colors, million color

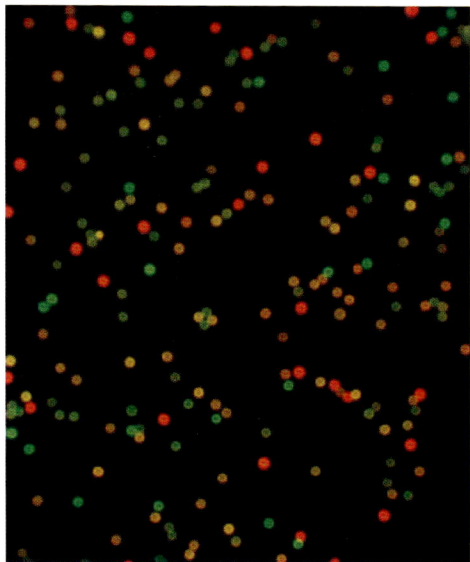

Fig. 19. Fluorescence image of a mixture of polymer microspheres loaded with CdTe NCs of different sizes. Reproduced with permission from [103], © 2002 Wiley–VCH

combinations can be established, which should hold immense promise in biological labeling and assay.

4. Outlook

Up to date, a great deal of methods has been successfully developed to create semiconductor NC–polymer composites. The integration of semiconductor NCs with various polymers can impart them with numbers of tailor-designed new properties derived from the polymers, including robust colloidal stability, robust chemical and photochemical stability, good compatibility, excellent processability, and immense flexibility to conjugate with a diversity of functional molecules to achieve multiple functionality. Once incorporating semiconductor NCs with polymers, we can initiate, develop, and even commercialize technical applications based on the unique size-dependent photoluminescence properties of NCs.

Nonetheless, the current work on fabrication of semiconductor NC–polymer composites mainly focus on using polymers to improve the properties of individual NCs or using NCs to improve the performance of polymers. Little attention is paid to manipulate the interplay between polymer function and semiconductor NC function with the intent of creating new collective functionality of NC ensembles. Most recently, a paradigmatic shift from development of new strategies to form semiconductor NC–polymer composites to use self-assembly behavior of polymers to direct self-assembly of NCs and control the NC interaction and thus their collective properties is rather obvious [20, 50]. The future intelligent devices should be based on intelligent self-assembly of various nanometer-sized building blocks [104, 105]. Extending enormously rich phase-separation behavior and stimuli-response behavior of polymers to supervise self-assembly of semiconductor NCs should be a promising way to make NC ensemble intelligent. One may be able to manipulate the interaction between NCs by tuning their separation spacing and the symmetry of their spatial association in a tailor-designed manner.

Abbreviations

AGET	Activator generated by electron transfer
AIBN	2,2′-azobisisobutyronitrile
ATRP	atom transfer radical polymerization
DMAEMA	dimethylaminoethyl methacrylate
LED	light-emitting diode
ME	mercaptoethanol
MEH-PPV	poly[2-methoxy-5-(2-ethylhexyloxy)-1,4-phenyl enevinylene]
MPA	3-mercaptopropionic acid
NC	nanocrystal
NMP	nitroxide-mediated living radical polymerization
NRD	nitro-containing diazoresin
OVDAC	octadecyl-p-vinyl-benzyl dimethylammonium chloride
PEG	polyethylene glycol
PEI	poly(ethylene imine)
PLMA	polylaurylmethacrylate
PNIPAM	poly(N-isopropylacrylamide)
ProDOT-CA	10-((3-methyl-3,4-dihydro-2H-thieno[3,4-b][1,4]dioxepin-3-yl)-methoxy)-10-oxo-decanoic acid
PS	polystyrene
PVP	poly(vinylpyrindine)
RAFT	reversible addition-fragmentation chain transfer
ROMP	ring opening metathesis polymerization
TGA	thioglycolic acid
TOPO	trioctylphosphine oxide

References

[1] Weller H (1993) Colloidal semiconductor Q-particles: chemistry in the transition region between solid state and molecules. Angew Chem Int Ed 32: 41–53
[2] Alivisatos AP (1996) Semiconductor clusters, nanocrystals, and quantum dots. Science 271: 933–937
[3] Kagan CR, Murray CB, Bawendi MG (1996) Lang-range resonance transfer of electronic excitations in close-packed CdSe quantum-dot solids. Phys Rev B 54: 8633–8643
[4] Nirmal M, Brus L (1999) Luminescence photophysics in semiconductor nanocrystals. Acc Chem Res 32: 407–414
[5] Medintz I, Uyeda T, Goldman E et al (2005) Quantum dot bioconjugates for imaging, labeling and sensing. Nat Mater 4: 435–446
[6] Rogach A, Franzl T, Klar T et al (2007) Aqueous synthesis of thiol-capped CdTe nanocrystals: state of the art. J Phys Chem C 111: 14628–14637
[7] Masala O, Seshadri R (2004) Synthesis routes for large volumes of nanoparticles. Ann Rev Mater Res 34: 41–81
[8] Dahl JA, Maddux BLS, Hutchison JE (2007) Toward greener nanosynthesis. Chem Rev 107: 2228–2269
[9] Kumar S, Nann T (2006) Shape-control of II–VI semiconductor nanomaterils. Small 2: 316–329
[10] Gerion D, Pinaud F, Williams S et al (2001) Synthesis and properties of biocompatible water-soluble silica-coated CdSe/ZnS semiconductor quantum dots. J Phys Chem B 105: 8861–8871
[11] Rogach A, Negesha D, Ostrander J et al (2000) "Raisin bun"-type composite spheres of silica and semiconductor nanocrystals. Chem Mater 12: 2676–2685

[12] Nann T, Mulvaney P (2004) Single quantum dots in spherical silica particles. Angew Chem Int Ed 43: 5393–5396
[13] Yang Y, Gao M (2005) Preparation of fluorescent SiO_2 particle with single CdTe nanocrystal core by the reverse microemulsion method. Adv Mater 17: 2354–2357
[14] Godovsky D (2000) Device application of polymer-nanocomposites. Adv Polym Sci 153: 163–205
[15] Abouraddy AF, Bayindir M, Benoit G et al (2007) Towards multimaterials multifunctional fibres that see, hear, sense and communicate. Nat Mater 6: 336–347
[16] Dubertret BD, Skourides P, Norris DJ et al (2002) In vivo imaging of quantum dots encapsulated in phospholipid micelles. Science 298: 1759–1762
[17] Wu X, Liu H, Liu J et al (2003) Immunoflurescent labeling of cancer marker Her2 and other cellular targets with semiconductor quantum dots. Nat Biotechnol 21: 41–46
[18] Nayak S, Lyon A (2005) Soft nanotechnology with soft nanoparticles. Angew Chem Int Ed 44: 7686–7708
[19] Hamley IW (2003) Nanotechnology with soft materials. Angew Chem Int Ed 42: 1692–1712
[20] Lin Y, Boeker A, He H et al (2005) Self-directed self-assembly of nanoparticle/copolymer mixtures. Nature 434: 55–59
[21] Kuang M, Wang D, Möhwald H (2005) Fabrication of thermoresponsive plasmonic microspheres with long-term stability from hydrogen spheres. Adv Funct Mater 15: 1611–1616
[22] Hofman-Caris CHM (1994) Polymers at the surface of oxide nanoparticles. New J Chem 18: 1087–1096
[23] Wang J, Chen W, Liu A et al (2002) Controlled fabrication of cross-linked nanoparticles/polymer composite thin films through the combined use of surface-initiated atom transfer radical polymerization and gas/solid reaction. J Am Chem Soc 124: 13358–13359
[24] Zhang J, Xu S, Kumacheva E (2004) Polymer microgels: reactors for semiconductor, metal, and magnetic nanoparticles. J Am Chem Soc 126: 7908–7914
[25] Peng X, Manna L, Yang W et al (2000) Shape control of CdSe nanocrystals. Nature 404: 59–61
[26] Puntes V, Krishnan K, Alivisatos AP (2001) Colloidal nanocrystal shape and size control: the case of cobalt. Science 291: 2115–2117
[27] Qu L, Peng Z, Peng X (2001) Alternative routes toward high quality CdSe nanocrystals. Nano Lett 1: 333–337
[28] Cao Y, Banin U (2000) Growth and properties of semiconductor core/shell nanocrystals with InAs cores. J Am Chem Soc 122: 9692–9702
[29] Gaponik N, Talapin D, Rogach A et al (2002) Thiol-capping CdTe nanocrystals: an alternative to organometallic synthesis routes. J Phys Chem B 106: 7177–7185
[30] Zhang H, Wang D, Yang B et al (2006) Manipulation of aqueous growth of CdTe nanocrystals to fabricate colloidally stable one-dimensional nanostructures. J Am Chem Soc 128: 10171–10180
[31] Zhang H, Wang D, Möhwald H (2006) Ligand-selective aqueous synthesis of one-dimensional CdTe nanostructures. Angew Chem Int Ed 45: 748–751
[32] Bao H, Gong Y, Li Z et al (2004) Enhancement effect of illumination on the photoluminescence of water-soluble CdTe nanocrystals: towards highly fluorescent CdTe/CdS core-shell structure. Chem Mater 16: 3853–3859
[33] Sastry M (2003) Phase transfer protocols in nanoparticle synthesis. Curr Sci 85: 1735–1745
[34] Fogg D, Radzilowski L, Dabbousi B et al (1997) Fabrication of quantum dot-polymer composites: semiconductor nanoclusters in dual-function polymer matrices with electron-transporting and cluster-passivating properties. Macromolecules 26: 8433–8439
[35] Fogg D, Radzilowski L, Blanski R et al (1997) Fabrication of quantum dot/polymer composites: phosphine-functionalized block copolymers as passivating hosts for cadmium selenide nanoclusters. Macromolecules 30: 417–426
[36] Lee J, Sundar V, Heine J et al (2000) Full color emission from II–VI semiconductor quantum dot–polymer composite. Adv Mater 12: 1102–1105
[37] Zhang H, Cui Z, Wang Y et al (2003) From water-soluble CdTe nanocrystals to fluorescent nanocrystal–polymer transparent composites using polymerizable surfactants. Adv Mater 15: 777–780
[38] Sudeep P, Emrick T (2007) Polymer–nanoparticle composites: preparative methods and electronically active materials. Polym Rev 47: 155–163
[39] Shallcross C, D'Ambruoso G, Korth B et al (2007) Poly(3,4-ethylene dioxythiophene)–semiconductor nanoparticle composite thin films tethered to indium tin oxide substrates via electropolymerization. J Am Chem Soc 129: 11310–11311
[40] Yang Y, Wen Z, Dong Y et al (2006) Incorporating CdTe nanocrystals into polystyrene microspheres: towards robust fluorescent beads. Small 2: 898–901

[41] Kim B, Taton A (2007) Multicomponent nanoparticles via self-assembly with cross-linked block copolymer surfactants. Langmuir 23: 2198–2202
[42] Duxin N, Liu F, Vali H et al (2005) Cadmium sulphide quantum dots in morphologically tunable triblock copolymer aggregates. J Am Chem Soc 127: 10063–10069
[43] Yu W, Chang E, Falkner J et al (2007) Forming biocompatible and noaggregated nanocrystals in water using amphiphlic polymers. J Am Chem Soc 129: 2871–2879
[44] Pellegrino T, Manna L, Kudera S et al (2004) Hydrophobic nanocrystals coated with an amphiphilic polymer shell: a general route to water soluble nanocrystals. Nano Lett 4: 703–707
[45] Zhang H, Wang C, Li M et al (2005) Fluorescent nanocrystal–polymer composites from aqueous nanocrystals: methods without ligand exchange. Chem Mater 17: 4783–4788
[46] Zhang H, Wang C, Li M et al (2005) Fluorescent nanocrystal–polymer complexes with flexible processability. Adv Mater 17: 853–857
[47] Sun H, Zhang J, Zhang H et al (2006) Preparation of carbozole-containing amphiphilic copolymers: an efficient method for the incorporation of functional nanocrystals. Macromol Mater Eng 291: 929–936
[48] Sun H, Zhang J, Zhang H et al (2006) Pure white-light emission of nanocrystal–polymer composites. ChemPhysChem 7: 2492–2496
[49] Qi X, Pu K, Fang C et al (2007) Semiconductor nanocomposites of emissive flexible random copolymers and CdTe nanocrystals: preparation, characterization, and optoelectronic properties. Macromol Chem Phys 208: 2007–2017
[50] Haryono A, Binder W (2006) Controlled arrangement of nanoparticle arrays in block-copolymer domains. Small 2: 600–611
[51] Wang X, Dykstra T, Salvador M et al (2004) Surface passivation of luminescent colloidal quantum dots with poly(dimethylaminoethyl methacrylate) through a ligand exchange process. J Am Chem Soc 126: 7784–8785
[52] Wang X, Oh J, Dykstra T et al (2006) Surface modification of CdSe and CdSe/ZnS semiconductor nanocrystals with poly(N-dimethylaminoethyl methacrylate). Macromolecules 39: 3664–3672
[53] Wang M, Felorzabihi N, Guerin G et al (2007) Water-soluble CdSe quantum dots passivated by a multidentate diblock copolymer. Macromolecules 40: 6377–6384
[54] Nikolic M, Krack M, Aleksandrovic V et al (2006) Tailor-made ligands for biocompatible nanoparticles. Angew Chem Int Ed 45: 6577–6580
[55] Kim S, Bawendi MG (2003) Oligometric ligands for luminescent and stable nanocrystal quantum dots. J Am Chem Soc 125: 14652–14653
[56] Guo W, Li J, Wang Y et al (2003) Luminescent CdSe/CdS core/shell nanocrystals in dendron boxes: superior chemical, photochemical and thermal stability. J Am Chem Soc 125: 3901–3909
[57] Lin Y, Skaff H, Böker A et al (2003) Ultrathin cross-linked nanoparticle membranes. J Am Chem Soc 125: 12690–12691
[58] Skaff H, Lin Y, Rangirala R et al (2005) Crosslinked capsules of quantum dots by interfacial assembly and ligand crosslinking. Adv Mater 17: 2082–2086
[59] Franzel U, Nuyken O (2002) Ruthenium-based metathesis initiators: development and use in ring-opening metathesis polymerization. J Polym Sci A: Polym Chem 40: 2895–2916
[60] Matyjaszewski K, Xia J (2001) Atom transfer radical polymerization. Chem Rev 101: 2921–2990
[61] Hawker C, Bosman A, Harth E (2001) New polymer synthesis by nitroxide mediated living radical polymerization. Chem Rev 101: 3661–3668
[62] Moad G, Rizzardo E, Thang S (2005) Living radical polymerization by the RAFT process. Aust J Chem 58: 379–410
[63] Braunecker W, Matyjaszewski K (2007) Controlled/living radical polymerization: feature, developments, and perspectives. Prog Polym Sci 32: 93–146
[64] Lu C, Cheng Y, Liu Y et al (2006) A facile route to ZnS-polymer nanocomposite optical materials with high nanophase content via γ-ray irradiation initiated bulk polymerization. Adv Mater 18: 1188–1192
[65] Skaff H, Ilker MF, Coughlin EB et al (2002) Preparation of cadmium selenide–polyolefin composites from functional phosphine oxides and ruthenium-based metathesis. J Am Chem Soc 124: 5729–5733
[66] Sill K, Emrick T (2004) Nitroxide-mediated radical polymerization from CdSe nanoparticles. Chem Mater 16: 1240–1243
[67] Skaff H, Emrick T (2004) Reversible addition fragmentation chain transfer (RAFT) polymerization fro unprotected cadmium selenide nanoparticles. Angew Chem Int Ed 43: 5383–5386
[68] Esteves A, Bombalski L, Trindade T et al (2007) Polymer grafting from CdS quantum dots via AGET ATRP in miniemulsion. Small 7: 1230–1236

[69] Edwards E, Chanana M, Wang D et al (2008) Stimuli-responsive resersible transport of nanoparticles across water/oil interfaces. Angew Chem Int Ed 47: 320–323
[70] Dabbousi B, Bawendi M, Onitsuka O et al (1995) Electroluminescence from CdSe quantum dot/polymer composites. Appl Phys Lett 66: 1316–1318
[71] Tessler N, Medvedev V, Kazes M et al (2002) Efficient near-infrared polymer nanocrystal light-emitting diodes. Science 295: 1506–1508
[72] Greenham NC, Peng X, Alivisatos AP (1996) Charge separation and transport in conjugated-polymer/semiconductor-nanocrystal composites studied by photoluminescence quenching and photoconductivity. Phys Rev B 54: 17628–17637
[73] Decher G (1997) Fuzzy nanoassemblies: toward layered polymeric multicomposites. Science 277: 1232–1237
[74] Gao M, Richter B, Kirstein S (1997) White light electroluminescence from self-assembled Q-CdSe/PPV multilayer structures. Adv Mater 9: 802–805
[75] Rogach A, Koktysh D, Harrison M et al (2000) Layer-by-layer assembled films of HgTe nanocrystals with strong infrared emission. Chem Mater 12: 1526–1528
[76] Zhang H, Zhou Z, Liu K et al (2003) Controlled assembly of fluorescent multilayers from an aqueous solution of CdTe nanocrystals and nonionic carbazole-containing copolymers. J Mater Chem 13: 1356–1361
[77] Gao M, Sun J, Dulkeith E et al (2002) Lateral patterning of CdTe nanocrystal films by the electric field directed layer-by-layer assembly method. Langmuir 18: 4098–4102
[78] Mamedov AA, Belov A, Giersig M et al (2001) Nanorainbows: graded semiconductor films from quantum dots. J Am Chem Soc 123: 7738–7739
[79] Franzl T, Klar T, Schietinger S et al (2004) Exciton recycling in graded gap nanocrystal structures. Nano Lett 4: 1599–1603
[80] Hao H, Lian T (2000) Layer-by-layer assembly of CdSe nanoparticles based on hydrogen bonding. Langmuir 16: 7879–7881
[81] Zhang H, Yang B, Wang R et al (2002) Fabrication of a covalently attached self-assembly multilayer film based on CdTe nanoparticles. J Colloid Interf Sci 247: 361–365
[82] Donath E, Sukhorukov G, Caruso F et al (1998) Novel hollow polymer shells by colloid-templated assembly of polyelectrolytes. Angew Chem Int Ed 37: 2201–2205
[83] Caruso F, Caruso R, Möhwald H (1998) Nanoengineering of inorganic and hybrid hollow spheres by colloidal templating. Science 282: 1111–1114
[84] Caruso F (2001) Nanoengineering of particle surfaces. Adv Mater 13: 11–22
[85] Peyratout C, Dähne L (2004) Tailor-made polyelectrolyte microcapsules: from multilayers to smart containers. Angew Chem Int Ed 43: 3762–3783
[86] Wang D, Rogach A, Caruso F (2002) Semiconductor quantum dot-labeled microsphere bioconjugates prepared by stepwise self-assembly. Nano Lett 2: 857–861
[87] Wang D, Möhwald H (2004) Template-directed colloidal self-assembly – the route to 'top-down' nanochemical engineering. J Mater Chem 14: 459–468
[88] Arsenault A, Fournier-Bidoz S, Hatton B et al (2004) Towards the synthetic all-optical computer: science fiction or reality? J Mater Chem 14: 781–794
[89] Susha A, Caruso F, Rogach A et al (2000) Formation of luminescent spherical core-shell particles by the consecutive adsorption of polyelectrolyte and CdTe(S) nanocrystals on latex colloids. Colloids Surf A 163: 39–44
[90] Rogach A, Susha A, Caruso F et al (2000) Nano- and microengineering: three-dimensional colloidal photonic crystals prepared from submicrometer-sized polystyrene latex spheres pre-coated with luminescent polyelectrolyte/nanocrystal shells. Adv Mater 12: 333–337
[91] Wang D, Rogach A, Caruso F (2003) Composite photonic crystals from semiconductor nanocrystals/polyelectrolyte-coated colloidal spheres. Chem Mater 15: 2724–2729
[92] Wang D, Caruso F (2001) Fabrication of heterogeneous macroporous materials based on a sequential electrostatic deposition process. Chem Commun 489–490
[93] Westenhoff S, Kotov NA (2002) Quantun dot on a rope. J Am Chem Soc 124: 2448–2449
[94] Crisp MT, Kotov NA (2003) Preparation of nanoparticle coating on surfaces of complex geometry. Nano Lett 3: 173–177
[95] Yin W, Liu H, Yates M et al (2007) Fluorescent quantum dot-polymer nanocomposite particles by emulsification/solvent evaporation. Chem Mater 19: 2930–2936
[96] Han M, Gao X, Su J et al (2001) Quantum-dot-tagged microbeads for multiplexed optical coding of biomoleucles. Nat Biotechnol 19: 631–635
[97] Kuang M, Wang D, Bao H et al (2005) Fabrication of multicolor-encoded microspheres by tagging semiconductor nanocrystals to hydrogel spheres. Adv Mater 17: 267–270

[98] Gong Y, Gao M, Wang D et al (2005) Incorporating fluorescent CdTe nanocrystals into a hydrogel via hydrogen bonding: toward fluorescent microspheres with temperature-responsive properties. Chem Mater 17: 2648–2653
[99] Duan H, Wang D, Sobal N et al (2005) Magnetic colloidosmes derived from nanoparticle interfacial self-assembly. Nano Lett 5: 949–952
[100] Li J, Hong X, Liu Y et al (2005) Highly photoluminescent CdTe/poly(N-isopropylacrylamide) temperature-sensitive gels. Adv Mater 17: 163–166
[101] Eichenbaum GM, Kiser PF, Dobrynin AV et al (1999) Investigation of the swelling response and loading of ionic microgels with drugs and proteins: the dependence on cross-link density. Macromolecules 32: 4867–4878
[102] Tekin E, Smith P, Hoeppener S et al (2007) Inkjet printing of luminescent CdTe nanocrystal–polymer composite. Adv Funct Mater 17: 23–28
[103] Gaponik N, Radtchenko I, Sukhorukov G et al (2002) Toward encoding combinatorial libraries: charge-driven microencapsulation of semiconductor nanocrystals luminescing in the visible and near IR. Adv Mater 14: 879–882
[104] Zhang H, Edwards E, Wang D et al (2006) Directing the self-assembly of nanocrystals beyond colloidal crystallization. Phys Chem Chem Phys 8: 3288–3299
[105] Edwards E, Wang D, Mohwald H (2007) Hierarchical organization of colloidal particles: from colloidal crystallization to supraparticle chemistry. Macromol Chem Phys 208: 439–445

Layer-by-layer (LBL) assembly with semiconductor nanoparticles and nanowires

By

Sudhanshu Srivastava, Nicholas A. Kotov

Departments of Chemical Engineering, Materials Science and Engineering,
Biomedical Engineering, University of Michigan,
Ann Arbor, MN, USA

1. Introduction

The length scales in nanometer range for both inorganic and organic materials have unique physical responses. The opto-electronic properties of metals [1] and semiconductors [2] strongly depend on their size, material used and crystalline shape in the nanometer size regime. New synthetic approaches have been applied to fabricate series of monodisperse nanometer size crystals, known as nanocrystals or nanoparticles (NPs). In particular, semiconductor NPs [3] have attracted tremendous attention due to their unique physical properties, which originate from the surface atoms of nanoscale objects and from the size quantization effect [4]. Group II–VI semiconductor NPs are currently of great technological interest as emitting materials for thin film electroluminescence devices [5, 6] and as optical amplifier media for telecommunication networks [7, 8], because of their strong bandgap luminescence. The incorporation of luminescent semiconductor NPs into photonic crystals [9] has also obtained substantial attention recently as a novel light source with controllable spontaneous emission. Thiol-capped semiconductor NPs with size-dependent luminescence in the visible spectral region have been synthesized in aqueous colloidal solutions [10] and used for fabrication of light-emitting diodes (LEDs) [11], and for impregnation of colloidal photonic crystals [12].

"Bottom–up" [13] and "top–down" [14] approaches have been employed to assemble NPs in controlled manner and to understand the scope of possible applications in the field of the biomedical and materials science. Polymers [15] and biomolecules [16] are used as linkers to assemble the NPs into various architectures [17]. Polymer/inorganic nanocomposites hold substantial promise for the production of novel materials in which optical, electrical, magnetic, and catalytic properties of inorganic nanostructures [18] are combined with optical, electrical, and mechanical properties of macromolecules [19]. The chemical and physical characteristics available for permutations from both classes of compounds provide a versatile platform for materials designed for various applications. Numerous prototype devices using patterning were designed to utilize the unique material properties of NPs and polymers for variety of applications. The potential integration of these materials with living systems lead to the development of promising technology for diagnostic and therapeutic applications in medicine and biology [20]. Layer-by-layer (LBL) assembly [21, 22] is a versatile technique that can be used for the preparation of such NPs/polymer nanocomposites. This technique appears to be convenient for constructing functional interfaces because (i) layers can be assembled with inorganic NPs for biocompatible coatings [23], (ii) the composition of the film can easily be changed from layer to layer to impart the desired properties and (iii) LBL allows for introduction of multiple nanocomponents in the self-assembled material.

Being originally developed for polyelectrolyte/polyelectrolyte systems [17a], the LBL technique is applicable for almost any type of charged species, including inorganic molecular clusters [24], NPs [21a], nanotubes and nanowires [25], nanoplates [26], organic dyes [27], dendrimers [28], porphyrins [29], biological polysaccharides, polypeptides [30], nucleic acids and DNA [31], and proteins. The universality of the LBL process has increased the rapid development of biomedical applications of polyelectrolyte multilayers and related nanostructured organic–inorganic composites. This translates into an exceptionally broad range of structural characteristics and, thus, functional properties. The forces promoting formation of LBL films are not only limited to electrostatic interactions. Assemblies based on hydrogen bonding [32], charge transfer [33], covalent bonding [34], biological recognition [35], and hydrophobic interactions [36] have also been studied. Besides charged inorganic substrates, hydrophobic polymer surfaces have also been used to offer good scaffolds for LBL growth based upon hydrophobic interactions [37]. Overall, availability of an extensive spectrum of fabrication components, variety of substrates, and versatility of assembly methods dramatically enriches the applications of LBL films.

A variety of functional films can be produced using NPs and polyelectrolytes within the LBL assembly technique [21a, 38]. LBL-based thin films are currently being evaluated for a variety of applications that include drug delivery [39], molecular sensing [40], solid battery electrolytes, and membranes [41]. In general, films are created by alternately exposing a substrate to positively and negatively charged molecules or particles, as shown in Fig. 1 [17a]. In this case, steps 1–4 are continuously repeated until the desired number of "bilayers" (or cationic–anionic pairs) is achieved. Each individual layer may be 1–100 nm thick depending on chemistry, molecular weight, charge density, temperature, deposition time, counterion, and pH of species being deposited. The ability to control coating thickness down to the nm-level, easily insert variable thin layers without altering the process, and economically use raw materials are advantages of LBL relative to other available techniques of deposition of comparable thick films ($\gg 1\ \mu m$). Furthermore, the LBL deposited films are often transparent, which opens up areas of use not available for comparable bulk films (e.g., bacterial sensors on food wrap or magnetic stripe over graphics). Introducing semiconductor NPs and nanowires (NWs) in the LBL films allows creating materials for new advanced applications.

In this chapter, we review recent efforts in processing semiconductor NP dispersions into high-quality thin films with thicknesses controllable on the nanometer scale using the LBL assembly method both on planar substrates and on colloidal

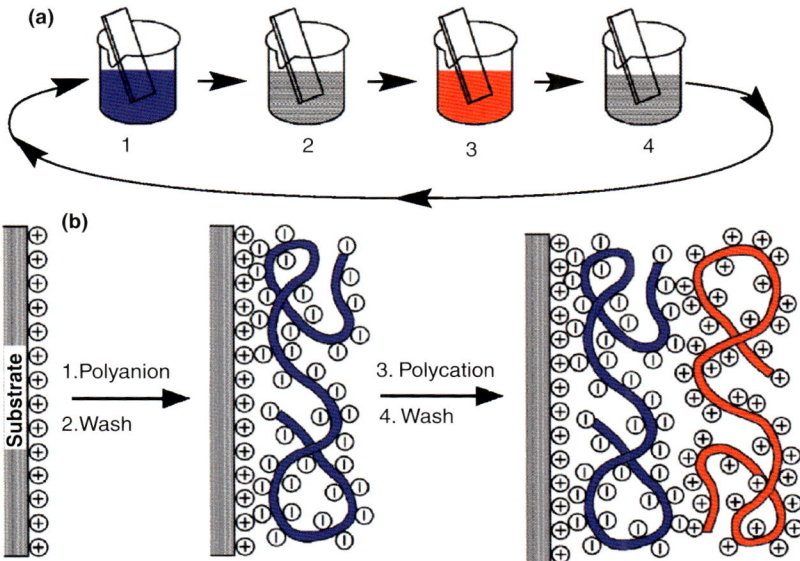

Fig. 1. a Scheme of the LBL film deposition using glass slides. Steps 1 and 3 represent the adsorption of a polyanion and polycation, respectively, and steps 2 and 4 are washing steps. The four steps are the basic build-up sequence for the simplest film architecture, $(A/B)n$. The construction of more complex film architectures requires only additional beakers and a different deposition sequence. **b** Simplified picture of the first two adsorption steps, depicting film deposition starting with a negatively charged substrate. The polyion conformation and layer interpenetration are an idealization of the surface-charge reversal with each adsorption step. Reproduced with permission from [17a], © 1997 AAAS

spheres. The LBL assembly discussed here is based on the alternating adsorption of oppositely charged species, such as positively and negatively charged polyelectrolyte pairs with semiconductor NPs and/or NWs. The technique for LBL can be applied universally to the coating of both macroscopically flat and non-planar (e.g. colloidal particles) surfaces for creating new photonic materials. LBL assemblies for semiconductor NWs also hold promise to create films with photoconductive and electronic applications. Besides discussing LBL assembly applications, the fabrication of different semiconductor NPs and NWs will also be shortly addressed.

2. Semiconductor nanoparticles

Semiconductor NPs possess attractive optical properties, as a result of size-dependent bandgaps, which determine the color of light emitted by the NPs. For fabrication of semiconductor NPs, the supersaturation and subsequent nucleation procedure can be initiated by the rapid addition of metal–organic precursors at 150–350°C into a flask containing hot coordinating solvents. The coordinating solvents initially used in the reaction were mixtures of long chain alkylphosphines and alkylphosphine oxides [42]. In the fabrication of II–VI semiconductor NPs (MX where $M = Zn$, Cd, Hg and $X = S$, Se, Te), metal alkyls (dimethylcadmium or diethylzinc) are selected from the group II sources. The group VI sources (where $X = S$, Se, Te) are often from salts of organophosphine chalcogenides or bistrimethylsilylchalcogenides, trimethylsilyl derivatives [43]. The organophosphine reagents (R3PX) are usually Se and Te sources because they are easy to synthesize; bistrimethylsilylchalcogenides are used as the source of S since they are more reactive than organophosphine derivatives. The use of mixed precursors, for example Se and S precursors, leads to formation of alloys and reflects the differential rate of precursor incorporation.

Synthesis of group II–VI semiconductor NPs is not limited to the use of alkylphosphines and alkylphosphine as high-boiling solvents in the synthesis. Injection of precursor reagents into hot alkylphosphites, alkylphosphates and alkylamines leads to formation of NPs. Mikulec showed that using alkylphosphoramide–tellurium precursors produces CdTe NPs with high-luminescence efficiencies [44]. Gaponik et al. summarized the advantages of CdTe NPs synthesis using short chain thiol ligands [45]: this kind of water-soluble NPs, considered in details in the Chapter by Gaponik and Rogach of this book are excellently suited for the use within the LBL approach. Guyot-Sionnest and Hines established that the use of alkylamines as the coordinating solvent considerably increases the rate of growth for ZnX NPs [46].

Similarly, synthesis of InP and InAs NPs has been carried out by mixing and vigorous heating of group III and V precursors in high-boiling solvents [47]. For groups III–V syntheses the precursor is normally present in the hot solvent prior to the addition of bis- or tris-trimethylsilyl derivatives. Growth of the NPs is very slow since Ostwald ripening takes about 6 days to reach the desired size. A lot of other potential organometallic precursors and high-boiling coordinating solvents are to be tested and thus provide opportunities for fabricating new semiconductor NP systems.

For many of the applications of semiconductor NPs, researchers aim to produce particles that are low cost, monodisperse, have high emission quantum yield, chemical functionality and stability in ambient conditions. CdTe NPs satisfying the most of these requirements can be synthesized using a wet chemical process that is relatively simple and allows one to functionalize the NPs using thiol type stabilizers in aqueous medium [10, 45, 48]. Thiol-stabilized CdTe NPs synthesized in water typically have a strong luminescence (with a quantum yield reaching 40–60% at room temperature). The synthesis starts with a solution that contains cadmium perchlorate and R-thiol stabilizer. Hydrogen telluride gas is passed through the solution to produce CdTe precursors that are not luminescent. The solution is then refluxed for minutes to several hours; during this time the particles increase in size through Oswald ripening. Within less than 1 h, green emission can be observed from the NP dispersion, and after few hours of boiling yellow and then red emission appear. The emission color change is an indication of the change in particle size.

The structure of the CdTe NPs has been shown to be zinc blend (cubic) through powder X-ray diffraction [45]. For small clusters it has been confirmed that the NP structures are cubic packed tetrahedrons with a surface layer of Cd–S–R [49]. However, for the larger particles (red emission) there is likely to be some CdS incorporated in gradient fashion throughout the NPs [50]. The functionality of the CdTe NPs can be achieved using different thiol-type stabilizers. The requirement for aqueous synthesis is that the thiol derivative must be soluble in water. The pH of the solution is typically modified so that, for negatively charged derivatives, particles are formed at pH > 11, whereas positively charged NPs are formed at pH ~ 6.

3. LBL assembly with semiconductor nanoparticles

In general, for LBL assembly the substrates are immersed in a solution of, e.g. positively charged polyelectrolyte, such as poly(diallyldimethylammonium chloride) (PDDA), poly(allylamine hydrochloride) (PAH) or polyethyleneimine (PEI), and subsequently rinsed by pure water (the aim of rinsing is the removal of loosely adsorbed polyelectrolyte from the substrate). The net charge of the substrate's surface becomes positive because of the adsorption of polyelectrolyte with positive charge. Subsequent execution of the analogous procedure with a negatively charged polyelectrolyte solution, such as poly(styrene sulfonate) (PSS), poly(vinyl sulfate) or poly(acrylic acid) (PAA), leads to the reversal of net charge on the substrate, bringing it back to the starting point. As a result, a double polyelectrolyte layer (bilayer) is built up on the substrate. With such cyclic depositions, one can obtain multilayer films on the substrates with desired structures and thicknesses (Fig. 1). The LBL technique allows for nanometer-scale control of the thickness of thin films [17a, 23]. In comparison to other assembly methods, the multilayer structure of LBL-deposited films allows for much higher loadings and results in more stable films.

Incorporation of semiconductor NPs in the LBL assembly based on alternating adsorption of oppositely charged species was investigated in detail [51]. Using the above approach to incorporate semiconductor NPs within polyelectrolytes films, HgTe NPs dispersions and PDDA were assembled into high-quality thin films with controllable thicknesses in nanometer scale. LBL films were prepared according to the standard cyclic procedure: (i) dipping the glass substrate in PDDA solution for 10 min; (ii) rinsing with water for 1 min; (iii) dipping into the dispersion of HgTe (pH = 10.0) for 20 min; (iv) rinsing with water again for 1 min. Each of such a procedure resulted in a "bilayer" consisting of a polymer/nanocrystal composite.

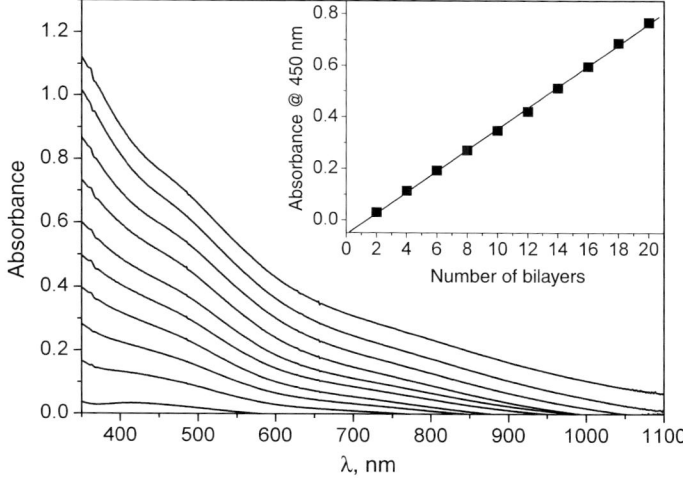

Fig. 2. Absorption spectra of LBL assembled film of HgTe NPs and PDDA as a function of a number of deposition cycles. Insert: absorbance at 450 nm vs. the number of deposited HgTe LBL layers. Reproduced with permission from [52], © 2000 ACS

The cycles were repeated according to the desired thickness of the films and were analyzed by UV–Vis (Fig. 2) [52].

The LBL assembled films with HgTe NPs displayed strong emission in the near-infrared, peaking around 1600 nm and covering the entire telecommunications spectral region of interest. LBL films of NPs can be viewed as very inexpensive yet versatile technology, enabling implementation of cost-effective high capacity optical networks. Due to exceptional versatility of LBL approach that can be effectively applied to the coating of planar substrates and highly curved surfaces such as optical fibers, this technique can be used in infrared opto-electronics for manufacturing all-fiber emitters, detectors, amplifiers, etc.

The LBL sequential adsorption technique was further employed for a variety of semiconductor NPs. It can be utilized as a deposition method for nanostructured coatings on highly curved micrometer to millimeter scale surfaces, which was demonstrated with strongly luminescent CdTe NPs. The preparation of uniform optical quality coatings made of highly luminescent NPs on optical fibers and tube interiors was done [53]. The LBL assembly was performed using CdTe NPs and PDDA. Confocal microscopy showed that CdTe/PDDA coatings were uniform and continuous (Fig. 3) [53]. Strong coupling of the NP luminescence into the 600 micron glass optical fiber attributed to high refractive index of CdTe/PDDA composite was observed. Two unexpected effects were detected: (i) gradual red shift of the luminescence spectrum as the layer thickness increases and (ii) a significant enhancement of the luminescence during first 100 min of UV illumination. Similar

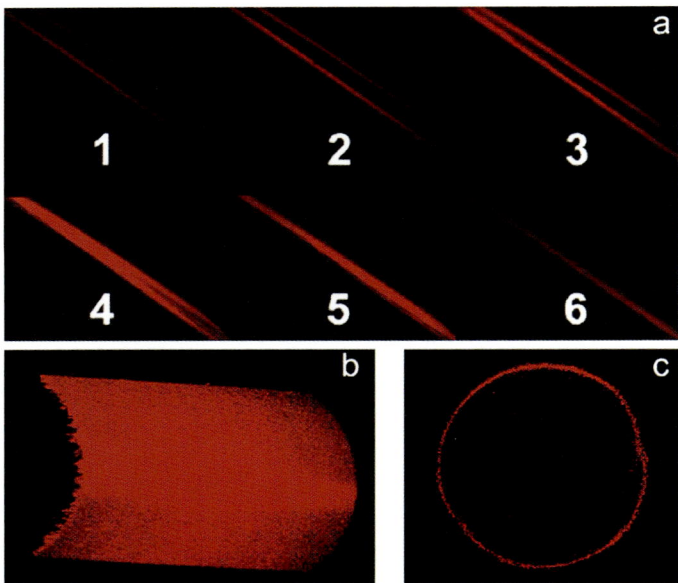

Fig. 3. Confocal microscopy examination of the optical fiber coated with $(PDDA/CdTe)_8$. **a** Depth profile image sequence. Numbers 1–6 denote gradual change of imaging depth from the middle of the fiber to the top surface. **b** Composite image of a piece of fiber surface. **c** Axial image of the fiber from the polished coated end. Reproduced with permission from [53], © 2003 ACS

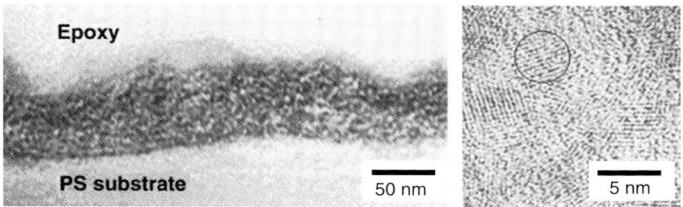

Fig. 4. Cross-sectional TEM image of a PAH/PAA/TiO$_2$[/PAA/PAH/PAA/TiO$_2$]$_7$ thin film. On the right side, a high-resolution image of TiO$_2$ NPs in the thin film is shown. Reproduced with permission from [54], © 2002 Elsevier

results were obtained for the coatings made in the glass tube interior, which can be considered as simple models of microfluidic devices with channels of complex geometry and sensing devices.

LBL films consisting of ionic polymers and positively charged TiO$_2$ NPs were demonstrated to possess photocatalytic properties [54]. Regular increase of the film thickness and the film weight by each deposition have been followed using ellipsometry, UV–vis spectroscopy, and a quartz crystal microbalance. Uniformly distributed TiO$_2$ NPs in the thin film were observed in scanning electron microscopy (SEM) and transmission electron microscopy (TEM) images (Fig. 4) [54]. Photocatalytic properties of the LBL thin films containing TiO$_2$ NPs were confirmed by oxidation of iodide and decomposition of methyl orange.

Other than electrostatic build-up, Wang and coworkers described a convenient LBL assembly approach to form multilayers of conjugated polymers and CdSe NPs based upon covalent coupling reactions [55]. This approach enables the formation of robust and smooth functional polymer thin films in common organic solvents. The stability of the resulting films was tested by sonication for several hours and no noticeable changes in the ellipsometric thickness of the films were observed. These properties, along with thickness control on the nanometer scale, render the resulting LBL structures very attractive as building blocks for organic device fabrication. Promising photocurrent responses have been found in the LBL self-assembled films. It has also been shown that the thin-film structures exert a great impact on the photovoltaic properties. The results of this study demonstrate that the covalent-based LBL method yields a unique approach for the fabrication of a variety of organic and organic–inorganic hybrid thin films.

4. LBL assembly of nanoparticles for biomedical applications

A critical emerging issue in the use of semiconductor NPs concerns their functionalization or modification to impart biocompatibility and target biospecificity. LBL assembly of NPs with polyelectrolytes provides the concept to impart biocompatibility and make use of NPs in the polymer films for further applications in the field of biomaterials. TiO$_2$ nanoshells were used to create ion-selective and biocompatible films using LBL technique. Hollow titania spheres were synthesized from the Ag and TiO$_2$ NPs [56]. In order to make nanoshells with ion-seiving properties, the silver core was removed by concentrated ammonia solution. As the titania shells were fabricated, LBL assembly was carried out by sequential dipping of the substrates in solutions of TiO_2 nanoshells (positive charge $+30\,mV$) and negatively charged PAA. TiO_2/PAA bilayers during the build-up were analyzed by ellipsometeric measurements and UV–vis absorbtion spectroscopy. The hollow sphere geometry of the nanoshells in the films provides high surface area and nanoscale porosity. The ion-permeable properties of the nanoshells were studied using the standard electrochemical method of cyclic voltammetery with potassium ferricyanide as the probe molecule. The change of pH from 11 to 2 resulted in a drastic increase of transparency of the nanoshell films toward the probe molecules, as indicated by the increase in the peak current in the voltammograms (Fig. 5) [56].

The high density of pores per unit volume of TiO_2 nanoshell-based LBL films resembles the qualities of biomembranes and tissues produced by biominerlization. In recent years there has been considerable interest in developing detection methods for the secretion of small molecules known as neurotransmitters, from the cells deep inside the brain. Understanding the mechanism of the secretion and metabolism is key for curing diseases like Parkinsons. Dopamine is one of the important chemicals found in the nervous system. In order to test the LBL films, the

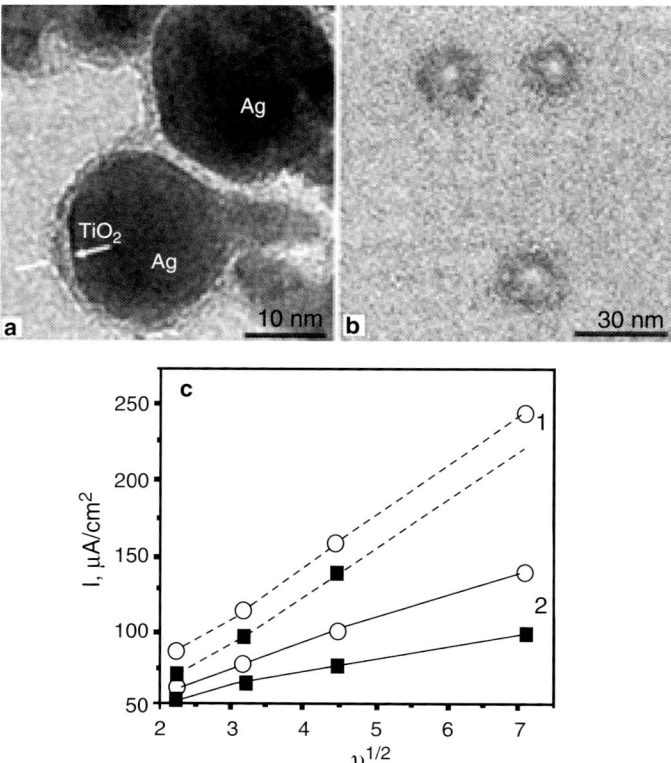

Fig. 5. TEM images of **a** Ag-core/TiO_2 shell NPs, **b** TiO_2 nanoshells, **c** Scan rate of cathodic and anodic currents for TiO_2 nanoshell LBL films at pH 2 (curves 1) and pH 11 (curves 2). Reproduced with permission from [56], © 2002 Wiley

electrochemical response to dopamine and ascorbic acid was investigated [56]. It was found that LBL coatings with nanoshells showed drastic difference to the magnitude of the electrochemical signal response of ascorbic acid relative to dopamine. The ion-selective capabilities of the nanoshells films for the successful detection of dopamine in the presence of ascorbic acid represents an important neurochemical problem and demonstrates a significant improvement in the detection sensitivity without any loss of sensitivity. The films were also found to be compatible with nerve cell line, which opens the way for their exploitations as coatings for the in vivo monitoring the brain activity. The ability to detect neurotransmitters directly inside the brain can facilitate the understanding of chemical communication between neurons and diagnoses the malfunction of the nervous system.

In another study [57], LBL assembly was carried out with absorption of positively charged PDDA in combination with negatively charged CdTe NPs on a glass substrate. The biocompatible coating of collagen on the surface of LBL assembled films of NPs was then built in the same cyclic manner with 0.5 wt% solutions of PAA at pH 4, which changes the surface charge to negative. Following the washing procedure, the substrate was exposed to positively charged 0.1% solution of collagen type IV at pH 4 for 20 min and rinsed. This procedure resulted in the deposition of a film with a layer sequence of $(PDDA/CdTe/)_n PDDA(/PAA/collagen)_m$, where n and m are the number of the corresponding deposition cycles. For testing the biocompatibility of the surfaces produced, mammalian C2C12 and PC12 cells in culture were used, which represent convenient models for expected interactions between the prepared NP/polyelectrolyte/collagen composite and tissues. When cells were cultured on $(PDDA/CdTe)_n$ multilayers, only clusters of dead cells were observed. When the CdTe NPs films were coated by a single PAA/collagen bilayer, the behavior of cells dramatically changed. They attached in large quantities and spread over the

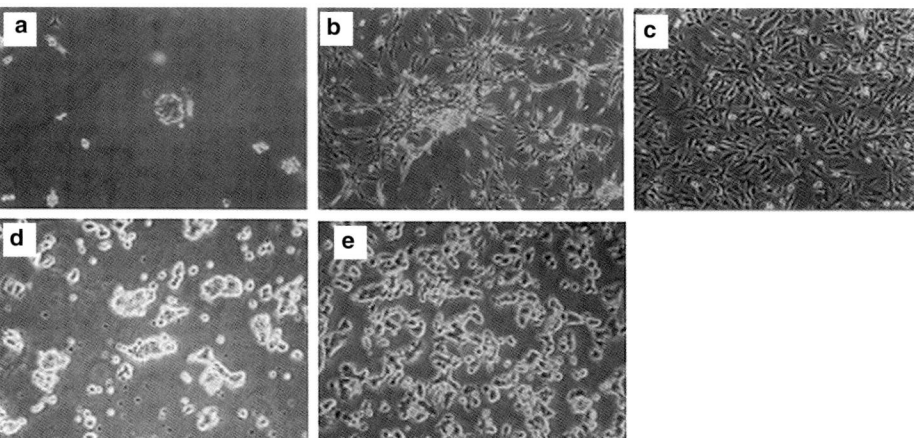

Fig. 6. Optical microscopy images of C2C12 myoblast (**a–c**) and PC12 (**d–e**) culture cells on the surface of (**a**) $(PDDA/CdTe)_3$, (**b**) $(PDDA/CdTe)_3 PDDA(PAA/collagen)_1$, (**c**) $(PDDA/CdTe)_3 PDDA(PAA/collagen)_5$, (**d**) $(PDDA/CdTe)_3 PDDA(PAA/collagen)_5$ and (**e**) glass. Reproduced with permission from [57], © 2003 ACS

surface, indicating that the cytotoxicity of the CdTe NPs was markedly screened (Fig. 6) [57]. These results indicate that biocompatibility factors must be engineered to the cell type used, as different cell surface molecules on various cell types determine attachment, survival and other functional interactions between the cell and the substrate. Although Cd-containing materials are unlikely candidates for in vivo use, ex vivo models with CdTe NPs can provide interesting information on fundamental problems at the interface of nanomaterials and biomaterials.

5. LBL assembly as a method to produce nanoparticle coatings on colloids

The coating technique using LBL is not limited to flat substrates, but can also be used for non-planar (e.g. colloidal particles) surfaces, resulting in core-shell materials. A new class of highly fluorescent, photostable and magnetic core-shell NPs in the submicrometer size range has been synthesized [58] and combined with the LBL assembly technique. Luminescent magnetic NPs [59] were fabricated using two main steps. The first step involved controlled addition of ligands to a dispersion of magnetic Fe_3O_4/Fe_2O_3 NPs, which were homogeneously incorporated in monodisperse silica spheres. The second, coating step involved the LBL assembly of polyelectrolytes and luminescent CdTe NPs onto the surfaces of the silica-coated magnetic NPs, which were finally covered with an outer shell of silica. These spherical particles had a typical diameter of ~220 nm with saturation magnetization of 1.34 emu/g and exhibited strong photoluminescence. Such multifunctional nanocomposites, addressable by a magnetic field and detectable by their emission, were prepared by the encapsulation of both magnetic iron oxide NPs and luminescent semiconductor NPs within composite silica spheres (Fig. 7).

Caruso et al. [60] utilized titanium dioxide, silica and laponite NPs as inorganic building block for multilayer formation on polystyrene (PS) sphere templates.

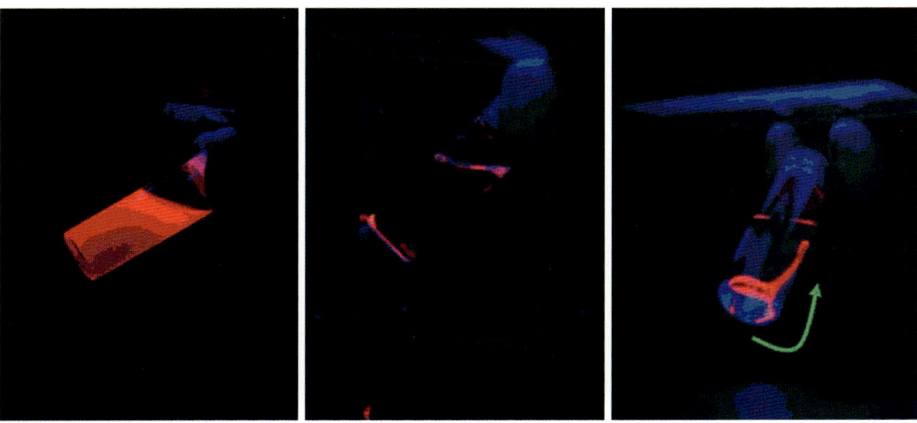

Fig. 7. Photographs of luminescent magnetic silica spheres driven by an external magnetic field. No magnetic field applied (left); a handheld magnet placed below the sample for 2 h (center); movement of the magnet dragging the concentrated particle phase (right). Reproduced with permission from [59], © 2006 Wiley

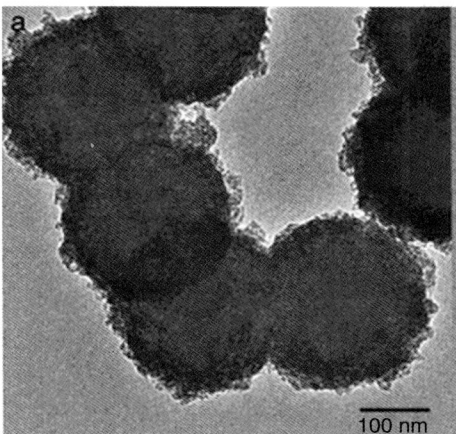

Fig. 8. TEM images of core-shell particles consisting of 210 nm diameter core PS spheres coated with three TiO_2 nanoparticle (positively charged)/PE_3 multilayers. Reproduced with permission from [60], © 2001 ACS

Composite organic–inorganic particles were formed by the controlled LBL assembly of the NPs in alternation with oppositely charged polyelectrolytes onto PS microspheres (Fig. 8). The influence of NPs type, shape (spherical to sheet-like), and size (3–100 nm), and the diameter of the PS sphere templates (210–640 nm) on the formation of multilayer shells was examined by TEM and SEM. In addition, the LBL technique for coating polymer spheres has been shown to be adaptable with small variations in the coating steps used to optimize the NP coatings of the different materials. These hybrid core-shell particles were subsequently calcinated to create hollow spheres with predetermined diameters. Such hollow spheres may find

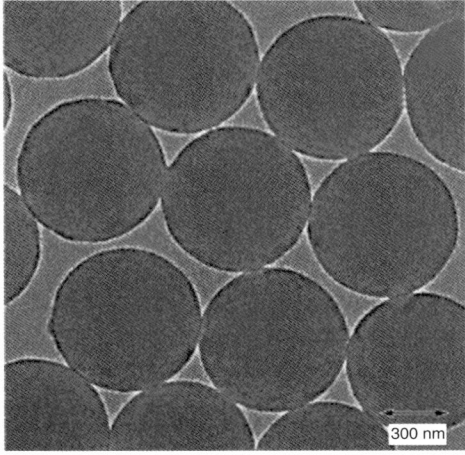

Fig. 9. TEM images of closely packed PS spheres coated with polyelectrolyte/CdTe NP shells. Reproduced with permission from [61], © 2002 Elsevier

application in diverse areas, ranging from photonics to fillers and pigments to microencapsulation.

In a similar study, highly luminescent thiol-capped CdTe and HgTe NPs were used for the deposition on submicron-sized monodisperse polystyrene spheres [61]. The nanocomposite shells on colloids were assembled by depositing negatively charged CdTe NPs on sulfate-stabilized PS spheres precoated with a three-layer film of the cationic polymer PAH and the anionic polymer PSS. The outermost surface layer (prior to the NP deposition) was PAH and hence the surface was positively charged. The procedure could be easily repeated, providing control over the composite multilayer film thickness. These nanocomposites made of nanocrystal/silica and core-shell latex/NP spheres have been used as building blocks for 3D colloidal photonic crystals (Fig. 9).

6. LBL assembly with nanowires and nanotubes

LBL assembly can also be extended to semiconductor or metallic nanowires (NWs) and carbon nanotubes. Use of NWs in the LBL assembly provides a useful approach to fabricate materials with unique and advanced opto-electronic responses. One-dimensional nanomaterials are expected to play a key role in fabricating nanoscale devices of the next generation. However, miniaturization in electronic and opto-electronic devices through improvements in conventional "top–down" technologies is approaching the limits of lithographic and etching processes. In contrast, bottom-up approaches to nanodevices, where the nanomaterials are assembled from basic building blocks, such as semiconductor NWs and carbon nanotubes, have the potential to go far beyond their limits. In general, one-dimensional nanostructures are synthesized by promoting the crystallization of solid-state structures along one direction [62]. The actual mechanisms of coaxing this type of crystal growth include (i) growth of an intrinsically anisotropic crystallographic structure, (ii) use of templates with one-dimensional morphologies for the formation of one-dimensional structures, (iii) introduction of a liquid/solid interface to reduce the symmetry, (iv) the use of an appropriate capping agent to control the growth rates of various facets, and (v) the self-assembly. There are a lot of ways to fabricate NWs using the above techniques; here we will only discuss in some more details the self-assembly based method leading to the formation of semiconductor NWs which have been used within the LBL approach. CdTe NWs can be formed by spontaneous self-assembly of individual negatively charged CdTe NPs [63]. For the NW formation to begin the excess stabilizer is removed through centrifugation. The precipitate is then dispersed in water at pH 9 and allowed to age at room temperature in the dark for several days. During the period of 6–8 days, the color of the solution gradually turns from orange to dark brown. Upon analyzing the respective solution with TEM, SEM and AFM solid CdTe NWs were detected (Fig. 10). NWs are formed irrespective of the CdTe particle size, as demonstrated using different CdTe dispersions with luminescence maxima at 520–530 nm (green), 550–565 nm (yellow), 590–605 nm (orange), and 610–625 nm (red) [64].

CdTe NWs display excellent uniformity in diameter, have a high aspect ratio, and significant photoluminescence quantum yield. The selection of the original particle

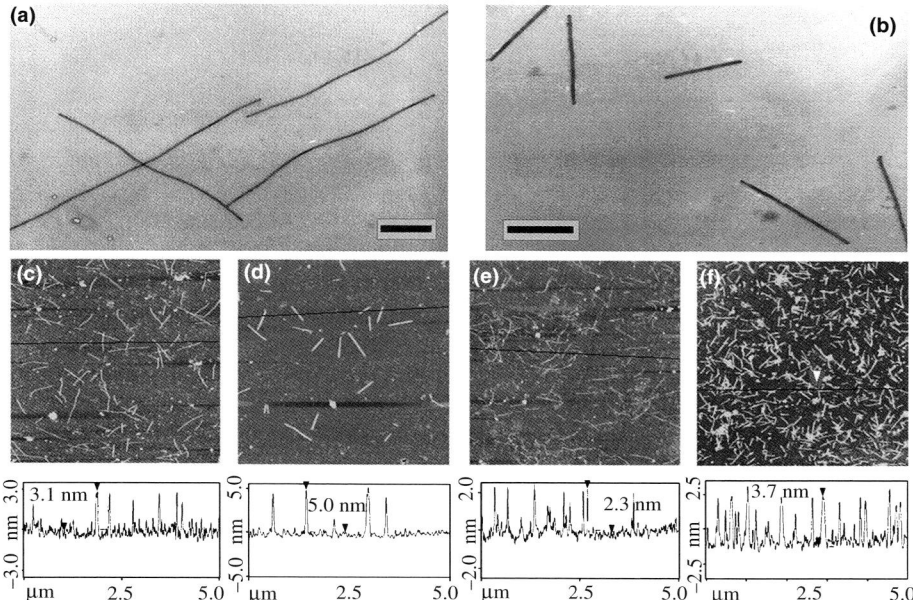

Fig. 10. TEM images of CdTe NWs made from 3.4-nm (**a**) and 5.4-nm (**b**) NPs. Bars, 100 nm. AFM images (top) and cross-sectional profiles (bottom) of NWs made from 3.4-nm (**c**), 5.4-nm (**d**), 2.5-nm (**e**), and 4.1-nm (**f**) CdTe NPs. A featureless cationic polyelectrolyte layer (poly(diallydimethylammonium) chloride), with surface features of 0.1 nm in height, was preadsorbed on the silica wafer to increase the adhesion of anionic NWs. Reproduced with permission from [63], © 2002 AAAS

sizes offers a convenient means for the control of the degree of quantum confinement and NW morphology. The described process also exemplifies the ability of the NPs to self-organize into other superstructures due to the intrinsic anisotropy of inter-NP interactions. NPs can be compared to proteins; self-organization processes in solution have been well established for proteins and other biomacromolecules [65], which have physical dimensions of several nanometers, i.e., the same scale as that of many inorganic nanocolloids. When positively charged CdTe NPs were allowed to self-assemble, TEM and AFM images showed free-floating sheets formed by a 2D network of assembled NPs. The experimentally observed template-free organization into free-floating particulate sheets resembled the assembly of S-proteins or chaperones [66].

A very general procedure for semiconductor NW formation has been introduced by Lieber and Duan [67]. Elemental Si and Ge semiconductor NWs were produced using the laser-assisted catalytic growth (LCG) method [68], which exploits laser ablation to generate nanometer diameter catalytic clusters that define the size and direct the growth of the crystalline NWs by a vapor–liquid–solid (VLS) mechanism. Importantly, it was shown that semiconductor NWs of the III–V materials GaAs, GaP, GaAsP, InAs, InP, InAsP, the II–VI materials ZnS, ZnSe, CdS, CdSe, and IV–IV alloys of SiGe can be synthesized in larger yields and high purity using this approach (Fig. 11).

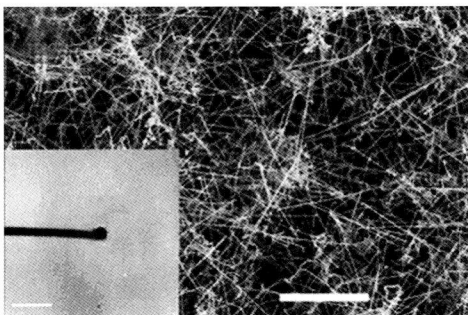

Fig. 11. FE-SEM image of CdSe nanowires prepared by LCG. The scale bar corresponds to 2 mm. The inset is the TEM image of an individual CdSe nanowire exhibiting nanocluster (dark feature) at the wire end. EDX shows that the nanocluster is composed primarily of Au. The scale bar is 50 nm. Reproduced with permission from [67], © 2000 Wiley

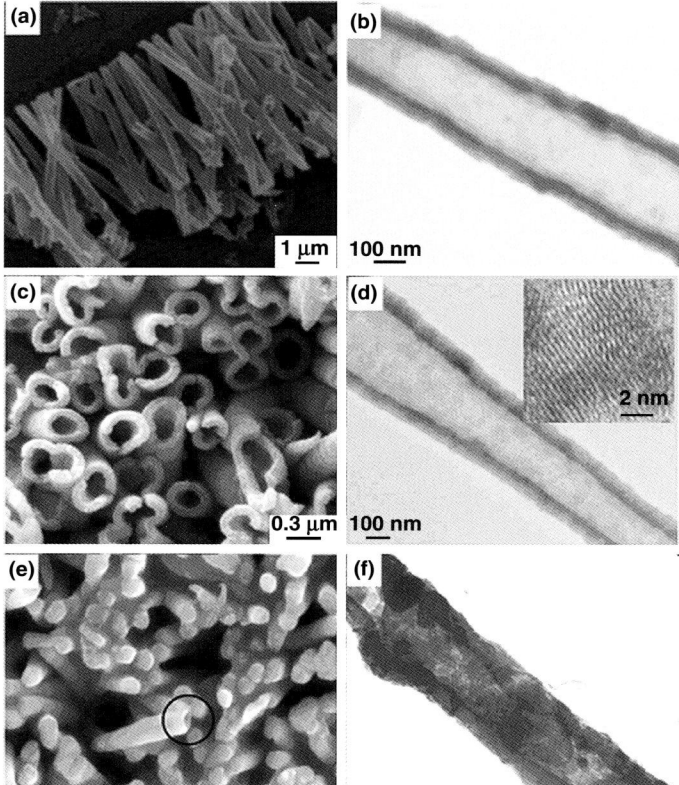

Fig. 12. SEM (**a**, **c**, **e**) and TEM (**b**, **d**, **f**) images of titania nanotubes obtained from (PEI/TALH)$_{12}$PEI-coated PC membrane after heating at 95°C for 24 h (**a**, **b**), 450°C for 10 h (**c**, **d**) and 950°C for 12 h (**e**, **f**). Reproduced with permission from [69], © 2007 Wiley

For NW formation, titania tubes were successfully prepared by LBL deposition of a water-soluble titania precursor, titanium(IV) bis(ammonium lactato) dihydroxide (TALH) and the oppositely charged poly(ethylenimine) (PEI) to form multilayer polyelectrolyte films (Fig. 12) [69]. Tubular structure was obtained by deposition of the above mentioned materials inside the cylindrical pores of a polycarbonate (PC) membrane template, followed by calcination at various temperatures. The as-prepared anatase titania tubes exhibit very promising photocatalytic properties, demonstrated by the degradation of the azodye methyl orange as a model molecule. They are also easily separated from the reaction system by simple filtration or centrifugation, allowing for straightforward recycling. The reported strategy provides a simple and versatile technique to fabricate titania based tubular nanostructures, which could easily be extended to prepare tubular structures of other materials and may find application in catalysis, chemical sensing, and nanodevices.

Cui and Liu [70] fabricated silica NWs by combining "top–down" e-beam lithography and "bottom–up" LBL assembly techniques. The simple and low-cost LBL method was used to construct silica NP thin films, while the e-beam lithography-based lift-off method was implemented for patterning the self-assembled thin films to nanometer scale lengths. The silica nanowires grown by this technique had an average width of 100 nm, while the minimum width obtained was 60 nm. These experimental data indicate a new approach to fabricate NWs that can be used in nanoelectronic devices and circuits.

CdTe NWs as those described above can be transformed to Te NWs using a Cd complexing agent EDTA [71]. The Te NWs exhibited a uniform diameter with lengths ranging from 100 to 800 nm. Solid thin films of Te NWs on glass (Fig. 13b) were prepared by using a standard LBL method by dipping the glass substrate alternatively into a negatively charged Te NW solution and positively charged polyelectrolyte, PDDA. Successful film formation was evidenced by a linear increase of the UV–vis absorbance with an increasing number of deposition cycles.

To characterize the conductive and photoconducting properties of the LBL assemblies made from Te NWs, a film of 20 bilayers, $(PDDA/NW)_{20}$, was prepared and electrical contacts were made by attaching silver wires to the surface of the Te. The electrical resistance of the Te NW films was measured in the dark and under

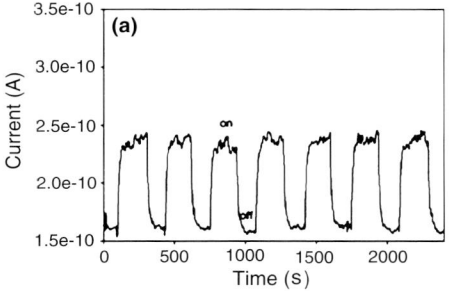 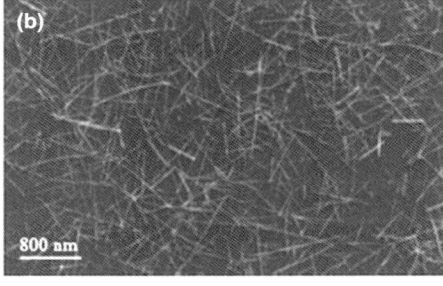

Fig. 13. a Photoresponse of the Te thin films for repetitive switching of the He:Ne laser between "on" and "off" states, **b** SEM image of $(PDDA/Te\ NW)_2$. Reproduced with permission from [71], © 2006 Wiley

illumination (Fig. 13a). The "light-on–light-off" cycle was demonstrated to be very stable under ambient conditions. The photocurrent rise-and-fall kinetics was also found to be relatively long [71]. At present, these results are interesting primarily from a fundamental point of view but also as a demonstration of the materials design potential as nanocomposites with high amounts of inorganic loading, made possible by LBL assembly. In addition, it was evidenced that the resistance can be adjusted by adding new chemical components. The photocurrent measurement of the thin films demonstrated that Te thin films can be reversibly switched between lower- and higher-conductivity states after exposure to the external light source. Such photoconducting NWs could serve as light detectors and switching devices for optoelectric applications, in which the binary states can be addressed optically.

Another important class of semiconducting quasi one-dimensional material is carbon nanotubes. Single-walled carbon nanotube (SWCNT) devices have been developed due to the outstanding structural, mechanical, thermal, electrical and chemical properties of SWCNTs [72]. An alternative approach is to use SWCNT thin films, in the form of either aligned arrays [73] or random networks [74], or as the semiconducting channel layers. In [72] the fabrication and characterization of high-mobility thin-film transistors (TFTs) using LBL assembled SWCNTs (Fig. 14) as the semiconducting material and SiO_2 NPs as the gate dielectric material was carried out. The channel length and the effective thickness of the SWCNT semiconductor layer were 50 microns and 38 nm, respectively. The effective thickness of the SiO_2 dielectric layer was 180 nm. The SWCNT TFT exhibited p-type semiconductor characteristics and operated in the accumulation mode. The combination of LBL assembly and microlithography provides a simple, low-temperature, and highly efficient approach to fabricate inexpensive TFT devices.

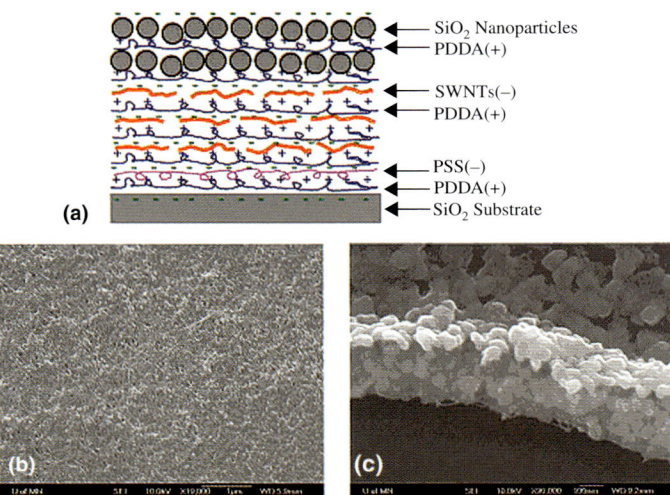

Fig. 14. **a** Model structure of LBL self-assembled SWNCT semiconducting layer and SiO_2 nanoparticle dielectric layer. **b** SEM image of the assembled SWNCTs. **c** Cross-sectional SEM image of the assembled SWCNT/SiO_2 layers. Reproduced with permission from [72], © 2006 AIP

7. Conclusions

LBL assembly based on sequential absorption of polyelectrolytes with semiconductor NPs or NWs is a unique way to introduce the properties from both the inorganic and organic components into resulting nanocomposite materials. This approach allows for high loading of NPs or NWs into the polyelectrolyte films with potential applications in biological and material science. In particular, the use of HgTe NPs holds a great promise for manufacturing infrared opto-electronic emitters, amplifiers and detectors. In the biomedical applications, LBL assemblies with TiO_2 nanoshells can be used for understanding the communication of neurons; CdTe NPs assemblies in combination with polyelectrolytes and collagen can be used as biocompatible surfaces. All these results indicate the scope for LBL technique to be employed on different kinds of semiconductor NPs for manufacturing new devices and materials for biological applications. Introducing semiconductor NWs and SWCNTs in the LBL assembly adds to the versatility of the approach for constructing transistors and conducting films. Modulation in any of the components used for the LBL assembly could lead to novel functional materials for such diverse areas as drug delivery, therapeutic biomaterials, sensors or transistors.

References

[1] Kubo R, Kawabata A, Kobayashi S (1984) Electronic properties of small particles. Ann Rev Mater Sci 14: 49–66
[2] Gaponenko SV (1998) Optical properties of semiconductor nanocrystals. Cambridge, Cambridge University Press
[3] Rogach AL, Talapin DV, Weller H (2004) Semiconductor nanoparticles. In: Caruso F (ed) Colloids and Colloid Assemblies. Weinheim, Wiley-VCH, pp 52–95
[4] Weller H (1993) Colloidal semiconductor Q-particles – chemistry in the transition region between solid-state and molecules. Angew Chem Int Ed 32: 41–53
[5] Schlamp MC, Peng X, Alivisatos AP (1997) Improved efficiencies in light emitting diodes made with CdSe(CdS) core/shell type nanocrystals and a semiconducting polymer. J Appl Phys 82: 5837–5842
[6] Mattoussi H, Radzilowski LH, Dabbousi BO, Thomas EL, Bawendi MG, Rubner MF (1998) Electroluminescence from heterostructures of poly(phenylene vinylene) and inorganic CdSe nanocrystals. J Appl Phys 83: 7965–7974
[7] Harrison MT, Kershaw SV, Burt MG, Rogach AL, Kornowski A, Eychmüller A, Weller H (2000) Colloidal nanocrystals for telecommunications. Complete coverage of the low-loss fiber windows by mercury telluride quantum dots. Pure Appl Chem 72: 295–307
[8] Kershaw SV, Harrison M, Rogach AL, Kornowski A (2000) Development of IR-emitting colloidal II–VI quantum-dot materials. IEEE J Sel Top Quant Electron 6: 534–543
[9] Miguez H, Blanco A, Lopez C, Meseguer F, Yates HM, Pemble ME, Lopez-Tejeira F, Garcia-Vidal FJ, Sanchez-Dehesa J (1999) Face centered cubic photonic bandgap materials based on opal-semiconductor composites. IEEE J Lightwave Technol 17: 1975–1981
[10a] Rogach AL, Katsikas L, Kornowski A, Su D, Eychmüller A, Weller H (1996) Synthesis and characterization of thiol-stabilized CdTe nanocrystals Ber. Bunsenges Phys Chem 100: 1772–1778
[10b] Gao MY, Kirstein S, Möhwald H, Rogach AL, Kornowski A, Eychmüller A, Weller H (1998) Strongly photoluminescent CdTe nanocrystals by proper surface modification. J Phys Chem B 102: 8360–8363
[11] Gao MY, Lesser C, Kirstein S, Möhwald H, Rogach AL, Weller H (2000) Electroluminescence of different colors from polycation/CdTe nanocrystal self-assembled films. J Appl Phys 87: 2297–2302
[12] Gaponenko SV, Bogomolov VN, Petrov EP, Kapitonov AM, Eychmüller A, Rogach AL, Kalosha II, Gindele F, Woggon U (2000) Spontaneous emission of organic molecules and semiconductor nanocrystals in a photonic crystal. J Lumin 87: 152–156

[13] Huang Y, Duan XF, Cui Y, Lauhon LJ, Kim KH, Lieber CM (2001) Logic gates and computation from assembled nanowire building blocks. Science 294: 1313–1317
[14] Alivisatos AP, Johnsson KP, Peng XG, Wilson TE, Loweth CJ, Bruchez MP, Schultz PG (1996) Organization of 'nanocrystal molecules' using DNA. Nature 382: 609–611
[15] Boal AK, Ilhan F, DeRouchey JE, Thurn-Albrecht T, Russell TP, Rotello VM (2000) Self-assembly of nanoparticles into structured spherical and network aggregates. Nature 404: 746–748
[16] Niemeyer CM (2001) Nanoparticles, proteins, and nucleic acids: biotechnology meets materials science. Angew Chem Int Ed 40: 4128–4158
[17a] Decher G (1997) Fuzzy nanoassemblies: toward layered polymeric multicomposites. Science 277: 1232–1237
[17b] Caruso F, Caruso RA, Möhwald H (1998) Nanoengineering of inorganic and hybrid hollow spheres by colloidal templating. Science 282: 1111–1114
[18a] Bruchez M, Moronne M, Gin P, Weiss S, Alivisatos AP (1998) Semiconductor nanocrystals as fluorescent biological labels. Science 281: 2013–2016
[18b] Sun SH, Murray CB, Weller D, Folks L, Moser A (2000) Monodisperse FePt nanoparticles and ferromagnetic FePt nanocrystal superlattices. Science 287: 1989–1992
[19a] Gangopadhyay R, De A (2000) Conducting polymer nanocomposites: a brief overview. Chem Mater 12: 608–622
[19b] Hawker CJ, Wooley KL (2005) The convergence of synthetic organic and polymer chemistries. Science 309: 1200–1205
[20] Panyam J, Labhasetwar V (2003) Biodegradable nanoparticles for drug and gene delivery to cells and tissue. Adv Drug Deliver Rev 55: 329–347
[21a] Kotov NA, Dekany I, Fendler JH (1995) Layer-by-layer self-assembly of polyelectrolyte–semiconductor nanoparticle composite films. J Phys Chem 99: 13065–13069
[21b] DeLongchamp DM, Kastantin M, Hammond PT (2003) High-contrast electrochromism from layer-by-layer polymer films. Chem Mater 15: 1575–1586
[22] Podsiadlo P, Kaushik AK, Arruda EM, Waas AM, Shim BS, Xu JD, Nandivada H, Pumplin BG, Lahann J, Ramamoorthy A, Kotov NA (2007) Ultrastrong and stiff layered polymer nanocomposites. Science 318: 80–83
[23] Tang ZY, Wang Y, Podsiadlo P, Kotov NA (2006) Biomedical applications of layer-by-layer assembly: from biomimetics to tissue engineering. Adv Mater 18: 3203–3224
[24] Ingersoll D, Kulesza PJ, Faulkner LR (1994) Polyoxometalate-based layered composite films on electrodes – preparation through alternate immersions on modification solutions. J Electrochem Soc 141: 140–147
[25] Jiang C, Ko H, Tsukruk VV (2005) Strain-sensitive Raman modes of carbon nanotubes in deflecting freely suspended nanomembranes. Adv Mater 17: 2127–2131
[26] Keller SW, Kim HN, Mallouk TE (1994) Layer-by-layer assembly of interclation compounds and heterostructures on surfaces – toward molecular beaker epitaxy. J Am Chem Soc 116: 8817–8818
[27] Cooper TM, Campbell AL, Crane RL (1995) Formation of polypeptide-dye multilayers by an electrostatic self-assembly technique. Langmuir 11: 2713–2718
[28] He JA, Valluzzi R, Yang K, Dolukhanyan T, Sung C, Kumar J, Tripathy SK, Samuelson L, Balogh L, Tomalia DA (1999) Electrostatic multilayer deposition of a gold-dendrimer nanocomposite. Chem Mater 11: 3268–3274
[29] Araki K, Wagner MJ, Wrighton MS (1996) Layer-by-layer growth of electrostatically assembled multilayer porphyrin films. Langmuir 12: 5393–5398
[30a] Mueller M (2001) Orientation of alpha-helical poly(L-lysine) in consecutively adsorbed polyelectrolyte multilayers on texturized silicon substrates. Biomacromolecules 2: 262
[30b] Boulmedais F, Ball V, Schwinte P, Frisch B, Schaaf P, Voegel JC (2003) Buildup of exponentially growing multilayer polypeptide films with internal secondary structure. Langmuir 19: 440–445
[31] Lvov Y, Decher G, Sukhorukov G (1993) Assembly of thin-films by means of successive deposition of alternate layers of DNA and poly(allylamine). Macromolecules 26: 5396–5399
[32] Stockton WB, Rubner MF (1997) Molecular-level processing of conjugated polymers. 4. Layer-by-layer manipulation of polyaniline via hydrogen-bonding interactions. Macromolecules 30: 2717–2725
[33] Shimazaki Y, Mitsuishi M, Ito S, Yamamoto M (1997) Preparation of the layer-by-layer deposited ultrathin film based on the charge-transfer interaction. Langmuir 13: 1385–1387
[34] Brynda E, Houska M (1996) Multiple alternating molecular layers of albumin and heparin on solid surfaces. J Colloid Interf Sci 183: 18–25
[35] Anzai JI, Kobayashi Y, Nakamura N, Nishimura M, Hoshi T (1999) Layer-by-layer construction of multilayer thin films composed of avidin and biotin-labeled poly(amine)s. Langmuir 15: 221–226

[36] Lojou E, Bianco P (2004) Buildup of polyelectrolyte-protein multilayer assemblies on gold electrodes. Role of the hydrophobic effect. Langmuir 20: 748–755
[37] Delcorte A, Bertrand P, Wischerhoff E, Laschewsky A (1997) Adsorption of polyelectrolyte multilayers on polymer surfaces. Langmuir 13: 5125–5136
[38a] Bertrand P, Jonas A, Laschewsky A, Legras R (2000) Ultrathin polymer coatings by complexation of polyelectrolytes at interfaces: suitable materials, structure and properties. Macromol Rapid Commun 21: 319
[38b] Decher G, Schlenoff JB (2003) Multilayer thin films – sequential assembly of nanocomposite materials. Wiley–VCH: Weinheim, Germany
[39] Nolan CM, Serpe MJ, Lyon LA (2004) Thermally modulated insulin release from microgel thin films. Biomacromolecules 5: 1940–1946
[40] Kim JH, Kim SH, Shiratori S (2004) Fabrication of nanoporous and hetero structure thin film via a layer-by-layer self assembly method for a gas sensor. Sensor Actuat B 102: 241–247
[41] DeLongchamp DM, Hammond PT (2004) Highly ion conductive poly(ethylene oxide)-based solid polymer electrolytes from hydrogen bonding layer-by-layer assembly. Langmuir 20: 5403–5411
[42] Murray CB, Norris DJ, Bawendi MG (1993) Synthesis and characterization of nearly monodisperse CdE (E = S, Se, Te) semiconductor nanocrystallites. J Am Chem Soc 115: 8706–8715
[43a] Murray CB, Kagan CR, Bawendi MG (2000) Synthesis and characterization of monodisperse nanocrystals and close-packed nanocrystal assemblies. Ann Rev Mater Sci 30: 545–610
[43b] Shiang JJ, Kadavanich AV, Grubbs RK, Alivisatos AP (1995) Shape control of CdSe nanocrystals. J Phys Chem 99: 17417–17422
[43c] Petit C, Taleb A, Pileni MP (1999) Cobalt nanosized particles organized in a 2D superlattice: synthesis, characterization, and magnetic properties. J Phys Chem B 103: 1805–1810
[44] Mikulec F (1999) Semiconductor nanocrystal colloids: manganese doped cadmium selenide, (core) shell composites for biological labeling, and highly fluorescent cadmium telluride. PhD Thesis. Massachusetts Institute of Technology, Cambridge, MA
[45] Gaponik N, Talapin DV, Rogach AL, Hoppe K, Shevchenko KV, Kornowski A, Eychmuller A, Weller H (2002) Thiol-capping of CdTe nanocrystals: an alternative to organometallic synthetic routes. J Phys Chem B 106: 7177–7185
[46] Hines MA, Guyot-Sionnest P (1998) Bright UV-blue luminescent colloidal ZnSe nanocrystals. J Phys Chem B 102: 3655–3657
[47] Micic OI, Sprague JR, Curtis CJ, Jones KM, Nozik AJ (1994) Synthesis and characterization of InP quantum dots. J Phys Chem 98: 4966–4969
[48] Rogach AL, Franzl T, Klar TA, Feldmann J, Gaponik N, Lesnyak V, Shavel A, Eychmüller A, Rakovich YP, Donegan JF (2007) Aqueous synthesis of thiol-capped CdTe nanocrystals: state-of-the-art. J Phys Chem C 111: 14628–14637
[49] Rockenberger J, Tröger L, Rogach AL, Tischer M, Grundmann M, Eychmüller A, Weller H (1998) The contribution of particle core and surface to strain, disorder and vibrations in thiolcapped CdTe nanocrystals. J Chem Phys 108: 7807–7815
[50] Rogach AL (2000) Nanocrystalline CdTe and CdTe(S) particles: wet chemical preparation, size-dependent optical properties and perspectives of opto-electronic applications. Mater Sci Eng B 69: 435–440
[51a] Gao MY, Richter B, Kirstein S, Möhwald H (1998) Electroluminescence studies on self-assembled films of PPV and CdSe nanoparticles. J Phys Chem B 102: 4096–4103
[51b] Aliev F, Correa-Duarte MA, Mamedov A, Ostrander JW, Giersig M, Liz-Marzan LM, Kotov NA (1999) Layer-by-layer assembly of core-shell magnetite nanoparticles: effect of silica coating on interparticle interactions and magnetic properties. Adv Mater 11: 1006–1010
[51c] Decher G, Schmitt J, Brand F, Lehr B, Oeser R, Losche M, Bouwman W, Kjaer K, Calvert J, Geer R, Dressik W, Shashidhar R (1998) Nonlinear optical properties of polyelectrolyte thin films containing gold nanoparticles investigated by wavelength dispersive femtosecond degenerate four wave mixing (DFWM). Adv Mater 10: 338
[51d] Gao MY, Kirstein S, Rogach AL, Weller H, Möhwald H (1999) Photoluminescence and electroluminescence of CdSe and CdTe nanoparticles. Adv Sci Technol 27: 347
[52] Rogach AL, Koktysh DS, Harrison M (2000) Layer-by-layer assembled films of HgTe nanocrystals with strong infrared emission. Chem Mater 12: 1526–1528
[53] Crisp MT, Kotov NA (2003) Preparation of nanoparticle coatings on surfaces of complex geometry. Nano Lett 3: 173–177
[54] Kim TH, Sohn BH (2002) Photocatalytic thin films containing TiO_2 nanoparticles by the layer-by-layer self-assembling method. Appl Surf Sci 201: 109–114

[55] Liang ZQ, Dzienis KL, Xu J, Wang Q (2006) Covalent layer-by-layer assembly of conjugated polymers and CdSe nanoparticles: multilayer structure and photovoltaic properties. Adv Func Mater 16: 542–548

[56] Koktysh DS, Liang XR, Yun BG, Pastoriza-Santos I, Matts RL, Giersig M, Serra-Rodriguez C, Liz-Marzan LM, Kotov NA (2002) Biomaterials by design: layer-by-layer assembled ion-selective and biocompatible films of TiO_2 nanoshells for neurochemical monitoring. Adv Func Mater 12: 255–265

[57] Sinani VA, Koktysh DS, Yun BG, Matts RL, Pappas TC, Motamedi M, Thomas SN, Kotov NA (2003) Collagen coating promotes biocompatibility of semiconductor nanoparticles in stratified LBL films. Nano Lett 3: 1177–1182

[58] Hong X, Li J, Wang MJ, Xu JJ, Guo W, Li JH, Bai YB, Li TJ (2004) Fabrication of magnetic luminescent nanocomposites by a layer-by-layer self-assembly approach. Chem Mater 16: 4022–4027

[59] Salgueirino-Maceira V, Correa-Duarte MA, Spasova M, Liz-Marzan LM, Farle M (2006) Composite silica spheres with magnetic and luminescent functionalities. Adv Func Mater 16: 509–514

[60] Caruso RA, Susha A, Caruso F (2001) Multilayered titania, silica, and Laponite nanoparticle coatings on polystyrene colloidal templates and resulting inorganic hollow spheres. Chem Mater 13: 400–409

[61] Rogach AL, Kotov NA, Koktysh DS, Susha AS, Caruso F (2002) II–VI semiconductor nanocrystals in thin films and colloidal crystals. Colloid Surf A: Physiochem Eng Asp 202: 135–144

[62] Law M, Goldberger J, Yang P (2004) Semiconductor nanowires and nanotubes. Ann Rev Mater Res 34: 83–122

[63] Tang ZY, Kotov NA, Giersig M (2002) Spontaneous organization of single CdTe nanoparticles into luminescent nanowires. Science 297: 237–240

[64] Li LS, Hu JT, Yang WD, Alivisatos AP (2001) Band gap variation of size- and shape-controlled colloidal CdSe quantum rods. Nano Lett 1: 349–351

[65] Jagannathan K, Chang RW, Yethiraj A (2002) A Monte Carlo study of the self-assembly of bacteriorhodopsin. Biophys J 83: 1902–1905

[66a] Tang ZY, Zhang ZL, Wang Y, Glotzer SC, Kotov NA (2006) Self-assembly of CdTe nanocrystals into free-floating sheets. Science 314: 274–278

[66b] Ellis RJ, Vandervies SM (1991) Molecular chaperones. Ann Rev Biochem 60: 321–347

[67] Duan XF, Lieber CM (2000) General synthesis of compound semiconductor nanowires. Adv Mater 12: 298–302

[68] Morales AM, Lieber CM (1998) A laser ablation method for the synthesis of crystalline semiconductor nanowires. Science 279: 208–211

[69] Yu A, Lu GQM, Drennan J et al (2007) Tubular titania nanostructures via layer-by-layer self-assembly. Adv Func Mater 17: 2600–2605

[70] Liu Y, Cui TH (2006) Silica nanowires fabricated with layer-by-layer self-assembled nanoparticles. J Nanosci Nanotechnol 6: 1019–1023

[71] Wang Y, Tang ZY, Podsiadlo P et al (2006) Mirror-like photoconductive layer-by-layer thin films of Te nanowires: the fusion of semiconductor, metal, and insulator properties. Adv Mater 18: 518–522

[72] Xue W, Liu Y, Cui TH (2006) High-mobility transistors based on nanoassembled carbon nanotube semiconducting layer and SiO_2 nanoparticle dielectric layer. Appl Phys Lett 89: 163512

[73] Xiao K, Liu Y, Hu P, Yu G, Wang X, Zhu D (2003) High-mobility thin-film transistors based on aligned carbon nanotubes. Appl Phys Lett 83: 150–152

[74] Zhou YX, Gaur A, Hur S, Kocabas C, Meitl MA, Shim M, Rogers JA (2004) p-channel, n-channel thin film transistors and p–n diodes based on single wall carbon nanotube networks. Nano Lett 4: 2031–2035

Exciton–phonon interaction in semiconductor nanocrystals

By

M. I. Vasilevskiy

Departamento de Física, Universidade do Minho, Campus de Gualtar, Braga, Portugal

1. Introduction

The interaction of electrons and holes with lattice vibrations in bulk semiconductors leads to a number of effects, such as a reduction in their mobility because of scattering, a (small) change in the effective mass, the introduction of phonon sidebands in the optical absorption and emission related to the excitons, and homogeneous broadening of spectral lines [1, 2]. It also determines the carrier dynamics in semiconductor devices. In particular, the emission of optical phonons is the main mechanism of relaxation of hot carriers (see Fig. 1), extremely important for the operation of semiconductor lasers [3]. This applies also to quantum wells, superlattices and quantum wires, nanostructures where the electron and hole energy spectra still are continuous.

The situation is different in semiconductor quantum dots (QDs). The three-dimensional (3D) quantum confinement of the electrons and holes results in discrete atomic-like energy spectra as shown schematically in Fig. 2. Owing to this electronic structure, the emission of optical phonons as the mechanism of relaxation to the ground state is apparently impossible because it would require strict energy conservation in the electron–phonon scattering, i.e. exact resonance between the optical phonon energy and level spacing, which should be rather accidental. On the other hand, the level spacing is rather big compared to the typical energy of acoustic phonons that may still form a continuum. This kind of argument justified the theoretical concept of 'phonon bottleneck', a very slow carrier relaxation which should be inherent to small QDs [4]. However, in the last years it has become clear that the electron and hole coupling to phonons in QDs leads to the formation of a polaron and, generally speaking, cannot be described within the Born approximation [5–7]. In other words, virtual transitions between different electronic levels, assisted by phonons and not requiring energy conservation may be important enough to guarantee a significant modification of the exciton energy spectrum and dynamics[1]. Precise quantitative knowledge of the mechanisms and intensity of the

---

[1] By exciton we understand an electron–hole pair, independently of the importance of the Coulomb interaction between them.

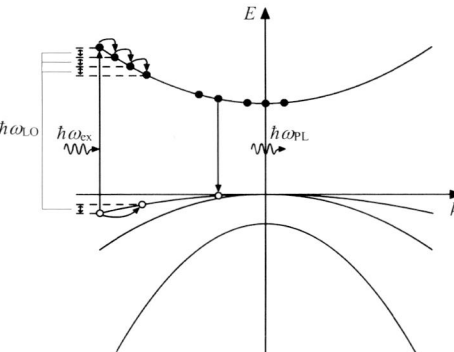

Fig. 1. A schematic illustration of hot electron thermalization and radiative recombination with holes. Incident photon of energy $\hbar\omega_{ex}$ excites an electron–hole pair and luminescence occurs at an energy $\hbar\omega_{PL}$

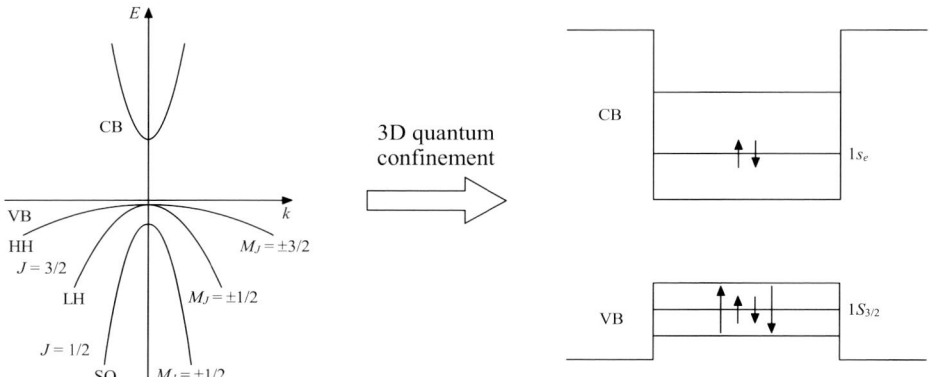

Fig. 2. A schematic illustration of the 3D confinement effect on the electron and hole spectra in a spherical QD made of a semiconductor material with zinc-blend structure

electron–phonon and hole–phonon interactions in QDs is therefore crucial for achieving a detailed understanding of their optical properties.

There exists a broad literature devoted to the electronic structure of the most studied II–VI, III–V and elemental QDs constituted by nearly spherical nanocrystals (NCs) that can be produced by melting, sputtering or colloidal chemistry techniques (see, e.g. [8] for review). Theoretical models that have been successfully applied to describe the electron and hole spectra of nearly spherical QDs range from the effective mass approximation (EMA) [9–11] to large scale *ab initio* calculations [12, 13]. There is a consensus that, if the complex valence band structure of the underlying material and the electron–hole exchange interaction are taken into account, the (semi-)analytical EMA calculations reproduce reasonably well the energy spectra and optical properties of confined excitons unless the QDs are too small (below 1 nm in radius).

The spatial confinement effect on optical phonons in semiconductor QDs has also been studied experimentally and theoretically (see [14] for a recent review). Again, continuum models provide a good understanding of experimental results obtained by means of Raman and FIR spectroscopies [15, 16]. However, the intensity of the exciton–phonon (ex–ph) interaction in QDs remains a controversial subject. Various theoretical studies led to different conclusions concerning the strength of the ex–ph coupling and its dependence on the QD size [17–20]. While calculations performed for II–VI and III–V dots generally agree that the ex–ph coupling strength is reduced in NCs compared to the bulk [21], they disagree regarding the numbers and trends in scaling of the intensity of this interaction with the QD size. Although the calculated values vary substantially depending on the approximations used, the Huang-Rhys parameter for the lowest exciton state, a measure of the intensity of the single-level coupling to optical phonons discussed in the next section, usually does not exceed $S \approx 0.1 - 0.2$ for II–VI spherical QDs [21–23] and is even smaller for the III–Vs [23]. Is then the polaron effect really important in semiconductor QDs? We shall discuss this in Sect. 3.

The situation is even more controversial as far as acoustic phonons are concerned. Experimental data obtained by low frequency Raman scattering studies of Ge [24], CdS [25] and CdSe [26] NCs, always embedded in a matrix, usually were interpreted [25–28] in terms of the Lamb's theory of elastic vibrations of a *freestanding* sphere [29]. This theory predicts quantised acoustic modes (see also [30] where the essential results are reproduced). However, as we will show in Sect. 2.3, the spectrum of the acoustic vibration modes of a QD embedded in an elastic matrix is continuous [31–33]. This is conceptually important for the exciton (polaron) dynamics (Sect. 4).

Turning to the experimental data, the exciton–optical-phonon coupling strength is most frequently obtained from photoluminescence (PL) spectra. Usually there is a large Stokes shift of the excitonic PL band with respect to the absorption in small II–VI [34, 35] and I–VII [36] QDs, which can be explained by strong exciton–phonon coupling ("experimental" Huang-Rhys parameter $S \approx 1$ for CuBr [36] and larger than unity for CdSe NCs [37, 38])[2]. However, as mentioned above, the calculated values happen to be at least one order of magnitude smaller. Furthermore, phonon replicas in the PL spectra caused by the recombination of excitons in QDs were found to disagree with the well-known Franck–Condon progression (Sect. 3.1), both in their spectral positions and relative intensities [21]. Although both the strong Stokes shift and apparently large intensity of the first phonon satellite (relative to the zero phonon line) can be characteristic of the QD size distribution and not of a single dot [35], these results indicate that the ex–ph interaction in QDs must be considered carefully and the adiabatic approximation (leading to the emission spectra described by the Franck–Condon progression) can be insufficient [21, 40].

For the sake of completeness, let us mention some works aimed at the determination of the ex–ph coupling strengths using other experimental techniques. Several

---

[2] Although self-assembled epitaxial QDs are beyond the scope of this book, it is worth noting that experimental and theoretical data on the Huang-Rhys parameter in such dots also vary by an order of magnitude [39].

authors have attempted to extract the Huang-Rhys parameter from the relative intensity of one- and two-phonon Raman scattering peaks [17, 41–43]. Quite high values of $S \approx 0.5$ for CdSe [17] and $S \approx 0.7$ for PbS [41] and CuBr [42] QDs have been reported. It should be pointed out that, for such a strong ex–ph coupling, the contribution of non-adiabatic transitions (i.e. the polaron effect) to the Raman scattering must be important and no single parameter can describe the relative intensity ratio between the one-phonon peak and its overtones [44]. We shall return to this point in Sect. 3.5.

In addition to these studies, exciton dephasing and homogeneous broadening of excitonic transitions has been measured for various semiconductor NCs as a function of the QD size and temperature [28]. The determination of the homogeneous broadening is possible by measuring PL spectra under size-selective excitation [45, 46] or by studying the spectral hole burning effect [36, 47, 48]. Alternatively, it can be deduced from the dephasing time extracted from the exponential decay of the polarisation in four-wave mixing [36, 49, 50] or photon echo [51] experiments. Intrinsic mechanisms of excitonic line broadening arise by coupling to phonons and, similarly to structures of higher dimensionality, the temperature dependence of the broadening is described by the following type of relation [36, 37, 45, 46]:

$$\Gamma(T) = \Gamma_0 + \gamma_{ac} N_{ac}(T) + \gamma_{LO} N_{LO}(T) \tag{1.1}$$

where Γ_0 is related to the radiation lifetime of the exciton, γ_{ac} and γ_{LO} are some constants, and $N_{ac}(T)$ and $N_{LO}(T)$ are the Bose factors for the acoustic and optical phonons, respectively. By fitting γ_{ac} and γ_{LO} one can estimate the strength of the ex–ph coupling. Although the variation of $(\Gamma(T)-\Gamma_0)$ with the temperature is mostly linear due to the second term in Eq. (1.1), there was found also a clear contribution of coupling to the optical phonons [36, 46]. This is not easy to understand for QDs, because the coupling of optical phonons to a single electron or exciton level should lead to the appearance of discrete satellites and not to the level broadening [2, 52]. Thinking in terms of the Fermi's Golden Rule (or, equivalently, within the Born approximation), the broadening could be a lifetime effect owing to the electron (or exciton) transition to another state with emission or absorption of an optical phonon. However, it would require strict energy conservation in the electron–phonon scattering, i.e. exact resonance between the optical phonon energy and level spacing, which should be rather accidental as already mentioned above. Once more, we come to the conclusion that the interpretation of experimental data within the models that are appropriate for bulk semiconductors and even for systems of reduced dimensionality, can be unsuitable for QDs. In particular, it means that multi-phonon (virtual) processes can be important and the ex–ph interaction in QDs must be treated in a non-perturbative way, even for the moderate values of the coupling constants coming out from the calculations. An important ingredient to be included in the theory is the non-adiabaticity of this interaction [21, 40] leading to a phonon-mediated coupling of different electronic levels, even if they are separated by an energy quite different from the optical phonon energy [7].

In this contribution we discuss the theoretical issues of the ex–ph interaction, polaron formation and its dynamics in nearly spherical QDs. It is organised as follows. In Sect. 2 we describe confined states of electrons, holes, and phonons and present the calculation of the ex–ph coupling parameters, within the EMA and continuum models for phonons. In Sect. 3, stationary polaron states are considered and the polaron effect on the steady state absorption, emission and Raman scattering is discussed. The polaron dynamics and hot carrier relaxation are discussed in Sect. 4, which is followed by conclusion.

2. Theory of coupling between confined electrons, holes and phonons in spherical quantum dots

2.1 Electron and hole states. The effect of quantum confinement on the electron and the hole in a dot competes with the Coulomb interaction between them. Two limiting cases can be distinguished [8], known as (i) strong ($R \ll a_{\text{ex}}$, a_{ex} is the bulk exciton Bohr radius) and (ii) weak ($R \gg a_{\text{ex}}$) confinement regimes. Here we shall concentrate on small QDs for which the strong confinement regime is realised (for instance, $a_{\text{ex}} = 5.6$ nm for CdSe). Then, in the first approximation, one can consider the spatial quantisation of the energy spectra of "free" electrons and holes. If one wishes to study the confinement effect departing from the electronic structure of bulk semiconductor, a relatively simple description of electron–hole pair states confined in a spherical QD is possible when the underlying material possesses cubic crystal lattice and direct band gap [8], as is the case of several III–Vs and II–VIs. Even though CdS and CdSe, probably the most studied NC materials, normally have hexagonal crystal structure, this (small) anisotropy effect can be taken into account as perturbation [9, 10]. It is more convenient to include this effect along with those arising from the electron–hole interaction which will be considered in the next section. In this section, we summarize the EMA results for cubic semiconductors that are necessary for the calculation of the ex–ph coupling rates (and the reader should refer to the original works [9, 10, 53–55] for details of derivation).

For wide-gap semiconductors, such as CdSe, the confined electron and hole levels may be treated independently[3]. The bulk electronic structure is shown schematically in Fig. 2. For electrons near the bottom of the conduction band, it is sufficient to consider a single parabolic band. Moreover, the problem can be simplified further by assuming infinitely high barriers at the semiconductor/matrix interface, which is a good approximation for NCs embedded in silica or similar dielectrics. This leads to the following simple expression [8–10] for the envelope wave functions for the electrons in a QD of radius R,

$$\psi_{lmn}(\vec{r}) = \theta(R - r)\sqrt{\frac{2}{R^3}}\frac{j_l(\xi_{ln}r/R)}{j_{l+1}(\xi_{ln})}Y_{lm}(\vartheta, \phi) \tag{2.1}$$

where $n = 1, 2, 3\ldots$ enumerates successive zeros, ξ_{ln}, of the l-th spherical Bessel function j_l, $\theta(R-r)$ is the Heaviside step function and $Y_{lm}(\vartheta, \phi)$ are the spherical harmonics. In addition to the three "spherical" quantum numbers, the fourth one

---
[3] See, however, [11] where the effects of band mixing on confined states in CdTe dots were considered.

characterising the confined electron states is spin projection, s_z. The full electron wave function, Ψ^e_{l,m,n,s_z}, is a product of ψ_{lmn} given by Eq. (2.1) with a Bloch function of the Γ_6 band, labelled by the electron spin projection $\sigma = \uparrow, \downarrow$ and denoted $|\sigma\rangle$. In the approximation considered here, $\sigma = s_z$ and

$$\Psi^e_{l,m,n,s_z}(\vec{r}) = \psi_{lmn}(\vec{r})\delta_{s_z,\sigma}|\sigma\rangle \qquad (2.2)$$

The double-degenerate ground state with $n=1$, $l=m=0$, $s_z = \uparrow, \downarrow$ is denoted $1s_e$ (Fig. 2). Its energy, with respect to the bottom of the bulk conduction band, E_C, is given by $\pi^2 \hbar^2/(2 m_e R^2)$ where m_e is the electron effective mass.

For holes, it is necessary to take into account the complex structure of the underlying valence band. This has been done by Al. Efros and co-workers who considered the limiting cases of very large [9, 10] and vanishing [53] spin-orbit splitting (Δ_{SO}), as well as the general one [54]. The hole states are classified by the total angular momentum quantum number, F_h, its projection, M, a radial quantum number, n_h, and the parity, P_h. In the case of $\Delta_{SO} \to \infty$, the holes can be considered as particles with spin 3/2 and, for a given F_h, there are four possible values of the orbital momentum, $l_h = F_h - 3/2$, $F_h + 1/2$ (for even states) and $l_h = F_h - 1/2$, $F_h + 3/2$ (for odd states). These values determine the order of the spherical harmonics that appear in the corresponding envelope functions, as exemplified below, and are important for establishing selection rules for phonons participating in the ex–ph coupling. For the four-fold degenerate even hole state with $F_h = 3/2$ and $n_h = 1$, usually denoted $1S_{3/2}$ (Fig. 2), the wave functions can be written as $\Psi^h_{3/2,M,1,+} = \psi^{(0)}_M + \psi^{(2)}_M$ with

$$\psi^{(0)}_M = R_0(r) Y_{0,0}(\vartheta,\phi)|M\rangle,$$

$$\psi^{(2)}_{1/2} = -R_2(r)\left\{\frac{1}{\sqrt{5}} Y_{2,0}(\vartheta,\phi)|1/2\rangle + \sqrt{\frac{2}{5}} Y_{2,-1}(\vartheta,\phi)|3/2\rangle + \sqrt{\frac{2}{5}} Y_{2,2}(\vartheta,\phi)|-3/2\rangle\right\},$$

$$\psi^{(2)}_{-1/2} = R_2(r)\left\{\frac{1}{\sqrt{5}} Y_{2,0}(\vartheta,\phi)|-1/2\rangle - \sqrt{\frac{2}{5}} Y_{2,1}(\vartheta,\phi)|-3/2\rangle - \sqrt{\frac{2}{5}} Y_{2,-2}(\vartheta,\phi)|3/2\rangle\right\},$$

$$\psi^{(2)}_{3/2} = R_2(r)\left\{-\frac{1}{\sqrt{5}} Y_{2,0}(\vartheta,\phi)|3/2\rangle + \sqrt{\frac{2}{5}} Y_{2,1}(\vartheta,\phi)|1/2\rangle - \sqrt{\frac{2}{5}} Y_{2,2}(\vartheta,\phi)|-1/2\rangle\right\},$$

$$\psi^{(2)}_{-3/2} = R_2(r)\left\{-\frac{1}{\sqrt{5}} Y_{2,0}(\vartheta,\phi)|-3/2\rangle + \sqrt{\frac{2}{5}} Y_{2,-1}(\vartheta,\phi)|-1/2\rangle \right.$$
$$\left. - \sqrt{\frac{2}{5}} Y_{2,-2}(\vartheta,\phi)|1/2\rangle\right\}. \qquad (2.3)$$

Here $M = \pm 1/2, \pm 3/2$ and $|\mu\rangle$ are the Bloch functions of the Γ_8 band ($\mu = \pm 1/2, 3/2$) which can be found, for instance, in [10]. The radial envelope functions are given by [9, 10]

$$R_0(r) = \theta(R-r)\frac{C}{R^{3/2}}\left[j_0(K_h r) - \frac{j_0(K_h R)}{j_0(\sqrt{\beta}K_h R)} j_0(\sqrt{\beta}K_h r)\right];$$

$$R_2(r) = \theta(R-r)\frac{C}{R^{3/2}}\left[j_2(K_h r) + \frac{j_0(K_h R)}{j_0(\sqrt{\beta}K_h R)} j_2(\sqrt{\beta}K_h r)\right] \qquad (2.4)$$

where C is a normalisation constant, $\beta = m_{lh}/m_{hh}$ is the ratio of the light and heavy hole masses, and $K_h = \chi_1/R$ with χ_1 denoting the first root of the equation

$$j_0(\sqrt{\beta}\chi)j_2(\chi) + j_0(\chi)j_2(\sqrt{\beta}\chi) = 0. \tag{2.5}$$

The energy of the $1S_{3/2}$ state, with respect to the top of the bulk valence band, E_V, is given by $\hbar^2 K_h^2/(2m_{hh})$.

In the case of negligible spin-orbit splitting ($\Delta_{SO} \approx 0$), relevant to CdS, the holes can be considered as particles with "pseudo-spin" 1. Although their ground state in a spherical QD is the odd p-type one [53, 54], it is the even s-type state that forms a "bright" exciton revealing itself in the optical spectra [54]. This state can be called $1S_1$, with six-fold degeneracy corresponding to $M=0, \pm 1$ and two projections of the true spin [53]. Again, the wavefunctions can be written in the form, $\Psi^h_{1,M,1,+} = \tilde{\psi}^{(0)}_M + \tilde{\psi}^{(2)}_M$, where

$$\tilde{\psi}^{(0)}_M = R_0(r)Y_{00}(\vartheta,\phi)|M\rangle,$$

$$\tilde{\psi}^{(2)}_{-1} = \sqrt{3}R_2(r)\left\{\frac{1}{\sqrt{30}}Y_{2,0}(\vartheta,\phi)|-1\rangle + \frac{1}{\sqrt{5}}Y_{2,-2}(\vartheta,\phi)|1\rangle - \frac{1}{\sqrt{10}}Y_{2,-1}(\vartheta,\phi)|0\rangle\right\},$$

$$\tilde{\psi}^{(2)}_0 = \sqrt{3}R_2(r)\left\{-\frac{2}{\sqrt{30}}Y_{2,0}(\vartheta,\phi)|0\rangle + \frac{1}{\sqrt{10}}Y_{2,-1}(\vartheta,\phi)|1\rangle + \frac{1}{\sqrt{10}}Y_{2,1}(\vartheta,\phi)|-1\rangle\right\},$$

$$\tilde{\psi}^{(2)}_1 = \sqrt{3}R_2(r)\left\{\frac{1}{\sqrt{30}}Y_{2,0}(\vartheta,\phi)|1\rangle + \frac{1}{\sqrt{5}}Y_{2,2}(\vartheta,\phi)|-1\rangle - \frac{1}{\sqrt{10}}Y_{2,1}(\vartheta,\phi)|0\rangle\right\}. \tag{2.6}$$

In Eq. (2.6) $|\mu\rangle$ ($\mu = 0, \pm 1$) are simply related to the Bloch functions of the Γ_4 valence band (see, e.g. [56]), $|0\rangle = i|Z\rangle$ and $|\pm 1\rangle = \mp i(|X\rangle \pm |Y\rangle)/\sqrt{2}$. The radial wavefunctions are different from those given by Eq. (2.4) just by a multiplicative factor of $\sqrt{2}$. Eq. (2.5) and the above expression for the hole energy also hold in this case.

The results presented above are also applicable, to a certain extent, to the holes confined in NCs of silicon and germanium. For Si, the $\Delta_{SO} \approx 0$ approximation is quite appropriate and Eq. (2.6) for the wavefunctions can be used. Calculations performed with the effect of finite barriers taken into account [57] show that the ground state is the six-fold degenerate one with total (orbital) angular momentum $\tilde{F}_h = 1$. For Ge dots, the spin-orbit splitting $\Delta_{SO} = 0.29$ eV can be comparable to the spatial quantisation energy. One can possibly use Eqs. (2.3)–(2.5) but only for a limited range of R.

Concerning the conduction band states, the situation is more complicated because of the location of the absolute minimum of this band outside the Brillouin zone centre for these materials. Within the EMA, it is difficult to include the mixing of the conduction band states of the absolute minimum with those corresponding to $\vec{k} = 0$. In the case of Si dots, given the large difference between the direct and indirect band gaps, it can be still acceptable to consider the confined electron states, relevant to the optical properties, in the framework of the envelope function approach on the basis of Bloch functions of the X-point of the Brillouin zone as it was done in [57, 58]. It is not the case of Ge dots where no reasonable analytical model exists. In a recent paper

[59], the confined states of both electrons and holes were considered in the simple parabolic band approximation, i.e. in the form of Eqs. (2.1) and (2.2), that seems to be an oversimplification.

There are a considerable number of published works devoted to large-scale numerical calculations of the electronic spectra of Si and Ge dots, both *ab initio* [60–62] and semi-empirical [63–66]. The well-known problem with such calculations is the difficulty in establishing simple qualitative trends, both for the electronic spectra and observable properties, so nicely seen from the continuous models. In this sense, the (semi-)empirical tight-binding methods [63–68] are a reasonable compromise. For instance, one can establish correspondence between the Luttinger parameters and Δ_{SO}, entering the EMA calculations of the hole spectra, and the hopping parameters of the tight-binding approach, thus allowing for a direct comparison between the two theories and assignment of the levels in terms of quantum numbers with clear physical meaning. Such a comparison shows, for instance, that there is virtually no difference between the EMA and tight-binding results for the highest hole level in CdSe dots [67]. As one might expect, the agreement is worse for small dots ($R < 2$ nm) where the scaling of the electron and hole confinement energies with the QD radius starts deviating from R^{-2} [13, 67, 69], mainly because of the band non-parabolicity [69] and finite barrier height [70] effects. The size-dependent energy level structure has been probed recently by means of the scanning tunnelling spectroscopy and it was concluded that the results are in good agreement with the tight-binding calculations (see [71] and references therein). Nevertheless, in spite of their limitations, EMA results are always useful, at least as a guideline.

For the sake of completeness, let us mention that the electronic properties of PbS and PbSe QDs were considered, within the EMA, in [72]. In bulk IV–VI semiconductors, both the minimum of the conduction band and the maximum of the valence band occur in the Δ point of the Brillouin zone and the parabolic band approximation is inadequate because of the strong non-parabolicity of the dispersion curves [72]. We do not reproduce here the rather complex expressions obtained in [72] to which the interested reader should refer. Some I–VII semiconductors that have been used as QD materials, possess the same crystal and (bulk) band structure as lead chalcogenides [73] but the exciton Bohr radius is so small (e.g. $a_{ex} = 0.7$ nm for CuCl [48]) that the strong confinement regime cannot be realised for such dots.

2.2 Exciton states. The Coulomb and exchange interactions between the electron and the hole forming an exciton depend on the QD size. In the weak confinement regime, the motion of the exciton's centre of mass is spatially quantised with the energy level spacing of the order of $\hbar^2/M_{ex}R^2$, where M_{ex} is the exciton mass. The corresponding wavefunctions are very similar to Eq. (2.1) and can be found, e.g. in [8]. Concerning the internal motion of the bound electron and hole, the situation is very similar to bulk excitons that, in a simple approximation, can be described by the hydrogen-like model [74, 75]. A more detailed theory, relevant to CuCl QDs, was developed in [76].

In the strong confinement regime, it is the structure of the few lowest exciton states that is important for the correct interpretation of PL and Raman spectra obtained under resonant excitation. In the first-order approximation, the exciton wave functions can be written as products of the electron and hole ones and the energy of the exciton ground state is given by

$$E_x(R) = E_g + \frac{\hbar^2 \pi^2}{2m_e R^2} + \frac{\hbar^2 \chi_1^2}{2m_{hh} R^2} - \zeta \frac{e^2}{\varepsilon_0 R} \qquad (2.7)$$

where E_g is the bulk band gap energy. The last term in Eq. (2.7) represents the Coulomb energy with ε_0 denoting the static dielectric constant of the QD material and ζ is a coefficient of the order of unity[4]. Note that the Coulomb attraction just shifts the energy of the lowest excitonic state, without lifting its degeneracy. Microscopic calculations [13, 69] have shown that the dielectric constant ε_0 decreases when the QD size is reduced, so that the Coulomb energy scales as $R^{-\alpha}$ with $\alpha = 0.82$, 0.90 and 0.86 for Si, GaAs and CdSe dots, respectively [69]. Also, the numerical coefficient should be modified because of the dielectric constant mismatch between the QD and the matrix [13, 71]. This mismatch originates an indirect "polarisation interaction" between the electron and the hole. Within the EMA, the polarisation shift of the exciton ground state scales as R^{-1} [13] and can be included in the last term of Eq. (2.7) with the coefficient ζ that depends also on the static dielectric constant of the matrix, ε_M, according to the relation $\zeta = [(0.86\varepsilon_0 + 2.72\varepsilon_M)/(\varepsilon_0 + \varepsilon_M)]$[5]. In the limit of $\varepsilon_0/\varepsilon_M \gg 1$, a common situation, the polarisation effect reduces the excitonic correction to about half its value. Recent first-principle and tight-binding calculations confirm this conclusion [13].

The electron–hole exchange interaction, proportional to the scalar product of the electron and hole spins, leads to a splitting of the ground state manifold that, within the EMA, scales as R^{-3} with the QD radius[6]. In the $\Delta_{SO} \to \infty$ limit, the $1s_e 1S_{3/2}$ octet splits into two groups of states, characterised by the total angular momentum $J = 1, 2$. Additionally, the degeneracy is lifted by the crystal field correction if the underlying QD material has hexagonal structure (e.g. for CdSe dots). These effects were considered in detail in [9, 10]. The result is that there are five energy levels corresponding to the exciton states with different total angular momentum, denoted 0^U, 0^L, $\pm 1^U$, $\pm 1^L$ and ± 2. The energies and the wavefunctions of these states are determined by two parameters,

$$\eta = \frac{\varepsilon_{exch} a_0^3}{3\pi R} \int_0^R \sin^2\left(\frac{\pi r}{R}\right) \left(R_0^2(r) + \frac{1}{5} R_2^2(r)\right) dr, \qquad (2.8)$$

and

$$\Delta = \Delta_{cr} \int_0^R \left(R_0^2(r) - \frac{3}{5} R_2^2(r)\right) r^2 dr, \qquad (2.9)$$

---

[4] This coefficient can be calculated using the wavefunctions (2.1) and (2.3). It varies between 1.77 and 1.91 depending on the hole mass ratio [14]. The number that usually appears in the literature is 1.79 as calculated in one of the first works devoted to the subject [77] assuming a simple parabolic band for holes.
[5] This result was obtained assuming a simple parabolic band for holes.
[6] Microscopic calculations predict exponents between 2 and 3 for Si, GaAs, InP and CdSe QDs [13, 66, 78].

where a_0 is the lattice constant and $\varepsilon_{\text{exch}}$ and Δ_{cr} are the exchange interaction and crystal field constants of the corresponding bulk material. The corrections, with respect to E_x given by Eq. (2.7), are:

$$E_0^{U,L} = \frac{1}{2}\eta + \frac{\Delta}{2} \pm 2\eta; \quad E_{\pm 1}^{U,L} = \frac{1}{2}\eta \pm \sqrt{4\eta^2 + \frac{\Delta^2}{4} - \eta\Delta}; \quad E_{\pm 2} = -\frac{3}{2}\eta - \frac{\Delta}{2}. \quad (2.10)$$

According to Eq. (2.10), there are 5 and 2 different exciton energy levels for dots made of a hexagonal and a cubic material, respectively. The splitting can reach some 20–30 meV in small CdSe and CdTe QDs [10]. Note that the lowest energy states (± 2) are optically inactive ("dark exciton"). The corresponding exciton wavefunctions are obtained from the products of the electron and hole ones, $\Psi^e_{l,m,n,s_z}(\vec{r}_e)$ $\Psi^h_{3/2,M,1,+}(\vec{r}_h)$, by using the following transformation matrix [44]:

$$\hat{T} = \begin{pmatrix} 0 & 0 & 1/\sqrt{2} & 0 & 0 & -i/\sqrt{2} & 0 & 0 \\ 0 & 0 & 1/\sqrt{2} & 0 & 0 & i/\sqrt{2} & 0 & 0 \\ 0 & a^- & 0 & 0 & -ia^+ & 0 & 0 & 0 \\ 0 & a^+ & 0 & 0 & ia^- & 0 & 0 & 0 \\ 0 & 0 & 0 & a^+ & 0 & 0 & -ia^- & 0 \\ 0 & 0 & 0 & a^- & 0 & 0 & ia^+ & 0 \\ 1 & 0 & 0 & 0 & 0 & 0 & 0 & 0 \\ 0 & 0 & 0 & 0 & 0 & 0 & 0 & 1 \end{pmatrix} \quad (2.11)$$

where

$$a^{\pm} = \sqrt{\frac{\sqrt{16\eta^2 + \Delta^2 - 4\eta\Delta} \pm (2\eta - \Delta)}{2\sqrt{16\eta^2 + \Delta^2 - 4\eta\Delta}}}.$$

The columns on the matrix (2.11) correspond to $(3/2, \uparrow)$, $(1/2, \uparrow)$, $(-1/2, \uparrow)$, $(-3/2, \uparrow)$, $(3/2, \downarrow)$, $(1/2, \downarrow)$, $(-1/2, \downarrow)$ and $(-3/2, \downarrow)$, while the rows are related to the 0^U, 0^L, $+1^U$, $+1^L$, -1^U, -1^L, $+2$ and -2 states (in this order).

In the opposite limit of $\Delta_{\text{SO}} \approx 0$ the exchange interaction splits the (12-fold degenerate) $1s_e 1S_1$ exciton manifold into a lower "triplet" state (nine-fold degenerate) and an upper "singlet" state (three-fold degenerate) [13]. The introduction of the spin-orbit interaction as perturbation leads to the splitting of the triplet state into three states with $J = 0, 1, 2$. Again, the lowest energy state ($J = 2$) is optically inactive[7]. Expressions for the corrections to the exciton energy, analogous to Eq. (2.10), can be found in [13].

For QDs made from indirect-gap materials like Si and Ge, the situation is perhaps too complex to be realistically described in the framework of a macroscopic model, even though the EMA consideration of the excitonic effect in such dots [80] led to some important qualitative conclusions. Excitons do exist in such dots and they can even recombine directly (i.e. without participation of phonons) via emission of a

---

[7] However, recent microscopic calculations performed for CdS QDs [79] do not support this prediction.

photon because the wave vector is not a good quantum number anymore and its conservation in optical transitions is "relaxed". However, it is not possible to construct the exciton states from those of free electrons and holes (which already can be obtained only with quite a limited validity) in a reasonably simple way because the Coulomb and exchange interactions are of the same order as the spin-orbit and intervalley couplings. In contrast with QDs made of a direct gap material, the electron–hole Coulomb interaction alone can give rise to a splitting of "dark" states even in the absence of exchange interaction [81]. The interested reader should refer to the book of Delerue and Lannoo [13] for discussion of the exciton spectra of Si QDs, calculated numerically within the tight-binding approach.

2.3 Phonons. Following the same strategy as in Sect. 2.2, we present here continuous models for both acoustic and optical phonons that lead to (semi-)analytical expressions for the atomic displacements and (in case of polar modes) for the associated electrostatic potential. Even though there exists a considerably broad literature devoted to phonons in nanostructures, including journal papers and even books [82], we shall present the equations and the final results, in order to clarify some doubtful points, approximations and limitations of the approach pursued.

Starting by *acoustic phonons*, the harmonic wave equation of the classical elasticity theory reads [83]:

$$-\omega^2 \rho(\vec{r}) u_j(\vec{r}) = \frac{\partial}{\partial x_k} \sigma_{jk}(\vec{r}) \qquad (2.12)$$

where ω is the frequency, ρ the mass density, u_j ($j=1$–3) are the components of the displacement vector and σ_{jk} is the elastic stress tensor. For a cubic crystal, neglecting elastic anisotropy, the components of the stress tensor are related to those of the strain tensor, ε_{ij}, through the Hooke law,

$$\sigma_{ij} = C_{12} \varepsilon_{ll} \delta_{ij} + 2 C_{44} \varepsilon_{ij} \qquad (2.13)$$

where C_{ij} are the elastic constants and ε_{ll} is the trace of the strain tensor. For a QD modelled by an elastic sphere, one has to use spherical coordinates (the strain components, expressed in terms of the displacement components, are given in [83]) and perform the covariant derivation in Eq. (2.12). A simplified version of the resulting equation (assuming $\rho=$const and $C_{ij}=$const) can be found, e.g. in [28, 30, 82]. This equation was solved by Lamb, more than 100 years ago, for the case of a *free-standing sphere* [29], i.e. by applying the zero-strain condition at the surface, $\sigma_{rr}(R) = \sigma_{r\theta}(R) = \sigma_{r\phi}(R) = 0$. The solution consists of a set of spheroidal modes[8] enumerated by the three "spherical" quantum numbers (l_p, m_p and n_p), with discrete frequencies, ω_{l_p,n_p}, that are ($2l_p + 1$)-fold degenerate with respect to m_p. The frequencies can be expressed as $\omega_{l_p,n_p} = \eta_{l_p,n_p}(c_L/R)$ where η_{l_p,n_p} is the n_p-th root of the equation,

$$2\xi j_{l_p+1}(\xi\eta)\{\eta^2 + (l_p-1)(l_p+2)[\eta j_{l_p+1}(\eta) - (l_p+1) j_{l_p}(\eta)]\} - j_{l_p}(\xi\eta)\{j_{l_p}(\eta)[\eta^4/2 \\ + (l_p-1)(2l_p+1)\eta^2] + \eta j_{l_p+1}(\eta)[\eta^2 - 2l_p(l_p-1)(l_p+2)]\} = 0, \qquad (2.14)$$

---
[8] There are also torsional modes, however, they are not Raman-active [84] and therefore difficult to observe.

$\xi = c_T/c_L$ and $c_L = \sqrt{C_{11}/\rho}$ and $c_T = \sqrt{C_{44}/\rho}$ are the longitudinal and transversal sound velocities, respectively (remember that $C_{11} = C_{12} + 2C_{44}$). Note that Eq. (2.14) is valid for $l_p > 0$ and the case of $l_p = 0$ is considered below. The Lamb's solution has been widely used to explain the experimentally observed structure in low-frequency Raman scattering spectra of semiconductor nanoparticles [24–28], even though the stress-free boundary conditions can hardly be a good approximation for a QD embedded in a matrix.

Let us consider the problem in more detail limiting ourselves by the mathematically simpler case of $l_p = 0$ where the displacement vector has only a radial component, u, depending only on r. The equation of motion (2.12) takes the following form:

$$-\omega^2 \rho(r) u(r) = \frac{d}{dr}\left[C_{11}\left(\frac{du}{dr} + 2\frac{u}{r}\right)\right] - 4\frac{u}{r}\frac{dC_{44}}{dr}. \tag{2.15}$$

Eq. (2.15) can be easily solved for a homogeneous sphere, with the solution given by the first spherical Bessel function,

$$u(r) = A j_1(kr), \tag{2.16}$$

where $k = \omega c_L$ and $A = \text{const}$. Applying the stress-free boundary condition,

$$\sigma_{rr}(R) = \left(C_{11}\frac{\partial u}{\partial r} + 2C_{12}\frac{u}{r}\right)\bigg|_{r=R} = 0,$$

one obtains the equation for the Lamb's $l_p = 0$ modes,

$$4(c_T/c_L)^2 j_1(kR) - (kR) j_0(kR) = 0. \tag{2.17}$$

Like Eq. (2.14), this equation predicts quantised phonon frequencies. The same happens in the opposite limit of the rigidly fixed surface, where the mode's frequencies are determined by the equation

$$j_1(kR) = 0. \tag{2.18}$$

A simple analytical solution of Eq. (2.15) is also possible for the case of a QD embedded in a homogeneous elastic matrix. The outgoing wave in the matrix can be written in the form [31, 33],

$$u(r)|_{r \geq R} = \frac{\sin(k_M r - \delta)}{(k_M r)^2} - \frac{\cos(k_M r - \delta)}{k_M r} \tag{2.19}$$

where $k_M = \omega/c_L^{(M)}$, $c_L^{(M)}$ is the longitudinal sound velocity in the matrix and δ is an unknown phase shift. By matching u and σ_{rr} corresponding to Eqs. (2.16) and (2.19) at the interface ($r = R$) one obtains two equations for two constants δ and A that have a solution *for any k*. In other words, the phonon frequency is *not* quantised. As an example, Fig. 3 shows the density of $l_p = 0$ phonon states calculated for a CdSe dot embedded in silica ($c_L = 3.59 \times 10^3$ m/s, $c_L^{(M)} = 5.95 \times 10^3$ m/s [85]), $g(E) = \sum_k \delta(E - \omega_k^2)$, and the local density of states (LDS) integrated over the QD,

$$g_{NC}(E) = 4\pi \sum_k \left\{\int_0^R C^{(k)} |u^{(k)}(r)|^2 \rho(r) r^2 dr\right\} \delta(E - \omega_k^2), \tag{2.20}$$

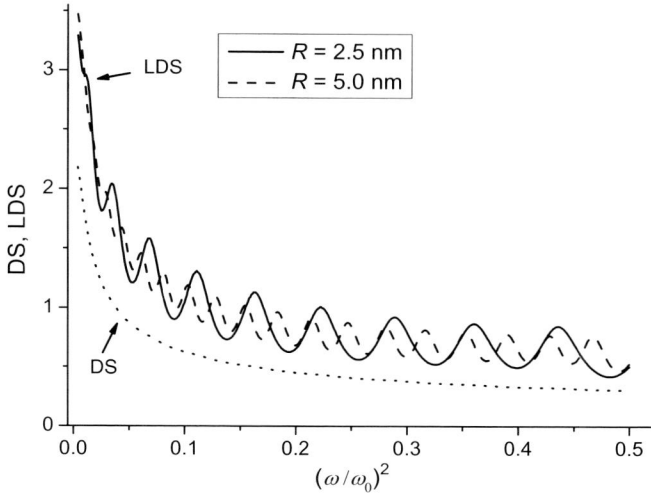

Fig. 3. Density of states and LDS (Eq. 2.20) of spherically symmetric acoustic phonon modes calculated for CdSe QDs of two different sizes, embedded in SiO_2 matrix

where $C^{(k)}$ is a normalisation constant defined by the normalisation condition,

$$4\pi C^{(k)} \int_0^\infty |u^{(k)}(r)|^2 \rho(r) r^2 dr = 1.$$

As one can see from Fig. 3, the total density of states (DS) does not show any QD related features and does not depend on R. The LDS shows oscillations originating from the interference phenomenon between the outgoing acoustic wave and the one reflected at the QD/matrix interface. Since this reflection is only partial, no complete confinement occurs. The LDS peaks correspond to the k numbers for which the (partial) interference inside the dot is constructive. This structure can be called "pseudo-quantisation" [31]. The peak's positions lie between the frequencies predicted by Eq. (2.17) (Lamb's modes) and those determined by Eq. (2.17), moreover, they depend on the matrix [85].

As it has been mentioned in Sect. 1, acoustic wave quantisation in nanoparticles embedded in different matrices has been investigated by several groups using low-frequency Raman scattering [24–27]. Based on the arguments presented above, we believe that the observed structure is characteristic of the pseudo-quantisation, rather than of truly discrete spectra of acoustic phonons in nanostructures. In fact, as pointed out in [86], confined modes in semiconductor nanoparticles were not directly observed as individual spectral peaks but rather as component ones obtained from the decomposition of broadened bands. This makes it difficult to establish the precise positions of the pseudo-quantised acoustic phonon modes, especially in the presence of the nanoparticle size dispersion. The authors of [86] performed their experiments on synthetic opals which are matrix-free arrays of closely packed SiO_2 spheres having diameters ranging from 200 to 340 nm and obtained nicely resolved spectral peaks whose position agree quite well with the Lamb's theory. The situation is more complex for semiconductor QDs.

Turning to the *optical phonons*, a number of continuum models have been described in the literature, overviewed in [14, 82]. In the simplest approximation for optical phonons in semiconductor nanostructures, known as dielectric continuum (DC) model, one solves the electrostatics equations for a (confined) polar medium with a frequency-dependent dielectric function describing its bulk phonon response. When applied to a spherical particle, this model predicts one surface (or interface) phonon mode for each angular momentum value $l_p > 0$ [87]. For a semiconductor heterostructure, the DC model yields interface and confined modes, both of which were observed experimentally [88]. However, further experiments on short-period superlattices showed that the confined and interface modes intermix. Moreover, microscopic calculations [89] have demonstrated that the relative ionic displacement field is continuous across the interface, contrary to the results obtained from the DC model. A number of alternative macroscopic approaches were proposed, in particular, the phonon dispersion was incorporated [90]. The most successful one is due to [91–93]. This model foresees the intermixing of the confined LO and TO as well as interface phonon modes and has been used to quantitatively describe the phonon-related properties of various semiconductor nanostructures including spherical QDs [14–16, 21, 93–99].

Following this approach, the spatially quantised optical phonon modes in a spherical QD made of an isotropic material (modelling a cubic crystal with two atoms per unit cell) are obtained by solving the following equation of motion [93],

$$\tilde{\rho}(\vec{r})(\omega^2 - \omega_{TO}^2(\vec{r}))\vec{u} = \vec{\nabla}[\tilde{\rho}(\vec{r})\beta_L(\vec{r})(\vec{\nabla} \cdot \vec{u})] - \vec{\nabla} \times [\tilde{\rho}(\vec{r})\beta_T(\vec{r})(\vec{\nabla} \times \vec{u})] + \alpha(\vec{r})\vec{\nabla}\varphi, \quad (2.21)$$

where $\vec{u}$ is the *relative* displacement vector, $\tilde{\rho}$ the atomic reduced mass density, ω_{TO} the bulk TO phonon frequency, β_L and β_T are phenomenological curvature parameters of the bulk phonon dispersion curves and α is the polarisability coefficient (equal to zero for non-polar materials like Si) that can be expressed as $\alpha = \sqrt{\tilde{\rho}(\omega_{LO}^2 - \omega_{TO}^2)/(4\pi)}$ with ω_{LO} denoting the bulk LO phonon frequency. Except for the last term, Eq. (2.21) is analogous to the acoustic wave equation[9], with the following correspondence: $-\tilde{\rho}\beta_L \to C_{11}$; $-\tilde{\rho}\beta_T \to C_{44}$; $\tilde{\rho} \to \rho$. General mechanical boundary conditions at an interface include the continuity of the displacement vector and the normal force, $\vec{\sigma} = (\sigma_{rr}, \sigma_{r\theta}, \sigma_{r\phi})$, with the "stress" components defined analogously to Eq. (2.13). However, if the optical phonon bands of the two materials do not overlap (as it is, for example, the case of all the most important semiconductors and silica glass), $\vec{u} = 0$ at the interface is a rather good approximation [15, 93].

The electrostatic potential, φ, associated with the optical phonon, is determined self-consistently through the Poisson equation,

$$\nabla^2 \varphi = \frac{4\pi\alpha(\vec{r})}{\varepsilon_\infty(\vec{r})}(\vec{\nabla} \cdot \vec{u}), \quad (2.22)$$

where ε_∞ is the high-frequency dielectric constant. Boundary conditions include the continuity of φ and $(\varepsilon_\infty \partial \varphi / \partial r)$ at the NC/matrix interface. Solving coupled

---

[9] There is, however, an important difference related to the non-zero TO phonon frequency, which is discussed below.

Eqs. (2.21) and (2.22) in spherical coordinates and applying the boundary conditions, one obtains a complete set of confined optical phonon modes (see [15, 93] for details). The frequencies of the spheroidal modes[10], ω_{l_p,n_p}, are determined by the successive roots of the following equation:

$$qRj''_{l_p}(qR)\{\gamma\varepsilon^S_\infty l_p[l_p g_{l_p}(kR) - kRg'_{l_p}(kR)] + \delta_{l_p}[g_{l_p}(kR) + kRg'_{l_p}(kR)]\} \\ - l_p(l_p+1)j_{l_p}(qR)\{\gamma\varepsilon^M_\infty[kRg'_{l_p}(kR) - l_p g_{l_p}(kR)] + \delta_{l_p}g_{l_p}(kR)\} = 0 \quad (2.23)$$

where $q = \sqrt{(\omega^2_{LO} - \omega^2)/\beta_L}$, $k = \sqrt{|(\omega^2_{TO} - \omega^2)/\beta_T|}$, $\gamma = (\omega^2_{LO} - \omega^2_{TO})/(\omega^2 - \omega^2_{TO})$, $g_{l_p}(x)$ is either $j_{l_p}(x)$ or $[i^{-1}j_{l_p}(ix)]$ depending on the sign of $(\omega^2_{TO} - \omega^2)/\beta_T$ (positive or negative, respectively), $\delta_{l_p} = l_p\varepsilon^S_\infty + (l_p+1)\varepsilon^M_\infty$, and ε^S_∞ and ε^M_∞ are the high-frequency dielectric constants of the sphere and the matrix, respectively. Equation (2.23) is valid for $l_p \geq 1$, while for $l_p = 0$ it reduces to

$$\tan(qR) = qR. \quad (2.24)$$

As an example, Fig. 4 shows the radial dependence of the (Raman-active) modes with $l_p = 0, 2$, confined in a CdSe QD embedded in glass.

The spherical components of the displacement vector are given by [23]:

$$u^r_{l_p,m_p,n_p}(\vec{r}) = \left(\frac{\hbar}{2\tilde{\rho}\omega_{l_p,n_p}}\right)^{1/2} A_{l_p,n_p} v_{l_p,n_p}(r) Y_{l_p,m_p}(\vartheta,\phi),$$

$$u^\vartheta_{l_p,m_p,n_p}(\vec{r}) = \left(\frac{\hbar}{2\tilde{\rho}\omega_{l_p,n_p}}\right)^{1/2} A_{l_p,n_p} w_{l_p,n_p}(r) \frac{\partial}{\partial\vartheta} Y_{l_p,m_p}(\vartheta,\phi), \quad (2.25)$$

$$u^\phi_{l_p,m_p,n_p}(\vec{r}) = \left(\frac{\hbar}{2\tilde{\rho}\omega_{l_p,n_p}}\right)^{1/2} A_{l_p,n_p} w_{l_p,n_p}(r) \frac{1}{\sin\vartheta} \frac{\partial}{\partial\phi} Y_{l_p,m_p}(\vartheta,\phi),$$

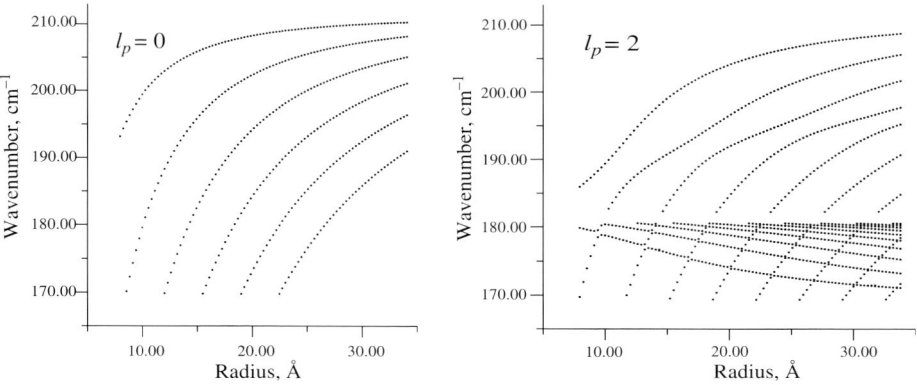

Fig. 4. Radial dependence of the frequencies of the confined optical phonon modes calculated for CdSe QDs embedded in SiO$_2$ matrix. Reproduced from [14] with permission of John Wiley & Sons Ltd.

---

[10] Again, there are also torsional modes but they are far-infrared and Raman-inactive.

where

$$v_{l_p,n_p}(r) = q_{l_pn_p}Rj'_{l_p}(q_{l_pn_p}r) + \zeta_{l_p,n_p}l_p(r/R)^{l_p-1} + \frac{\eta_{l_p,n_p}l_p(l_p+1)R}{r}g_{l_p}(k_{l_pn_p}r),$$

$$w_{l_p,n_p}(r) = \frac{R}{r}\{j_{l_p}(q_{l_pn_p}r) + \zeta_{l_p,n_p}(r/R)^{l_p} + \eta_{l_p,n_p}[g_{l_p}(k_{l_pn_p}r) + k_{l_pn_p}rg'_{l_p}(k_{l_pn_p}r)]\},$$

$$\zeta_{l_p,n_p} = \frac{q_{l_pn_p}Rj'_{l_p}(q_{l_pn_p}R)[g_{l_p}(k_{l_pn_p}R) + k_{l_pn_p}Rg'_{l_p}(k_{l_pn_p}R) - l_p(l_p+1)j_{l_p}(q_{l_pn_p}R)g_{l_p}(k_{l_pn_p}R)]}{l_p[l_pg_{l_p}(k_{l_pn_p}R) - k_{l_pn_p}Rg'_{l_p}(k_{l_pn_p}R)]},$$

$$\eta_{l_p,n_p} = \frac{l_pj_{l_p}(q_{l_pn_p}R) - q_{l_pn_p}Rj'_{l_p}(q_{l_pn_p}R)}{l_p[l_pg_{l_p}(k_{l_pn_p}R) - k_{l_pn_p}Rg'_{l_p}(k_{l_pn_p}R)]}$$

and the normalization constant, A_{l_p,n_p}, is determined by

$$A_{l_p,n_p} = \left\{\int_0^R [v_{l_p,n_p}^2(r) + l_p(l_p+1)w_{l_p,n_p}^2(r)]r^2 dr\right\}^{-1/2}.$$

The corresponding electrostatic potential (vanishing for non-polar modes) is given by

$$\varphi_{l_p,m_p,n_p}(\vec{r}) = \frac{C_F}{e}\Phi_{l_p,n_p}(r)Y_{l_p,m_p}(\vartheta,\phi), \qquad (2.26)$$

where $C_F = e\sqrt{2\pi\hbar\omega_{LO}[(\varepsilon_\infty^S)^{-1} - (\varepsilon_0^S)^{-1}]/R}$ is the Fröhlich constant, ε_0^S is the static dielectric constant of the sphere and the dimensionless function Φ in Eq. (2.26) is

$$\Phi_{l_p,n_p}(r) = A_{l_p,n_p}[j_{l_p}(q_{l_pn_p}r) + \gamma_{l_pn_p}\zeta_{l_p,n_p}(r/R)^{l_p}].$$

Note that the above expressions are valid for $l_p > 0$. For $l_p = 0$, one has much simpler formulae [91],

$$u^r_{0,0,n_p}(r) = \left(\frac{\hbar}{4\pi R^3 \tilde{\rho}\omega_{0,n_p}}\right)^{1/2} \frac{j_1(q_{0n_p}r)}{|j_0(q_{0n_p})|},$$

$$\Phi_{0,n_p}(r) = \sqrt{2}\frac{[j_0(q_{l_pn_p}) - j_0(q_{l_pn_p}r)]}{|\sin(q_{l_pn_p})|}. \qquad (2.27)$$

It is important to note that the confined optical phonon modes, in general, have a mixed longitudinal-transverse-interface nature. The displacement vector of a mode with certain n_p, l_p and m_p contains each of these three components. This is an effect of the confined geometry. The only exception is the $l_p = 0$ mode which is purely radial, i.e. longitudinal. Contrary to the acoustic phonons, the optical phonon modes in QD nanostructures are truly quantised. This is due to the presence of $\omega_{TO} \neq 0$ in Eq. (2.21), otherwise very similar to that for acoustic phonons. The term $\tilde{\rho}(\vec{r})\omega_{TO}^2(\vec{r})$ plays the role of the potential energy for electrons and leads to the confinement of optical phonons.

The spatial quantisation of the optical phonon modes in QDs has been nicely demonstrated by means of far-infrared spectroscopy [94–97]. It is more difficult with

Raman spectroscopy because the Raman-active modes with $l_p=0$, 2 and small n_p numbers have frequencies that differ from ω_{LO} just by few cm^{-1} and the QD size dispersion hinders the discrete structure of the spectrum of individual dots. Nevertheless, there are published works where the experimentally observed Raman spectra were successfully modelled by integrating the individual dot spectra, calculated within the formalism presented above, with appropriate QD size distributions [14, 16, 98, 99].

2.4 Interaction mechanisms and coupling constants. The (one-particle) electronic states and the excitonic states discussed in Sects. 2.1 and 2.2, respectively, were obtained within the Born–Oppenheimer approximation assuming that the atoms are frozen. Similarly, the consideration of phonons in Sect. 2.3 ignored the electronic degrees of freedom. Now we are going to consider the linear coupling between the atomic and electronic motions, leading to the ex–ph interactions. The mechanisms and the Hamiltonians of these interactions have been derived for bulk semiconductors many years ago [100, 101]. Within the spirit of our approach, we shall consider the same Hamiltonians, however, involving the electronic states and phonon modes characteristic of spherical QDs.

The most universal mechanism of coupling between the electrons and *acoustic phonons* is through the volume deformation potential. The bottom of a non-degenerate band (e.g. Γ_6 conduction band in materials with zinc-blend structure) is shifted proportionally to the (local) relative variation of the volume,

$$\widehat{H}'_{e-AP} = a_c \sum_\nu (\vec{\nabla} \cdot \vec{u}_\nu(\vec{r})), \qquad (2.28)$$

where a_c is the bulk deformation potential constant and $\vec{u}_\nu$ denotes the atomic displacement for a phonon mode $\nu \equiv \{l_p, m_p, n_p\}$. For degenerate conduction band minima located outside the Γ point in one of the high-symmetry directions in the Brillouin zone ([1 0 0] for Si, [1 1 1] for Ge), the Hamiltonian involves two deformation potential constants, Ξ_d and Ξ_u, and has the form [102]

$$\widehat{H}'_{e-AP} = \sum_\nu [\Xi_d \varepsilon^{(\nu)}_{ll} + \Xi_u \varepsilon^{(\nu)}_{ij} q_i^{\min} q_j^{\min}], \qquad (2.29)$$

where $\varepsilon^{(\nu)}_{ij}$ is the strain tensor corresponding to the phonon mode ν, $\varepsilon^{(\nu)}_{ll} \equiv (\vec{\nabla} \cdot \vec{u}_\nu(\vec{r}))$ and $\vec{q}^{\min}$ is the location of the conduction band minimum in the Brillouin zone. The situation is more complex for holes because the strain produces not only a shift but also a splitting and a mixing of the valence bands. The corresponding Hamiltonian for the $J=3/2$ bands, $\widehat{H}'_{h-AP}$, was derived by Bir and Pikus [100], involves three different deformation potential constants and can be found also in [102]. However, if we consider only spherically symmetric $l_p=0$ phonon modes for which $\varepsilon_{ij} = \varepsilon \delta_{ij}$, then $\widehat{H}'_{h-AP}$ takes the simple form of Eq. (2.28). Then the ex–ph interaction Hamiltonian can be written as [75]

$$\widehat{H}'_{xe-AP}(\vec{r}_e, \vec{r}_h) = \sum_\nu [a_c(\vec{\nabla}_e \cdot \vec{u}_\nu(\vec{r}_e)) - a_v(\vec{\nabla}_h \cdot \vec{u}_\nu(\vec{r}_h))]. \qquad (2.30)$$

Assuming bulk-like acoustic phonons[11], i.e. considering longitudinal plane waves of the form $u_{\vec{k}}^z(\vec{r},t) = \sqrt{2\hbar/(\rho\omega V)}\,\text{Re}\exp(ikz - i\omega t)$, diagonal matrix elements of the Hamiltonian (2.30) for each of the exciton states constituting the $1s_e 1S_{3/2}$ octet, calculated using the wavefunctions (2.1)–(2.4), are given by

$$g_k = \left(\frac{2\hbar k}{c_L \rho V}\right)^{1/2} [a_c f_1(kR) - a_v f_2(kR)] \quad (2.31)$$

where

$$f_1(kR) = 2\pi^2 \int_0^1 j_0(kRx) j_0^2(\pi x) x^2\, dx,$$

$$f_2(kR) = \int_0^R j_0(kr)[R_0^2(r) + R_2^2(r)] r^2\, dr.$$

The functions f_1 and f_2, both equal to unity for $kR = 0$, decrease rapidly as the argument increases. A more elaborate calculation performed using the solution (2.19), taking into account the partial reflection of the acoustic waves at the QD/matrix interface, leads to the result similar to Eq. (2.31), however, with an oscillatory modulation of the variation of g_k with kR [31], similar to Fig. 3.

The result presented above corresponds to the strong confinement regime. In the opposite case of weak confinement, the situation is quite similar to bulk excitons. The deformation potential interaction between acoustic phonons and bulk exciton was considered some 50 years ago [104] and the main results can be found also in [75]. The matrix element can be expressed in the form (2.31) with the functions f_1 and f_2 replaced by $f_{1,2}(ka_{\text{ex}}) = [1 + (\beta_{1,2} ka_{\text{ex}})^2]^{-1}$ where $\beta_{1,2} = m_{e,h}/(m_e + m_h)$, m_e and m_h are the electron and hole masses.

Beyond the short-range deformation potential interaction, there exists a long-range coupling between excitons and acoustic phonons in non-centrosymmetric polar crystals (e.g. wurtzite CdSe and CdS) because of the piezoelectric effect. The corresponding Hamiltonian and calculation of the matrix elements can be found in [105].

As far as *optical phonons* are concerned, also there is a universal short-range ex–ph interaction mechanism mediated by a deformation potential. In bulk semiconductors with cubic structure, for long-wavelength optical phonons, the optical deformation potential (ODP) coupling vanishes by symmetry for any non-degenerate band (e.g. Γ_2 conduction band in Ge or Γ_6 conduction band in materials with zinc-blend structure) [106] but it is non-zero for holes near the top of the valence band. The short-range ODP interaction is proportional to the relative displacement of two atoms in the same unit cell, with the corresponding operator given by [107],

$$\widehat{H}'_{\text{ex-OP}} = \frac{\sqrt{3}d_0}{2a_0}\sum_\nu (\vec{D}\cdot\vec{u}_\nu(\vec{r})), \quad (2.32)$$

---

[11] This assumption, yet simple, is not quite consistent with the form (2.30) of the ex–ph interaction Hamiltonian. Still it has been used by several authors [7, 103].

Table 1. Diagonal ex–ph coupling constants g_ν calculated for different $1S_{3/2}$ hole states and confined phonon modes with different angular momenta and $m_p = 0$

	M			
l_p	3/2	1/2	−1/2	−3/2
0	J_{0,n_p}	J_{0,n_p}	J_{0,n_p}	J_{0,n_p}
1	$C_{1,n_p} I^+_{1,0,n_p}$	$C_{1,n_p} I^-_{1,0,n_p}$	$C_{1,n_p} I^-_{1,0,n_p}$	$C_{1,n_p} I^+_{1,0,n_p}$
2	$-J_{2,n_p}$	J_{2,n_p}	J_{2,n_p}	$-J_{2,n_p}$
3	$C_{3,n_p} I^+_{3,0,n_p}$	$C_{3,n_p} I^-_{3,0,n_p}$	$C_{3,n_p} I^-_{3,0,n_p}$	$C_{3,n_p} I^+_{3,0,n_p}$

where a_0 is the lattice constant and d_0 is the ODP constant for the Γ^v_{15} valence band [106] and $\vec{u}_\nu$ is the relative displacement of two ions in the same unit cell for a phonon mode $\nu \equiv \{l_p, m_p, n_p\}$. Considering only the light and heavy hole sub-bands ($\Delta_{SO} \to \infty$), the vector $\vec{D}$ in Eq. (2.28) is constituted by numerical 4×4 matrices, (D_x, D_y, D_z), with columns and rows corresponding to the Bloch states with different angular momentum projection $J_z = 3/2, 1/2, -1/2, -3/2$ (in this order),

$$D_x = \begin{pmatrix} 0 & -1/\sqrt{3} & 0 & 0 \\ -1/\sqrt{3} & 0 & 0 & 0 \\ 0 & 0 & 0 & 1/\sqrt{3} \\ 0 & 0 & 1/\sqrt{3} & 0 \end{pmatrix}; \quad D_y = \begin{pmatrix} 0 & i/\sqrt{3} & 0 & 0 \\ -i/\sqrt{3} & 0 & 0 & 0 \\ 0 & 0 & 0 & -i/\sqrt{3} \\ 0 & 0 & i/\sqrt{3} & 0 \end{pmatrix};$$

$$D_z = \begin{pmatrix} 0 & 0 & i/\sqrt{3} & 0 \\ 0 & 0 & 0 & i/\sqrt{3} \\ -i/\sqrt{3} & 0 & 0 & 0 \\ 0 & -i/\sqrt{3} & 0 & 0 \end{pmatrix}. \quad (2.33)$$

In the limit of $\Delta_{SO} \approx 0$, the matrices have the dimensions 3×3, with columns and rows corresponding to the Bloch functions $|0\rangle$ and $|\pm 1\rangle$ appearing in Eq. (2.6). They are quite simply related to the matrices (D_x, D_y, D_z) corresponding to the Γ_4 band [107]. The general case of the 6×6 Konh–Luttinger Hamiltonian is considered in [108].

Calculation of the ODP matrix elements, $\langle M, s_z | \widehat{H}'^{(\nu)}_{\text{ex-OP}} | M, s_z \rangle$, for the $1s_e 1S_{3/2}$ exciton states was performed in [23] by expressing the scalar product in Eq. (2.32) in terms of spherical components of the displacement vector and expanding three terms containing D_x, D_y and D_z in series of spherical harmonics. The diagonal matrix elements, non-zero for phonons with $l_p = 1, 3$, are summarized in Table 1, expressed in terms of the overlap integrals[12],

$$I^\pm_{1,0,n_p} = \frac{2d_0}{5\sqrt{3}\pi} \int_0^R \left(R_0(r) R_2(r) \pm \frac{1}{7} R_2^2(r) \right) [v_{l_p,n_p}(r) - w_{l_p,n_p}(r)] r^2 dr; \quad (2.34)$$

---

[12] Some unimportant phase factors have been omitted in Table 1. Note also that the definition of spherical harmonics may vary by a factor of i^l. Here we used the spherical harmonics as defined in [109] since the explicit expressions for the hole wavefunctions, Eqs. (2.3) and (2.6), were obtained using the 3j-symbols from the same source. There is some discrepancy between the previously published results [23] and those presented here.

$$I^{\pm}_{3,0,n_p} = \frac{2d_0}{5\sqrt{7\pi}} \int_0^R \left\{ R_0(r)R_2(r)[v_{l_p,n_p}(r) + 4w_{l_p,n_p}(r)] \right.$$
$$\left. \pm R_2^2(r)\left[\frac{1}{3}v_{l_p,n_p}(r) - 2w_{l_p,n_p}(r)\right] \right\} r^2 dr; \qquad (2.35)$$

The functions appearing in the above expressions have been defined in the previous sections and the dimensionless constants C_{l_p,n_p} are given by

$$C_{l_p,n_p} = \frac{\sqrt{3}}{2a_0}\left(\frac{\hbar}{2\tilde{\rho}\omega_{l_p,n_p}}\right)^{1/2} A_{l_p,n_p}$$

with corresponding l_p and n_p. Note that only holes are involved in this interaction and the ODP matrix elements scale with the QD radius approximately as $R^{-3/2}$.

In addition to $\hat{H}'_{\text{ex-OP}}$, the long-range electrostatic field associated with the optical phonons in polar materials introduces a stronger coupling mechanism, the Fröhlich-type interaction described by the operator

$$\hat{H}''_{\text{ex-OP}} = e\sum_\nu [\varphi_\nu(\vec{r}_h) - \varphi_\nu(\vec{r}_e)]. \qquad (2.36)$$

The corresponding matrix elements were calculated for the $1s_e 1S_{3/2}$ exciton states in [21] and are non-zero only for phonons with $l_p = 0, 2$. The diagonal ones, $\langle M, s_z | \hat{H}''^{(\nu)}_{\text{ex-OP}} | M, s_z \rangle$, are given in Table 1, expressed in terms of the following two integrals,

$$J_{0,n_p} = \frac{C_F}{\sqrt{4\pi}} \int_0^R [R_0^2(r) + R_2^2(r) - \psi_{001}^2(r)] \Phi_{0,n_p}(r) r^2 dr, \qquad (2.37)$$

and

$$J_{2,n_p} = \frac{C_F}{\sqrt{5\pi}} \int_0^R R_0(r)R_2(r) \Phi_{2,n_p}(r) r^2 dr. \qquad (2.38)$$

One can also take into account the effects of electron–hole exchange interaction and hexagonal structure of the underlying material that mix the $|M, s_z\rangle$ pair states and lift their degeneracy as discussed in Sect. 2.2. In order to obtain the ex–ph interaction matrix in the basis of the $0^U, 0^L, +1^U, +1^L, -1^U, -1^L, +2$ and -2 states, one can use the matrix elements defined above (written in the $|M, s_z\rangle$ representation) and the transformation matrix (2.11). The result can be found in [23].

The dependence of the coupling constants (2.37) and (2.38) on the QD radius for CdSe dots is shown in Fig. 5. While these parameters are much larger than those corresponding to the ODP mechanism for II–VI QDs, the two coupling mechanisms are of comparable importance for III–V dots. This has been confirmed by means of Raman spectroscopy, which is one of the techniques most directly related to the ex–ph interaction. While resonant Raman scattering spectra obtained from nearly spherical CdS, CdSe and CdTe QDs normally present a single asymmetric peak centred slightly below ω_{LO} [14], those recorded from InP QDs prepared in a similar

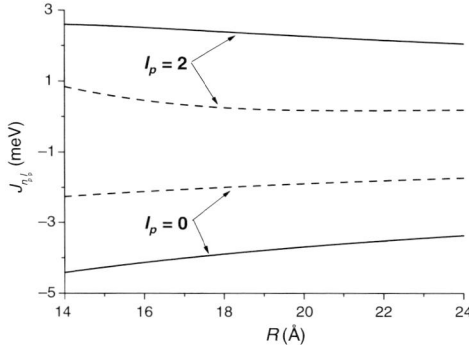

Fig. 5. Fröhlich-type coupling constants between exciton and confined optical phonons as functions of the dot radius, calculated for CdSe QDs embedded in SiO$_2$ matrix. The solid lines represent the phonon modes with $n_p = 1$ and the dashed lines the $n_p = 2$ modes. Adapted from [44] with permission of the American Physical Society

way contain an extra peak situated near the TO phonon frequency of bulk InP [110–112], which can be explained only if the ODP-mechanism-mediated scattering by $l_p = 1$ and especially $l_p = 3$ phonon modes is taken into account. The experimentally observed trends in the variation of the relative intensity of the "anomalous" TO-type Raman peak with the QD size has been shown to agree quite well with the theoretical predictions based on the above consideration of the ex–ph interaction occurring through the ODP and Fröhlich-type mechanisms [112].

Let us finish this section with a remark concerning the weak confinement regime where the effect of spatial confinement on phonons is negligible and can be included into consideration rather easily for excitons. The reader can refer to the papers [113, 114] where detailed calculations of the ex–ph coupling rates were performed for excitons in quantum wells. These results, with small modifications, can be applied also to quantum dots.

3. Polaron states

3.1 Franck–Condon progression.
Since excitons and phonons interact, their consideration as independent entities, even with modified energy spectra, is an approximation. Strictly speaking, they form a new quasi-particle called polaron. In bulk semiconductors, free excitons should be considered as free polarons, however, the polaron energy spectrum differs from that of the exciton only by a (small) renormalisation of the effective mass [2, 74]. The situation is different if the "bare" exciton spectrum is discrete, the case of an exciton bound to an impurity or confined in a QD. The δ-like density of bare states leads to an increased probability of multiphonon processes even for the modest values of the ex–ph coupling constants. Consequently, the polaron spectrum cannot be obtained within a (finite-order) perturbation theory.

An exact solution of the many-body problem is possible for the case that the interaction with the vibrational degrees of freedom (hereafter designated by "lattice", L) is described by a diagonal coupling to a single non-degenerate bare exciton state, i.e. if the interaction gives place only to virtual transitions. In the second quantisation formalism, the Hamiltonian of such a system is written as [2]

$$\hat{H}^{(1)}_{\text{ex}-L} = E_0 \hat{c}^+ \hat{c} + \sum_\nu \hbar\omega_\nu \left(\hat{b}_\nu^+ \hat{b}_\nu + \frac{1}{2}\right) + \sum_\nu g_\nu \hat{c}^+ \hat{c} (\hat{b}_\nu + \hat{b}_\nu^+). \tag{3.1}$$

Here, E_0 is the bare exciton state energy, $\hat{c}^+(\hat{c})$ and $\hat{b}_\nu^+(\hat{b}_\nu)$ denote the exciton and phonon creation (annihilation) operators, respectively, and g_ν are the ex–ph coupling constants. If we assume that there can be not more than one exciton per QD, the excitons are equivalent to fermions. The Hamiltonian (3.1) can be diagonalised by making a canonical transformation first proposed by Lang and Firsov [115], that has the form $\exp(\hat{s}) \hat{H}^{(1)}_{\text{ex}-L} \exp(-\hat{s})$ where

$$\hat{s} = \hat{c}^+ \hat{c} \sum_\nu \frac{g_\nu}{\hbar\omega_\nu} (\hat{b}_\nu^+ - \hat{b}_\nu).$$

The result is called "independent boson model" and the new (polaronic) energy levels are [2]:

$$E(\{m_\nu\}) = (E_0 - \Delta) n_{\text{ex}} + \sum_\nu \hbar\omega_\nu \left(m_\nu + \frac{1}{2}\right) \tag{3.2}$$

where $\{m_\nu\}$ are the phonon occupation numbers, $n_{\text{ex}} = 0, 1$ is the exciton occupation number and the ground state energy shift is given by

$$\Delta = \sum_\nu g_\nu^2 / (\hbar\omega_\nu). \tag{3.3}$$

Within the independent boson model, the polaron spectral function, that is the imaginary part of the retarded Green's function, calculated for Einstein phonons (with $\omega_\nu = \omega_0$ for any ν) takes, at zero temperature, the following simple form:

$$A(E) = 2\pi e^{-S} \sum_{m=0}^\infty \frac{S^m}{m!} \delta(E - E_0 + \Delta - m\hbar\omega_0) \tag{3.4}$$

where $S = \sum_\nu g_\nu^2 / (\hbar\omega_0)^2$, a dimensionless measure of coupling between the exciton and optical phonons, is called Huang-Rhys parameter (HRP). The spectral function (3.4) is a series of δ-functions, spaced by $\hbar\omega_0$ apart (note that the null vibrations energy has been omitted for simplicity), which is called Franck–Condon progression. The coefficients of the δ-functions follow a Poisson distribution with respect to the number of optical phonons (m) forming the polaron states. It should be borne in mind that the relative intensities of the m-phonon satellites change with temperature (T). A more elaborate expression for the spectral function at $T \neq 0$ can be found in [2].

As we know, the exciton ground state in a typical QD is degenerate, therefore the HRP must be defined for each sub-state $|M, s_z\rangle$ constituting the bare exciton octet $1s_e 1S_{3/2}$,

$$S_{s_z, M} = (\hbar\omega_{\text{LO}})^{-2} \sum_{l_p, m_p, n_p} |\langle M, s_z | H_{\text{ex-ph}} | M, s_z \rangle|^2. \tag{3.5}$$

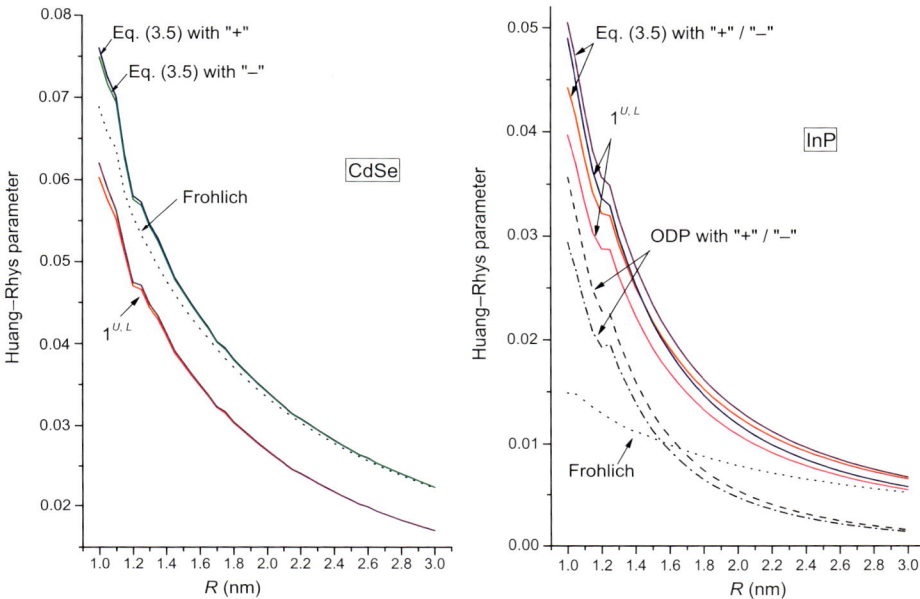

Fig. 6. HRP calculated for CdSe (left) and InP (right) QDs of different radii, with and without exchange interaction and crystal field (for CdSe) effects. In the latter case, "$\pm$" corresponds to the upper and lower sign in the matrix elements from Table 1, respectively. The curves labelled $1^{U,L}$ represent the exchange interaction and crystal field effects that affect only these four states. The individual contributions of the Fröhlich-type and ODP mechanisms also are shown (calculated without exchange interaction effect). Reproduced from [23] with permission of IOP Publishing Ltd.

Let us first neglect the effects of exchange interaction and hexagonal crystal field. It turns out that, if only the Fröhlich-type interaction is taken into account, $S_{S_z,M}$ is exactly the same for all eight states, within the approximation considered here. The ODP interaction introduces some difference between the states with $M = \pm 3/2$ and $M = \pm 1/2$, respectively, since the corresponding matrix elements are different (Table 1). The dependence of the HRP calculated according to Eq. (3.5) for CdSe and InP[13] dots on the QD radius is shown in Fig. 6.

As it can be seen from this figure, the difference between these two groups of states is fairly small in the case of CdSe dots where the Fröhlich-type interaction is dominant and the ODP contribution to the HRP does not exceed 10% (consequently, S scales approximately as R^{-1}). The case of InP QDs is different. This material is characterised by a lower degree of ionicity and a larger ODP interaction constant (see Table 2). As a result, the ODP and Fröhlich-type mechanisms produce comparable contributions to the HRP and the difference between the exciton states involving $M = \pm 1/2$ and $M = \pm 3/2$ holes becomes significant. There is no simple scaling law for S versus R in this case. With the effects of exchange interaction and

---

[13] The value of the spin-orbit splitting energy, $\Delta_{SO} = 0.11$ eV, probably is not sufficiently large to justify the use of the hole wavefunctions given by Eq. (2.3). However, calculations performed with the wavefunctions given by Eq. (2.6) (valid in the limit $\Delta_{SO} \approx 0$) give similar results.

Table 2. Material parameters used in the calculations

Parameter	CdSe	InP
m_e/m_0	0.13	0.08
m_{lh}/m_0	0.26	0.09
ε_0	9.7	12.4
ε_∞	6.2	9.6
Δ_{cr} (meV)	25	0
$\varepsilon_{exch} a_0^3$ (meV nm^3)	36	36
ω_{LO} (cm^{-1})	211	345
ω_{TO} (cm^{-1})	169	305
d_0 (eV)	8.9	35.6
a_0 (eV)	6.05	5.87
β_{LO} (cm^{-1} Å^2)	70.55	29.43
β_{TO} (cm^{-1} Å^2)	−35.08	−15.6

crystal field taken into account, the results given by Eq. (3.5) hold for the states 0^U and 0^L [with the lower sign in the expressions (2.34) and (2.35)], and ±2 (with the upper sign). However, the HRP has a smaller value for the remaining states [23]. This leads to an additional (phonon-related) splitting between these groups of states.

The polaron spectral function at $T \neq 0$ is affected by including the acoustic phonons into consideration. This leads to a small shift and a temperature-dependent broadening of all optical phonon satellites, however, the individual spectral weights of the optical phonon replicas do not change. Let us briefly discuss the effect of the acoustic phonon continuum on the zero-optical-phonon line (ZPL) of the spectral function [116]. It is now structured into a true δ–like zero-acoustic-phonon line (ZAPL) surrounded by two broad sidebands. There is a downward shift of the ZAPL with respect to the bare exciton energy, additional to Eq. (3.3) and given by a similar expression. The shape of the sidebands is determined by the coupling function g_k [Eq. (2.31)] times the density of acoustic phonon states times either N_k (for the upper energy sideband) or $(N_k + 1)$ (for the lower energy one), where $N_k = [\exp(-\hbar c_L k/k_B T) - 1]^{-1}$ is the Bose function and $c_L k \approx E - E_0 + \Delta$. Thus, the sidebands are rather asymmetric, especially at low temperatures. As the temperature increases, the central part of the ZPL becomes looking like a slightly asymmetric Lorentzian whose width increases linearly with T [50, 52].

3.2 Effect of non-adiabaticity.
It is necessary to bear in mind that the independent boson model discussed above assumes the existence of only *one* exciton state[14]. That corresponds to a hypothetical case where the phonon-mediated coupling of the lowest energy exciton state to all other states can be neglected. Only under such circumstances the Huang-Rhys parameter describes entirely the polaron spec-

---

[14] Since no phonon-mediated transitions into other states are allowed, one can say that it is still an *adiabatic* approximation.

trum. As it has been mentioned in Sect. 1, there are several important experimental results which cannot be explained in terms of the independent boson model, even if one assumes unrealistically large HRP values[15], namely:

(i) Observation of a forbidden optical transition to the 1P exciton state in CuCl QDs, interpreted as a result of the formation of a hybrid ex–ph state [117];

(ii) "Strange" phonon replicas in single QD PL excitation spectra [39, 117, 118], i.e. spectral features separated from the ZPL by energies considerably smaller that the LO phonon energy (e.g. 18 meV for InAs QDs [118] while the LO phonon energy in InAs is 32 meV);

(iii) Anomalously strong nLO phonon satellites ($n=2, 3$), stronger than the 1LO replica, observed for NC ensembles [21] and in single self-assembled QDs [119–121];

(iv) Optical-phonon-related component of the homogeneous broadening of absorption lines, already discussed in Sect. 1 (Eq. 1.1).

These experimental results indicate that it is essential to include into consideration the non-adiabaticity of the ex–ph interaction [21, 40] leading to a phonon-mediated coupling of different bare exciton levels, even if they are separated by energies quite different from the optical phonon energy. A step towards understanding the new effects arising from such coupling was made in recent works [7, 122–124] and it seems that it is possible to explain the essence of the above experimental findings even within the minimal model considered in these works, illustrated in Fig. 7.

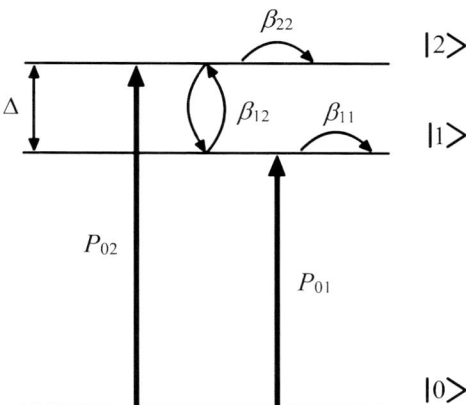

Fig. 7. Bare exciton states included in the minimal model. Two exciton levels, both allowed for optical transitions from exciton vacuum (0), coupled by phonon-assisted transitions. Virtual phonon-assisted intra-level transitions are also allowed

---

[15] Even though some of these results were obtained for self-assembled QDs, the physics involved is general for systems with discrete energy spectra.

The Hamiltonian of this model is an obvious generalisation of Eq. (3.1),

$$\widehat{H}_{\text{ex}-L}^{(2)} = \sum_{i=1,2} E_i \hat{c}_i^+ \hat{c}_i + \sum_\nu \hbar\omega_\nu \left(\hat{b}_\nu^+ \hat{b}_\nu + \frac{1}{2} \right) + \sum_\nu \hbar\omega_\nu (\hat{b}_\nu + \hat{b}_\nu^+) \sum_{i,j=1}^{2} \beta_{ij}^{(\nu)} \hat{c}_i^+ \hat{c}_j \tag{3.7}$$

where dimensionless ex–ph coupling constants, $\beta_{ij}^{(\nu)} = \langle i | H_{\text{ex-OP}}^{(\nu)} | j \rangle / \hbar\omega_\nu$, have been introduced and they have been assumed real. As proposed in [7], it can be diagonalised (numerically) exactly including a finite number of multi-phonon processes, large enough to guarantee that the result can be considered exact in the physically important polaron energy region. By rewriting Eq. (3.7) in the basis constituted by independent ex–ph states, $|n_1, n_2, \{m_v\}\rangle$ (n_1 and n_2 are the occupation numbers of two exciton states, $n_1 + n_2 = 1$) one obtains an infinite Hamiltoinian matrix with elements,

$$H_{ij}(\{m_v\}, \{m'_v\}) = [E_i \delta_{ij} + \sum_\nu m_\nu \hbar\omega_\nu] \prod_\nu \delta_{m_v m'_v}$$
$$+ \sum_\nu \hbar\omega_\nu \beta_{ij}^{(\nu)} [\sqrt{m_v + 1} \delta_{m_v+1,m'_v} + \sqrt{m_v} \delta_{m_v-1 1, m'_v}] \prod_{\mu \neq \nu} \delta_{m_\mu m'_\mu}.$$

It can be truncated by allowing a certain maximum number of phonons for each mode and then diagonalised numerically. The numerical results have been checked against analytical solutions available for some particular cases of the Hamiltonian (3.7), such as $\beta_{ij}^{(\nu)} = 0$ for $i \neq j$ considered in the previous section, and $\beta_{11} = \beta_{11} = 0$ for a single phonon mode [125], and it was shown that they are very accurate in a limited range of energy that depends on the maximum number of phonons included in the calculation [7].

Once the polaron states have been found, it is straightforward to calculate any (one-exciton) observable property. Some of these properties will be discussed in the next sections. The formation of non-adiabatic electron–polaron states has been convincingly demonstrated using far-infrared spectroscopy of intra-band electron transitions in a doped QD in magnetic field [124]. One of the polaron effects produced by the inter-level coupling is known as "Rabi splitting", in analogy with the resonant absorption of electromagnetic radiation by a two-level atom [125]. It consists in the appearance of new peaks in the spectral functions, with positions determined by all the involved parameters that have the dimension of energy, namely, $\{\hbar\omega_\nu\}$, $\Delta E = E_2 - E_1$, and the ex–ph coupling constants.

In practical terms, the calculation scheme [7] outlined above can include 2–3 most important optical phonon modes and it can be extended to a larger number of bare exciton states such as the $1s_e 1S_{3/2}$ octet [44]. As shown in [7], the continuum of acoustic phonons can also be included in these calculations if their inter-level coupling is neglected. If the interaction parameter g_k is the same for all the bare exciton states involved, each of the polaron states simply is broadened in the same way as discussed in the end of the previous section. A difference between g_k corresponding to different exciton levels would lead to acoustic-phonon-mediated transitions between the polaron states. We shall return to this point in Sect. 4.

3.3 Optical absorption and emission. Within the polaron concept, the optical absorption and emission spectra of a QD can be calculated using the Kubo formula for the frequency-dependent dielectric function [2]. The following expression was derived [7] for its imaginary part:

$$\text{Im } \varepsilon(\omega) = \left(\frac{2\pi e}{m_0 \omega}\right)^2 \frac{1}{VZ} \sum_{i,j} p_{0i}^* p_{0j} \sum_{\{m_\nu\},\kappa} \left\{ \exp\left(-\frac{\sum_\nu m_\nu \hbar \omega_\nu}{k_B T}\right) (C_{i,\{m_\nu\}}^\kappa)^* C_{j,\{m_\nu\}}^\kappa \right.$$
$$\times \left[\delta(\hbar\omega - E_\kappa + \sum_\nu m_\nu \hbar \omega_\nu) + \exp\left(-\frac{E_\kappa - \sum_\nu m_\nu \hbar \omega_\nu}{k_B T}\right) \right.$$
$$\left. \left. \times \delta(\hbar\omega + E_\kappa - \sum_\nu m_\nu \hbar \omega_\nu) \right] \right\} \quad (3.8)$$

where m_0 is the free electron mass, i and j denote bare exciton states, κ enumerates the polaron states (with energies E_κ), $C_{j,\{m_\nu\}}^\kappa$ are the polaron wavefunctions in the basis of independent ex–ph states that are obtained by numerical diagonalisation of the Hamiltonian matrix, and $Z = \sum_{\{m_\nu\}} \exp(-\sum_\nu m_\nu \hbar \omega_\nu/(k_B T))$. Two terms in the second line of Eq. (3.8) represent the absorption and emission, respectively, of a photon of frequency ω.

From the structure of the function Im $\varepsilon(\omega)$ it is not difficult to see how the optical spectra are determined by the polaron states. If E_κ are equidistant, as within the independent boson model (Eq. 3.2), the same applies to the absorption and emission peaks. Moreover, the absorption and emission spectra are mirror-symmetric with respect to the ZPL, a well-known property of the FC progression [2] that does not take place in the general case where the polaron states are not separated by any regular intervals. This naturally explains the "strange" phonon replicas, not matching any known phonon energy, that have been observed by single dot spectroscopy [39, 118]. These odd peaks may be generated by a single phonon mode that couples different exciton levels [7]. Resonant (Rabi-type) coupling of a phonon mode to a pair of exciton states can be the reason for the observed enhancement of certain (phonon-assisted) optical transitions that otherwise would be very weak (given the small HRP value) or even not allowed in the dipole approximation. In [7] the reader can find a more extended discussion of this issue.

Now we would like to revisit the point concerning the homogeneous broadening of excitonic transitions in QDs, already discussed in Sect. 1. As we have seen explicitly in Sect. 3.1 (Eq. 3.4), the interaction of an exciton confined in a QD with optical phonons leads to the appearance of satellite peaks, not to a broadening. Then the presence of the last term in Eq. (1.1) seems to have no physical reason as far as QDs are concerned. Nevertheless, it can be justified, at least in a phenomenological way. Figure 8 shows room temperature absorption and emission spectra of a CdSe QD calculated including the most important ($l_p = 0, 2; n_p = 1$) optical phonon modes and acoustic phonons, within the approximations discussed in the end of the previous section. The ex–ph coupling constants were calculated as explained in Sect. 2.4, however, the deformation potential parameter for acoustic phonons was taken approximately five times smaller than the real value in order to reveal the underlying structure due to the interaction with optical phonons.

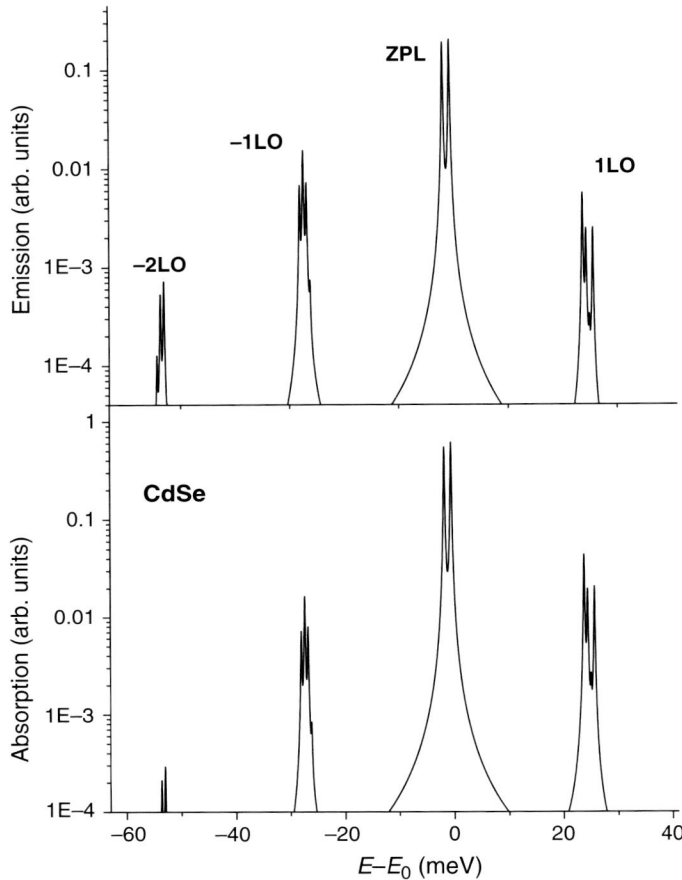

Fig. 8. Absorption and emission spectra calculated for a CdSe QD of $R = 1.9$ nm embedded in SiO_2 matrix. The principal optical phonon modes (with both intra- and inter-level coupling) and acoustic phonons (with only intra-level coupling and reduced DP constant) were included in the calculation

It can be seen from Fig. 8 that the broadened bands designated ZPL, 1LO, etc. are formed by *many* optical-phonon-related polaron states whose structure is smeared by the interaction with the continuum of acoustic phonons. As the temperature increases, the intensity of the (δ-like) peaks produced by the interaction with solely optical phonons increases, leading to an apparent broadening of the bands that, approximately, can be described by the term $\gamma_{LO} N_{LO}(T)$. Of course, this is not valid at low temperatures where the acoustic-phonon-related structure of a polaron state by no means can be described by a smooth function like Lorentzian. At liquid helium temperatures, the central part of each polaron band (the ZAPL) is very narrow, both experimentally [50] and theoretically [126]. The ZAPL broadening, experimentally estimated at 1.5 μeV/K [50], is related to so called "pure dephasing" [126, 127], an interesting effect beyond the considered ex–ph interaction model.

3.4 Anti-Stokes photoluminescence.

The up-converted or anti-Stokes photoluminescence (ASPL) is the emission of photons with energies higher than the excitation energy (E_{exc}). This effect was observed for colloidal II–VI QDs (see the following Chapter by Rakovich/Donegan) and discussed in [128–131]. The ASPL occurs when an ensemble of QDs is excited at the very edge of the absorption spectrum, below the normal (i.e., excited with high-energy photons) PL band. The principal experimental facts concerning this effect, as summarised in [7], are the following:

(i) The ASPL intensity increases linearly with the excitation power, which can be rather low.

(ii) The blue (anti-Stokes) shift between the ASPL peak and E_{exc} does not significantly depend upon the QD size, if E_{exc} is chosen proportionally to the absorption peak energy (which depends upon the size). However, the shift increases with temperature and can range from 20 to 150 meV.

(iii) If E_{exc} increases (approaching the absorption peak energy), the ASPL also moves continuously towards higher energies. Its intensity increases and finally the spectrum transforms into the normal PL band.

(iv) The ASPL intensity increases strongly with temperature.

Possible ASPL mechanisms were discussed in [128–131], but no convincing conclusion was made. In virtue of (i), processes like two-photon absorption and Auger excitation can be excluded in this case. It looks most plausible that incident photons excite electrons to some intermediate sub-gap states from which they eventually proceed to the higher energy (luminescent) states through a thermal activation effect. Since ASPL has been observed for several different types of NCs, the sub-gap states involved in the mechanism should be intrinsic to QDs, independently of their material and covering shell. It was suggested [7] that the polaron states lying

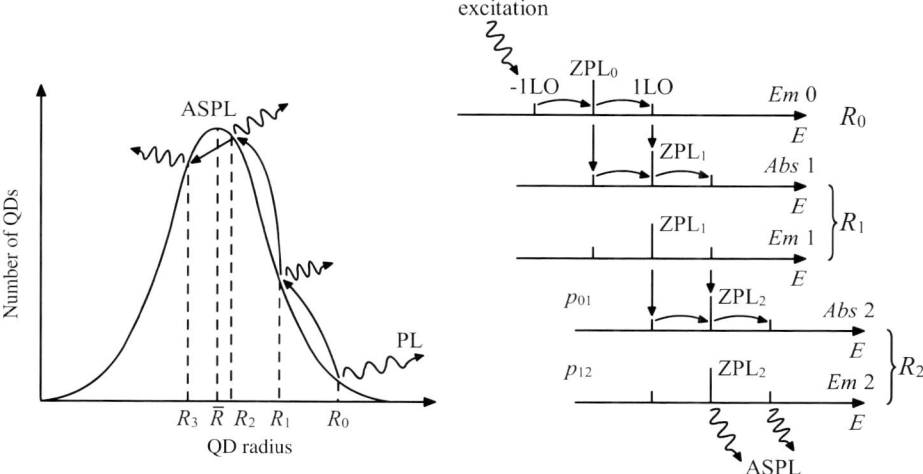

Fig. 9. Cascade energy transfer process leading to the ASPL. It begins by the excitation of a large QD, of a size R_0, through its "-1LO" polaron state. At each step, some photons are emitted and some of them are re-absorbed. A certain fraction of the latter will be re-emitted with a higher energy

below the ground exciton level (designated "-nLO" in Fig. 8) would be the natural candidate to this role. After being excited through one of these states, the dot can emit a photon, most likely with the frequency corresponding to the ZAPL. The energy gain of $\hbar\omega_{LO}$ comes from the thermostat that maintains the dot at a given temperature. This explains why the anti-Stokes shift (i.e. the gap between the intermediate and emitting states) does not depend on the QD radius [129]. Spectral shifts larger than $\hbar\omega_{LO}$ can be explained by a cascade excitation as schematically shown in Fig. 9.

Indeed, the ASPL is excited through the largest (and rare) dots in the ensemble. In some cases, the emission occurs at a higher energy and such a photon has a larger chance to be re-absorbed because there are more dots of the appropriate size. (Of course, many of the emitted photons just leave the sample contributing to the unshifted or slightly shifted luminescence.) The cascade re-absorption and re-emission process occurs until the number of dots able to re-absorb becomes too small. Then, beyond the usual (unshifted) luminescence, a spectral maximum is formed at a higher frequency. Monte-Carlo modelling of the processes illustrated in Fig. 9 shows that, in accordance with experiment, the observation of the ASPL requires high optical density of the QD solution. It also confirms the experimental findings (ii–iv) summarised above.

3.5 Multi-phonon Raman scattering.

Another steady-state optical phenomenon that reveals the polaron effect in QDs is the resonant Raman scattering (RRS). In an K-th order Stokes RRS process, an incident photon (of frequency Ω_I) resonantly creates an exciton that recombines after being inelastically scattered K times, each time creating an optical phonon. Then the emitted photon has a smaller frequency, $\Omega_S = \Omega_I - \sum_\nu r_\nu \hbar \omega_\nu$ where $\sum_\nu r_\nu = K$. Within the polaron concept, this process should be seen as a two-step one involving three states: (i) QD with a phonon configuration $\{m_\nu\}$ plus a photon of frequency Ω_I, (ii) excited QD whose state is a linear combination of various polaron states, (iii) QD with a phonon configuration $\{m'_\nu\}$ plus a photon of frequency Ω_S. The final phonon configuration is such that $m'_\nu = m_\nu - r_\nu$ for each phonon mode ν. In simple terms, the difference is that in the intermediate state the exciton is scattered many times by emitting and absorbing all possible combinations of phonons, with the only condition that the final net result is represented by the numbers r_ν. The scattering intensity, in the polaronic picture, is given by the following expression [44],

$$I_S(\Omega_I, \omega) \sim \frac{1}{Z} \sum_{\{m_\nu\}} e^{-\sum_\nu m_\nu \hbar \omega_\nu / (k_B T)} \sum_{\{r_\nu\}} \left| \sum_\kappa \frac{\sum_{i,j} p_{0i} p_{0j}^* \left(c_{i,\{m_\nu\}}^\kappa\right)^* c_{j,\{m'_\nu\}}^\kappa}{E_\kappa - \sum_\nu m_\nu \hbar \omega_\nu - \hbar \Omega_I} \right|^2$$
$$\times \delta\left(\omega - \sum_\nu r_\nu \omega_\nu\right) \tag{3.9}$$

where $\omega \equiv \Omega_I - \Omega_S$ is the Raman shift.

Usually RRS studies of QDs are limited to the first-order scattering [14]. Even though the absolute value of I_S predicted within the polaron concept is different from that obtained in the framework of the standard perturbation theory, this is very

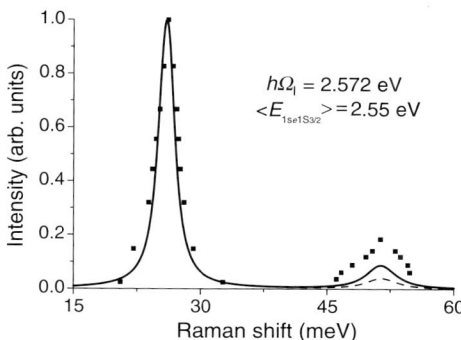

Fig. 10. RRS spectra calculated at $T=77$ K in the perturbation theory (dashed line) and polaron (solid line) approaches for CdSe QDs with a mean radius of 1.9 nm and a size dispersion of 10%. The squares are experimental data from [17]. Each curve is normalised to the height of the main (1LO) peak whose absolute value is approximately 30% lower in the polaron approach. Adapted from [44] with permission of the American Physical Society

hard to measure experimentally [132]. The lineshape, however, is not significantly influenced by the polaron effect [133]. What is indeed sensitive to it is the relative intensity of the higher order Raman peaks. RRS spectra of CdSe QDs [17] were modelled in [134] using the standard perturbation theory, considering a complete set of confined exciton states and ex–ph coupling constants calculated in a way similar to the formalism presented in Sect. 2. While the lineshape of the experimental spectra was reproduced quite well by the theory, it was necessary to artificially multiply the second and the third order peak's intensities by some factors (of the order of 10) in order to attain the experimental values. As demonstrated in [44], it is possible to reach a much better agreement with experiment using the polaron approach.

Figure 10 shows normalised RRS spectra calculated using Eq. (3.9) and the perturbation theory of the ex–ph interaction, with no fitting parameters used. In both cases, the same set of exciton states ($1s_e 1S_{3/2}$) and optical phonon modes ($l_p = 0, 2$; $n_p = 1$) was included in the calculation. The comparison between the Raman spectra calculated using the two approaches and experimental data [17] strongly supports the importance of the polaron effect on the multi–phonon Raman scattering, even for moderated values of the exciton–phonon coupling rates. The relative intensity of the second peak predicted within the polaron concept is still too low but this is probably due to the small number of exciton states and phonon modes considered, the limitation implied by the method of calculation of polaron states described in Sect. 3.2.

4. Polaron decay and hot carrier relaxation

As it was demonstrated in the previous section, the interaction of electrons and holes with lattice vibrations leads to the formation of a quasiparticle called polaron. Polaron states are the true stationary states of the quantum mechanical system constituted by an exciton and optical phonons confined in a QD. They determine the steady-state properties of the dot. What about non-stationary phenomena, such as time-resolved luminescence? These phenomena involve relaxation of electrons or electron–hole pairs that initially are not at thermal equilibrium. If their energy is continuous, the relaxation occurs via emission of optical phonons (Fig. 1). In a QD, the spectrum is discrete and the phonons are already incorporated into the polaron concept. Then how does the relaxation occur? Once again, we come across the "phonon bottleneck" problem mentioned Sect. 1.

Does the "phonon bottleneck" in QDs really exist? An efficient phonon-mediated relaxation of optically created electron–hole pairs has been reported in a number of works studying self-assembled QDs [118, 135], with a characteristic time of the order of several picoseconds [135]. A recent study [136] performed on chemically grown NCs, where the exciton energy level spacings are much larger than in self-assembled dots, also revealed ultra-fast intra-band exciton relaxation, although the mechanism seems to be different for CdSe and PbSe NCs. On the other hand, there are several published experimental results that support the existence of a phonon bottleneck effect in the relaxation of optically created electron–hole pairs [137–139], for instance, a relaxation time of 7.7 ns was measured for InAs/GaAs QDs [137]. The so called "hot luminescence" (i.e. the emission from excited states) under low power excitation, observed in several studies, is also manifest of slow carrier relaxation into their ground state in QDs.

Whether the intra-band relaxation of carriers in QDs is fast or slow, the polaron effect itself cannot explain it. Electron-polaron is a stationary state of an electron coupled to optical phonons. Exciton–polaron exists because of the electron–hole, electron–phonon and hole–phonon interactions. If there were no further interactions, the polaron states would be everlasting, as pointed out in [5, 6, 140]. Some further interactions are responsible for the relaxation. Several possible mechanisms of hot carrier relaxation in QDs have been proposed:

(i) Polaron has a rather short lifetime because of the anharmonic effects that lead to a fast decay of confined optical phonons (forming the polaron), into a couple of new phonons [6, 140].

(ii) Polaron can relax by Coulomb coupling to electrons that are present in the wetting layer where the energy spectrum is continuous [120, 141, 142]. This Auger-type mechanism can be possible even in chemically grown QDs where no wetting layer exists. If the hole energy spectrum is sufficiently dense, the relaxation can occur just because of the electron–hole Coulomb interaction within QD [143, 144].

(iii) Acoustic phonons can provide the possibility of transitions between different polaron states [7] as it was briefly discussed in the end of Sect. 3.2. If the acoustic phonon spectrum is (at least partially) continuous and the polaron spectrum is sufficiently dense, this interaction would allow for the polaron dynamics towards equilibrium.

The mechanism (i) seems to be the most popular one. Calculations performed for self-assembled QDs in [140], based on the experimental studies [145] of the anharmonic decay of long-wavelength optical phonons in bulk semiconductors, predict fast or slow polaron decay depending on the dot parameter. If we extrapolate the results of this work to NCs, with much stronger confinement and, generally, much larger energy intervals between adjacent polaron states, we should not expect fast exciton relaxation for such dots. The mechanism (ii) involves a continuum of electronic (or excitonic) states and therefore is intrinsically limited to self-assembled QDs. Last, there are no concrete calculations relative to the mechanism (iii). Thus, the establishing of a universal relaxation mechanism of the QD polaron relaxation is still an open question and further theoretical and experimental efforts are necessary to solve it.

5. Conclusions

We have presented an overview of analytical models aimed at the understanding of the exciton and phonon states and their interaction in nanocrystalline semiconductor QDs. These models allow for establishing the key features which are important for applications of QDs in optoelectronics, such as the dependence of the stationary state energies and transition rates on the dot size, shape, type of the underlying material and its surrounding (e.g. capping the dot with another semiconductor). The results obtained within this approach generally agree with the predictions of more elaborate large-scale numerical methods, unless the nanoparticles are extremely small. Calculations of the exciton–phonon coupling parameters, based on the same mechanisms and interaction constants known for bulk semiconductors, allow for a satisfactory description of Raman lineshapes [14] and lead to the experimentally verified important conclusion that the radiative recombination of excitons in silicon QDs usually remains phonon-assisted [57].

Strictly speaking, the exciton–phonon interaction in QDs leads to the formation of the polaron and must be treated non-perturbatively for coupling strengths typical for II–VI and III–V dots. A number of many-body theory approaches have been proposed for this including a direct method [7], however, limited to a small number of exciton levels and phonon modes. It should be stressed that both intra- and inter-level interactions are important, leading to a rather complex polaron energy spectra. The polaron effect allows for the explanation of the fine structure of the absorption and emission spectra of a single self-assembled QD, and of the anti-Stokes PL of colloidal solutions of QDs. The polaron effect can also be important for the resonance behaviour of the one-phonon Raman scattering intensity [133]. Although not changing significantly the one-phonon Raman lineshape, the polaron effect can drastically enhance the relative intensity of the multi-phonon scattering.

The coupling between the different vibrational and electronic degrees of freedom is certainly relevant to the excitation, relaxation and dephasing in QDs, where some novel dynamical effects may arise in the limit of very small clusters consisting of a hundred of atoms [146]. However, for dots with size $R \geq 1.5-2$ nm containing, at least, several hundred of atoms, one hopes that the well established concepts of semiconductor physics and the properly treated quantum confinement effect should be sufficient to explain the dynamical effects that may look puzzling or contradictory so far, such as the "phonon bottleneck". It should be borne in mind that the polaron effect by itself does not explain (neither contradicts) fast or slow carrier relaxation in QDs. Some further interactions are responsible for this. Which one(s)? So far, it is not completely clear.

Acknowledgements

The collaboration and valuable discussions with E.V. Anda, V.A. Burdov, S.S. Makler, R.P. Miranda, and C. Trallero-Giner are gratefully acknowledged. The original work was supported by the Portuguese Foundation for Science and Technology (FCT).

References

[1] Kuper CG, Whitfield GD (eds) (1963) Polarons and excitons. Plenum, New York
[2] Mahan GD (1990) Many-particle physics. Plenum, New York
[3] Ridley BK (1982) Quantum processes in semiconductors. Oxford University Press, Oxford
[4] Bockelmann U, Bastard G (1990) Phonon scattering and energy relaxation in two-, one-, and zero-dimensional electron gases. Physical Review B 42: 8947–8951
[5] Li XQ, Arakawa Y (1997) Ultra-fast energy relaxation in quantum dots through defect states: a lattice-relaxation approach. Physical Review B 56: 10423–10427
[6] Verzelen O, Ferreira R, Bastard G (2002) Excitonic polarons in semiconductor quantum dots. Physical Review Letters 88: 146803
[7] Vasilevskiy MI, Anda EV, Makler SS (2004) Electron–phonon interaction effects in semiconductor quantum dots: a non-perturbative approach. Physical Review B 70: 035318
[8] Woggon U (1999) Optical properties of semiconductor quantum dots. Springer, Berlin Heidelberg New York Tokyo
[9] Efros AlL (1992) Luminescence polarization of CdSe micro-crystals. Physical Review B 46: 7448–7458
[10] Efros AlL, Rosen M, Kuno M, Nirmal M, Norris DJ, Bawendi M (1996) Band-edge exciton in quantum dots of semiconductors with a degenerate valence band: dark and bright exciton states. Physical Review B 54: 4843–4856
[11] Prado SJ, Trallero-Giner C, Alcalde AM, López-Richard V, Marques GE (2003) Optical transitions in a single CdTe spherical quantum dot. Physical Review B 68: 235327
[12] Wang L-W, Zunger A (1996) Pseudopotential calculations of nano-scale CdSe quantum dots. Physical Review B 53: 9579–9582
[13] Delerue C, Lannoo M (2004) Nanostructures. Theory and modelling. Springer, Berlin Heidelberg New York Tokyo
[14] Rolo AG, Vasilevskiy MI (2007) Raman spectroscopy of optical phonons confined in semiconductor quantum dots and nano-crystals. Journal of Raman Spectroscopy 38: 618–633
[15] Vasilevskiy MI (2002) Dipolar vibrational modes in spherical semiconductor quantum dots. Physical Review B 66: 195326
[16] Trallero-Giner C, Debernardi A, Cardona M, Menendez-Proupin E, Ekimov AI (1998) Optical vibrons in CdSe dots and dispersion relation of the bulk material. Physical Review B 57: 4664–4669
[17] Klein MC, Hache F, Ricard D, Flytzanis C (1990) Size dependence of electron-phonon coupling in semiconductor nano-spheres: the case of CdSe. Physical Review B 42: 11123–11132
[18] Marini JC, Stebe B, Kartheuser E (1994) Exciton–phonon interaction in CdSe and CuCl polar semiconductor nano-spheres. Physical Review B 50: 14302–14308
[19] Alcalde AM, Weber G (2000) Scattering rates due to electron–phonon interaction in CdS_xSe_{1-x} quantum dots. Semiconductor Science and Technology 15: 1082–1086
[20] Ajiki H (2001) Exciton–phonon interaction in a spherical quantum dot: effect of electron–hole exchange interaction. Physica Status Solidi (b) 224: 633–637
[21] Fomin VM, Gladilin VN, Devreese JT, Pokatilov EP, Balaban SN, Klimin SN (1998) Photoluminescence of spherical quantum dots. Physical Review B 57: 2415–2425
[22] Garcia-Cristobal A, Minnaert AWE, Fomin VM, Devreese JT, Silov AYu, Haverkort JEM, Wolter JH (1999) Electronic structure and phonon-assisted luminescence in self-assembled quantum dots. Physica Status Solidi B-Basic Research 215: 331–336
[23] Hamma M, Miranda RP, Vasilevskiy MI, Zorkani I (2007) Calculation of the Huang-Rhys parameter in spherical quantum dots: the optical deformation potential effect. Journal of Physics Condensed Matter 19: 346215
[24] Ovsyuk NN, Gorokhov EB, Grishchenko VV, Shebanin AP (1988) Low-frequency Raman scattering by small semiconductor particles. JETP Letters 47: 298–300
[25] Tanaka A, Onari S, Arai T (1993) Low-frequency Raman scattering from CdS micro-crystals embedded in a germanium dioxide glass matrix. Physical Review B 47: 1237–1243
[26] Savoit L, Champagnon B, Duval E, Kudryavtsev IA, Ekimov AI (1996) Size dependence of acoustic and optical vibrational modes of CdSe nano-crystals in glasses. Journal of Non-Crystalline Solids 197: 238–246
[27] Bragas AV, Aku-Leh C, Merlin R (2006) Raman and ultra-fast optical spectroscopy of acoustic phonons in $CdTe_{0.68}Se_{0.32}$ quantum dots. Physical Review B 73: 125305
[28] Takagahara T (2002) Electron–phonon interactions in semiconductor quantum dots. In: Masumoto Y, Takagahara T (eds) Semiconductor quantum dots. Springer, Berlin, pp. 115–147
[29] Lamb H (1881) On the vibrations of an elastic sphere. Proceedings of the London Mathematical Society 13: 189–212

[30] Cheng W, Ren S-F, Yu PY (2005) Microscopic theory of the low frequency Raman modes in germanium nano-crystals. Physical Review B 71: 174305
[31] Goupalov SV, Merkulov IA (1999) Theory of Raman light scattering by nano-crystal acoustic vibrations. Physics of the Solid State 41: 1349–1358
[32] Goupalov SV, Suris RA, Lavallard P, Citrin DS (2001) Homogeneous broadening of the zero-optical-phonon spectral line in semiconductor quantum dots. Nanotechnology 12: 518–522
[33] Grosse F, Zimmermann R (2007) Electron–phonon interaction in embedded semiconductor nanostructures. Physical Review B 75: 235320
[34] Bányai L, Koch SW (1993) Semiconductor quantum dots. World Scientific, Singapore
[35] Bawendi MG, Carroll PJ, Wilson WL, Brus LE (1992) Luminescence properties of CdSe quantum crystallites: resonance between interior and surface localized states. Journal of Chemical Physics 96: 946–954
[36] Valenta J, Moniatte J, Gilliot P, Hönerlage B, Grun JB, Levy R, Ekimov AI (1998) Dynamics of excitons in CuBr nano-crystals: spectral-hole burning and transient four-wave-mixing measurements. Physical Review B 57: 1774–1783
[37] Jungnickel V, Henneberger F (1996) Luminescence related processes in semiconductor nanocrystals: the strong confinement regime. Journal of Luminescence 70: 238–252
[38] Empedocles SA, Norris DJ, Bawendi MG (1996) Photoluminescence spectroscopy of single CdSe nanocrystallite quantum dots. Physical Review Letters 77: 3873–3876
[39] Bissiri M, von Högersthal GB, Bhatti AS, Capizzi M, Frova A, Frigeri P, Franchi S (2000) Optical evidence of polaron interaction in InAs/GaAs quantum dots. Physical Review B 62: 4642–4646
[40] Devreese JT (2007) Fröhlich polaron from 0D to 3D: concepts and recent developments. Journal of Physics Condensed Matter 19: 255201
[41] Krauss TD, Wise FW (1997) Raman-scattering study of exciton–phonon coupling in PbS nanocrystals. Physical Review B 55: 9860–9865
[42] Baranov AV, Yamaguchi S, Masumoto Y (1997) Exciton–LO-phonon interaction in CuCl spherical quantum dots studied by resonant hyper-Raman spectroscopy. Physical Review B 56: 10332–10337
[43] Scamarcio G, Spagnoto V, Ventruti G, Lugará M, Righini GC (1996) Size dependence of electron-LO-phonon coupling in semiconductor nano-crystals. Physical Review B 53: 10489–10492
[44] Miranda RP, Vasilevskiy MI, Trallero-Giner C (2006) Non-perturbative approach to the calculation of multi-phonon Raman scattering in semiconductor quantum dots: polaron effect. Physical Review B 74: 115317
[45] Nomura S, Kobayashi T (1992) Exciton–LO-phonon couplings in spherical semiconductor micro-crystallites. Physical Review B 45: 1305–1316
[46] Rakovich YuP, Vasilevskiy MI, Artemiev MV, Filonovich SA, Rolo AG, Barber DJ, Gomes MJM (2001) Temperature dependence of the absorption lines in very small quantum dots: the role of electron–phonon interaction. In: Miura N, Ando T (eds) Proceeding of 25th Interational Conference on the Physics of Semiconductors. Springer, Berlin, part II, pp 1203–1204
[47] Woggon U, Gaponenko S, Langbein W, Uhrig A, Klingshirn C (1993) Homogeneous line width of confined electron–hole-pair states in II–VI quantum dots. Physical Review B 47: 3684–3689
[48] Naoe K, Zimin LG, Masumoto Y (1994) Persistent spectral hole burning in semiconductor nanocrystals. Physical Review B 50: 18200–18210
[49] Banin U, Cerullo G, Guzelian AA, Bardeen CJ, Alivisatos AP, Shank CV (1997) Quantum confinement and ultrafast dephasing dynamics in InP nano-crystals. Physical Review B 55: 7059–7067
[50] Besombes L, Kheng K, Marsal L, Mariette H (2001) Acoustic phonon broadening mechanism in single quantum dot emission. Physical Review B 63: 155307
[51] Mittleman DM, Schoenlein RW, Shiang JJ, Colvin VL, Alivisatos AP, Shank CV (1994) Quantum size dependence of femtosecond electronic dephasing and vibrational dynamics in CdSe nanocrystals. Physical Review B 49: 14435–14447
[52] Schmitt-Rink S, Miller DAB, Chemla DS (1987) Theory of the linear and nonlinear optical properties of semiconductor micro-crystallites. Physical Review B 35: 8113–8125
[53] Efros AlL, Rodina AV (1989) Confined excitons, trions and biexcitons in semiconductor microcrystals. Solid State Communications 72: 645–649
[54] Grigorian GB, Kazaryan EM, Efros AlL, Yazeva TV (1990) Hole energy quantisation and absorption edge in spherical micro-crystals with complex valence band structure. Soviet Physics Solid State 32: 1031–1038

[55] Einevoll GT (1992) Confinement of excitons in quantum dots. Physical Review B 45: 3410–3417
[56] Cardona M (1996) Fundamentals of semiconductors. Springer, Berlin Heidelberg New York Tokyo, pp. 21–77
[57] Moskalenko AS, Berakdar J, Prokofiev AA, Yassievich IN (2007) Single-particle states in spherical Si/SiO_2 quantum dots. Physical Review B 76: 085427
[58] Burdov VA (2002) Electron and hole spectra of silicon quantum dots. Journal of Experimental and Theoretical Physics 94: 411–418
[59] Huntzinger J-R, Mlyah A, Paillard V, Wellner A, Combe N, Bonafos C (2006) Electron–acoustic-phonon interaction and resonant Raman scattering in Ge quantum dots: matrix and quantum confinement effects. Physical Review B 74: 115308
[60] Vasiliev I, Öğut S, Chelikowsky JR (2001) *Ab initio* absorption spectra and optical gaps in nanocrystalline silicon. Physical Review Letters 86: 1813–1816
[61] Weissker H-Ch, Furthmüller J, Bechstedt F (2002) Optical properties of Ge and Si nano-crystallites from *ab initio* calculations. I. Embedded nano-crystallites. Physical Review B 65: 155327
[62] Weissker H-Ch, Furthmüller J, Bechstedt F (2003) Structural relaxation in Si and Ge nano-crystallites: influence on the electronic and optical properties. Physical Review B 67: 245304
[63] Ren SY (1997) Quantum confinement in semiconductor Ge quantum dots. Solid State Communications 102: 479–484
[64] Ren SY (1997) Quantum confinement of edge states in Si crystallites. Physical Review B 55: 4665–4669
[65] Delerue C, Allan G, Lannoo M (2001) Electron–phonon coupling and optical transitions for indirect-gap semiconductor nano-crystals. Physical Review B 64: 193402
[66] Nishida Y (2004) Electronic state calculations of Si quantum dots: oxidation effects. Physical Review B 69: 165324
[67] von Grünberg HH (1997) Energy levels of CdSe quantum dots: wurtzite versus zinc-blende structure. Physical Review B 55: 2293–2302
[68] Trani F, Ninno D, Iadonisi G (2007) Tight-binding formulation of the dielectric response in semiconductor nano-crystals. Physical Review B 76: 085326
[69] Franceschetti A, Zunger A (1997) Direct pseudopotential calculation of exciton Coulomb and exchange energies in semiconductor quantum dots. Physical Review Letters 78: 915–918
[70] Burdov VA (2002) Dependence of the optical gap in Si quantum dots on the dot size. Semiconductors 36: 1154–1158
[71] Jdira L, Liljeroth P, Stoffels E, Vanmaekelberg D, Speller S (2006) Size-dependent single-particle energy levels and inter-particle Coulomb interactions in CdSe quantum dots measured by scanning tunnelling spectroscopy. Physical Review B 73: 115305
[72] Kang I, Wise FW (1997) Electronic structure and optical properties of PbS and PbSe quantum dots. Journal of the Optical Society of America B 70: 1632–1646
[73] Tsidilkovskii IM (1978) Band Structure of Semiconductors (in Russian). Izdatel'stvo Nauka, Moscow, pp. 21–77
[74] Anselm AI (1978) Introduction to Semiconductor Theory (in Russian). Izdatel'stvo Nauka, Moscow, pp. 319–325
[75] Basu PK (1997) Theory of optical processes in semiconductors. Clarendon Press, Oxford, pp. 124–152
[76] Ekimov AI, Onuschenko AA, Raikh ME, Efros AlL (1986) Size quantization of excitons in microcrystals with large longitudinal-transverse splitting. Soviet Physics JETP 63: 1054–1062
[77] Brus LE (1984) Electron–electron and electron–hole interactions in small semiconductor crystallites: the size dependence of the lowest excited electronic states. Journal of Chemical Physics 80: 4403–4409
[78] Franceschetti A, Wang L-W, Fu H, Zunger A (1998) Short-range versus long-range electron–hole exchange interactions in semiconductor quantum dots. Physical Review B 58: 13367–13370
[79] Demchenko DO, Wang L-W (2006) Optical transitions and nature of Stokes shift in spherical CdS quantum dots. Physical Review B 73: 155326
[80] Takagahara T, Takeda K (1992) Theory of the quantum confinement effect on excitons in quantum dots of indirect-gap materials. Physical Review B 46: 15578–15581
[81] Reboredo FA, Franceschetti A, Zunger A (1999) Excitonic transitions and exchange splitting in Si quantum dots. Applied Physics Letters 75: 2972–2974
[82] Stroscio MA, Dutta M (2001) Phonons in nanostructures. Cambridge University Press, Cambridge
[83] Landau LD, Lifshitz EM (1986) Theory of elasticity. Pergamon Press, Oxford

[84] Duval E (1992) Far-infrared and Raman vibrational transitions of a solid sphere: selection rules. Physical Review B 46: 5795–5797
[85] Rufo S, Dutta M, Stroscio MA (2003) Acoustic modes in free and embedded quantum dots. Journal of Applied Physics 93: 2900–2905
[86] Kuok MH, Lim HS, Ng SC, Liu NN, Wang ZK (2003) Brillouin study of the quantization of acoustic modes in nano-spheres. Physical Review Letters 90: 255502
[87] Ruppin R, Englman R (1970) Optical phonons in small crystals. Reports on Progress in Physics 33: 149–196
[88] Dumelow T, Parker TJ, Smith SRP, Tilley DR (1993) Far-infrared spectroscopy of phonons and plasmons in semiconductor superlattices. Surface Science Reports 17: 151–212
[89] Rücker H, Molinari E, Lugli P (1991) Electron–phonon interaction in quasi-two-dimensional systems. Physical Review B 44: 3463–3466
[90] Babiker M (1986) Longitudinal polar optical modes in semiconductor quantum wells. Journal of Physics C-Solid State Physics 19: 683–697
[91] Trallero-Giner C, Garcia-Moliner F, Velasco VR, Cardona M (1992) Analysis of the phenomenological models for long-wavelength polar optical modes in semiconductor layered systems. Physical Review B 45: 11944–11948
[92] Chamberlain MP, Cardona M, Ridley BK (1993) Optical modes in GaAs/AlAs superlattices. Physical Review B 48: 14356–14364
[93] Roca E, Trallero-Giner C, Cardona M (1994) Polar optical vibrational modes in quantum dots. Physical Review B 49: 13704–13711
[94] Krauss TD, Wise FW, Tanner DB (1996) Observation of coupled vibrational modes of a semiconductor nano-crystal. Physical Review Letters 76: 1376–1379
[95] Vasilevskiy MI, Rolo AG, Artemyev MV, Filonovich SA, Gomes MJM, Rakovich YuP (2001) FIR absorption in CdSe quantum dot ensembles. Physica Status Solidi (b) 224: 599–603
[96] Milekhin A, Friedrich M, Zahn DRT, Sveshnikova L, Repinsky S (1999) Optical investigation of CdS quantum dots in Langmuir-Blodgett films. Applied Physics A 69: 97–100
[97] Vasilevskiy MI, Rolo AG, Gaponik NP, Talapin DV, Rogach AL, Gomes MJM (2002) Dipole-active vibrations confined in InP quantum dots. Physica B 316–317: 452–454
[98] Vasilevskiy MI, Rolo AG, Gomes MJM, Vikhrova OV, Ricolleau C (2001) Impact of disorder on optical phonons confined in CdS nano-crystallites embedded in a SiO_2 matrix. Journal of Physics Condensed Matter 13: 3491–3508
[99] Pokatilov EP, Klimin SN, Fomin VM, Devreese JT, Wise FW (2002) Multi-phonon Raman scattering in semiconductor nanocrystals: importance of non-adiabatic transitions. Physical Review B 65: 075316
[100] Bir GL, Pikus GE (1972) Symmetry and strain-induced effects in semiconductors (in Russian). Izdatel'stvo Nauka, Moscow, pp. 374–448
[101] Stoneham AM (1996) Theory of defects in solids. Clarendon, Oxford, pp. 271–341
[102] Cardona M (1996) Fundamentals of semiconductors. Springer, Berlin Heidelberg New York Tokyo, pp. 113–147
[103] Sirenko AA, Belitsky VI, Ruf T, Cardona M, Ekimov AI, Trallero-Giner C (1998) Spin-flip and acoustic-phonon Raman scattering in CdS nano-crystals. Physical Review B 58: 2077–2087
[104] Anselm AI, Firsov YA (1955) Mean free path of delocalized exciton in an atomic crystal (in Russian). Zh Exp Theor Fiz 28: 152–159
[105] Mahan GD, Hopfield JJ (1964) Piezoelectric polaron effects in CdS. Physical Review Letters 12: 241–243
[106] Blacha A, Presting H, Cardona M (1984) Deformation potentials of $k=0$ states of tetrahedral semiconductors. Physica Status Solidi (b) 126: 11–36
[107] Woerner M, Elsaesser T (1995) Ultra-fast thermalization of non-equilibrium holes in p-type tetrahedral semiconductors. Physical Review B 51: 17490–17498
[108] Dargys A (2005) Hole spin relaxation: optical deformation potential scattering. Semiconductor Science and Technology 20: 733–739
[109] Landau LD, Lifshitz EM (1991) Quantum mechanics. Non-relativistic theory. Pergamon, Oxford
[110] Guzelian AA, Katari JEB, Kadavanich AV, Banin U, Hamad K, Juban E, Alivisatos AP, Wolters RH, Arnold CC, Hearth JH (1996) Synthesis of size-selected, surface-passivated InP nanocrystals. Journal of Physical Chemistry 100: 7212–7218
[111] Seong MJ, Mićić OI, Nozik AJ, Mascarenhas A, Cheong HM (2003) Size-dependent Raman study of InP quantum dots. Applied Physics Letters 82: 185–187
[112] Rolo AG, Vasilevskiy MI, Talapin DV, Rogach AL (2005) Resonant Raman scattering in spherical InP QDs: the role of the optical deformation potential interaction. In: Physics of Semiconductors

(Menendez J, Van der Walle CG, eds). ICPS27 Proceedings. AIP Publishing, Melville, NY; pp. 747–748
[113] Spector HN, Lee J, Melman P (1986) Exciton linewidth in semiconducting quantum-well structures. Physical Review B 34: 2554–2560
[114] Rudin S, Reinecke TL (1986) Temperature-dependent exciton linewidths in semiconductor quantum wells. Physical Review B 41: 3017–3027
[115] Lang IG, Firsov YA (1962) Kinetic theory of semiconductors with low mobility (in Russian). Zh Exp Theor Fiz 43: 1843–1860
[116] Duke CB, Mahan GD (1965) Phonon-broadened impurity spectra. I. Density of states. Physical Review 139: A1965–A1982
[117] Itoh T, Nishijima M, Ekimov AI, Gourdon C, Efros AlL, Rosen M (1995) Polaron and exciton–phonon complexes in CuCl nano-crystals. Physical Review Letters 74: 1645–1648
[118] Ignatiev IV, Kozin IR, Davydov VG, Nair SV, Lee JS, Ren H-W, Sugou S, Masumoto Y (2001) Phonon resonances in photoluminescence spectra of self-assembled quantum dots in an electric field. Physical Review B 63: 075316
[119] Heitz R, Veit M, Ledentsov NN, Hoffmann A, Bimberg D, Ustinov VM, Kop'ev PS, Alferov ZhI (1997) Energy relaxation by multi-phonon processes in InAs/GaAs quantum dots. Physical Review B 56: 10435–10445
[120] Toda Y, Moriwaki O, Nishioka M, Arakawa Y (1999) Efficient carrier relaxation mechanism in InGaAs/GaAs self-assembled quantum dots based on the existence of continuum states. Physical Review Letters 82: 4114–4117
[121] Lemaitre A, Ashmore AD, Finley JJ, Mowbray DJ, Skolnik MS, Hopkinson M, Krauss TF (2001) Enhanced phonon-assisted absorption in single InAs/GaAs quantum dots. Physical Review B 63: 161309
[122] Stauber T, Zimmermann R, Castella H (2000) Electron–phonon interaction in quantum dots: a solvable model. Physical Review B 62: 7336–7343
[123] Jacak L, Machnikowski P, Krasnyi J, Zöller P (2003) Coherent and incoherent phonon processes in artificial atoms. European Physical Journal D 22: 319–331
[124] Hameau S, Guldner Y, Verzelen O, Ferreira R, Bastard G, Zeman J, Lemaitre A, Gerard JM (1999) Strong electron–phonon coupling regime in quantum dots: evidence for everlasting resonant polarons. Physical Review Letters 83: 4152–4155
[125] Meystre P, Surgent III M (1998) Elements of quantum optics. Springer, Berlin, p. 287
[126] Krummheuer B, Axt VM, Kuhn T (2002) Theory of pure dephasing and the resulting absorption line shape in semiconductor quantum dots. Physical Review B 65: 195313
[127] Muljarov EA, Zimmermann R (2004) Dephasing in quantum dots: quadratic coupling to acoustic phonons. Physical Review Letters 93: 237401
[128] Poles E, Selmarten DC, Micic OI, Nozik AJ (1999) Anti-Stokes photoluminescence in colloidal semiconductor quantum dots. Applied Physics Letters 75: 971–973
[129] Rakovich YuP, Filonovich SA, Gomes MJM, Donegan JF, Talapin DV, Rogach AL, Eychmüller A (2002) Anti-Stokes photoluminescence in II–VI colloidal nano-crystals. Physica Status Solidi (b) 229: 449–452
[130] Rakovich YuP, Gladyshchuk AA, Rusakov KI, Filonovich SA, Gomes MJM, Donegan JF, Talapin DV, Rogach AL, Eychmüller A (2002) Anti-Stokes luminescence of cadmium telluride nano-crystals. Applied Spectroscopy 69: 444–449
[131] Wang X, Yu WW, Zhang J, Aldana J, Peng X, Xiao M (2003) Photoluminescence up-conversion in colloidal CdTe quantum dots. Physical Review B 68: 125318
[132] Cantarero A, Trallero-Giner C, Cardona M (1989) Excitons in one-phonon resonant Raman scattering: deformation-potential interaction. Physical Review B 39: 8388–8397
[133] Vasilevskiy MI, Miranda RP, Anda EV, Makler SS (2004) Polaron effect on Raman scattering in semiconductor quantum dots. Semiconductor Science and Technology 19: S312–S315
[134] Rodríguez-Suárez R, Menéndez-Proupin E, Trallero-Giner C Cardona M (2000) Multiphonon resonant Raman scattering in nano-crystals. Physical Review B 62: 11006–11016
[135] Müller T, Schrey FF, Strasser G, Unterrainer K (2003) Ultra-fast intra-band spectroscopy of electron capture and relaxation in InAs/GaAs quantum dots. Applied Physics Letters 83: 3572–3174
[136] Schaller RD, Petryga JM, Goupalov SV, Petrushka MA, Ivanov SA, Klimov VI (2005) Breaking the phonon bottleneck in semiconductor nano-crystals via multi-phonon emission induced by intrinsic non-adiabatic interactions. Physical Review Letters 95: 196401
[137] Heitz R, Born H, Guffarth F, Stier O, Schliwa A, Hoffmann A, Bimberg D (2001) Existence of a phonon bottleneck for excitons in quantum dots. Physical Review B 64: 241305

[138] Sauvage S, Boucaud P, Lobo RPSM, Bras F, Fishman G, Prazeres R, Glorin F, Ortega JM, Gérard J-M (2002) Long polaron lifetime in InAs/GaAs self-assembled quantum dots. Physical Review Letters 88: 177402
[139] Urayama J, Norris TB, Singh J, Bhattacharya P (2001) Observation of phonon bottleneck in quantum dot electronic relaxation. Physical Review Letters 86: 4930–4933
[140] Verzelen O, Ferreira R, Bastard G (2002) Energy relaxation in quantum dots. Physical Review B 66: 081308
[141] Wetzler R, Wacker A, Schöll E (2004) Non-local Auger effect in quantum dot devices. Semiconductor Science and Technology 19: S43–S44
[142] Seebeck J, Nielsen TR, Gartner P, Jahnke F (2005) Polarons in semiconductor quantum dots and their role in the quantum kinetics of carrier relaxation. Physical Review B 71: 125327
[143] Efros AlL, Kharchenko VA, Rosen M (1995) Breaking the phonon bottleneck in nanometer quantum dots: role of Auger-like processes. Solid State Communications 93: 281–284
[144] Narvaez GA, Bester G, Jahnke F (2006) Carrier relaxation mechanisms in self-assembled (In,Ga)As/GaAs quantum dots: efficient $p \to s$ Auger relaxation of electrons. Physical Review B 74: 075403
[145] Vallée F (1994) Time-resolved investigation of coherent LO-phonon relaxation in III–V semiconductors. Physical Review B 49: 2460–2468
[146] Stoneham AM, McKinnon BA (1998) Excitation, dynamics and dephasing in quantum dots. Journal of Physics Condensed Matter 10: 7665–7677

Anti-Stokes photoluminescence in semiconductor nanocrystal quantum dots

By

Yury P. Rakovich, John F. Donegan

School of Physics and CRANN Research Centre, Trinity College Dublin, Dublin, Ireland

1. Introduction

George Gabriel Stokes, while studying the process of light emission from solid materials, discovered that the wavelength of the emitted light is generally longer than that of the light which is exciting the emission process. This excitation–emission shift bears his name, as an observed "Stokes-shift" between the excited and emitted light [1]. In our modern quantum mechanical description of the process, the emitted photons have a lower energy than those exciting the photoluminescence, the excess energy being delivered into the material, usually by phonon excitation. Within this review, we call the process the Stokes-shifted photoluminescence (SSPL).

It has been shown that some materials emit light at shorter wavelengths than that with which the material was illuminated because of thermal (phonon) interactions with the excited atoms [2]. This up-conversion process is termed anti-Stokes photoluminescence (ASPL), as opposed to the more common Stokes emission process. ASPL has been observed in a variety of systems such as atoms and molecules [3, 4], polymers [5, 6], fullerenes [7, 8], semiconductor macrocrystals [9–14] and structures [15–23].

Recent research interest in the field of up-conversion processes has been sparked by the development of multi-colour displays [24], dynamical imaging microscopy [25], bio-imaging systems [26, 27], unconventional lasers [28] and solid-state optical refrigeration devices [29, 30]. Here, we will concentrate on one particular up-conversion process, phonon-assisted anti-Stokes emission, in one particular type of semiconductor nanostructures, namely in nanocrystal quantum dots (QDs) fabricated via colloidal chemistry. These QDs represent the ultimate in semiconductor-based quantum-confined systems with atom-like energy levels, large optical transition dipole moment and high photoluminescence (PL) quantum efficiency. The interest developed in nanocrystal QDs has been fueled by the high degree of reproducibility and control that is currently available in the fabrication and manipulation of these quantum-confined structures [31]. It is worth mentioning the utilization of QDs in LEDs [32, 33], photonic [34, 35] and core-shell structures [36, 37], and as biological labels [38, 39].

Apart from phonon-assisted up-conversion [40] a number of different mechanisms have been suggested to explain the anti-Stokes emission process. The suggested microscopic mechanisms are Auger recombination [15, 41], direct two-photon absorption [42, 43], and two-step/two-photon absorption [44]. Below, we will discuss the plausibility of these mechanisms in the ASPL of semiconductor nanocrystals.

2. Phonon-assisted ASPL in nanocrystal QDs

The ASPL occurs when the emission spectrum with intensity I_{em} is obtained at higher energy than the excitation (P) (Fig. 1) while the energy gap ΔE between the excitation energy and the excited electronic level is comparable or even large than the maximum phonon energy in the material. It is instructive to start the consideration of the problem with a simple three-level model where two upper energy levels are separated by an energy gap of value ΔE (Fig. 1a).

To set up rate equations, we first consider the excitation transition from the ground state ("0") to the state "1", from where two pathways are possible. First, the resonant radiative transition from state "1" may return the system to its ground state with rate γ_1. Alternatively, a thermally induced population of level "2" may occur with a radiative transition, which is the ASPL process, from state "2" with rate γ_2. At elevated temperatures this process may prevail over the resonant recombination. The excited carriers may relax nonradiatively from level "2" to level "1" with rate γ_r followed by the resonant radiative recombination.

The dynamic equations, which describe this model for population densities n_1 and n_2 are

$$\frac{dn_1}{dt} = P - [\gamma_1 + \gamma_r \exp(-\Delta E/kT)]n_1 + \gamma_r n_2$$
$$\frac{dn_2}{dt} = \gamma_r \exp(-\Delta E/kT)n_1 - \gamma_r n_2 - \gamma_2 n_2 \quad (1)$$

where γ_i is the recombination rate; ΔE is the activation energy; k is Boltzmann's constant; T is the temperature.

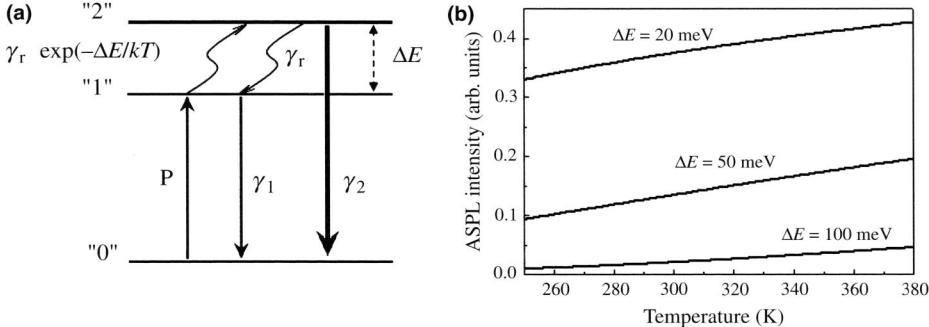

Fig. 1. Energy scheme for phonon-assisted up-converted emission (**a**) and temperature dependence of ASPL intensity calculated using Eq. (2) for three values of ΔE supposing that $\gamma_r = \gamma_1 = \gamma_2$ (**b**)

Solving system (1) for continuous-wave excitation one obtains equations for both the resonant emission ($I_1 = \gamma_1 n_1$) and up-converted emission intensity ($I_2 = \gamma_2 n_2$):

$$I_1 = P \frac{c_1}{c_1 + A \exp(-\Delta E/kT)}$$
$$I_2 = P \frac{A \exp(-\Delta E/kT)}{c_2 + A \exp(-\Delta E/kT)} \quad (2)$$

where coefficients $c_1 = \gamma_1/\gamma_r$; $c_2 = \gamma_2/\gamma_r$; $A = \gamma_2/(\gamma_2 + \gamma_r)$) describe the ratios of recombination rates. It is noteworthy that experimentally these parameters can be estimated from analysis of the PL decay curve [31, 45].

The first implication of this analysis is that the intensity of up-converted emission is a linear function of excitation power. However, the most striking feature of this model is that the intensity of the up-converted emission increases with temperature (Fig. 1b) gaining energy from the thermal bath in contrast to the conventional quenching of resonant or Stokes-shifted luminescence with increasing temperature. This anomalous temperature behaviour of ASPL may therefore be used as an indicator of phonon-assisted processes while analysing mechanisms of up-converted luminescence in materials.

To date ASPL caused by one-photon phonon-assisted carrier excitation has been reported for InP [46], CdSe [46–48], CdTe [47, 49, 50], PbS [51] and PbSe [52] nanocrystals.

3. Dependence on excitation wavelength and the efficiency of ASPL

A general feature of this kind of up-conversion process is that the ASPL signal can be detected only for excitation energies that are well below the maximum energy position of the normal Stokes emitted PL signal.

Figure 2a shows PL spectra of 4-nm size CdSe/ZnS QDs measured at room temperature varying the excitation energy (E_{exc}) (2.0–3.10 eV) in the spectral region from the high energy region of the absorption spectra to the tail far below the first absorption peak (Fig. 2b). When $E_{exc} = 2.43$ eV (Fig. 2a, Region I) the position of the Stokes-shifted emission peak is almost independent of the excitation wavelength. For this sample, the value of the "nonresonant Stokes shift" (the difference between the lowest-energy peak in the absorption spectra and the emission peak) is about 67 meV. The PL linewidth shows only a very small additional broadening with increasing excitation energy in this region. This weak dependence of the linewidth and the nonresonant Stokes shift is due to the fact that the higher the excitation energy – the better the PL spectrum reflects the entire size distribution of nanocrystals in the sample. While the data presented in Fig. 2a are normalized for comparison, on an absolute scale, a 57% reduction in the integrated PL intensity has been observed as the excitation energy decreases from 3.10 eV to 2.43 eV.

Providing excitation between 2.30 and 2.43 (Fig. 2a, Region II), the emission profile becomes more complex. A weak shoulder appears on the blue edge of PL band as the photon energy decreases down to 2.38 eV, whereas the main PL peak shifts to the low-energy side. For lower energy excitation (2.34 eV), the PL spectrum has one

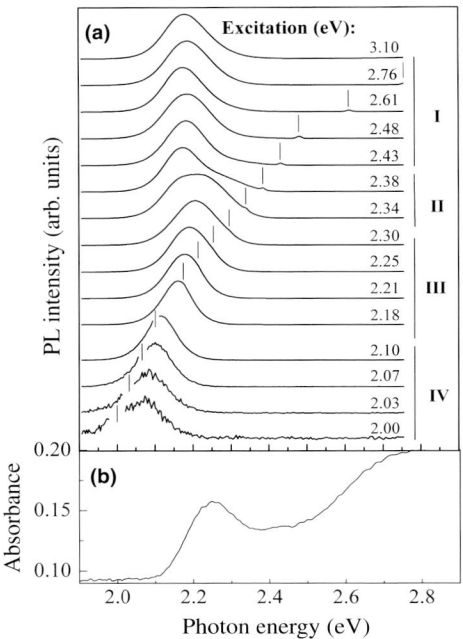

Fig. 2. a PL spectra of CdSe/ZnS nanocrystals in toluene for various excitation wavelengths at room temperature. For each spectrum the excitation energy is marked by a vertical line. All PL spectra are normalized for comparison. **b** The absorption spectrum of the sample

very broad band, which can be well de-convoluted into two peaks. This double peak structure can be explained as a result of size-selective excitation within the population of QDs and is expected for an inhomogeneously broadened system. Since there are multiple states present, there is a photon energy (2.38–2.34 eV) where the largest QDs can be excited into their second excited state, while a smaller size is excited into their first excited state. Since each nanocrystal emits at one photon energy, regardless of excitation energy, at this point the emission spectrum shows two peaks, one from the smaller set of particles absorbing into their first state and the second from the larger QDs excited into their second state [53].

As the excitation photon energy is tuned further below the first absorption maximum of 2.25 eV down to 2.18 eV (Fig. 2a, Region III), one PL peak is observed again, whose emission energy shifts to the red with decreasing excitation energy. The width of the PL peak decreases considerably in this spectral region (from ∼137 meV at $E_{exc} = 2.30$ eV to ∼96 meV at $E_{exc} = 2.18$ eV) demonstrating PL line narrowing, because only larger nanocrystals within the finite size distribution are excited on the low-energy side of the absorption profile. On the other hand, a 52% reduction in the integrated PL intensity has been observed as the excitation energy decreases in Region II because only a very small fraction of the size distribution is selected to be excited.

A distinctly different behaviour is observed in the spectral region far below the absorption peak (Fig. 2a, Region IV). A tail of ASPL can be seen with 2.10 eV excitation, ranging up to ∼200 meV above the excitation energy. A similar decaying anti-Stokes tail in PL has been also observed in InP [46] and CdTe nanocrystals [54]. At still lower excitation energies, a pronounced ASPL peak appears in the high-energy region, which shifts to lower energy following the excitation wavelength. This behaviour is very similar to that of the SSPL in region III. Actually, the progressive transition from SSPL into ASPL is evident in Fig. 2a, Regions III–IV. A similar observation was reported recently for PbS QDs [51]. The gradual change in the band shape suggests that the physical process involved in QD emission is the same throughout the whole range of excitation energies. However, in contrast to SSPL, the ASPL does not reflect the size distribution of the QDs. As mentioned above, the effect of inhomogeneous broadening caused by the distribution of QDs sizes can be clearly seen in the excitation wavelength dependence of the SSPL spectra (Fig. 2a, Regions II and III). When the excitation is restricted to the onset region of the absorption spectra, then QDs of a much narrower size range are excited; these QDs are the largest size in the ensemble. In this spectral region the SSPL spectra of QDs show a decrease of the width of the PL band demonstrating pronounced line narrowing [53, 55, 56]. In the spectra presented in Fig. 2a, the full width at half maximum (FWHM) of the SSPL band decreased from 95 meV at $E_{ex} = 2.32$ eV to 76 meV at $E_{ex} = 2.16$ eV. In contrast, the ASPL linewidth shows extra broadening with decreasing excitation energy: from 83 meV at $E_{ex} = 2.12$ eV to 150 meV at $E_{ex} = 1.95$ eV.

It is noteworthy that all spectra presented in Fig. 2a were obtained by exciting the samples with a Xenon lamp (output power of 40 μW to 0.1 mW, depending on the spectral region). This demonstrates that phonon-assisted excitation of the ASPL process in QDs is a highly efficient process since even for samples with moderate quantum yield (∼20%) there is no need for laser excitation [57]. Also it can be seen that the efficiency of ASPL is comparable with that of SSPL at least for small magnitudes of the up-converted blue shift ΔE (at excitation energy $E_{ex} = 2.10$ eV). At lower excitation energies (i.e. bigger magnitudes of ΔE) the efficiency of ASPL rapidly declines in accordance with Vavilov's law [50, 58].

4. Dependence on excitation power

In light of the above model for population densities, photon energy up-conversion is a linear process of the excitation intensity in nanocrystal QDs. The linear behaviour of the ASPL intensity (I_{ASPL}) has been experimentally verified for various QD materials [46, 47, 49, 50, 59] with a slope depending on the quantum yield of the sample (Fig. 3).

It is known [46, 60] that in some cases, the analysis of the dependence of the ASPL intensity on the excitation intensity (I_{exc}) can give information on the mechanism of excitation energy transfer in the high-energy spectral region. Thus, with two-photon excitation or two-stage excitation of electrons from the valence to the conduction band through deep impurity levels, a quadratic dependence $I_{ASPL} \sim I_{exc}^2$ should be observed [60]. For an ASPL process induced by Auger recombination, the dependence $I_{ASPL} \sim I_{exc}^3$ is characteristic [22], whereas in the case of mixed mechanisms, the power-law dependence of I_{ASPL} on I_{exc} becomes more complicated [13].

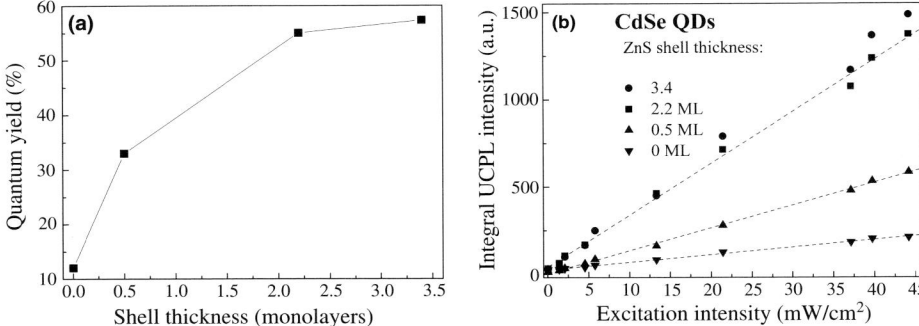

Fig. 3. a The PL quantum yield of CdSe QDs as a function of the ZnS shell thickness (in monolayers (ML)). The dashed line is a guide to the eye. **b** The dependence of the ASPL intensity in CdSe QDs on the excitation power density at room temperature. The lines show a linear fit to the data. The enhancement of up-conversion efficiency is observed with increasing of thickness of the ZnS shell, i.e. with improving passivation of surface dangling bonds. Reproduced with permission from MAIK Nauka/Interperiodica [48]

Therefore, when analyzing an up-converted PL signal, an observed linear dependence of ASPL on excitation intensity alone cannot be taken as an indication of the participation of phonons in the excitation process. As will be discussed in the following section, the same dependence can be observed, for example, for two-photon or two-step excitation under conditions of saturation [61]. A definitive conclusion for the mechanisms of ASPL must be supported by a series of independent measurements. In this respect the temperature dependence investigations of up-converted PL are the most conclusive [46, 47, 49, 61, 62].

5. Temperature dependence of ASPL

As mentioned above, the ASPL process with a linear dependence on excitation intensity can be observed as a result of energy transfer to the excited electron–hole pair from the phonon bath. In this case, the ASPL intensity should grow with increasing temperature because of the increase of the population of phonons. Indeed, this behavior was reported for CdSe [46, 47] and CdTe [47, 49, 50] QDs. As can be seen from Fig. 4 an increase of up to 12 times in the ASPL intensity of CdSe QDs was achieved, when the sample was heated in the temperature range 283–353 K, while the width of ASPL band is gradually reduced [63]. At the same time the SSPL shows thermal induced quenching and broadening. It turns out that the thermally stimulated increase in the ASPL is almost independent of the size of the QDs (Fig. 5, inset), although it is more efficient when the QD surface is better passivated (Fig. 4, inset) [48, 49, 57].

The spectral position of the ASPL peak shows little dependence on temperature (Figs. 4 and 5) whereas the temperature variation of the peak energy of the SSPL was found to be practically coincident with that of the ($1S_e \rightarrow 1S_{3/2}$) absorption peak energy. These experimental findings demonstrate an important role of the electron–phonon interaction in ASPL processes in QDs and this will be considered later.

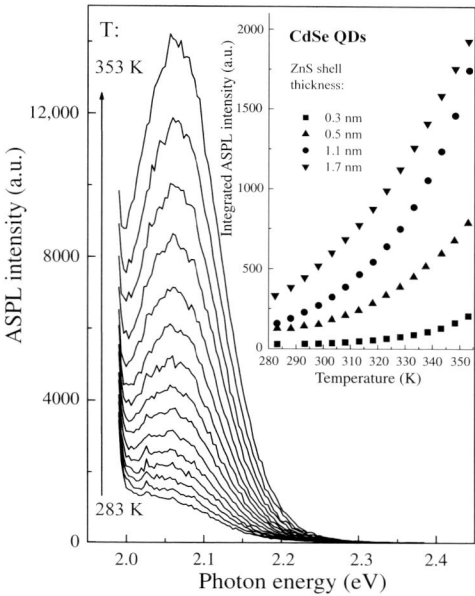

Fig. 4. Temperature dependence of the ASPL in 4-nm CdSe QDs with 1.1-nm ZnSe shell. Inset: Temperature stimulated enhancement of ASPL intensity for increasing shell thickness (i.e. increasing quantum yield of the sample). Reproduced with permission from MAIK Nauka/Interperiodica [48]

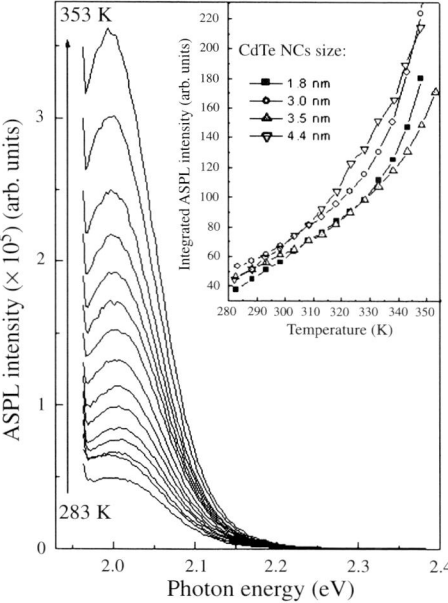

Fig. 5. Temperature behaviour of the ASPL band in 3.5-nm CdTe NCs. Inset: Variation of the integrated intensity of the ASPL with temperature for all samples studied. Reproduced with permission from Materials Research Society [57]

Table 1. The maximum magnitude of the anti-Stokes shift for different QD materials

Material	ΔE^{max} (meV)	References
CdSe	335	[47]
	413	[59]
CdTe	285	[47]
	350	[57]
	360	[49]
PbS	330	[51]

As can be seen from Fig. 1b, an increase in ASPL intensity with temperature strongly depends on the magnitude of the up-converted blue shift demonstrating a sub-linear growth at small (20 meV) values of ΔE; almost linear dependence is obtained for $\Delta E = 50$ meV and a super-linear exponential-like dependence for $\Delta E = 100$ meV.

For a phonon-assisted mechanism, the minimum magnitude of the up-converted blue shift is expected to be about the typical energy of optical phonons (24.8 meV for the bulk CdSe) [64] which is comparable to the thermal energy ($\sim k_B T$) at room temperature. In data presented in Fig. 4, ΔE is about 97 meV giving rise to a well-resolved ASPL band.

As can be seen from Fig. 2, the value of the anti-Stokes blue shift increases with decreasing excitation energy. The maximum magnitude of the up-converted shift (ΔE^{max}) can be defined as the difference between the excitation energy and the energy value at which an exponential fit of the ASPL high-energy wing crosses the average background noise level [46]. The reported values of ΔE^{max} estimated in this way are summarized in Table 1.

6. Mechanism of ASPL: thermally populated defect states versus electron–phonon interaction

Let us summarize the reported experimental results on the ASPL in colloidal QDs:

(i) The ASPL intensity increases linearly with excitation power.

(ii) The ASPL intensity grows strongly with increasing temperature. The efficiency of the thermally stimulated ASPL growth is independent of the size of the QDs. The up-conversion process is more efficient in samples with higher quantum yield.

(iii) If E_{ex} increases (approaching the absorption peak) the ASPL peak moves continuously towards higher energy accompanied by a narrowing of the ASPL band. Its efficiency increases approaching the efficiency of SSPL and finally the spectrum switches to the Stokes-shifted regime.

(iv) The blue (anti-Stokes) shift between the ASPL peak and E_{ex} does not depend significantly upon QDs size. The shift can range approximately from 20 meV to 400 meV.

Possible ASPL mechanisms were discussed in [46–50, 57], but no definite conclusions were made. According to point (i) above, nonlinear optical processes such

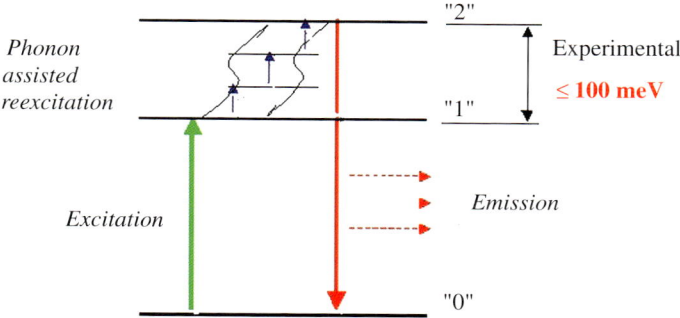

Fig. 6. Schematic of phonon-assisted up-conversion mechanism proposed in [47]

as two-photon absorption and (since the possibility of emergence of more than one-exciton per QD is negligible) the Auger excitation can be excluded in this case. Experimental findings (ii) and (iii) strongly suggest the involvement of phonons in the re-excitation process. Finally, in order to explain (iv), it was suggested that up-conversion occurs via thermally populated states. This mechanism requires that the defect states are populated via phonons and then absorption of the incident photon leads to the excitation into the conduction band states followed by the higher energy photon emission. Depending on the electronic structure of the conduction and valence energy levels in particular QDs, electron [47, 49] or hole [46] up-conversion is believed to be driven by phonon absorption.

Although theoretical data reported in support of this mechanism are consistent with the surface energy levels for dangling bonds or other sub-bandgap defect states, there is no direct experimental evidence that these states are responsible for the ASPL process [46, 49, 59]. Also the fact that the ASPL efficiency increases with quantum yield of the QDs provides strong proof that the ASPL emission cannot originate from sub-band gap states.

In order to account for all the experimental features of ASPL, an alternative model has been recently suggested based on direct re-excitation of QDs by longitudinal optical (LO) phonons without resorting to the surface (defect) states [65]. Taking advantage of a nonperturbative approach for the calculation of the polaronic effects in QDs, it was shown that red-shifted optical phonon replicas can be involved in up-conversion and that the polaronic effects are significant, even when the interlevel spacing in the QDs is quite far away from resonance with the optical phonon energies.

Figure 7a presents the lower-energy side of absorption spectrum of CdSe QDs calculated using this approach and showing two sub-bandgap bands (indicated as "−1LO" and "−2LO") through which the QDs can be excited. The QDs will then emit a photon most likely having the energy of the zero phonon line. In a sense 1LO and 2LO optical phonon replicas represent virtual sub-gap states, which are separated from the fundamental absorption line (i.e. zero phonon line) by energies that are only weakly dependent on the QD size. The probability of such an up-conversion process increases with temperature because so does the integrated intensity of the "−1LO" and "−2LO" absorption bands due to increased population of LO phonons (Fig. 7b). The experimental temperature dependence of the ASPL

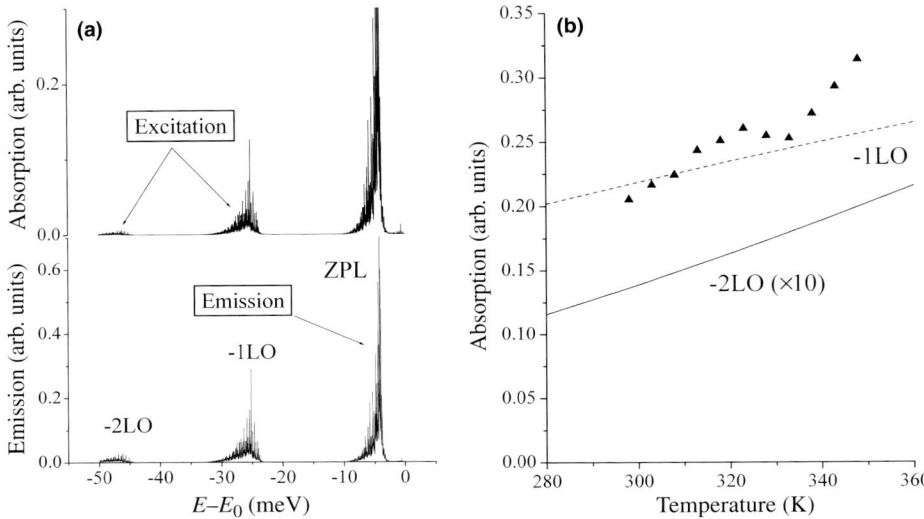

Fig. 7. a Low-energy part of the absorption and emission spectra calculated for a hypothetical CdTe QD considering three optical phonon modes with parameters given in the figure. The level spacing $\Delta E = 100$ meV, and the temperature is 300 K. **b** Temperature dependence of the integrated intensity of two sub-bandgap bands in the absorption spectrum (**a**) (lines) and experimental data of [47] (points) showing the temperature dependence of the ASPL peak amplitude. Reproduced with permission from American Physical Society [65]

intensity described above can be understood by taking into account that, at a certain temperature, further (higher order) red-shifted satellites (whose intensity depends more strongly on temperature) become more efficient.

In reality the situation is complicated by the distribution of the QD size, so that E_{ex} can match different "$-n$LO" bands of QDs of several different size subsets within the ensemble. It is also necessary to bear in mind that inter-level interaction, multiple confined optical phonon modes, and interaction with acoustic phonons will broaden the satellite band for each QDs in the ensemble. Consequently, the resulting ASPL band will not appear as a sum of narrow features as it appears from the above figure. Another complication arises from the fact that the calculated emission spectrum (Fig. 7a) assumes thermal equilibrium. This will be the case only if carrier thermalization processes are fast. Recent studies of carriers dynamics in QDs [55, 66, 67] at above bandgap excitation show that immediately after photoexcitation, the initially formed hot carriers thermalize quickly to the bottom of the conduction and valence bands and subsequently decay either into shallow trap states [55, 66] or an intrinsic "dark" exciton state [68]. This thermalization is an extremely fast process occurring in the 300–500 fs range [67, 69] providing fast establishment of thermal equilibrium. It is noteworthy that the reported time-resolved studies of ASPL demonstrate a much longer, ns-scale decay [49]. However, there is no reason why such relaxation should be slower at room temperature and/or if the carrier is "cool".

Another and more serious problem with the proposed polaronic mechanism of up-conversion is that using this approach it is difficult to explain the observed values of the up-converted shift (Table 1) which are much larger than the LO phonon energy.

Of course in spectra of individual QDs there are further "$-n$LO" satellites (with larger n), but their intensities are very low for realistic values of the electron–phonon coupling constants. In order to explain the experimentally observed large anti-Stokes shifts, one has to consider a cascaded mechanism of the ASPL. More detailed discussion of this particular mechanism is given in the chapter of M. Vasilevskiy of this book.

7. Availability of phonon modes

Whichever model is used to describe the phonon-assisted ASPL excitation in the QDs, the availability of vibrational modes is crucial to provide efficient up-conversion. To this end, recent experimental studies of Raman spectra in semiconductor QDs [70–75] may provide fresh insight into understanding the mechanism of phonon-assisted up-conversion.

The evolution of the optical phonon spectra of colloidal core/shell CdSe/ZnS QDs which demonstrate efficient ASPL (Figs. 2 and 4) has been recently reported [70]. These QDs were studied by resonant Raman spectroscopy with an increase of the shell thickness from 0.5 to 3.4 monolayers. A significant improvement of the PL efficiency has been observed with increase of the ZnS shell thickness.

As an example of the Raman spectra of the CdSe/ZnS nanocrystals, the spectrum of CdSe QDs with the thickest (3.4 ML) ZnS shell is presented in Fig. 8a. Raman lines of the LO and 2LO phonons of the CdSe core are clearly seen in the region of

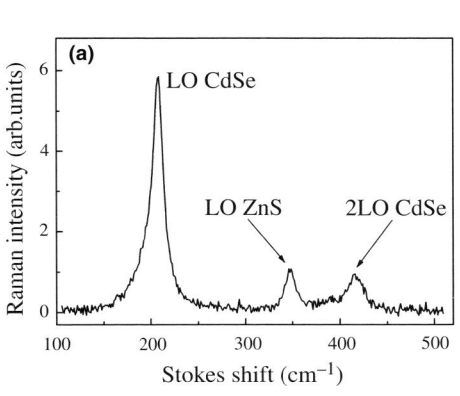
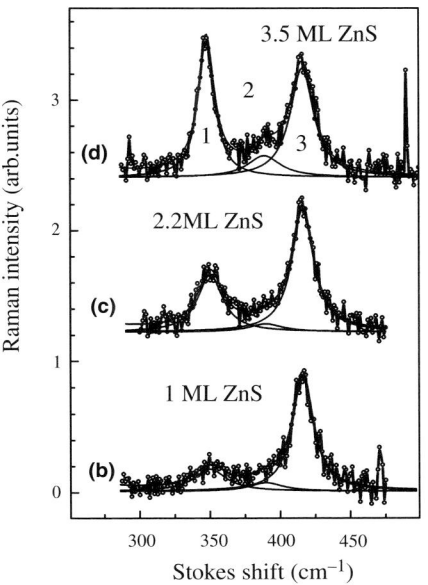

Fig. 8. a Raman spectrum of CdSe/ZnS QDs with a shell thickness 3.4 ML. **b–d** The parts of the Raman spectra of CdSe/ZnS QDs in the region of the ZnS LO phonon for different shell thicknesses. The regions of the ZnS LO phonons and 2SO and 2LO phonons of CdSe are denoted by the numbers 1, 2, and 3, respectively. Reproduced with permission from American Physical Society [70]

200 cm^{-1} and 400 cm^{-1}, respectively. Apart from these modes a pronounced peak associated with the LO phonons of the ZnS shell can be seen in Fig. 8a at about 350 cm^{-1}. The intensity and line shape of the ZnS LO line are determined by the shell crystallographic structure [70].

It has turned out that the ZnS LO phonon line at 350 cm^{-1}, which partly overlaps the second order Raman lines of the CdSe core, can be distinguished, even at a shell thickness of 0.5 ML. The increase in shell thickness results in the increase in the line intensity, which is roughly proportional to the ZnS volume. A remarkable decrease in linewidth from 30 cm^{-1} for the 0.5 ML shell down to 12.5 cm^{-1} for the 3.4 ML manifests to the substantial improvement in the shell crystallographic structure. It is particularly remarkable that the increase in the shell thickness results in an enhanced efficiency of the ASPL process (Figs. 3 and 4). The observed peak between the ZnS LO and CdSe 2LO phonon lines (Fig. 8) was the subject of extensive discussion in the last several years [72, 76–79]. In most of the studies it was suggested that the asymmetry in the low-frequency part of the LO phonon Raman peak of the CdSe QDs is caused mainly by surface optical (SO) phonon modes. In spite of the fact that the Raman scattering by the SO modes is forbidden for an ideal spherical shape of the QDs, the appearance of the SO peak in Raman spectra can be explained by the relaxation of the angular momentum phonon selection rule because of the lack of wave-vector conservation in QDs [78]. It has been also predicted that the SO mode can be observed in the Raman spectra in the case of a nonspherical shape of the QDs, as well as due to the effect of impurities or interface imperfections [72, 78].

The theory of Raman scattering by spherical particles predicts the participation of phonons with angular momenta $l = 0$, 1, 2 and 3 (0 and 2 through the Fröhlich mechanism while 1 and 3 through the optical deformation potential hole–phonon interaction). Normally only LO-type phonons are observed in the spectra of II–VI

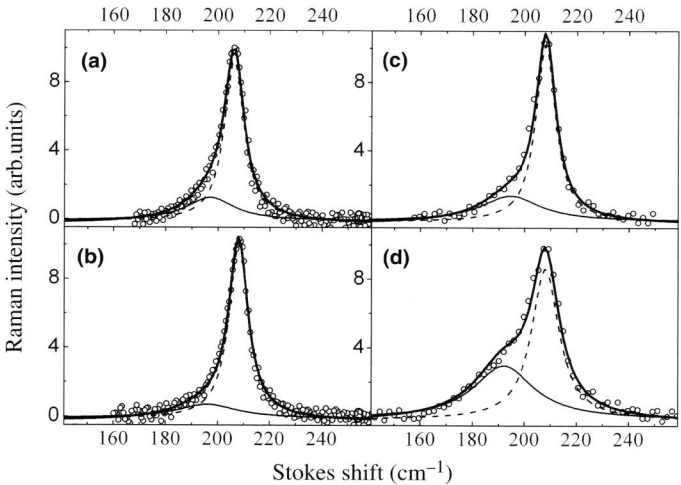

Fig. 9. The parts of the Raman spectra of CdSe/ZnS QDs in the region of the CdSe LO phonon for different shell thickness: 0 ML (**a**); 0.5 ML (**b**) 2.2 ML (**c**) and 3.4 ML (**d**), respectively. Reproduced with permission from American Physical Society [70]

QDs, and it was suggested that the polar Fröhlich-type interaction dominates the Raman scattering in CdSe/ZnS QDs. Therefore the peak labeled 2 in Fig. 8 can be assigned to the scattering from the interface phonon modes with $l = 2$ (although "breathing" $l = 0$ modes can also contribute to the polarised scattering).

A closer look at the evolution of the first order Raman spectra of the CdSe core as the ZnS shell thickness is varied reveals an enhanced contribution of SO modes (Fig. 9). The most striking result presented in Fig. 9 is the fact that with increasing shell thickness (i.e. with increase in quantum yield) the SO phonon band continues to shift to lower energy and grows in intensity. This was suggested to arise from an incoherent epitaxial growth of the ZnS shell at high coverage [70]. However, in the context of the present review, this fact clearly indicates the involvement of surface states in the up-conversion process, which is mediated by the interaction with optical surface phonons as well as with LO modes of the CdSe core and the ZnS shell. Due to spectral tunability of the SO mode energy (depending on shell thickness) optical phonon modes of various energies can contribute to the re-excitation of carriers causing efficient ASPL.

8. Applications of ASPL and further research directions

In the previous sections of this chapter, recent advances in the understanding of the fundamental properties of the phonon-assisted ASPL in semiconductor QDs are reviewed. In view of the potential applications, ASPL has a lot to offer in the fields of nanotechnology, lasing, optical cooling, bio-imaging and information technology. One of the applications suggested recently is the development of an all-optical temperature sensor based on nanoparticles [82]. Measurement of temperature changes in

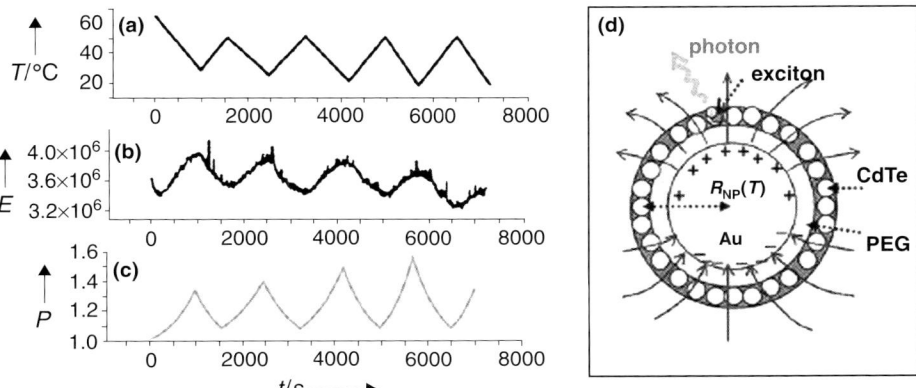

Fig. 10. Variation in the ASPL intensity E (**b**) of the PEG-tethered Au and CdTe QDs depending on the temperature T (**a**); **c** calculated photon-field enhancement factor P of the CdTe QDs as a function of time. **d** Schematic representation of the dielectric model used for calculating the curve in (**c**) as well as the plasmon excitation with associated field lines; the plasmon excitation inside an Au nanosphere interacts with excitons in the CdTe QDs through electric fields. The distance $R_{NP}(T)$ varies with the temperature. The curve in **c** also represents a theoretical dielectric model of the QDs assembly in which the CdTe QDs form a continuous spherical shell around the Au nanoparticles. Reproduced with permission from Wiley–VCH [82]

supersmall (nanoliter) volumes is a difficult problem, especially if both high precision and nanosecond time resolution are required.

One of the proposed approaches is to use the plasmon–exciton interaction in superstructures formed by metal nanoparticles and semiconductor QDs connected by a polymer acting as a molecular spring. A higher temperature leads to a modification in the luminescence due to the more extended conformation of the polymer chain (Fig. 10). In the reported experiments [82] the superstructure was excited in the anti-Stokes regime in order to eliminate scattered light from the excitation source.

Another promising direction is to use the unique properties of ASPL in hybrid photonic structures. In this respect whispering gallery mode (WGM) microcavities (e.g., microspheres, microcylinders, microrings and microtoroids) are ideally suited for observation of enhanced optical effects with extremely low excitation intensities [83]. High quality factors (Q) and large field densities associated with WGMs result in resonant enhancement of linear and nonlinear interactions of various kinds [36, 71, 84, 85]. Up-conversion of semiconductor QDs combined with photon confinement in three-dimensional microcavities has strong potential to be useful in microlaser technology, optical data storage, lighting and bio-imaging applications.

Figure 11b shows a room temperature spectrum of a single polystyrene microsphere of 70 μm size covered by a monolayer of CdTe QDs on a Si substrate using low intensity nonresonance excitation by a He–Ne laser. The QD monolayer was formed

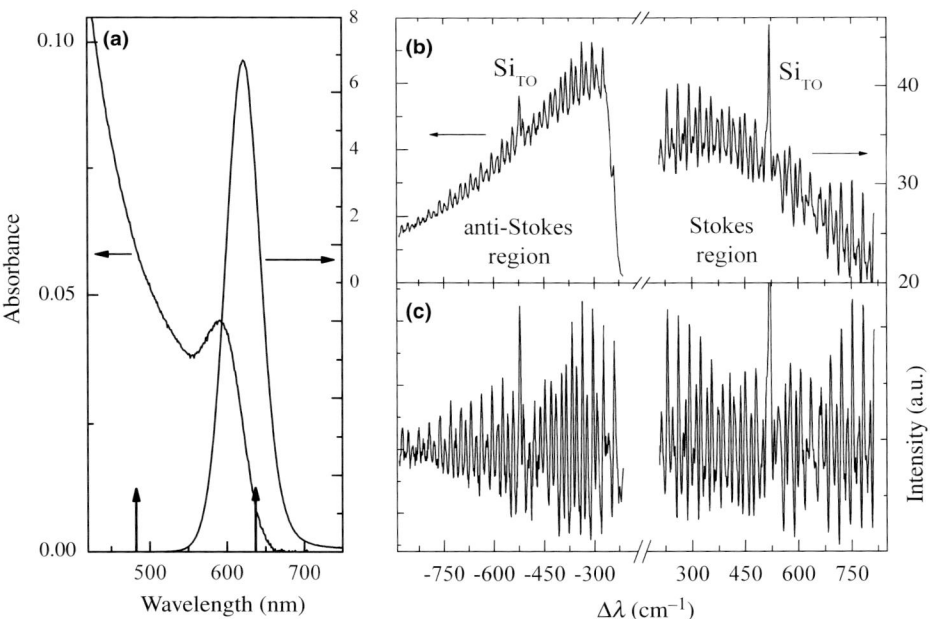

Fig. 11. Room temperature absorption and PL spectra of CdTe QDs in water (**a**). Arrows indicate the excitation wavelength used in micro-PL and Raman experiments. Raman spectrum of a single polystyrene/CdTe microsphere on a Si substrate. Excitation by HeNe laser $\lambda = 632.8$ nm before (**b**) and after (**c**) PL background subtraction. Reproduced with permission from Springer [86]

using the layer-by-layer deposition technique [37]. In this case, strong coupling between the WGM of the spherical microcavity and the electronic states of the CdTe QDs results in an enhanced luminescence contribution to the Raman signal simultaneously in both the Stokes and anti-Stokes spectral regions. The ASPL process is certainly highly efficient having an intensity comparable to the SSPL as seen from Fig. 11b. A similar effect was recently reported in small (2 μm) microspheres with a thin shell of semiconductor QDs [71]. The observation of ASPL from the polymer–CdTe microsphere can be attributed to the optical feedback via the microcavity with a WGM structure which leads to an increased probability of energy transfer to the emitting species.

A further major step in the application of ASPL using QDs is the development of nanoscale optical refrigerators. The concept of laser cooling (optical refrigeration) by luminescence up-conversion in solids dates back to 1929 [87], when it was recognized that thermal vibrational energy can be removed by anti-Stokes fluorescence creating a local cooling if a material is excited with photons having energy below the mean fluorescence energy. Material purity problems prevented observation of this type of laser cooling until 1995, when it was first demonstrated in ytterbium-doped glass [88]. This was followed soon after by reports of cooling in dye solutions [89] and thulium-doped glass [90]. A primary advantage of semiconductors compared to rare-earth doped solids is their potential for achieving temperatures down to ~10 K and below [91]. This is due to the difference of the ground state populations in the two systems. Also colloidal semiconductor QDs have a number of advantages over dyes such as tunable absorption and emission wavelengths, better photostability and longer excited state lifetimes [92].

Although the theory of semiconductor cooling has been tackled previously [91, 93, 94], the critical issue hindering laser cooling applications is the limited quantum yield of semiconductor materials due to nonradiative carrier recombination. It was recently demonstrated that in order to achieve efficient optical cooling, the quantum yield must be higher than 90% [95] which is unachievable for bulk semiconductor materials or epitaxial heterolayers. However, recent advances in the synthesis of highly efficient colloidal QDs with suppressed nonradiative transitions [31] may allow for the realization of semiconductor cooling even below 10 K [91].

9. Conclusions

The unique optical properties of semiconductor colloidal QDs have enabled a comprehensive study of photon energy up-conversion mediated by the interaction with optical phonons. This has allowed for an improved understanding of the anti-Stokes luminescence in nanostructures. The study of phonon-assisted excitation of ASPL, reviewed here, reveals many fascinating questions and fundamental problems that appear when merging the main concepts of the electron–phonon interaction and electronic and photonic confinement, in one structure. The dependence of the ASPL parameters on the excitation intensity and temperature suggests a manifold of photonic applications, in particular in the fields of sensing and optical cooling.

Acknowledgements

We thank Prof. A. Eychmüller, Dr. N. Gaponik and Prof. M. I. Vasilevskiy for helpful discussions. This work was supported by Science Foundation Ireland (SFI) under its CRANN CSET Project PR23.

References

[1] Stokes GG (1858) On the change of refrangibility of light. Abstracts of the Papers Communicated to the Royal Society of London 6: 195–200
[2] Wood RW (1928) Anti-Stokes radiation of fluorescent liquids. Phil Mag 6: 310–315
[3] Shen YR (2003) The principles of nonlinear optics. Wiley-Interscience, Hoboken, NJ
[4] Pope M, Swenberg CE (1999) Electronic processes in organic crystals and polymers. Oxford University Press, Oxford
[5] Soos ZG, Kepler RG (1991) Two-photon-absorption spectrum of poly(di-n-hexylsilane) films. Phys Rev B 43: 11908–11912
[6] Lemmer U, Rischer R, Feldmann J, Mahrt RF, Yang J, Greiner A, Bassler H, Gobel EO, Heesel H, Kurz H (1993) Time-resolved studies of two-photon absorption processes in poly(p-phenylenevinylene)s. Chem Phys Lett 203: 28–32
[7] Pichler K, Graham S, Gelsen OM, Friend RH, Romanow WJ, McCauley JP, Coustel N, Fischer JE, Smith AB (1991) Photophysical properties of solid films of fullerene, C_{60}. J Phys Condens Mater 3: 9259–9270
[8] Feldmann J, Fischer R, Guss W, Gobel EO, Schmitt-Rink S, Kratschmer W (1992) White luminescence from solid C_{60}. Europhys Lett 20: 553–558
[9] Litton CW, Reynolds DC, Collins TL, Park YS (1970) Exciton–LO-phonon interaction and anti-Stokes emission line in CdS. Phys Rev Lett 25: 1619–1621
[10] Halsted RE, Apple EF, Prener JS (1959) Two-State optical excitation in sulfide phosphors. Phys Rev Lett 2: 420–421
[11] Brown MR, Cox AFJ, Orr DS, Williams JM, Woods J (1970) Anti-Stokes excited edge emission in cadmium sulphide. J Phys C 3: 1767–1779
[12] Gundersen M (1974) Conversion of 28-mkm far-infrared radiation to visible light using bound excitons in CdS. Appl Phys Lett 24: 591–592
[13] Ivanov VY, Semenov YG, Surma M, Godlewski M (1996) Anti-Stokes luminescence in chromium-doped ZnSe. Phys Rev B 54: 4696–4701
[14] Ganichev D, Raab W, Zepezauer E, Prettl W, Yassievich IN (1997) Storage of electrons in shallow donor excited states of GaP:Te. Phys Rev B 55: 9243–9246
[15] Seidel W, Titkov A, Andre JP, Voisin P, Voos M (1994) High-efficiency energy up-conversion by an "Auger fountain" at an InP-AlInAs type-II heterojunction. Phys Rev Lett 73: 2356–2359
[16] Cho Y-H, Kim DS, Choe B-D, Lim H, Lee JL, Kim F (1997) Dynamics of anti-Stokes photoluminescence in type-II $Al_xGa_{1-x}As$ – $GaInP_2$ heterostructures: the important role of long-lived carriers near the surface. Phys Rev B 56: R4375–R4378
[17] Schrottke L, Grahn HT, Fujiwara K (1997) Enhanced anti-Stokes photoluminescence in a GaAs/$Al_{0.17}Ga_{0.83}As$ single quantum well with growth islands. Phys Rev B 56: R15553–R15556
[18] Finkeissen E, Potemski M, Wyder P, Vina L, Wiemann G (1999) Cooling of a semiconductor by luminescence up-conversion. Appl Phys Lett 75: 1258–1260
[19] Kral K, Zdenek P, Khas Z (2004) Transient processes and luminescence upconversion in zero-dimensional nanostructures. Surf Sci 566: 321–326
[20] Ignatiev IV, Kozin IE, Wen RN, Sugou S, Masumoto Y (1999) Anti-Stokes photoluminescence of InP self-assembled quantum dots in the presence of electric current. Phys Rev B 60: R14001–R14004
[21] Yamamoto A, Sasao T, Goto T, Arai K, Lee H-Y, Makino H, Yao T (2003) Anti-Stokes photoluminescence in CdSe self-assembled quantum dots. Phys Stat Sol (c) 0: 1246–1249
[22] Paskov PP, Holtz PO, Monemar B, Garcia JM, Schoenfeld WV, Petroff PM (2000) Photoluminescence up-conversion in InAs/GaAs self-assembled quantum dots. Appl Phys Lett 77: 812–814
[23] Kammerer C, Cassabois G, Voisin C, Delalande C, Roussignol P, Gerard JM (2001) Photoluminescence up-conversion in single self-assembled InAs/GaAs quantum dots. Phys Rev Lett 87: 207401-1/4
[24] Downing E, Hesselink L, Ralston J, Macfarlane R (1996) A three-color, solid-state, three-dimensional display. Science 273: 1185–1189
[25] Fujino T, Fujima T, Tahara T (2005) Femtosecond fluorescence dynamics imaging using a fluorescence up-conversion microscope. J Phys Chem B 109:15327–15331

[26] Kachynski AV, Kuzmin AN, Pudavar HE, Prasad PN (2005) Three-dimensional confocal thermal imaging using anti-Stokes luminescence. Appl Phys Lett 87: 023901–1/3
[27] Pena A-M, Strupler M, Boulesteix T, Schanne-Klein M-C (2005) Spectroscopic analysis of keratin endogenous signal for skin multiphoton microscopy. Opt Express 13: 6268–6274
[28] Macfarlane RM, Tong F, Silversmith AJ, Lenth W (1998) Violet cw neodymium upconversion laser. Appl Phys Lett 52: 1300–1302
[29] Luo X, Eisaman MD, Gosnell TR (1998) Laser cooling of a solid by 21 K starting from room temperature. Opt Lett 23: 639–641
[30] Thiede J, Distel J, Greenfield SR, Epstein RI (2005) Cooling to 208 K by optical refrigeration. Appl Phys Lett 86:154107–1/3
[31] Rogach AL, Franzl T, Klar TA, Feldmann J, Gaponik N, Lesnyak V, Shavel A, Eychmüller A, Rakovich YP, Donegan JF (2007) Aqueous synthesis of thiol-capped CdTe nanocrystals: state-of-the-art. J Phys Chem C 111: 14628–14637
[32] Colvin V, Schlamp M, Alivisatos A (1994) Light-emitting diodes made from cadmium selenide nanocrystals and a semiconducting polymer. Nature 370: 354–357
[33] Gao MY, Lesser C, Kirstein S, Mohwald H, Rogach AL, Weller H (2000) Electroluminescence of different colors from polycation/CdTe nanocrystal self-assembled films. J Appl Phys 87: 2297–2302
[34] Kershaw SV, Harrison MT, Burt MG (2003) Putting nanocrystals to work: from solutions to devices. Phil Trans 361: 331–343
[35] Rakovich YP, Donegan JF, Gerlach M, Bradley AL, Connolly TM, Boland JJ, Gaponik N, Rogach A (2004) Fine structure of coupled optical modes in photonic molecules. Phys Rev A 70: 051801 (R)–1/4
[36] Rakovich YP, Yang L, McCabe EM, Donegan JF, Perova T, Moore A, Gaponik N, Rogach A (2003) Whispering gallery mode emission from a composite system of CdTe nanocrystals and a spherical microcavity. Sem Sci Technol 18: 914–918
[37] Susha AS, Caruso F, Rogach AL, Sukhorukov GB, Kornowski A, Möhwald H, Giersig M, Eychmüller A, Weller H (2000) Formation of luminescent spherical core-shell particles by the consecutive adsorption of polyelectrolyte and CdTe(S) nanocrystals on latex colloids. Colloid Surf A 163: 39–44
[38] Alivisatos AP, Weiwei GW, Larabell C (2005) Quantum dots as cellular probes. Ann Rev Biomed Eng 7: 55–76
[39] Byrne SJ, Corr SA, Rakovich TY, Gun'ko YK, Rakovich YP, Donegan JF, Mitchell S, Volkov Y (2006) Optimisation of the synthesis and modification of CdTe quantum dots for enhanced live cell imaging. J Mater Chem 16: 2896–2902
[40] Auzel F (1976) Multiphonon-assisted anti-Stokes and Stokes fluorescence of triply ionized rare-earth ions. Phys Rev B 13: 2809–2817
[41] Potemski M, Stepniewski R, Maan JC, Martinez G, Wyder P, Etienne B (1991) Auger recombination within Landau levels in a two-dimensional electron gas. Phys Rev Lett 66: 2239–2242
[42] Cingolani R, Ploog K (1991) Frequency and density dependent radiative recombination processes in III-V semiconductor quantum-wells and superlattices. Adv Phys 40: 535–623
[43] Bhawalkar JD, Guang SH, Park C-K, Chan FZ, Ruland G, Prasad PN (1996) Efficient, two-photon pumped green upconverted cavity lasing in a new dye. Opt Commun 124: 33–37
[44] Baltramiejunas R, Vaitkus J, Gavryushin V (1978) Two-photon and two-step absorption of light in II–VI semiconductors. Sov Phys – Collection 18: 46–49
[45] Byrne SJ, Corr SA, Rakovich TY, Gun'ko YK, Rakovich YP, Donegan JF, Mitchell S, Volkov Y (2006) Optimisation of the synthesis and modification of CdTe quantum dots for enhanced live cell imaging. J Mater Chem 16: 2896–2902
[46] Poles E, Selmarten DC, Micic OI, Nozik AJ (1999) Anti-Stokes photoluminescence in colloidal semiconductor quantum dots. Appl Phys Lett 75: 971–973
[47] Rakovich YP, Filonovich SA, Gomes MJM, Donegan JF, Talapin DV, Rogach AL, Eychmüller A (2002) Anti-Stokes photoluminescence in II–VI colloidal nanocrystals. Phys Stat Sol (b) 229: 449–452
[48] Rusakov KI, Gladyshchuk AA, Rakovich YP, Donegan JF, Filonovich SA, Gomes MJM, Talapin DV, Rogach AL, Eychmüller A (2003) Control of efficiency of photon energy up conversion in CdSe/ZnS quantum dots. Opt Spectr 94: 921–925
[49] Wang X, Yu WW, Zhang J, Aldana J, Peng X, Xiao M (2003) Photoluminescence upconversion in colloidal CdTe quantum dots. Phys Rev B 68: 125318–1/6
[50] Rakovich YP, Gladyshchuk AA, Rusakov KI, Filonovich SA, Gomes MJM, Talapin DV, Rogach AL, Eychmüller A (2002) Anti-Stokes luminescence of cadmium telluride nanocrystals. J Appl Spectr 69: 444–449

[51] Fernee MJ, Jensen P, Rubinsztein-Dunlop H (2007) Unconventional photoluminescence upconversion from PbS quantum dots. Appl Phys Lett 91: 043112–1/3
[52] Harbold JM, Wise FW (2007) Photoluminescence spectroscopy of PbSe nanocrystals. Phys Rev B 76: 125304–125306
[53] Hoheisel W, Colvin VL, Johnson CS, Alivisatos AP (1994) Threshold for quasicontinuum absorption and reduced luminescence efficiency in CdSe nanocrystals. J Chem Phys 101: 8455–8460
[54] Talapin DV, Haubold S, Rogach AL, Kornowski A, Haase M, Weller H (2001) A novel organometallic synthesis of highly luminescent CdTe nanocrystals. J Phys Chem B 105: 2260–2263
[55] Bawendi MG, Carroll PJ, Wilson W, Brus L (1992) Luminescence properties of CdSe quantum crystallies: resonance between interior and surface localized states. J Chem Phys 96: 946–954
[56] Rakovich YP, Walsh L, Bradley L, Donegan JF, Talapin D, Rogach AL, Eychmüller A (2000) Size selective photoluminescence excitation spectroscopy in CdTe quantum dots. Proc SPIE 4876: 432–437
[57] Filonovich SA, Gomes MJM, Rakovich YP, Donegan JF, Talapin DV, Gaponik NP, Rogach AL, Eychmüller A (2003) Up-conversion luminescence in colloidal CdTe nanocrystals. MRS Proc 737: 157–162
[58] Vavilov S (1946) Photoluminescence and thermodynamics. J Phys 10: 499–502
[59] Rakovich YP, Donegan JF, Filonovich SA, Gomes MJM, Talapin DV, Rogach AL, Eychmüller A (2003) Up-conversion luminescence via a below-gap state in CdSe/ZnS quantum dots. Phys E 17: 99–100
[60] Carlone C, Beliveau A, Rowell NL (1991) On the anti-Stokes fluorescence in $Cd_{1-x}Zn_xS$ crystals. J Lumin 47: 309–317
[61] Joly AG, Chen W, McCready DE, Malm J-O, Bovin J-O (2005) Upconversion luminescence of CdTe nanoparticles. Phys Rev B 71: 165304–1/9
[62] Chen W (2005) Upconversion luminescence from CdSe nanoparticles. J Chem Phys 122: 224708–1/7
[63] Rakovich YP, Donegan JF, Filonovich SA, Gomes MJM, Talapin DV, Rogach AL, Eychmüller AA (2003) Photon energy up-conversion in CdSe quantum dots. In: Long AR, Davies JD (eds). Proceedings of the 26th International Conference on the Physics of Semiconductors. Edinburgh, UK: Institute of Physics Conference Series pp R2.7.1–R2.7
[64] Widulle F, Kramp S, Pyka NM, Gobel A, Ruf T, Debernardi A, Lauck R, Cardona M (1999) The phonon dispersion of wurtzite CdSe. Phys B 263–264: 4448–4451
[65] Vasilevskiy MI, Anda EV, Makler SS (2004) Electron–phonon interaction effects in semiconductor quantum dots: a nonperturabative approach. Phys Rev B 70: 035318–1/14
[66] Nirmal M, Murray CB, Bawendi MG (1994) Fluorescence-line narrowing in CdSe quantum dots: surface localization of the photogenerated exciton. Phys Rev B 50: 2293–2300
[67] Underwood DF, Kippeny T, Rosenthal SJ (2001) Ultrafast carrier dynamics in CdSe nanocrystals determined by femtosecond fluorescence upconversion spectroscopy. J Phys Chem B 105: 436–443
[68] Efros AL, Rosen M, Kuno M, Nirmal M, Norris DJ, Bawendi M (1996) Band-edge exciton in quantum dots of semiconductors with a degenerate valence band: dark and bright exciton states. Phys Rev B 54: 4843–4856
[69] Klimov VI, McBranch DW (1998) Femtosecond 1P-to-1S electron relaxation in strongly confined semiconductor nanocrystals. Phys Rev Lett 80: 4028–4031
[70] Baranov AV, Rakovich YP, Donegan JF, Perova TS, Moore RA, Talapin DV, Rogach AL, Masumoto Y, Nabiev I (2003) Effect of ZnS shell thickness on the phonon spectra in CdSe quantum dots. Phys Rev B 68: 165306–1/7
[71] Rakovich YP, Donegan JF, Gaponik N, Rogach AL (2003) Raman scattering and anti-Stokes emission from a single spherical microcavity with a CdTe quantum dot monolayer. Appl Phys Lett 83: 2539–2541
[72] Trallero-Giner C, Debernardi A, Cardona M, Menéndez-Proupín E, Ekimov AI (1998) Optical vibrons in CdSe dots and dispersion relation of the bulk material. Phys Rev B 57: 4664–4669
[73] Tamulaitis G, Rodrigues PAM, Yu PY (1995) Screening of longitudinal optical phonons by carriers in quantum dots. Sol Stat Commun 95: 227–231
[74] Rolo AG, Vasilevskiy MI, Gaponik NP, Rogach AL, Gomes MJM (2002) Confined optical vibrations in CdTe quantum dots and clusters. Phys Stat Sol (b) 229: 433–437
[75] Vasilevskiy MI (2002) Dipolar vibrational modes in spherical semiconductor quantum dots. Phys Rev B 66: 195326–1/9
[76] Hwang Y-N, Shin S, Park HL, Park S-H, Kim U, Jeong HS, Shin E-J, Kim D (1996) Effect of lattice ontraction on the Raman shifts of CdSe quantum dots in glass matrices. Phys Rev B 54: 15120–15124

[77] Hwang Y-N, Park S-H, Kim D (1999) Size-dependent surface phonon mode of CdSe quantum dots. Phys Rev B 59: 7285–7288
[78] Comas F, Trallero-Giner C, Studart N, Marques GE (2002) Interface optical phonons in spheroidal dots: Raman selection rules. Phys Rev B 65: 073303-1/3
[79] Rodriguez-Suarez R, Menendez-Proupin E, Trallero-Giner C, Cardona M (2000) Multiphonon resonant Raman scattering in nanocrystals. Phys Rev B 62: 11006-1/16
[80] Ruppin R, Englman R (1970) Optical phonons of small crystals. Rep Prog Phys 33: 146–196
[81] Fedorov AV, Baranov AV, Inoue K (1997) Exciton–phonon coupling in semiconductor quantum dots: resonant Raman scattering. Phys Rev B 56: 7491–7502
[82] Lee J, Govorov AO, Kotov NA (2005) Nanoparticle assemblies with molecular springs: a nanoscale thermometer. Angew Chem Int Ed 2005 44: 7439–7442
[83] Vahala KJ (2003) Optical microcavities. Nature 424: 839–846
[84] Braginsky VB, Gorodetsky ML, Ilchenko VS (1989) Quality factor and nonlinear properties of optical whispering-gallery modes. Phys Lett A 137: 393–397
[85] Hill SC, Benner RE (1986) Morphology-dependent resonances associated with stimulated processes in microspheres. J Opt Soc Am B 3: 1509–1514
[86] Gaponik N, Rakovich YP, Gerlach M, Donegan JF, Savateeva D, Rogach AL (2006) Whispering gallery modes in photoluminescence and Raman spectra of a spherical microcavity with CdTe quantum dots: anti-Stokes emission and interference effects. Nanoscale Res Lett 1: 68–73
[87] Pringsheim P (1929) Zwei Bemerkungen über den Unterschied von Lumineszenz- und Temperaturstrahlung. Z Physik 57: 739–746
[88] Epstein RI, Buchwald MI, Edwards BC, Gosnell TR, Mungan CE (1995) Observation of laser-induced fluorescent cooling of a solids. Nature 377: 500–503
[89] Clark JL, Rumbles G (1996) Laser cooling in the condensed phase by frequency up-conversion. Phys Rev Lett 76: 2037–2040
[90] Hoyt CW, Sheik-Bahae M, Epstein RI, Edwards BC, Anderson JE (2000) Observation of anti-Stokes fluorescence cooling in thulium-doped glass. Phys Rev Lett 87: 3600–3603
[91] Sheik-Bahae M, Epstein RI (2004) Can laser light cool semiconductors? Phys Rev Lett 92: 247403-1/4
[92] Alphandéry E, Walsh LM, Rakovich Y, Bradley AL, Donegan JF, Gaponik N (2004) Highly efficient Förster resonance energy transfer between CdTe nanocrystals and Rhodamine B in mixed solid films. Chem Phys Lett 388: 100–104
[93] Oraevsky AN (1996) Cooling of semiconductors by laser radiation. Quantum Electron 26: 1018–1022
[94] Rivlin LA, Zadernovsky AA (1997) Laser cooling of semiconductors. Opt Commun 139: 219–222
[95] Zander C, Drexhage KH (1995) Cooling of dye solution by anti-Stokes fluorescence. In: Neckers DC, Volman DH, von Bunau G (eds) Advances in photochemistry. Wiley, New York, pp 59–78

Exciton dynamics and energy transfer processes in semiconductor nanocrystals

By

Andries Meijerink

Condensed Matter and Interfaces, Department of Chemistry, Debye Institute, Utrecht, The Netherlands

1. Introduction

Exciton dynamics provide unique information on both the nature of optical transitions and the local environment of an optically active species. However, experimental facilities for measuring (fast) luminescence decay dynamics have been developed long after techniques for time-averaged optical spectroscopy (absorption, excitation and emission spectroscopy). Studies on the dynamics of excited states therefore lag behind of steady state spectroscopy. This situation is also true for research on the optical properties of semiconductor nanocrystals. The first scientific record relating the change in optical properties of semiconductor nanocrystals to the particle size dates back to 1926 when Jaeckel explained the red-shift of the absorption onset in glasses containing CdS particles to a change in the CdS particle size [1]. It took until the 1980's that a fundamental understanding of the effects underlying the size-dependent optical properties of semiconductor nanocrystals were explained by quantum confinement and the name quantum dots (QDs) was introduced for these nanocrystals [2–4]. Since then the work on semiconductor nanocrystals has developed to an active and still growing field of research. Initially information on the optical properties of QDs was obtained from luminescence (excitation and emission) and absorption spectra. The results from the optical spectra could be related to theoretical calculations on the energy level structure. As for other optically active systems, additional information was later obtained from the dynamics of the excited state. The decay kinetics of the luminescence provide information on for example the nature of excited states, the quality of nanocrystals, competition between radiative and non-radiative recombination processes and interactions between QDs. At present, time resolved studies play a prominent role in the understanding of the optical properties of QDs. The time range involved varies between fs/ps (governed by non-radiative relaxation processes of charge carriers and energy transfer processes) to ns/µs (for radiative recombination processes).

The rapid increase in fast dynamics studies in the field of semiconductor nanocrystals, and in optical spectroscopy in general, has not only been triggered by the realization that decay kinetics provide important additional information on their optical and electronic properties. It is also the spectacular development of experimentally available systems for fast (fs to ns) optical spectroscopy. Equipment for ps and ns lifetime measurements is now commonplace while even 20 years ago the building of a ps laser was a research project in its own. The development of ps diode lasers has resulted in cheap commercially available systems for sub-ns life time measurements, using time-correlated-single-photon counting techniques. For probing the faster dynamics of electron and hole relaxation processes within the conduction or valence band, Ti-sapphire fs laser-systems are available. In the past decade pump-probe systems for measuring fs carrier dynamics, for example by transient absorption spectroscopy, are becoming available in a growing number of laboratories. Alternatively, fast time-resolved emission spectra can be recorded on a ps time scale using a streak camera in combination with a Ti-sapphire laser.

In this chapter the exciton and charge carrier dynamics in quantum dots will be discussed with a focus of colloidal semiconductor nanocrystals. In the first part of this chapter, the recombination dynamics in isolated QDs will be considered. For different types of QDs the radiative and non-radiative relaxation processes are discussed in relation to theoretical work on energy level calculations and relaxation processes. The focus will be on the widely studied model systems CdSe and CdTe, but also results on other systems will be reviewed. The second part of this chapter is devoted to energy transfer processes between QDs. A short theoretical basis will be provided on energy transfer and energy migration after which exciton dynamics in a variety of QD systems will be treated. Again, the well-known model systems will be used to illustrate the state of the art of the knowledge on energy transfer processes in semiconductor nanocrystals. In addition to energy transfer between QDs, also energy transfer to other chromophores and to metal nanoparticles (NPs) will be covered since these processes are becoming increasingly important in many applications of QDs.

2. Exciton dynamics in quantum dots

2.1 General. The life time of the excitonic emission is an important parameter and contains information on the nature of the ground state and excited states of QDs. The theory for calculating the spontaneous emission probability for electric dipole transitions is well-established (Fermi's golden rule). The general equation for the transition probability between an excited state b and a ground state a is [5]:

$$A_{ba}(\text{ED}) = \frac{1}{\tau_{\text{rad}}} = \frac{1}{4\pi\varepsilon_0} \frac{4n\omega^3}{3\hbar c^3} \left(\frac{E_{\text{loc}}}{E}\right)^2 \frac{1}{g_b} \sum_{a_n, b_m} |<a_n|\mu_c|b_m>|^2 \qquad (1)$$

where A_{ba} (ED) is the transition probability (s^{-1}) and is the inverse of the (radiative) life time τ_{rad}, n is the refractive index, ω is the transition frequency, $(E_{\text{loc}}/E)^2$ is the local field correction factor and g_b is the degeneracy of the excited state. The summation is over all levels in the ground state a_n and excited state b_m for the transition dipole moments μ connecting the levels. For an allowed electric dipole transition the transition dipole moment $\mu = e \cdot r$ and with $r = 10^{-10}$ m (approximately the Bohr radius of an hydrogen atom) and $n = 1.7$ this gives $A(\text{ED}) = 10^8 \text{ s}^{-1}$ in the visible spectral region. For QDs this equation is often rewritten as [6]:

$$A_{ba}(\text{ED}) = \frac{2e^2 \omega \varepsilon_1^{1/2} f}{3 m_0 c^3} \qquad (2)$$

where the oscillator strength f is given by

$$f = \frac{2P^2}{m_0 E_{1S_h 1S_e}}$$

P is the momentum transition moment of the exciton and is known as the Kane parameter. The expressions do not account for the screening of the electric field inside the QD. The high optical dielectric constants of semiconductors strongly influence the radiative lifetime. The internal electric field is reduced by dielectric screening and lengthens the life time by a factor $[3\varepsilon_0/(\varepsilon_1 + 2\varepsilon_0)]^{-2}$ where ε_1 is the (wavelength-dependent) dielectric constant of the semiconductor and ε_0 is the dielectric constant of the surrounding medium [6]. Semiconductor materials with large dielectric constants, like PbSe and PbS, have therefore long radiative life time.

In addition to radiative decay, non-radiative decay processes influence the decay kinetics of excitons in QDs. An important non-radiative relaxation process is trapping of an exciton or charge carriers at a defect or impurity site in the QD or at the QD surface. For measurements on an ensemble of QDs these non-radiative decay processes give rise to non-exponential decay curves. The distribution of defects or impurities over the QD population is inhomogeneous which results in a wide variety of (non-radiative) decay rates for different QDs in the ensemble. The exciton emission from QDs with a defect will decay faster with a total rate $A_{\text{tot}} = A_{\text{r}} + A_{\text{nr}}$. The non-radiative decay rate A_{nr} will depend on the nature of the defect or impurity and the number of defects. QDs showing a single exponential decay are of high quality and exhibit a high luminescence quantum yield. For many studies it is crucial to use high quality QDs exhibiting a single exponential radiative decay. By studying the deviation of or changes in the single exponential decay, it will be possible to obtain quantitative information on for example energy transfer rates or factors influencing the radiative decay rate. If the reference decay curves are non-exponential due contributions from non-radiative decay processes, it is much harder to observe and analyze this influence. In addition to radiative and non-radiative recombination processes, also hot carrier relaxation and bi-exciton decay are interesting processes in exciton dynamics studies. One of the interesting topics in the past decades has been the question whether or not a phonon-bottleneck affects the exciton dynamics in QDs and more recently exciton dynamics studies have provided evidence for multiple exciton generation (MEG). In the sections below the various decay processes for different classes of QDs will be discussed.

2.2 Exciton dynamics in CdSe and CdTe quantum dots

2.2.1 *Radiative and non-radiative decay.*
CdSe and CdTe nanocrystals are the most widely studied colloidal QDs and serve as the work horse for investigations

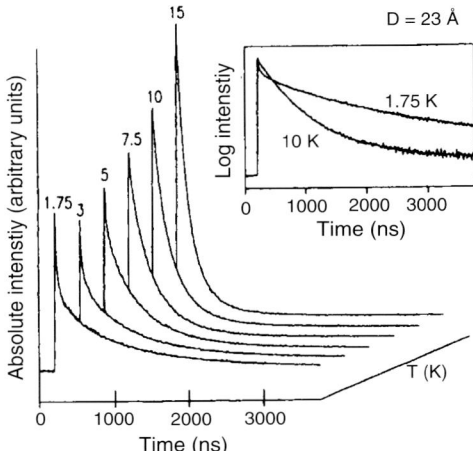

Fig. 1. Temperature-dependent decay curves for the exciton emission in 2.3 nm CdSe QDs. Reproduced with permission from [8], © 1994 APS

on quantum size effects. It is therefore not surprising that studies on the exciton dynamics have been done most extensively for CdSe and CdTe QDs and the understanding is the most advanced for these systems. First the exciton decay kinetics will be treated for CdSe QDs. Early studies on the exciton dynamics in CdSe QDs date back to 1994 shortly after the introduction of the hot-injection method for high quality CdSe nanocrystals [7]. Figure 1 shows the temperature-dependent decay profiles for 2.3 nm CdSe QDs. At the lowest temperatures a strongly non-exponential decay curve is observed with an initial decay in the ns range and long time decay in the μs range. Upon heating, the decay becomes single exponential (the short lived component disappears) and the long time component becomes faster and decreases from μs to ns between 2 and 15 K.

Later more extensive experiments and a theoretical analysis were applied to clarify the temperature-dependent decay behavior and relate the observations to the energy level scheme of CdSe QDs (wurtzite structure). Calculations show that the energy level structure is strongly dependent on the size and shape of the CdSe nanocrystal. In the effective mass approximation the $1S_e$ electron interacts with the $1S_{3/2}$ hole giving rise to an 8-fold degenerate state. The degeneracy is lifted by electron–hole exchange interaction, the crystal field and crystal shape asymmetry into five states which are denoted $0^L, 0^U, \pm 1^L, \pm 1^U$ and ± 2 [9]. The order of the states depends on the shape of the QD and the energy separation between the different states increases as the size of the QD decreases. The energy level structure of the $1S(e)–1S_{3/2}(h)$ is shown in Fig. 2, as it was calculated by Efros et al. [9]. Since then, more refined calculations and different models have been applied to explain the energy level structure but the main features are the same and are consistent with the experimentally observed temperature dependence of the decay time [10–12]. The transition from the lowest ±2 excited state of the CdSe QD to the ground state is formally forbidden. This results in a long lived emission at low temperatures. In addition, because of the low transition

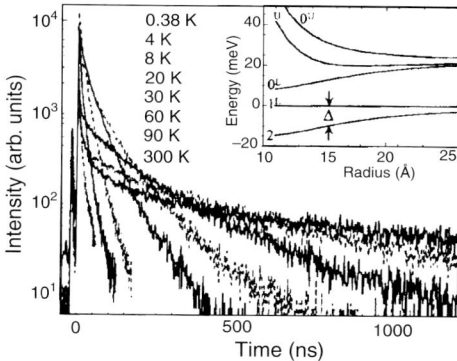

Fig. 2. Exciton emission decay curves for 2.6 nm core/shell CdSe/ZnS QDs between 0.38 K and 300 K. Inset shows results on theoretical calculations on the size-dependent energy level structure. Reproduced with permission from [14], © 2003 AIP

probability, transitions to this level cannot be observed in excitation or absorption spectra, so that this level has been named 'dark state'. Note however that the term 'dark state' is somewhat misleading since emission from this state at low temperatures can be very bright! Transitions from the $\pm 1^L$ level are allowed (this state is therefore called 'bright state') and at higher temperatures when the $\pm 1^L$ level is thermally occupied the exciton decay time drops from μs to ns. From the temperature dependence of the decay time the energy difference between the two lower excited states can be determined. The splitting is small, typically of the order of a meV.

A more careful analysis of the size and temperature dependence of the exciton life times for both ensemble of CdSe QDs (e.g. in [13, 14]) and single CdSe QDs [15] confirms the early observations. Figure 2 shows exciton decay measurements over a wide temperature range (0.38–300 K) for 1.3 nm CdSe QDs. Especially at the lowest temperatures, the very fast initial decay (ns) followed by a slow μs decay component can now be observed more clearly. Analysis of the results show that there is a size dependence of both the energy difference between the dark state and the bright state and the life time of the dark state. The energy difference and the dark state life time

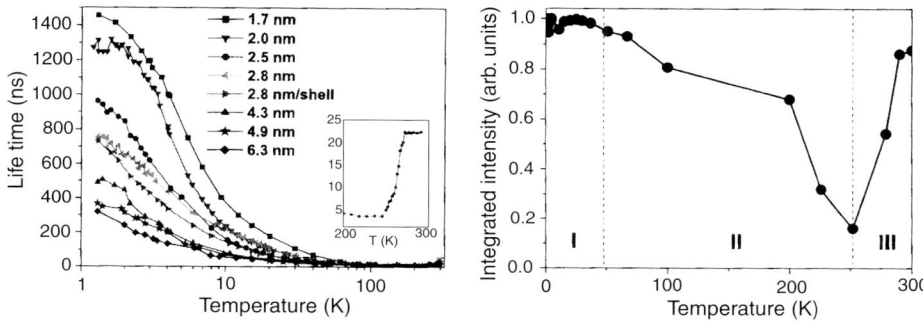

Fig. 3. (Left) Temperature dependence of the exciton emission life time for CdSe QDs between 1.7 nm and 6.3 nm, and (right) integrated emission intensity for CdSe QDs as a function of temperature. Reproduced with permission from [13], © 2006 APS

both increase for smaller particles. In Fig. 3 the temperature dependence of the long lived component of the excitonic emission is shown over a wide variety of CdSe QD sizes. In the low temperature regime the temperature dependence can be fitted fairly well to a three level model:

$$\frac{1}{\tau} = \frac{1}{\tau_{\text{dark}}} \left(\frac{e^{\Delta E/kT}}{1 + e^{\Delta E/kT}} \right) + \frac{1}{\tau_{\text{bright}}} \left(\frac{e^{\Delta E/kT}}{1 + e^{\Delta E/kT}} \right) \quad (3)$$

where τ_{dark} and τ_{bright} are the life times for the transition from the dark (± 2) and bright exciton state (± 1). Analysis of the size dependence shows that the splitting of the dark–bright state excitons increases from 0.7 meV to 1.7 meV upon decreasing the CdSe QD size from 6 nm to 1.5 nm [13]. The life time for the dark state decreases from 1.3 μs for the smallest particles to 0.3 μs for the largest (6 nm) CdSe QDs. In the literature there are various reports confirming this size dependence qualitatively although there are differences in the experimentally determined values in similar systems. Theoretical calculations on the size-dependent exciton splitting and life times, also confirm the observed trends as a function of particle size but usually give larger absolute values for the splitting and also the life times [9, 12]. The differences

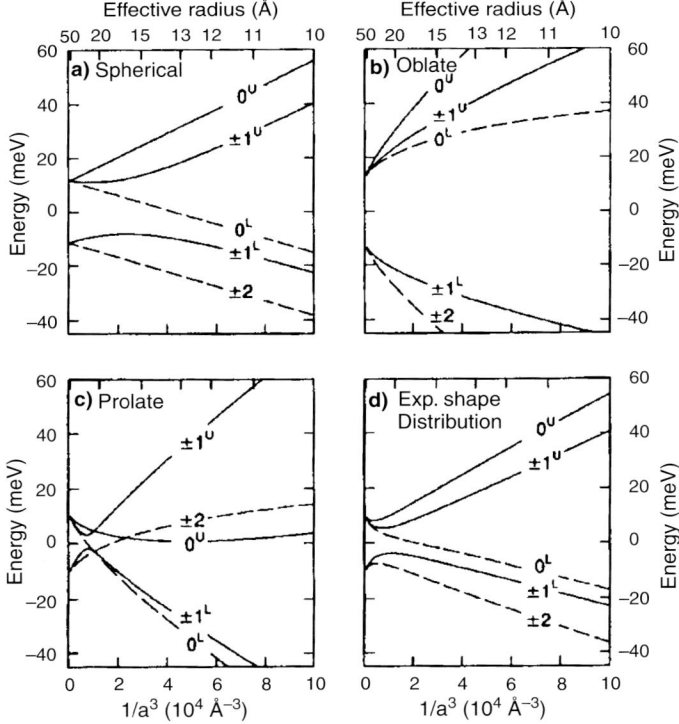

Fig. 4. Size dependence of the exciton band edge structure in CdSe QDs of various shapes. Solid lines represent states for which transitions to the ground state are allowed ('bright' states) while the dashed lines represent states for which transition to the ground state are formally forbidden ('dark states'). Reproduced with permission from [9], © 1998 APS

between observations and between observation and calculation are not surprising. Not only the size, but also the shape (and aspect ratio) of the QDs will strongly influence the energy level structure (see also Fig. 4). Small variations in shape for QDs of similar sizes can explain differences in the details. Also surface states have been suggested to strongly influence the energy level splitting and the life time of the dark state [9, 11, 12]. Interestingly, experiments on similar CdSe QDs with different cappings (organic ligands vs. ZnS) do not show a difference in the low-temperature decay behavior which suggests that the size-dependent and temperature-dependent exciton decay kinetics are intrinsic to the CdSe QDs and not caused by surface states.

At higher temperatures (above 50 K) temperature induced quenching of the exciton emission is observed. In Fig. 3, right hand side, this temperature region is labeled II. The quenching may be related to thermally activated trapping of charge carriers in surface states but the exact nature of the quenching process is not known. In this temperature region the decay curves become non-exponential which confirms that the shortening of the exciton decay time is not due to changes in the radiative decay rate but to quenching processes which have different rates for different QDs within the ensemble. The quenching behavior in this temperature regime is also dependent on the synthesis conditions. An interesting observation is the increase in life time in temperature region III. It is rather unusual that a luminescence decay time increases upon heating since most non-radiative decay processes becomes faster at higher temperatures. Not only the luminescence life time but also the luminescence intensity (quantum yield) increases around the transition temperature. This phenomenon has been called luminescence temperature anti-quenching and is explained by a phase transition in the capping layer [16]. At low temperatures the capping layer is 'frozen' and the rigid configuration of the surface capping molecules hinders relaxation of the Cd and Se surface atoms. It is well known that relaxation is required to prevent surface trapping states situated in the bandgap [12]. Upon 'melting' the surface capping layer, surface relaxation can occur and the disappearance of the surface quenching states causes an increase in the luminescence life time and luminescence intensity. Luminescence temperature anti-quenching has also been observed for CdTe QDs in ice. Here the quenching is related to local freezing of the solvent (water) around the CdTe QDs [17].

2.2.2 *Influence of the environment.*
The discussion above beautifully illustrates an important aspect in research on QDs: the sensitivity of the luminescence properties of QDs to the surface. Subtle changes at the surface can strongly influence the luminescence properties and exciton dynamics of QDs. The surface to volume ratio is large and especially for the smaller QDs a significant fraction of the total number of atoms in the QDs is at surface sites. It is therefore no surprise that surface passivation is crucial. Synthesis procedures have been optimized to yield QDs with quantum yields close to unity and single exponential decay curves [18–20]. These highly efficient QDs are ideal probes to study the influence of the surroundings on the luminescence properties. Deviations from the exponential decay and changes in the luminescence life time can be related to changes in the environment. Here we will discuss two effects: the introduction of quenching molecules at the surface which causes non-radiative decay and changes in the refractive index or distribution of

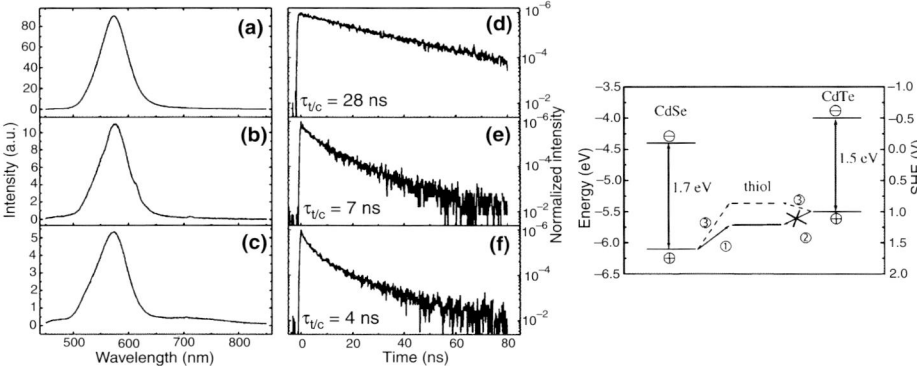

Fig. 5. (Left) Emission spectra and luminescence decay curves of the exciton emission of CdSe QDs capped with TOP/TOPO/HDA in chloroform (**a**, **d**), allylmercaptam in chloroform (**b**, **e**), and amino-ethane-thiol in water (**c**, **f**). (Right) Positions of bulk CdSe and bulk CdTe band edges with respect to a standard hydrogen electrode reference. Hole trapping can occur from CdSe (process ①) but not from CdTe (process ②). The dashed line indicates the assumed position for the standard potential of a thiol that quenches the luminescence of both CdSe and CdTe (process ③). Reproduced with permission from [21]. © 2004 ACS

density of states in the surroundings of the QDs which influence the radiative life time.

As an example, Fig. 5 shows emission spectra and luminescence decay curves for CdSe QDs in chloroform [21]. Upon addition of allylmercaptam (Fig. 5B and E) or amino-ethane-thiol (Fig. 5C and F), quenching centers are introduced at the QD surface. As a result, the decay curves become non-exponential and show a faster decay ($\tau_{1/e}$ is 7 ns or 4 ns) while the emission spectra do not change. The single exponential decay curve in Fig. 5D shows that for this ensemble of QDs the decay is dominated by radiative relaxation. From the curve the radiative decay time (in this case 28 ns) can be determined. The quenching of the CdSe by thiols is related to trapping of holes in the valence band of CdSe by thiols (Fig. 5, right hand side). For CdTe the higher energy position of the valence inhibits this quenching and the luminescence life time of the CdTe emission remains single exponential with a $\tau_{1/e}$ of 17 ns [21].

A more subtle influence of the local environment is to change the radiative decay rate of the QD emission. Changing the nature of the surrounding medium (for example the solvent) causes a variation of the local field correction factor and this influences the radiative decay rate. In fact, QDs are a sensitive probe to test theoretical models on the influence of changes in the local surroundings on radiative decay processes. For CdSe and CdTe QDs it was shown that there is a weak but significant increase in the radiative decay rate in solvents with a higher refractive index [22]. For QDs in photonic bandgap structures the variation of the local density of states at specific wavelengths can be probed. The radiative decay rate was shown to be either enhanced or inhibited in line with a calculated increase or decrease of the density of states inside the photonic crystal [23]. In Fig. 6 on the left hand side an inverse opal of titania shells is shown which has a strong wavelength-dependent variation in the local density of states as a result of the periodic structure. On the right hand side the decay rates are plotted as a function of emission frequency (measured

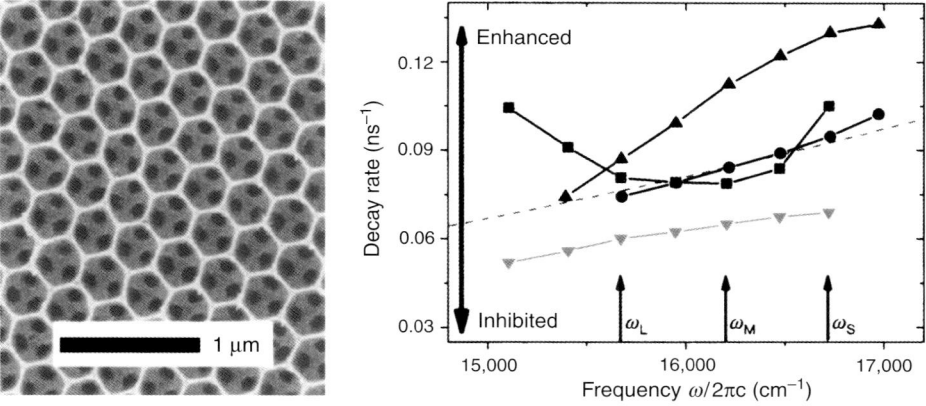

Fig. 6. (Left) Scanning electron microscope image of a (1 1 1) face of a titania inverse opal. (Right) Measured decay rates for CdSe/ZnSe QDs in photonic crystals with different lattice parameters (370 nm, filled dots; 420 nm, squares; 500 nm, upside-down triangles; 580 nm, triangles). The dashed curve gives the calculated decay rate in a homogeneous medium. Three sets of QDs (small, medium and large, central frequencies indicated by arrows) were used. Reproduced with permission from [23], © 2004 Nature

using a set of three size distributions of QDs in titania inverse opals). For example, in a photonic crystal with a lattice parameter of 580 nm a significant increase (up to a factor of 3) in the decay rate is measured and predicted in the energy range 16,000 cm^{-1} to 17,000 cm^{-1} due to the higher photonic density of states. Also note that the decay rate is not strongly influenced by the size of the QD. The dashed line in Fig. 6 gives the decay rate in a homogeneous medium (not photonic). There is an increase of the decay rate for the smaller sized QDs. The supra-linear increase has been explained by including variations in the thermal population over the various exciton states as a result of a size-dependent splitting of the energy levels [24].

2.2.3 Fast relaxation processes.
A final aspect where carrier dynamics have received considerable attention is the relaxation of high-energy charge carriers to the band edges. In this process they loose their excess energy by interaction with phonons. The time scale of these relaxation processes is much faster than the time scale for the radiative and non-radiative recombination processes discussed above [25]. Typically these phonon relaxation processes take place in the 100 fs to 10 ps time regime. In bulk semiconductors typical rates are of the order of 0.5 eV/ps [26]. However, in a QD the discrete energy level structure and the change in the phonon spectrum can be expected to considerably reduce the phonon relaxation rates [27]. This phenomenon is known as the "phonon bottleneck". In the literature there are a number of beautiful demonstrations on how a phonon bottleneck can slow down the relaxation from a higher energy level, the most convincing experiments being on the work horse in solid state spectroscopy: ruby (Al_2O_3:Cr^{3+}) [28, 29]. The phonon bottleneck can even be used to create a "phonon laser" in ruby [29]. However, for semiconductor QDs contradictory results have been reported [30, 31]. In some papers reduced relaxation rates have been reported in QD structures (e.g. InGaAs

QDs) and ascribed to phonon bottleneck effects [25]. In CdSe and CdTe QDs the relaxation rates always show a fast ~ps component while in some cases also a slower (~100 ps) component has been observed and ascribed to a phonon bottleneck [25, 32]. The interest in a phonon bottleneck reducing the relaxation rate of hot charge carriers is not purely academic. If indeed the carrier relaxation is slowed down in QDs as compared to bulk semiconductors, this will be beneficial for the application of QDs in solar cells. Extraction of hot charge carriers (at a higher potential) gives the possibility of obtaining a higher voltage than the bandgap after absorption of high energy photons. A long life time of the hot charge carriers is required to make this feasible and this could be promoted by phonon bottleneck effects.

Most recent experiments on CdSe and CdTe indicate that there is no convincing evidence for a phonon bottleneck. This is not surprising. The justification for a phonon bottleneck is that the limited number of atoms in a semiconductor nanocrystal reduces the density of phonon modes. Since these are the accepting modes in the phonon relaxation process a reduced relaxation rate is plausible. However, the phonon density of states, especially in the higher energy region involved in the relaxation of hot charge carriers, is still sufficiently high to allow for fast relaxation while also coupling with high-energy vibrational modes in surface ligands contributes to fast charge carrier relaxation. There is however one exception: a phonon bottleneck does explain the slow relaxation process from the bright state to the dark state in CdSe QDs. These states are separated by a few cm^{-1}. At low temperatures (below 10 K) relaxation from the higher energy bright state to the lower energy dark is so slow that fast emission from the higher bright state is observed with a ns decay time (see Figs. 2 and 18). This shows that the relaxation from the bright $\pm 1^L$ state to the dark ± 2 dark state is slow (ns) which can be explained by the absence of accepting acoustic phonon modes that can bridge the meV gap between the two states. Due to the small size, the wavelength of the acoustic phonons is cut off, which effectively eliminates all low energy (~meV) acoustic phonons [33].

To monitor the fast relaxation pump-probe techniques like transient absorption (TA) are very suitable. A fs pump pulse creates a hot electron–hole pair and subsequently a probe pulse can monitor the relaxation of the hot charge carriers to the band edges [34, 35]. By varying the photon energy (wavelength) of the pump pulse, the excess energy of the hot electrons and holes can be tuned. The absorption of a fs probe pulse is measured as a function of wavelength and time delay between the pump and probe pulse. The change in the absorption spectrum induced by the pump pulse is ascribed to two effects. First, the presence of the (hot) exciton influences the absorption due to Coulomb interactions of the electron–hole pair created by the pump pulse (which can also be considered as a shift in the absorption spectrum due to a Stark effect induced by the electron–hole pair). The second effect is the filling of states due to relaxation of the hot electrons and holes. For example, when the electrons have relaxed to the edge of the conduction band, absorption at wavelengths corresponding to transitions to the conduction band edge will be reduced due to the fact that the density of 'available' states is reduced (since the relaxed charge carriers occupy these states). In Fig. 7 an example of a transient absorption spectrum for 4.1 nm CdSe QDs is shown. From the transient absorption spectra at different delay times a wealth of information is obtained. For example, after an initial change in the

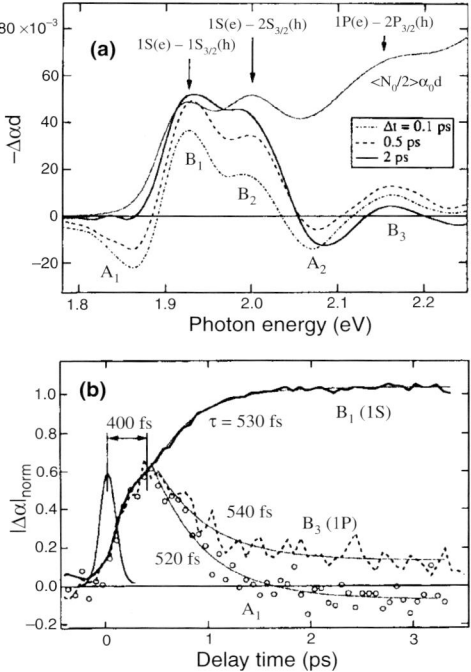

Fig. 7. a Transient absorption spectra of 4.1 nm CdSe QDs recorded with a 0.1 ps; 0.5 ps and 2 ps time delay, **b** Transient absorption kinetics at spectral position B_1 (thick solid line), B_3 (thick dashed line) and A_1 (circles). The thin solid line is the pump-probe cross-correlation profile. Reproduced with permission from [34], © 2000 ACS

absorption strength due to the Stark effect, it is observed that the absorption at the wavelength corresponding to B1 (absorption $1S(e)$–$1S_{3/2}(h)$) decreases with a time constant of 540 fs while the B3 absorption ($1P(e)$–$1P_{3/2}(h)$) increases with the same time constant. From this it can be concluded that the 1P–1S relaxation rate of conduction band electrons is about 540 fs. Further research has provided a fairly detailed understanding of the relaxation rates for different types of QDs and different sizes of QDs. Fast relaxation of electrons is mediated by energy transfer to holes (Auger process) which can relax through a large number of closely spaced energy levels in the valence band. The influence of the Auger cooling process is evident from experiments on QDs capped with molecules that efficiently trap holes (e.g. pyridine or thiols). A lengthening of the relaxation time from the sub-ps regime to several ps is observed when the holes are trapped thus blocking the Auger quenching process.

Further evidence for the fast intraband relaxation is obtained from single particle excitation spectra [36]. Experimentally it is very challenging to obtain a single particle excitation spectrum. The spectrum shown in Fig. 8 shows the energy level structure for a single CdSe QD. Even though the spectrum may seem to only contain information on the energy level structure, also information on relaxation rates can be obtained. The linewidth of excitation (and emission) lines of single particles reflect the homogeneous linewidth which is determined by the coherence life time of the

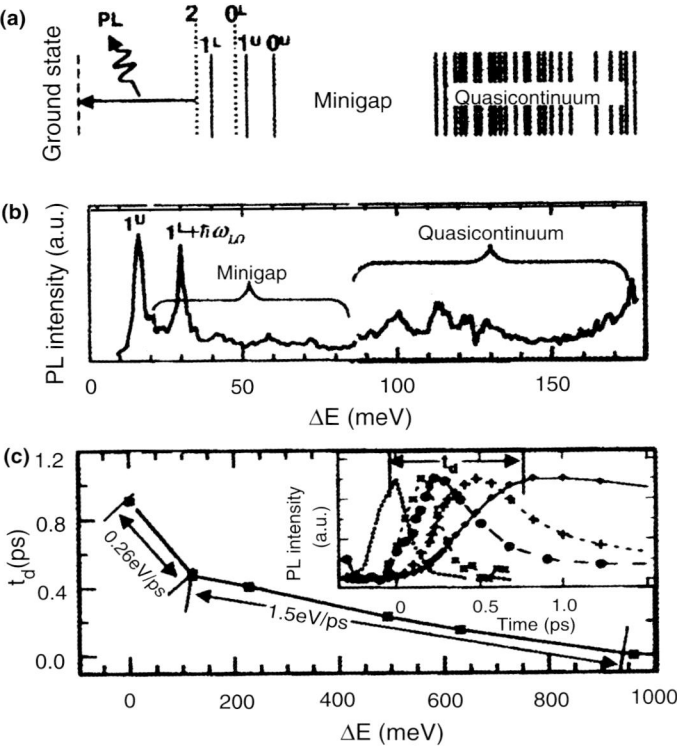

Fig. 8. a Schematic overview of the lower energy optical transitions in CdSe QDs, **b** single-QD excitation spectrum and **c** time delay of the hot photoluminescence maximum with respect to the arrival time of a pump pulse plotted as a function of the difference between detection energy and the energy of the lowest emitting transition (at 300 K). The relaxation rates are also indicated (in eV/ps). Reproduced with permission from [36], © 2004 APS

initial and final state through the Heisenberg uncertainty principle ($\Delta \nu = 5.3$ cm$^{-1}/\tau$ (ps)) [37]. The narrow excitation line corresponding to the 1^U transition indicates that the coherence lifetime for this excited state is relatively long, while the broader structures in quasicontinuum reflect fast relaxation from the states to the band edge states. The spectral width of ~ 10 cm^{-1} is consistent with ps relaxation times.

The mechanism for the fast ps relaxation rate of hot electrons in QDs where Auger cooling has been suppressed by hole trapping shows that in this situation there is a strong influence of the vibrational modes of the capping molecules on the relaxation rate [26]. Figure 9 shows the 1P–1S relaxation profile for 5 nm CdSe QDs capped with different types of ligands. The relaxation time increases from 6 ps (in TOP/TOPO capped QDs) to 18 ps in *n*-dodecanethiol-capped QDs. The large variation in relaxation times is explained by energy transfer to resonant vibrational modes of the surface ligands. The faster relaxation rates are consistent with stronger surface ligand absorption peaks in the infrared region that is resonant with the 1P–1S transition.

Under high excitation powers the probability for the generation of multiple excitons in a single QD increases. The relaxation rate of these multi-exciton states

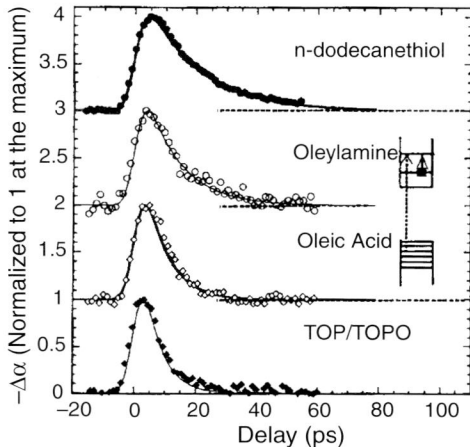

Fig. 9. Influence of capping molecules on the 1P–1S relaxation rate in CdSe QDs, measured by TA spectroscopy. Reproduced with permission from [26], © 2005 ACS

is extremely fast, again due to Auger quenching processes. The decay of the blue-shifted emissions takes place on a 100 ps timescale, the exact rate depends on the particle size (150 ps for 3.4 nm QDs and 50 ps for 2.3 nm QDs) [38, 39]. Also the radiative decay rate for bi-exciton and tri-excitons have been determined. Studies on the dynamics of the bi-excitons and tri-excitons in single CdSe core-shell QDs (5.1 nm) yielded values of 8.4 ns and 6.8 ns for radiative decay of bi-excitons and tri-excitons, respectively [39]. This is in line with theoretical predictions on the bi-exciton and tri-exciton decay.

2.3 Exciton dynamics in other types of quantum dots. Research on radiative and non-radiative relaxation processes in other QD systems is limited in comparison to the widely studied CdSe and CdTe model systems discussed above, although there is rapid increase in research on carrier dynamics in IV–VI PbSe and PbS QDs. These infrared emitting QDs have gained considerable interest since reports on efficient multiple exciton generation MEG upon excitation with photons exceeding $2.5\,E_g$ [40, 41]. Hot carrier relaxation plays an important role in explaining the high efficiency of this process since it is competing with MEG. The radiative life time of the emission from PbSe and PbS QDs is long, typically in the µs regime [6, 42–44]. As an example a decay curve of PbSe QD emission is shown in Fig. 10 together with the absorption and emission spectra for these nanocrystals [6]. The quantum yield can be high (up to 85%) and the decay curve shown in the inset represents the radiative decay of these QDs. The luminescence decay time of 0.88 µs is considerably longer than the ns decay times observed for CdSe or CdTe QDs. Similarly long luminescence life times are reported for PbS QDs. The origin for the long life time is explained by dielectric screening as discussed above and is in good agreement with theoretical calculations [6, 44].

An interesting question is whether or not a lengthening of the radiative decay time will occur upon cooling. The most recent theoretical calculations predict a dark state

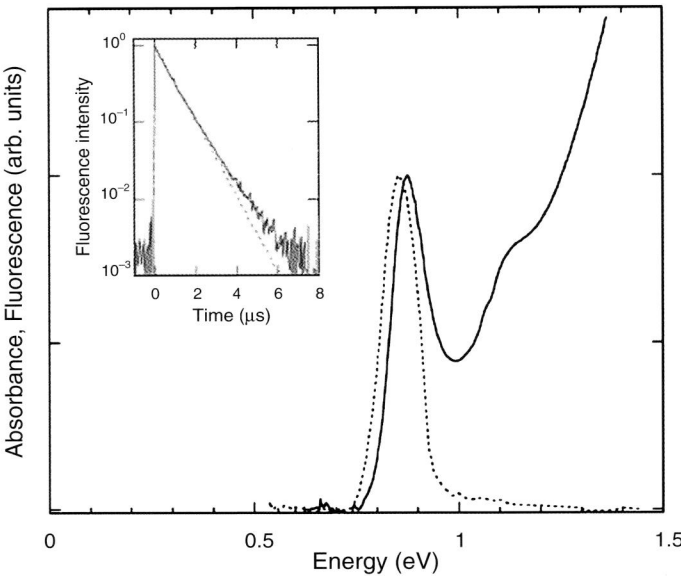

Fig. 10. Emission (dotted line) and absorption (drawn) spectrum of PbSe QDs. The inset shows the luminescence decay curve. Reproduced with permission from [6], © 2002 ACS

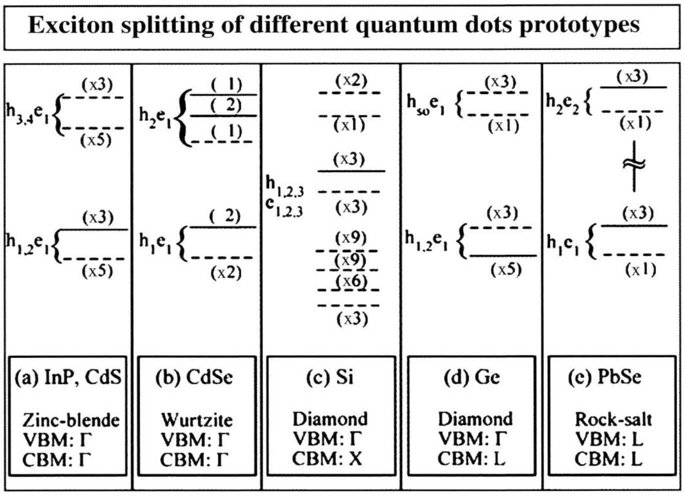

Fig. 11. Energy level diagram of the splitting of the lowest exciton state from pseudopotential calculations. Solid lines represent bright states (allowed transition to the ground state) and broken lines represent dark states (forbidden transitions). Reproduced with permission from [44], © 2007 ACS

to be lowest energy exciton state in PbSe QDs (see Fig. 11). Just as for CdSe and CdTe QDs this is expected to result in a long decay time upon cooling. At this point, no experimental results have been published on the temperature dependence of the life time of the exciton emission in PbS or PbSe QDs. Initial (unpublished) results do not provide evidence for substantial lengthening of the decay time at temperatures down to 4 K suggesting that either the lowest exciton state is not a dark state or the splitting between the dark and bright state is much smaller than 1 meV [45].

A beautiful illustration of how the measurement of the dynamics of excitons can be used is the recent observation of multiple exciton generation in PbS and PbSe QDs. Upon excitation of these IV–VI QDs with photons of energies exceeding ~2.5 E_g multiple excitons are generated in a single QD [40, 41]. The fingerprint for multiple exciton generation is the observation of fast decaying components in the transient absorption spectra resonant with the 1S(e)–1S$_{3/2}$(h) transition: only when bi-excitons are created, a fast filling of these band-edge states is expected (for a single exciton state the time constant for filling is the (ns) single exciton decay rate). From the ratio of the TA signal for the single exciton decay and the fast component, the number of excitons generated after absorption of a single high-energy photon can be estimated. Analysis of the results shown in Fig. 12 indicates that the efficiency of multiple exciton generation is surprisingly high (700% for photons of 8 E_g). The mechanism responsible for MEG is unclear but, if the reported efficiencies for MEG are correct, it must take place on a fs time scale. To explain the high efficiency the creation of

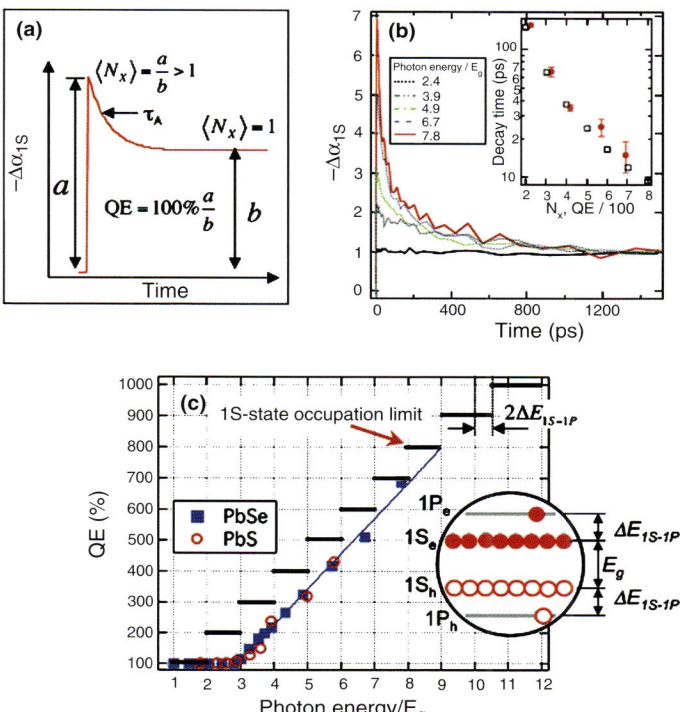

Fig. 12. a Schematic picture of the method to derive MEG yields from the transient absorption (TA) curve. **b** Experimentally measured transient absorption at the band edge transition after (low power) excitation with various photon energies (indicated in the figure). **c** Efficiency of exciton generation (number of excitons per absorbed photon)

multiple excitons has to occur much faster than competing processes (like hot carrier relaxation) which take place on a ps time scale.

After the observations of MEG in PbS and PbSe, reports appeared on MEG in other systems like CdSe, InP and Si QDs [46–48]. However, later the claim for MEG in InP QDs was withdrawn while for CdSe QDs contradictory results were reported for seemingly the same experiment. The difference in the results are possibly related to a trivial aspect as stirring (no MEG) or not stirring (MEG) the QD solution [49]. The controversy shows that fast dynamic measurements do not only provide a wealth of information, but are at the same time extremely sensitive for several external parameters and the interpretation is not always straightforward.

For III–V QDs extensive research has been done on exciton dynamics in epitaxially grown QD structures. For I–VII QDs like CuCl exciton dynamics have also been extensively studied. This chapter focuses on colloidal QD systems and the reader is referred to [50] and [51] for further information on exciton dynamics in III–V and I–VII QD structures. InP is an exception since these III–V QDs have been widely studied as colloidal nanocrystals. The radiative life time of the exciton emission is similar to that of CdSe and CdTe QDs. For efficient red-emitting InP QDs a decay time of 65 ns has been reported at room temperature [52]. The II–VI ZnO QDs are also widely studied, but the radiative life time of the exciton emission is hard to determine since trapping of charge carriers in defects occurs (giving rise to the well-known green emission from ZnO) and hinders the observation of the radiative decay rate. From the multitude of studies on ZnO nanostructures it is clear that exciton relaxation is fast and typically sub-ns life times are observed, even in systems where trapping of charge carriers is prevented by trap filling [53, 54]. These results suggest that the radiative rate for the exciton emission in ZnO is an order of magnitude faster than in CdSe and CdTe QDs. Finally, an important class of QDs are Si and Ge QDs. For Si nanocrystals the discussion on the nature of the blue-shifted emission has been long debated (and the debate is still going on) [55, 56]. From many studies it is however clear that Si nanocrystals can show blue-shifted emission due to quantum confinement. The life of the emission is long, µs-ms and is related to the forbidden nature of the indirect bandgap transition [57, 58].

3. Energy transfer processes with quantum dots

3.1 Energy transfer and energy migration.

In diluted systems the optical properties and time response are determined by the properties of the single (isolated) QD. When the concentration of QD is increased or when clusters of QDs are formed, energy transfer between QDs in close proximity becomes an important alternative pathway for de-excitation of a QD. Energy transfer between optically active ions and molecules is well known [5]. The initially excited species is called the donor and the excitation energy can be transferred from a donor to an acceptor in close proximity. In the case of QDs energy transfer will be directed from the smaller QDs (higher energy bandgap) to the larger QDs (smaller energy bandgap).

Quantitative modeling and understanding of energy transfer processes between luminescent ions and molecules has greatly benefited from the work by Förster and Dexter in the 1950's. Equations were derived for energy transfer rates for various

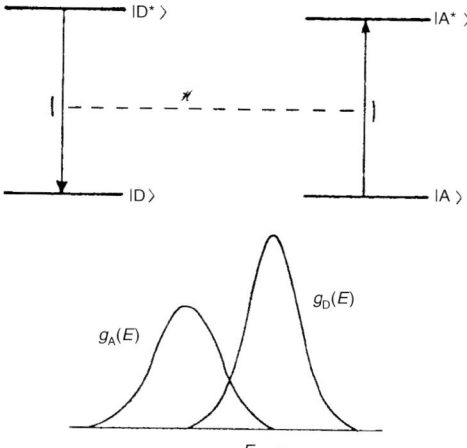

Fig. 13. Schematic representation of energy transfer from a donor D to an acceptor A. Both interaction (top picture) and spectral overlap (bottom picture) are required. Reproduced with permission from [5], © 1989 Oxford University Press

mechanisms, i.e. multipole–multipole interaction and exchange interaction. In both cases energy transfer does not involve the emission of a photon. It concerns non-radiative energy transfer where the energy transfer is mediated by electrostatic coupling, magnetic coupling or exchange coupling between donor and acceptor. The concept is schematically depicted in Fig. 13. The general expression for the probability of energy transfer is given by [5]:

$$W_{DA} = \frac{2\pi}{\hbar} |<D, A^*|H'|D^*, A>|^2 \int g_D(E) g_A(E) dE \qquad (4)$$

Here the left hand side represents the interaction between the donor and acceptor ion involved in the energy transfer process (multipole–multipole or exchange) and the right hand side term gives the overlap integral between the donor emission spectrum and the acceptor absorption spectrum (resonance condition). In the case of dipole–dipole interaction (the dominant mechanism for energy transfer in most cases discussed here) the equation becomes [5]:

$$W_{DA}^{dd} = \left(\frac{1}{4\pi\varepsilon_0}\right)^2 \frac{3\pi\hbar e^4}{n^4 m^2 \omega^2} \frac{1}{R^6} f_D(ED) f_A(ED) \int g_D(E) g_A(E) dE \qquad (5)$$

where W_{DA}^{dd} is the energy transfer rate (through dipole–dipole interaction), R is the distance between the (point) dipoles, f_A and f_D give the (electric dipole) oscillator strengths for the acceptor and donor transition, the integral is the spectral overlap while the other symbols have their usual meaning.

Often the Förster equation is rewritten to give the energy transfer rate k [5]:

$$k_{\text{Förster}} = \frac{Q_D}{\tau_D} \left(\frac{8.785 \times 10^{-25} I}{n^4 R^6}\right) \quad \text{with} \quad I = \int_0^\infty \alpha_A(\lambda) f_D(\lambda) \lambda^4 d\lambda \qquad (6)$$

where Q_D is the dipole moment of the donor transition, τ_D is emission life time, R is the distance between the dipoles and I is overlap integral over acceptor extinction coefficient (α_A) and the donor oscillator strength (f_D) over the full wavelength interval. An interesting question concerns the validity of the (widely applied) Förster model to quantitatively describe energy transfer between QDs. The Förster model has been derived for the limit of point dipoles (i.e. the distance between the dipoles is much larger than the spatial extension of the dipoles). In the case of energy transfer between QDs this boundary condition is not valid. The oscillating dipoles in the QDs cannot be considered as point dipoles and also local fields can play an important role. In general the transfer rates calculated using the Förster equation are nevertheless in good agreement with the observed rates. In a recent theoretical paper on the validity of Försters theory for energy transfer between QDs it was concluded that Förster theory cannot be applied for energy transfer between indirect bandgap semiconductor nanocrystals like Si [59].

In Fig. 14 the luminescence decay profiles are sketched for different types of energy transfer processes [5]. In the absence of energy transfer processes (curve a), a single exponential decay is observed and the decay rate (radiative and non-radiative) can be determined from the emission life time. In the case of one-step energy transfer (curve b), the excitation energy is transferred from a donor to one or more acceptors in the immediate surroundings but there is no energy transfer between donors. In the luminescence decay curves this leads to an initial non-exponential decay that is faster than the radiative decay rate. The faster non-exponential decay reflects the different acceptor configurations around the donor. In the long time regime the decay time approaches the radiative decay rate for those donors that do not have an acceptor in the immediate surroundings. In case of concentrated systems of donors, energy

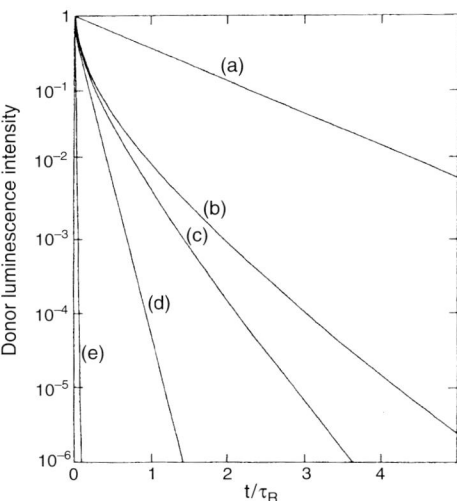

Fig. 14. Shape of luminescence decay curves of the donor emission for **a** isolated donors, **b** single step energy transfer to acceptors, **c** diffusion limited energy migration to acceptors and **d** and **e** fast diffusion. Reproduced with permission from [5], © 1989 Oxford University Press

transfer between donors can occur and energy transfer to acceptors or quenching sites can occur via diffusion of the excitation energy over the donor sublattice. If the energy transfer probability between donors is lower than the transfer probability to the traps, one speaks of diffusion limited energy migration and the luminescence decay is non-exponential (curve c). The initial fast non-exponential component reflects the single step energy transfer from donors to nearby acceptors (just as in curve b) but in the long time regime the decay is faster than the single exponential decay for isolated donors due to energy migration over the donor sublattice to acceptors, thus providing an additional decay channel. Finally, there is the regime of fast diffusion where the donor–donor transfer rate is much higher than the donor–acceptor transfer rate. In this case, a single exponential decay is observed (curves d and e) since all donors sense the same environment of acceptors due to the very rapid diffusion of the excitation energy over the donor sublattice.

The theory for analysis of the donor emission decay profiles and extracting information for donor–donor and donor–acceptor transfer rates is extensive in the literature on luminescence of transition metal ions and fluorescent molecules. In the case of random distributions of optically active species (glass like systems) decay profiles are fitted using the Inokuti–Hirayama model (for single step D–A energy transfer) and the Yokoto–Tanimoto model (in case of diffusion limited energy migration). For crystalline systems where energy transfer is dominated by short range energy transfer, the better models are shell models where the actual (discrete) distribution of acceptors is modeled based on the discrete donor–acceptor distances that are possible in the given crystal structure [60, 61].

3.2 Energy transfer between quantum dots. In the field of energy transfer processes involving QDs the analysis is more complicated and often more phenomological models are used to describe the observed decay behavior. An important difference between energy transfer processes between transition metal ions or fluorescent molecules and QDs is the homogeneity of the system. The inhomogeneous broadening for optical transitions is larger for QDs, usually much larger than the homogeneous broadening. This gives rise to a variable energy mismatch for energy transfer processes between QDs and it is hard to derive a single transfer rate for QDs at a certain distance. The energy mismatch needs to be made up by phonon assistance and the energy transfer probability for a given donor–acceptor pair at the same distance R will vary due to differences in the energy mismatch. This gives rise to a distribution in the donor–acceptor transfer rates. A second aspect is the inhomogeneity in the distribution of distances between QDs. There is a wide variety of systems where energy transfer between QDs has been studied, as will be discussed below, but rarely one can determine the distribution in distances between donors and acceptors with the same accuracy as it can be done for transition metal ion-doped crystals where the crystal structure imposes discrete donor–acceptor distances. Only in well-ordered self-assembled structures of QDs one can obtain more quantitative information on the number of acceptors at specific distances. In spite of the more complex nature of energy transfer processes in QD systems, it is useful to be aware of the analysis methods developed for energy transfer processes between optically active transition metal ions since they can serve as a starting point for the analysis.

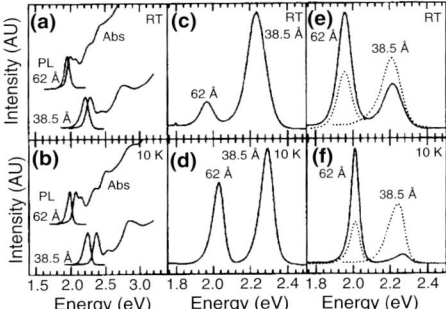
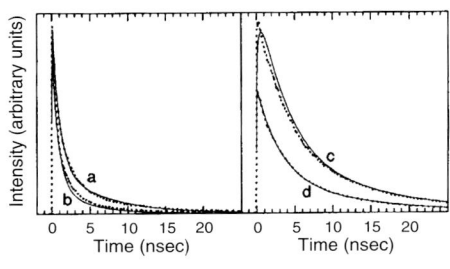

Fig. 15. (Left) Absorption and emission spectra for a mixture of 3.9 nm and 6.2 nm CdSe QDs at RT **a** and 10 K **b**. In **c** and **d** the photoluminescence spectra for a (frozen in (d)) solution with 18% 6.2 nm (acceptor) and 82% 3.9 nm (donor) QDs are shown at RT and 10 K. In **e** and **f** the RT and 10 K emission spectra of close packed solid films are shown. (Right) Luminescence decay curves of the emission from the 3.9 nm QDs in a pure film of 3.9 nm QDs (a) and a mixed film with 18% of 6.2 nm QDs (b). Curves (c) and (d) show the decay curves of the emission from the 6.2 nm QDs in the same mixed film upon excitation in blue (c) and red (d) absorption edge of the 3.9 nm CdSe QDs. Reproduced with permission from [62], © 1996 APS

In two pioneering papers on energy transfer between QDs, Kagan et al. studied the energy transfer from small 3.9 nm CdSe QDs to large 6.2 nm CdSe QDs in thin solid films of QDs with 18% of the large QDs and 82% of the smaller (donor) QDs [62, 63]. Spectroscopic evidence for the presence of efficient energy transfer from the smaller to the larger QDs is provided by comparison of Fig. 15c and d to e and f: the 2.3 eV emission from the smaller QDs is clearly quenched in the solid films. The luminescence decay curves were analyzed using a model similar to the Inokuti–Hirayama model:

$$n_{D,\text{mixed}}(t) = n_{D,\text{pure}}(t)\exp\left[-\gamma\left(\frac{\pi t}{\tau_D}\right)^{1/2}\right] \quad (7)$$

where $\gamma = C(4/3\pi R_0^3)$. From the fit (drawn line for the right hand side of Fig. 15, curve (a)) the critical distance for energy transfer was determined to be 4.8 nm and was found to be in agreement with the critical distance for energy transfer derived using the Förster equation.

This gives a nearest neighbor transfer rate of $1\times 10^8\,\text{s}^{-1}$ at RT and $0.6\times 10^8\,\text{s}^{-1}$ at 10 K. The nearest neighbor distance between can be estimated as 6.1 nm (the sum of the radii of the QDs (5 nm) and the thickness of the capping layer (1.1 nm)). This value for the nearest neighbor separation distance was confirmed by SAXS measurements. Finally, the nearest neighbor energy transfer rates were also determined using the Förster equation for dipole–dipole interaction and found to be an excellent agreement with the experimentally determined values showing that dipole–dipole interaction is the dominant mechanism for energy transfer between nearest neighbor QDs.

In addition to the model resembling the Inokuti–Hirayama model, also more exact shell models have been applied, in analogy with the analysis of energy transfer between luminescent ions in solids [64]. Langmuir–Blodgett (LB) assemblies of

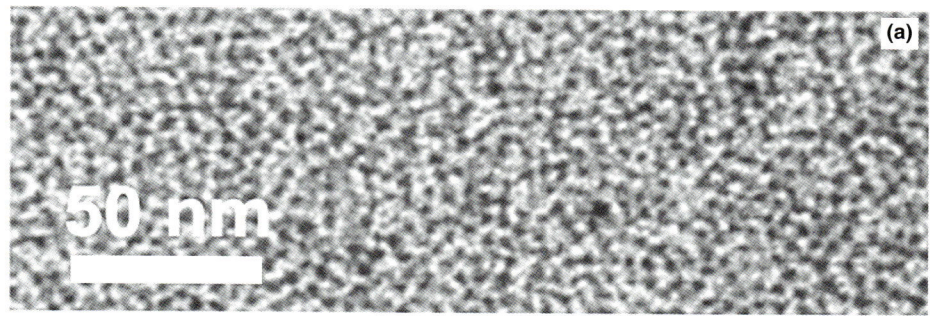

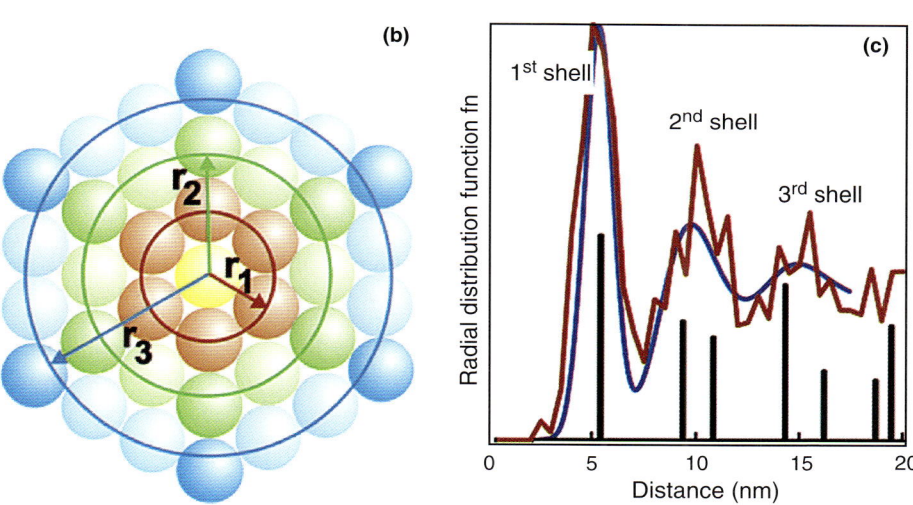

Fig. 16. a A TEM image of an LB monolayer of octylamine-capped CdSe NQDs with radius $R = 2.2$ nm. **b** An ideal hexagonal 2D array of spheres. Different colors show three main shells. Different shadings within the main shells show different subshells. **c** Radial distribution function (RDF) of the LB sample shown in panel a (red line). Black bars represent the RDF of an ideal hexagonal array composed of spheres with a center-to-center distance of 5.4 nm. The blue line is the RDF of a "non-ideal" hexagonal array with fluctuation in the sphere-to-sphere distances of 13%. Reproduced with permission from [64], © 2003 ACS

octylamine-capped 2.2 nm CdSe QDs were prepared and the luminescence decay curves were analyzed at different wavelengths within the inhomogeneously broadened line. In Fig. 16a TEM image of the Langmuir–Blodget film is shown, together with the distribution of donor–acceptor distances within this ordered film (Fig. 16c). For this situation the luminescence decay curves can be modeled for different wavelengths within the inhomogeneously broadened line. The nearest neighbor transfer time was determined to be 50 ps, yielding transfer times of 0.75 ns and 10 ns for transfer to acceptors in the 2nd and 3rd shell. A good agreement between the experimentally determined transfer rates at different emission energies was obtained using these three rates and statistical determinations of the number of acceptors in each shell as a function of emission energy. The transfer time of 50 ps is considered to

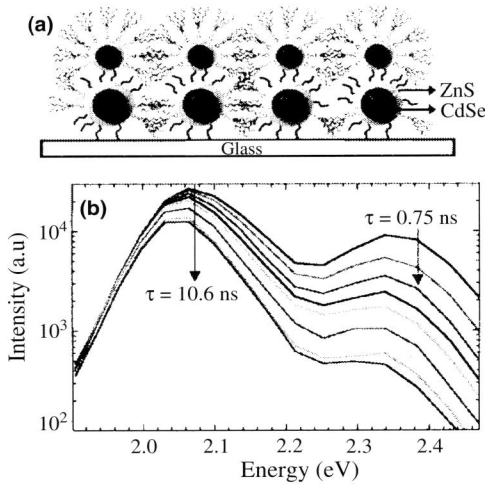

Fig. 17. (Top) Schematic picture of a QD bilayer in which directional energy flow from the smaller to larger QDs can occur. (Bottom) Time-resolved photoluminescence spectra recorded with 500 ps time intervals. Reproduced with permission from [66], © 2002 APS

represent an upper limit for the transfer rate that can be expected for energy transfer between CdSe QDs.

Energy transfer between QDs of different sizes is not only of scientific interest, it may also be applied for light harvesting purposes. By constructing layered structures of QDs where the size of the particles is decreased or increased in every subsequent layer, a system in which a directional flow of energy occurs to the layer with the largest QDs [65, 66]. In this way all the excitation energy that is absorbed by a large number of QD layers can be concentrated in a single layer. To demonstrate the efficiency of this light harvesting concept, directional energy transfer in a bi-layer is demonstrated in Fig. 17. A bilayer of 1.3 nm on top of 2.1 nm core-shell CdSe/ZnS QDs was fabricated. The lower part of Fig. 17 shows emission spectra at 500 ps time interval. The emission from the smaller QDs decay with a time constant of 750 ps due to fast energy transfer to the larger dots which decay with a time of 10.6 ns, close to the radiative decay time. The large ratio (20) between the interlayer transfer rate and the radiative decay rate allows for efficient transfer up to 20 layers of QDs of continuously increasing sizes.

Evidence that the mechanism for the energy transfer process is dipole–dipole (or in general: multipole–multipole) interaction is demonstrated by studying the temperature dependence of the energy transfer rate and the efficiency of the energy transfer process [67]. In the case of dipole–dipole interaction, the energy transfer rate is proportional to the oscillator strengths of the transitions involved on the donor and acceptor. Since both the transfer rate and the radiative decay of the donor are proportional to the oscillator strength of the transition on the donor, the transfer efficiency (or the ratio of the donor/acceptor emission intensity) is expected to be independent on the oscillator strength of the transition on the donor. In the case of CdSe or CdTe QDs the oscillator strength of the donor transition can be easily tuned

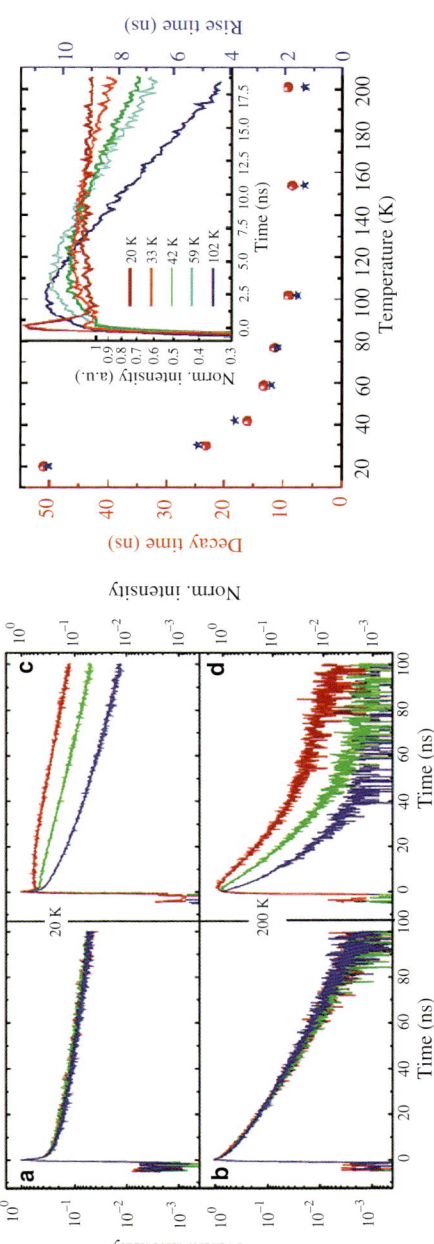

Fig. 18. (Left) Normalized luminescence decay curves of orange emitting CdTe QDs dissolved in chloroform (**a**, **b**) and as a QD solid on quartz (**c**, **d**) at 20 K and 200 K measured at the peak maximum (green) 20 nm red-shifted (red) and 20 nm blue-shifted (blue) from the maximum. (Right) Luminescence decay (●) and rise time (★) of a QD solid on quartz, prepared from orange-luminescing hexanethiol-capped CdTe QDs recorded as function of temperature, measured 20 nm red-shifted of the excitonic peak maximum. The inset shows an enlargement of the normalized luminescence decay curves at different temperatures in the first 18 ns. Note the fast component at low temperatures which is due to emission from the bright state. Reproduced with permission from [67], © 2005 ACS

by cooling the QDs and freezing the system partly in the dark state. The spectral shift involved is small and will not influence the transitions involved on the (larger) acceptor QDs. In Fig. 18 results on temperature dependence of the donor emission decay curves is shown for orange emitting CdTe QDs (3 nm) in chloroform at low concentration (no energy transfer expected) and as a QD solid on quartz (energy transfer possible). Decay curves were recorded as a function of temperature in the maximum and on the short wavelength side and long wavelength side of the inhomogeneously broadened emission band. From the faster decay curves in the solid for the short wavelength side (donors) and the rise time on the long wavelength side (acceptors) the energy transfer rate can be determined. The decay curves in the solvent (no energy transfer) serve as a reference. The results in Fig. 18b show that the energy transfer rate is proportional to the radiative decay rate on the donor, consistent with a dipole–dipole energy transfer mechanism. Further evidence was obtained from the temperature dependence of the transfer rate in mixed solids of green (donor) and orange (acceptor) QDs systems. Also in this system the transfer rates are observed to scale with donor emission decay rate while the transfer efficiency is not influenced by the temperature.

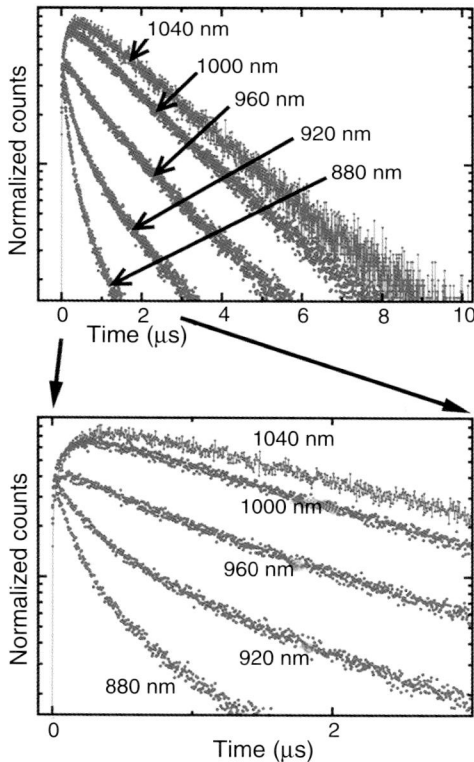

Fig. 19. Temporally resolved fluorescence of a film of 2.5 nm PbS QDs at different wavelengths within the emission band for the 0–10 μs window (top) and the 0–3 μs window (bottom) at RT. Reproduced with permission from [43], © 2007 ACS

Similar conclusions are obtained by studying energy transfer processes in PbS QD solids [43]. The RT radiative decay time of the PbS exciton emission is of the order of 1 μs, about two orders of magnitude slower than the ∼20 ns decay time of the CdSe or CdTe emission at RT. Since the dipole–dipole transfer rate scales with donor oscillator strength, also the energy transfer rates are expected to be two orders of magnitude slower. The luminescence decay curves of QD films of 2.5 nm PbS QDs are shown in Fig. 19. The QDs in the film emit around 960 nm (920 nm in the solvent). The fast emission of the short wavelength side (donors) and the build-up on the long wavelength side (acceptors) indicate that the donor–acceptor transfer time for PbS QDs is about 200–400 ns, indeed also two orders of magnitude slower than observed for CdSe or CdTe QDs.

3.3 Energy transfer from quantum dots to dye molecules and metal nanoparticles.
Energy transfer between QDs is interesting mainly from a fundamental point of view and has no immediate prospect for application. The main field of application of QDs is in the area of imaging in living organisms. QDs are perfect labels for bio-imaging due to the high stability, tunability of emission color, broad excitation range and narrow emission band [68–70]. Also in this field energy transfer processes play an important role, but now it involves not only energy transfer between QDs but also energy transfer to dye molecules and interaction with metal nanoparticles. In this last section some examples will be discussed where interaction between QDs and dye molecules or metal nanoparticles are used.

An elegant method to probe interaction between proteins (or other biomolecules) is to take advantage of the strong distance dependence of fluorescence (or Förster) resonance energy transfer (FRET). Energy transfer commonly proceeds through dipole–dipole interaction (both for fluorescent dyes and QDs) with the well-known R^{-6} distance dependence. By labelling two different biomolecules with fluorescent probes, energy transfer from the higher energy emitting probe (donor) to the probe emitting at a lower energy takes place, so that the information on the proximity of the two probes (typically within 10 nm) can be obtained. This technique is widely used with different types of dye molecules and is a very sensitive method to study receptor–ligand interactions or conformational changes of a biomolecule. For example, a conformational change where two areas of a protein form an active center can be probed by binding different chromophores that are capable of FRET on specific positions in the protein molecule that come together or apart. This process can be measured on a single molecule level and the dynamics of the folding and unfolding can be monitored. The application of the QDs instead of dye molecules in this type of studies offers the same advantages as mentioned above for the straightforward labelling of biomolecules with QDs for fluorescence imaging where especially the high stability of chromophore's emission is important [71–73]. For dye molecules, changes in the relative intensity of donor and acceptor emission is severely affected by quenching of donor and acceptor molecules which complicates the analysis of changes in the intensity due to FRET in closely linked donor–acceptor pairs.

A classic study showing the great potential of QDs in a quantitative analysis of protein binding is based on binding maltose binding proteins (MBP-zb) with an

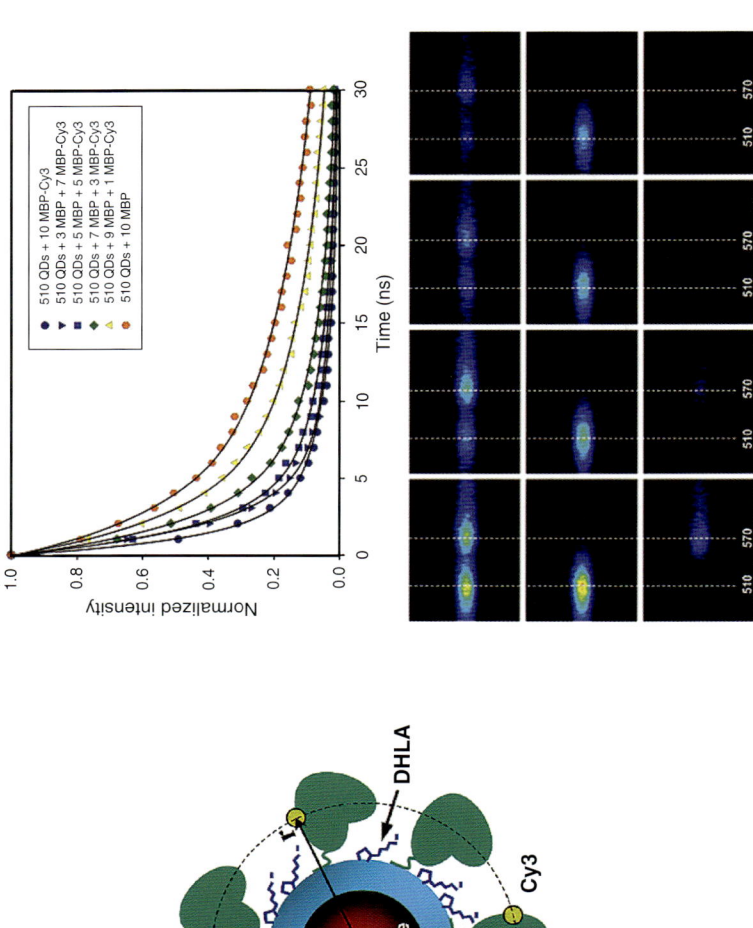

Fig. 20. (Left) Schematic representation of the QD–MBP-dye nanoassembly. The distance r represents the radius or average distance between the QD center and location of the Cy3-labeled residue on MBP. (Right) Plot of the 510 nm emitting QD photoemission intensity vs. time immediately after a short pulse excitation signal. Data are shown for various QD–MBP–Cy3 conjugate configurations where the number of Cy3-labeled proteins was increased from 0 to 10, while maintaining the total number of proteins fixed at 10 MBP/QD. Series of images showing the intensity of 510 nm QDs with 3 MBP/7 MBP-Cy3 (top row), 510 nm QDs with 10 MBP per QD (middle row), and free MBP-Cy3 equivalent to 10 MBP-Cy3 per QD (no QDs present, bottom row) as recorded by the CCD camera at 2 ns intervals. Reproduced with permission from [74]. © 2004 ACS

electrostatic attachment domain irreversibly onto surfaces of CdSe/ZnS core shell QDs capped with dihydrolipoic acid ligands [74]. By attaching a cyanine dye (Cy3) to MPD dye-labeled proteins could be bound at a well-defined distance from the QD which acts as a donor due to the spectral overlap of the QD emission with the Cy3 absorption spectrum. The QD–MBP-dye complex is shown schematically in Fig. 20. Up to 15 MPD-proteins could be bound to the QD surface. By varying the number of MPD proteins with a dye molecule, the number of acceptors around the QD could be accurately varied (thus changing the acceptor concentration). Alternatively, the donor acceptor distance and the spectral overlap between the donor emission and the acceptor absorption could be varied by changing the size of the CdSe core. In Fig. 20 (right hand side) the QD decay curves are shown for different numbers of Cy3 acceptor, while in the images below the emission spectra recorded with a streak camera at different time intervals after the excitation pulse are shown. The decay curves show that the donor emission decay becomes progressively faster as the number of Cy3 acceptors around the QD donor increases. Analysis of the results allowed an accurate determination of the Förster distance for energy transfer ($\sim$6–7 nm) that was in excellent agreement (within 0.3 nm) with the distance determined using structural characterization techniques.

The control of energy transfer between asymmetric CdSe/CdS quantum rods (QRs) and dye molecules has been demonstrated at low temperatures and using an electric field [75]. At low temperatures the emission of the QRs becomes narrow, as do the absorption spectra of dye molecules. An important property of this particular kind of QRs is the much larger shift of the emission in an external electric field (as compared to QDs) [76]. This so-called quantum confined Stark effect (QCSE) in QRs was used to shift the emission of the QR in or out of resonance with the absorption spectrum of a nearby dye molecule (a Cy5 derivative). The QRs and dye molecules are incorporated (at well-chosen concentrations so that the distance between QRs and dye molecules is 20 nm on average) in a polystyrene film. The idea of the experiment is presented in Fig. 21, left hand side. Below the TEM-images showing the asymmetric CdSe/CDs QRs with a CdSe core located at one end of the nanostructure, the shift of the QR emission in resonance with the dye absorption upon applying an electric field is schematically depicted. On the right hand side of Fig. 21 it is demonstrated that the concept works beautifully: in the top graph the QR emission is shown for a film without dye molecules, demonstrating a 10 nm Stark shift upon cyclic application of a bias. In the lower images (b and c) turning on (b) and turning off (c) of Förster energy transfer by the electric field are shown. Depending on the resonance conditions for an individual QR-dye pair the electric field can shift the QR emission in or out of resonance with the dye absorption, thus electrically switching energy transfer.

Finally, energy transfer between QDs and metal nanoparticles is a hot topic. In spite of an exploding number of papers in this field (often called plasmonics), the basic understanding of the interactions that govern the response of excitons in QDs (or fundamental excitations in order systems) in the proximity of a metal nanoparticle is still lacking. In addition, the practical implications of plasmonic interactions seem to be only limited by ones imagination. To illustrate this, the influence of plasmonic coupling on the exciton life time is shown to be able to probe with a nm accuracy the

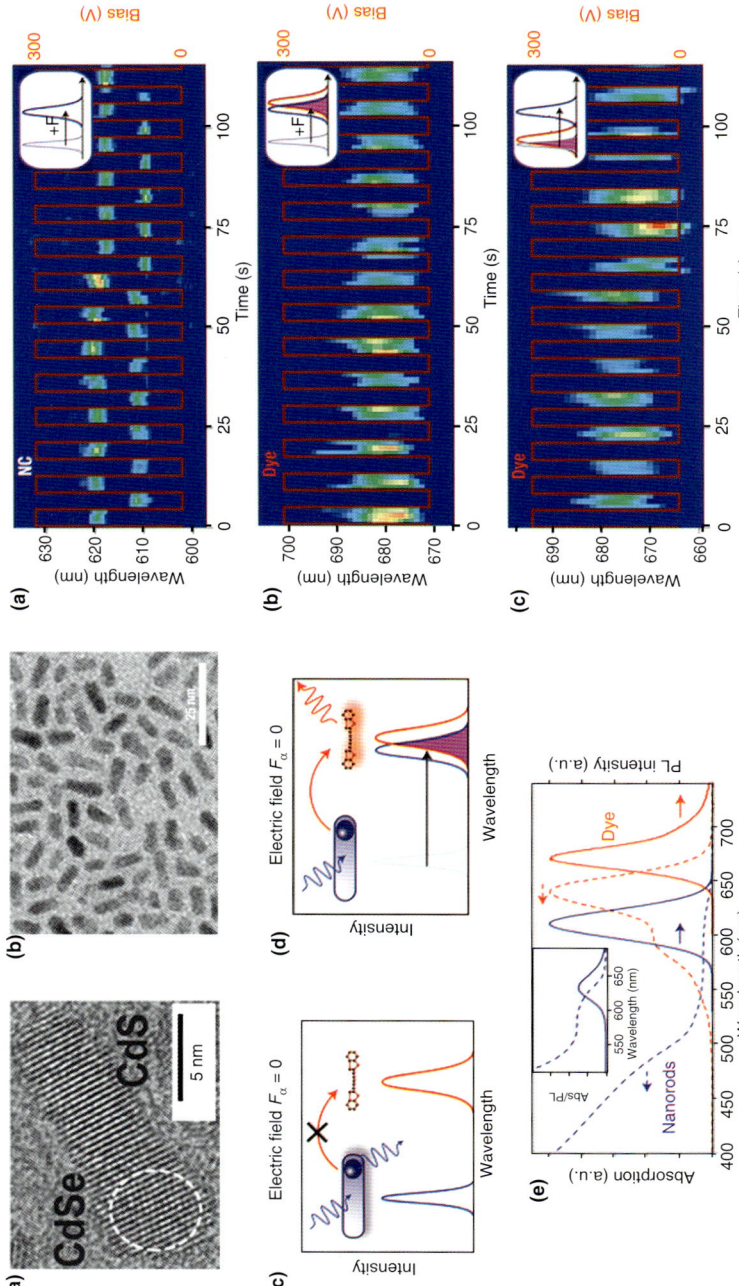

Fig. 21. (Left) **a, b** High-resolution TEM image and overview of CdSe/CdS quantum rods (QRs). **c, d** Schematic representation of spectral shifting the QR emission to overlap with the dye absorption by applying an electric field. **e** Ensemble excitation and emission spectra of the CdSe/CdS QRs and dye molecules at 50 K. (Right) Electrical tuning of resonant energy transfer from nanocrystals to single dye molecules using the QCSE at 50 K. **a** Dependence of the PL spectrum of a single nanocrystal on electric field in the absence of dye molecules. **b, c** Modulation of the fluorescence of two different nanocrystal–dye couples by cyclic application of a bias. Depending on the spectral position of the nanocrystal, FRET occurs either with (**b**) or without (**c**) an electric field, so that the dye emission switches on and off with a change in field. Reproduced with permission from [75], © 2006 Nature Publishing Group

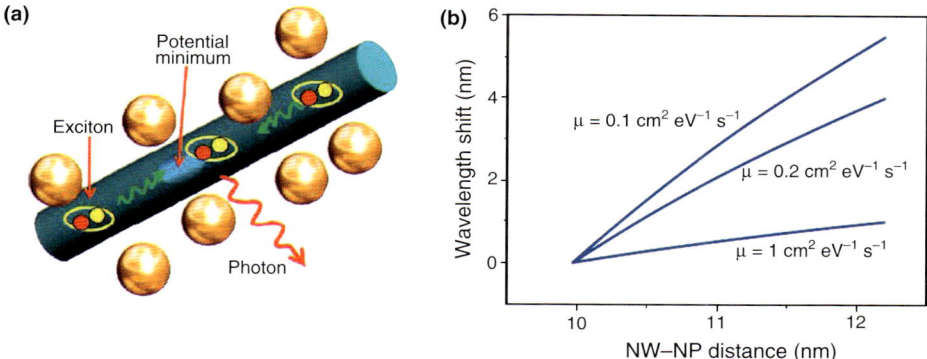

Fig. 22. **a** Schematic diagram of excitons diffusing towards the potential minima where they recombine. This leads to red-shifted emission. **b** Calculated wavelength shift of exciton emission as a function of the NW–NP distance for the specified exciton mobilities. Reproduced with permission from [77], © 2006 Nature Publishing Group

distance between a semiconductor nanowire (NW) and a gold nanoparticle [77]. This very sensitive distance probe can potentially be used in living cells to monitor the movement of biomolecules with nm accuracy. The basic idea is that the exciton life time in a semiconductor nanowire determines the emission wavelength. As the exciton diffuses in the nanowire it will be trapped in regions of lower bandgap separation and the longer the exciton lives, the larger the red-shift will be. The exciton life time will be shortened by energy transfer to a nearby metal nanoparticle and for simplicity the energy transfer rate is described by the Förster formula (giving a R^{-6} distance dependence). Based on these considerations the wavelength shifts to the blue as the metal nanoparticle approaches the wire (shortening the exciton life time) and shifts to the red as the metal nanoparticle moves away. The wavelength shift as a function of NW–NP is shown in Fig. 22 for different exciton mobilities.

To illustrate that this concept does indeed work, a molecular spring assembly was constructed as shown in Fig. 23 in which a CdTe nanowire is linked via PEG–aB–PEG strings to gold nanoparticles (PEG being poly-ethyleneglycol, aB being an antibody, here streptavidine SA, further details can be found in [77]). Upon adding an antigen (aG) that binds to the aB, the PEG chain expands and the NW–NP distance increases. As a result, the exciton lifetime increases due to slower energy transfer to the gold NPs and the emission shifts to the red. Upon adding free aB that competes with the aG bound on the PEG string, the aG is released from the string, the string contracts and the expected blue-shift is observed. This fully reversible wavelength shift demonstrates the viability of this approach. From the shift also the exciton mobility was estimated to be 0.1–0.2 cm^2 eV^{-1} s^{-1}.

Research on the influence of metal nanoparticles on the exciton dynamics in QDs is an area where new results have emerged in the past years [78–80] and some of the most promising new schemes for applications of (enhanced) QD fluorescence have been proposed by taking advantage of exciton–plasmon interaction. The near future will show which of these will actually work, but even if it is only a fraction it will be a successful avenue.

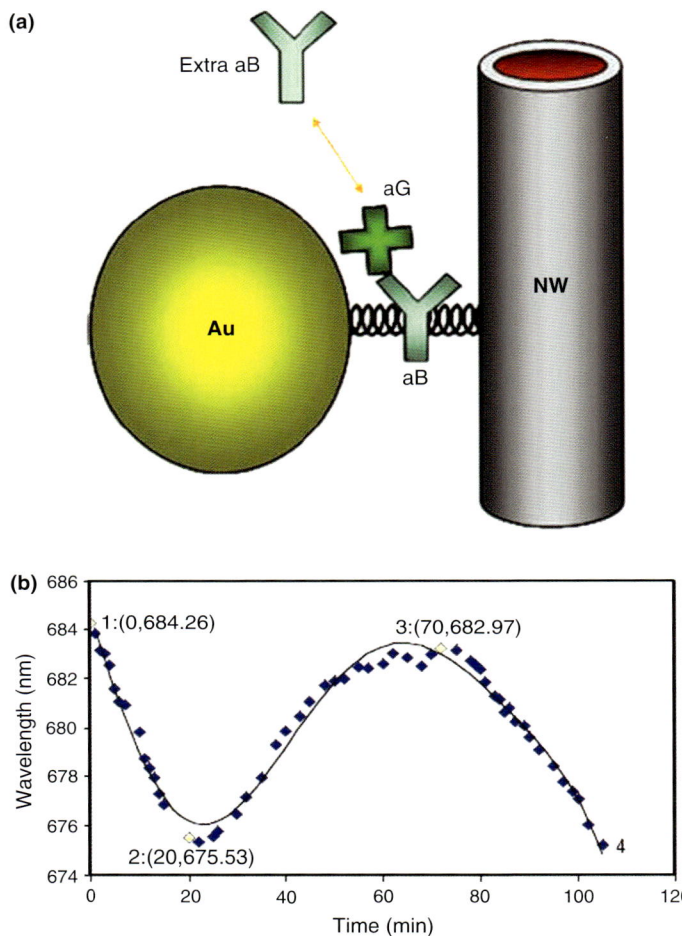

Fig. 23. a Schematic illustration of a molecular spring assembly of a CdTe nanowire linked via PEG–aB–PEG to a gold nanoparticle. The reversible coupling aG–aB (SA) changes the chain length. The addition of extra amounts of aB will disrupt the aG–aB reaction in the NP–PEG–aB–PEG–NW superstructure. **b** Reversible shift of the peak luminescence wavelength: 1, attachment of a NP to a NW; 2, after adding 20 μl SA; 3, after adding free aB to the media; 4, after adding 20 μl SA. Excitation wavelength: 420 nm. Reproduced with permission from [77], © 2006 Nature Publishing Group

4. Conclusions

Studies on the exciton dynamics in QDs have greatly contributed to the understanding of the energy level structure and dynamic processes in QDs in a broad time window (fs–ns). The research has benefited from the rapid development of generally accessible equipment for fast time resolved spectroscopy. Especially for the colloidal II–VI CdSe and CdTe QDs dynamic processes like exciton recombination, charge carrier relaxation and energy transfer are extensively studied and well understood. Nevertheless, these model systems will continue to be of great interest for the discovery of new phenomena. An important new focus in exciton dynamics studies is aimed at controlling and understanding changes in the exciton dynamics through external influences, for example plasmonic coupling or embedding in photonic structures. For the infrared emitting IV–VI QDs PbSe and PbS a rising interest in the exciton dynamics has been triggered by potential application for telecommunication and reports on multiple exciton generation. Energy transfer processes involving QDs have been widely studied and can be well explained using the Förster theory for energy transfer. Exciting developments in this area involve controlling energy transfer on a nm scale where especially coupling with metal nanoparticles or dye molecules offer new avenues for detection schemes that can be used in bio-imaging and possibly other areas. In the past 10 years of research on exciton dynamics has developed into a mature and exciting field that continues to attract a growing number of (young) researchers.

References

[1] Jaeckel G (1926) Some modern absorption glasses. Z Tech Phys 7: 301–304
[2] Rossetti R, Nakahara S, Brus LE (1983) Quantum size effects in the redox potentials, resonance Raman spectra, and electronic spectra of cadmium sulfide crystallites in aqueous solution. Journal of Chemical Physics 79: 1086–1088
[3] Henglein A (1982) Photodegradation and fluorescence of colloidal cadmium sulfide in aqueous solution. Berichte der Bunsen-Gesellschaft Physikalische Chemie 86: 301–305
[4] Ekimov AI, Onushchenko AA (1981) Quantum dimensional effect in three-dimensional microcrystals of semiconductors. Pis'ma v Zhurnal Eksperimental'noi i Teoreticheskoi Fiziki 34: 363–366
[5] Henderson B, Imbusch GF (1989) Optical spectroscopy of solids. Oxford University Press
[6] Wehrenberg BL, Wang C, Guyot-Sionnest P (2002) Interband and intraband optical studies of PbSe colloidal quantum dots. Journal of Physical Chemistry B 106: 10634–10640
[7] Murray CB, Norris DJ, Bawendi M (1993) Synthesis and characterization of nearly monodisperse CdE (E = sulfur, selenium, tellurium) semiconductor nanocrystallites. Journal of the American Chemical Society 115: 8706–8715
[8] Nirmal M, Murray CB, Bawendi M (1994) Fluorescence-line narrowing in CdSe quantum dots: surface localization of the photogenerated exciton. Physical Review B: Condensed Matter 50: 2293–2300
[9] Efros AL, Rosen M, Kuno M, Nirmal M, Norris DJ, Bawendi M (1996) Band-edge exciton in quantum dots of semiconductors with a degenerate valence band: dark and bright exciton states. Physical Review B: Condensed Matter 54: 4843–4856
[10] Bryant GW, Jaskolski W (2003) Tight-binding theory of quantum – dot quantum wells: single-particle effects and near-band-edge structure. Physical Review B: Condensed Matter and Materials Physics 67(20): 205320
[11] Califano M, Franceschetti A, Zunger A (2005) Temperature dependence of excitonic radiative decay in CdSe quantum dots: the role of surface hole traps. Nano Letters 5: 2360–2364
[12] Leung K, Pokrant S, Waley KB (1998) Exciton fine structure in CdSe nanoclusters. Physical Review B 57: 12291–12301
[13] Mello Donegá de C, Bode M, Meijerink A (2006) Size- and temperature-dependence of exciton lifetimes in CdSe quantum dots. Physical Review B 74: 085320/1–085320/9
[14] Crooker SA, Barrick T, Hollingsworth JA, Klimov VI (2003) Multiple temperature regimes of radiative decay in CdSe nanocrystal quantum dots: intrinsic limits to the dark-exciton lifetime. Applied Physics Letters 82: 2793–2795
[15] Labeau O, Tamarat P, Lounis B (2003) Temperature dependence of the luminescence. Lifetime of single CdSe/ZnS quantum dots. Physical Review Letters 90: 257404
[16] Wuister SF, Houselt van A, de Mello Donegá C, Vanmaekelbergh D, Meijerink A (2004) Temperature antiquenching of the luminescence from capped CdSe quantum dots. Angewandte Chemie-International Edition 43: 3029–3033
[17] Wuister SF, Mello Donegá de C, Meijerink A (2004) Luminescence temperature antiquenching of water-soluble CdTe quantum dots: role of the solvent. Journal of the American Chemical Society 126: 10397–10402
[18] Qu L, Peng X (2002) Control of photoluminescence properties of CdSe nanocrystals in growth. Journal of the American Chemical Society 124: 2049–2055

[19] Reiss P, Carayon S, Bleuse J (2003) Large fluorescence quantum yield and low size dispersion from CdSe/ZnSe core/shell nanocrystals. Physica E: Low-Dimensional Systems & Nanostructures 17: 95–96
[20] Mello Donegá de C, Hickey SG, Wuister SF, Vanmaekelbergh D, Meijerink A (2003) Single-step synthesis to control the photoluminescence quantum yield and size dispersion of CdSe nanocrystals. Journal of Physical Chemistry B 107: 489–496
[21] Wuister SF, de Mello Donegá C, Meijerink A (2004) Influence of thiol capping on the exciton luminescence and decay kinetics of CdTe and CdSe quantum dots. Journal of Physical Chemistry B 108: 17393–17397
[22] Wuister SF, de Mello Donegá C, Meijerink A (2004) Local-field effects on the spontaneous emission rate of CdTe and CdSe quantum dots in dielectric media. Journal of Chemical Physics 121: 4310–4315
[23] Lodahl P, Driel van AF, Nikolaev IS, Irman A, Overgaag K, Vanmaekelbergh D, Vos WL (2004) Controlling the dynamics of spontaneous emission from quantum dots by photonic crystals. Nature 430: 654–657
[24] Driel van AF, Allan G, Delerue C, Lodahl P, Vos WL, Vanmaekelbergh D (2005) Frequency-dependent spontaneous emission rate from CdSe and CdTe nanocrystals: influence of dark states. Physical Review Letters 95: 236804/1–236804/4
[25] Nozik AJ (2001) Spectroscopy and hot electron relaxation dynamics in semiconductor quantum wells and quantum dots. Annual Reviews in Physical Chemistry 52: 193–231
[26] Guyot-Sionnest P, Wehrenberg B, Yu D (2005) Intraband relaxation in CdSe nanocrystals and the strong influence of the surface ligands. Journal of Chemical Physics 123: 074709
[27] Boudreaux DS, Williams F, Nozik AJ (1980) Hot carrier injection at semiconductor-electrolyte junctions. Journal of Applied Physics 51: 2158–2163
[28] Dijkhuis JI, Pol van der A, Wijn de HW (1976) Spectral width of optically generated bottlenecked (29 cm^{-1}) phonons in ruby. Physical Review Letters 37: 1554–1557
[29] Tilstra LG, Arts AFM, Wijn de HW (2007) Optically excited ruby as a saser: experiment and theory. Physical Review B 76: 024302
[30] Heitz R, Born H, Guffarth F, Stier O, Schliwa A, Hoffmann A, Bimberg D (2001) Existence of a phonon bottleneck for excitons in quantum dots. Physical Review B: Condensed Matter and Materials Physics 64: 241305
[31] Kral K, Khas P (1998) Absence of phonon bottleneck and fast electronic relaxation in quantum dots. Physica Status Solidi B: Basic Research 208: R5–R6
[32] Guyot-Sionnest P, Shim M, Matranga C, Hines M (1999) Intraband relaxation in CdSe quantum dots. Physical Review B: Condensed Matter and Materials Physics 60: R2181–R2184
[33] Rufo S, Dutta M, Stroscio MA (2003) Acoustic modes in free and embedded quantum dots. Journal of Applied Physics 93: 2900–2905
[34] Klimov V (2000) Optical nonlinearities and ultrafast carrier dynamics in semiconductor nanocrystals. Journal of Physical Chemistry B 104: 6112–6123
[35] Wang H, Mello Donegá de C, Meijerink A, Glasbeek M (2006) Ultrafast exciton dynamics in CdSe quantum dots studied from bleaching recovery and fluorescence transients. Journal of Physical Chemistry B 110: 733–737
[36] Htoon H, Cox PJ, Klimov VI (2004) Structure of excited-state transitions of individual semiconductor nanocrystals probed by photoluminescence excitation spectroscopy. Physical Review Letters 93: 187402
[37] Salvador MR, Hines MA, Scholes GD (2003) Exciton-bath coupling and inhomogeneous broadening in the optical spectroscopy of semiconductor quantum dots. Journal of Chemical Physics 118: 9380–9388
[38] Caruge J-M, Chan Y, Sundar V, Eisler HJ, Bawendi MG (2004) Transient photoluminescence and simultaneous amplified spontaneous emission from multiexciton states in CdSe quantum dots. Physical Review B 70: 085316
[39] Fisher B, Caruge JM, Zehnder D, Bawendi M (2005) Room-temperature ordered photon emission from multiexciton states in single CdSe core-shell nanocrystals. Physical Review Letters 94: 087403
[40] Schaller RD, Klimov VI (2004) High efficiency carrier multiplication in PbSe nanocrystals: implications for solar energy conversion. Physical Review Letters 92: 186601
[41] Schaller RD, Sykora M, Pietryga JM, Klimov VI (2006) Seven excitons at a cost of one: redefining the limits for conversion efficiency of photons into charge carriers. Nano Letters 6: 424–429
[42] Warner JH, Thomsen E, Watt AR, Heckenberg NR, Rubinsztein-Dunlop H (2005) Time-resolved photoluminescence spectroscopy of ligand-capped PbS nanocrystals. Nanotechnology 16: 175–179

[43] Clark SW, Harbold JM, Wise FW (2007) Resonant energy transfer in PbS quantum dots. Journal of Physical Chemistry C 111: 7302–7305
[44] An JM, Franceschetti A, Zunger A (2007) The excitonic exchange splitting and radiative life time in PbSe quantum dots. Nano Letters 7: 2129–2135
[45] Rijssel van J, Koole R, Mello Donegá de C, Meijerink A, unpublished
[46] Schaller RD, Petruska MA, Klimov VI (2005) The effect of electronic structure on carrier multiplication efficiency: a comparative study of PbSe and CdSe nanocrystals. Applied Physics Letters 87: 253102
[47] Pijpers JJH, Hendry E, Milder MTW, Fanciulli R, Savolainen J, Herek JL, Vanmaekelbergh D, Ruhman S, Mocatta D, Oron D, Aharoni A, Banin U, Bonn M (2007) Carrier multiplication and its reduction by photodoping in colloidal InAs quantum dots. Journal of Physical Chemistry C 111: 4146–4152
[48] Timmerman D, Izeddin I, Stallinga P, Yassievich IN, Gregorkiewicz T (2008) Space-separated quantum cutting with silicon nanocrystals for photovoltaic applications. Nature Photonics 2: 105–109
[49] Nair G, Bawendi MG (2007) Carrier multiplication yields of CdSe and CdTe nanocrystals by transient photoluminescence spectroscopy. Physical Review B 76: 081304(R)
[50] Itoh T, Iwabuchi Y, Ikehara T, Furumiya M, Katagiri N, Yano S, Iwai S, Edamatsu K, Gourdon C, Ekimov A (1999) Fundamental and nonlinear optical properties of semiconductor mesoscopic particles. Mesoscopic Materials and Clusters 31: 31–46
[51] Berstermann T, Auer T, Kurtze H, Schwab M, Yakovlev DR, Bayer M, Wiersig J, Gies C, Jahnke F, Reuter D, Wieck AD (2007) Systematic study of carrier correlations in the electron–hole recombination dynamics of quantum dots. Physical Review B: Condensed Matter and Materials Physics 76: 165318
[52] Lucey DW, MacRae DJ, Furis M, Sahoo Y, Cartwright AN, Prasad PN (2005) Monodispersed InP quantum dots prepared by colloidal chemistry in a noncoordinating solvent. Chemistry of Materials 17: 3754–3762
[53] Xiong G, Pal U, Serrano JG (2007) Correlations among size, defects, and photoluminescence in ZnO nanoparticles. Journal of Applied Physics 101: 024317
[54] Hirai T, Harada Y, Hashimoto S, Itoh T, Ohno N (2005) Luminescence of excitons in mesoscopic ZnO particles. Journal of Luminescence 112: 196–199
[55] Calcott PDJ (1998) The mechanism of light emission from porous silicon: where are we 7 years on? Materials Science & Engineering B: Solid-State Materials for Advanced Technology B51: 132–140
[56] Godefroo S, Hayne M, Jivanescu M, Stesmans A et al (2008) Classification and control of the origin of photoluminescence from Si nanocrystals. Nature Nanotechnology 3: 174–178
[57] Dovrat M, Goshen Y, Jedrzejewski J, Balberg I, Sa'ar A (2004) Radiative versus nonradiative decay processes in silicon nanocrystals probed by time-resolved photoluminescence spectroscopy. Physical Review B: Condensed Matter and Materials Physics 69: 155311
[58] Walters RJ, Kalkman J, Polman A, Atwater HA, de Dood MJA, Watson TJ (2006) Photoluminescence quantum efficiency of dense silicon nanocrystal ensembles in SiO_2. Physical Review B: Condensed Matter and Materials Physics 73: 132302
[59] Allan G, Delereu C (2007) Energy transfer between semiconductor nanocrystals: validity of Forster's theory. Physical Review B: Condensed Matter and Materials Physics 75: 195311
[60] Siebold H, Heber J (1981) "Discrete shell model" for analyzing time-resolved energy transfer in solids. Journal of Luminescence 22: 297–319
[61] Vergeer P, Vlugt TJH, Kox MHF, Hertog den MI, Eerden van der JPJM, Meijerink A (2005) Quantum cutting by cooperative energy transfer in $YbxY_{1-x}PO_4$: Tb^{3+}. Physical Review B 71: 014119
[62] Kagan CR, Murray CB, Nirmal M, Bawendi MG (1996) Electronic energy transfer in CdSe quantum dot solids. Physical Review Letters 76: 1517–1520
[63] Kagan CR, Murray CB, Bawendi MG (1996) Long-range resonance transfer of electronic excitations in close-packed CdSe quantum-dot solids. Physical Review B: Condensed Matter 54: 8633–8643
[64] Achermann M, Petruska MA, Crooker SA, Klimov VI (2003) Picosecond energy transfer in quantum dot Langmuir-Blodgett nanoassemblies. Journal of Physical Chemistry B 107: 13782–13787
[65] Franzl T, Koktysh DS, Klar TA, Rogach AL, Feldmann J, Gaponik N (2004) Fast energy transfer in layer-by-layer assembled CdTe nanocrystal bilayers. Applied Physics Letters 84: 2904–2906
[66] Crooker SA, Hollingsworth JA, Tretiak S, Klimov VI (2002) Spectrally resolved dynamics of energy transfer in quantum-dot assemblies: toward engineered energy flows in artificial materials. Physical Review Letters 89: 186802

[67] Wuister SF, Koole R, Mello Donegá de C, Meijerink A (2005) Temperature-dependent energy transfer in cadmium telluride quantum dot solids. Journal of Physical Chemistry B 109: 5504–5508
[68] Bruchez M, Moronne M, Gin P, Weiss S, Alivisatos P (1998) Semiconductor nanocrystals as fluorescent biological labels. Science 281: 2013–2016
[69] Chan WCW, Nie SM (1998) Quantum dot bioconjugates for ultrasensitive nonisotopic detection. Science 281: 2016–2018
[70] Alivisatos P (2004) The use of nanocrystals in biological detection. Nature Biotechnology 22: 47–52
[71] Willard DM, Carillo LL, Jung J, Van Orden A (2001) CdSe-ZnS quantum dots as resonance energy transfer donors in model protein-rotein binding assay. Nano Letters 1: 469–474
[72] Ebenstein Y, Mokari T, Banin U (2004) Quantum dot functionalized scanning probes for fluorescence-energy-transfer-based microscopy. Journal of Physical Chemistry B 108: 93–99
[73] Hohng S, Ha T (2005) Single-molecule quantum dot fluorescence resonance energy transfer. Chem Phys Chem 6: 956–960
[74] Clapp AR, Medintz IL, Mauro M, Fisher BR, Bawendi MG, Mattoussi H (2004) Fluorescence resonance energy transfer between quantum dot donors and dye-labeled protein acceptors. Journal of the American Chemical Society 126: 301–310
[75] Becker K, Lupton JM, Müller J, Rogach AL, Talapin DV, Weller H, Feldmann J (2006) Electrical control of Förster energy transfer. Nature Materials 5: 777–781
[76] Müller J, Lupton JM, Lagoudakis P, Koeppe R, Rogach AL, Feldmann J, Talapin D, Weller H (2005) Wave function engineering in elongated semiconductor nanocrystals with heterogeneous carrier confinement. Nano Letters 5: 2044–2049
[77] Lee J, Hernandez P, Lee J, Govorov AO, Kotov NA (2007) Exciton–plasmon interactions in molecular spring assemblies of nanowires and wavelength-based protein detection. Nature Materials 6: 291–295
[78] Nikoobakht B, Burda C, Braun M, Hun M, El-Sayed MA (2002) The quenching of CdSe quantum dot photoluminescence by gold nanoparticles in solution. Photochemistry and Photobiology 75: 591–597
[79] Fu A et al. (2004) Discrete nanoctructures of quantum dots/Au with DNA Journal of American Chemical Society 126: 10832–10833
[80] Shevchenko EV, Ringler M, Schwemer A, Talapin DV, Klar TA, Rogach AL, Feldmann J, Alivisatos P (2008) Self-assembled binary superlattices of CdSe and Au nanocrystals and their fluorescence properties. Journal of the American Chemical Society 130: 3274–3275

Fluorescence spectroscopy of single CdSe nanocrystals

By

John M. Lupton[1], Josef Müller[2]

[1]Department of Physics, University of Utah, Salt Lake City, Utah, USA
[2]Photonics and Optoelectronics Group, Physics Department and Center for NanoScience (CeNS), Ludwig-Maximilians-Universität München, Munich, Germany

1. Introduction

The true beauty of nature reveals itself in the delicate balance between order and disorder, between large and small scales. The leaves on a tree may appear to us as highly disordered at first glance, certainly once they have fallen to the ground; but their making relies on a high degree of order, a subtle interplay between physical processes and photochemical reactions in the light-harvesting complexes responsible for photosynthesis [1–3]. Indeed, this very process of photosynthesis has long caught the attention of researchers. Light from the sun is absorbed by pigments within such a complex, passed from molecule to molecule arranged in a particular order imposed by a protein scaffold, and funnelled to a sequence of reaction centres in which charge carrier separation, protonation, and ultimately the formation of glucose take place. At first sight these complexes appear suitably rigid and ordered [4]. Yet detailed theoretical investigations [2] along with high-resolution spectroscopy in the time [3] and frequency domain (i.e. single molecule spectroscopy) [5, 6] have revealed that a high degree of order – the spatial arrangement of pigments – is not the only horse nature places its bet on.

Consider the simplest case of fluorescence energy transfer from one molecule to another, a straightforward dipole–dipole coupling problem. As in antenna theory, the maximum coupling efficiency occurs if both the transmitter and receiver are in resonance. In a molecular picture, this translates into a spectral overlap between absorption and emission of donor and acceptor. If the individual pigments experience a high degree of local order, their optical transitions will be extremely narrow. However, some change in transition energies is inevitable from molecule to molecule due to miniscule differences in the local environment, isotopic effects, isomeric effects, and the likes [7]. The net result is that dipolar coupling breaks down, and excitation energy becomes localised [8]. This effect is akin to disorder localisation, well-known from transport theory. As in multiple trapping transport theory, however, dynamic disorder can help overcome the effect of static disorder. Temporal fluctuations in the local electronic structure of the individual molecular entity can actually promote microscopic fluorescence resonance energy transfer (FRET) [2]. In a light-harvesting complex, the protein scaffold ensures that all pigments are situated in the correct geometry, optimized by evolution. On the other hand, proteins are dielectrically dynamic objects, imposing a rapidly varying dielectric environment on the local pigments. Consequently, the effective electronic transitions broaden [6], improving FRET between adjacent pigments [2, 3].

The temporal fluctuations of nanoscale emitters and absorbers – spectral diffusion – constitutes the central theme of this chapter. Semiconductor nanocrystals are excellent systems to investigate random fluctuations in the emission properties of a light source in the single photon limit [9, 10]. Along with the intensively studied fluorescence intermittency [11–26] – fluorescence blinking – the seemingly random spectral fluctuations follow a well-defined pattern which can provide insight into either the immediate dielectric environment and nature of the emitting species or the overall nanoparticle shape [27, 28]. Indeed, suitable choice of the immediate surrounding of the semiconductor surface can even virtually suppress blinking [29, 30]. As nanocrystals are significantly larger than most dye molecules studied in single molecule investigations [31], spectral fluctuations correspond to changes in the properties of the primary excitonic species. These properties are controlled by surface states of the nanocrystal, which can acquire a net charge [32–35], in turn polarising the exciton. Whereas most blinking studies have been carried out at room temperature in the past, the most profound insight into these subtle fluctuations is available at cryogenic temperatures, where the elementary transitions become orders of magnitude narrower than the ensemble spectrum [36].

This chapter serves to review some of the recent progress made in studying and exploiting the spectral characteristics of individual CdSe nanocrystals. After discussing some of the basic experimental techniques, the spectral fluctuations of a particularly interesting class of CdSe nanocrystals are highlighted. Building on the seminal demonstration of shape control in inorganic semiconductor nanoparticles [37], these CdSe spherical quantum dots are surrounded by an elongated CdS shell and support, in contrast to purely spherical particles, linearly polarised emission [38]. The broken symmetry of the system results in spatial separation of the electron and hole wave functions, which can be controlled by external electric fields. Such fields are created by surface charges, which in turn control the optical properties such as the transition energy, transition line width, intensity, and electron–phonon coupling strength of the single emitter. The spectral jitter observed follows a well-defined pattern and exhibits either Gaussian or Lorenzian statistics. It is illuminating to draw parallels to seemingly unrelated classes of nanoscale light sources, conjugated polymers. Although the electronic properties of these carbon-based systems are very different, near identical spectral fluctuations are observed, which indicate the role of highly local dielectric effects. Finally, an example is presented on how to put the spectral properties of single, highly polarisable nanoscale light sources to work. Electric fields enable a tuning of the excitonic transition so that the dipole–dipole coupling strength in a microscopic, single particle FRET couple can be controlled. This electrical control of energy transfer constitutes a nanophotonic field-effect switch, or FRET gate, which, with suitable self assembly techniques, may well provide a building block for future nanophotonic circuitry. Most importantly, however, this electrical tuning of the elementary transition opens a window to control order effects in nominally disordered systems, by driving a microscopic transition in and out of resonance.

2. Why single particle spectroscopy?

Every nanocrystal is different. Sizes vary from particle to particle, shapes differ ever so slightly, and even the atomic composition, the isotopic distribution, becomes significant for extremely small particles. Most importantly, every particle experiences a slightly different dielectric environment. CdSe nanocrystals are stabilised internally through ionic bonds. However, the smaller the particle becomes, the greater the percentage of atoms forming the crystal sitting at the surface. These surface atoms are characterised by only partially saturated bonds and can therefore form dangling bonds. Many of these surface states are pacified sufficiently by organic ligands, which also serve to stabilise the nanocrystals in solution and prevent aggregation and precipitation [39]. The crucial influence of these ligands on the optical properties is well documented: removal of the ligands by, e.g. exposure of the particles to the electron beam of a transmission electron microscope results in non-emissive particles in which the fluorescence is quenched [40]. However, these ligands form a thermodynamic equilibrium with the particles in solution, continuously binding and unbinding, which necessitates a surplus of ligands. A simple experiment reveals this effect. When compared to organic dye molecules, nanocrystals are reasonably stable to environmental influences such as oxidation. However, the temporal stability of the fluorescence of a solution of nanocrystals depends on concentration. The stability of the emission intensity under constant illumination in part depends on the process of statistical aging, which constitutes the ensemble average of single particle blinking [41–43]. As this phenomenon follows a scale-invariant power-law distribution of the probabilities of detecting a particular duration over which the emission of the particle remains "on" or "off" under constant excitation [44], the probability for increased "off" periods increases within an ensemble of nanocrystals with time. In the absence of illumination, serial dilution of a nanocrystal solution by addition of pure solvent reduces the ratio of free ligands to particles, thereby ultimately lowering the overall surface coverage. The effective result is that nanocrystals can degrade in highly dilute solutions, a phenomenon which can be counterbalanced through the addition of further ligands.

In any case, both the charged surface of the nanocrystal and the surrounding ligand shell constitute a variable nanoscale dielectric environment, which can influence the elementary electronic transition and therefore lead to a difference in transition energy from one nanocrystal to the next. This distribution of transition energies between particles within the ensemble is referred to as inhomogeneous broadening [45]. Each particle, however, has its own intrinsic spectral width. The quantum mechanical limitation of spectral width Γ is given by the Heisenberg uncertainty principle through the radiative transition time τ, $\Gamma \geq h/4\pi\tau$. This relation between transition line width and transition time only applies in the limit of coherence loss through population loss, i.e. through an electronic transition. There are other conceivable routes to reducing the electronic coherence time – that is, the time, in which the excited electron wave function retains its phase – in an individual emitter, such as through electron–phonon scattering or any other form of nuclear motion. This homogeneous line width, controlled only by the loss of coherence in the excited state, constitutes the primary limitation to the spectral width of the emission and absorption of a single particle [46]. Long-phase coherence is particularly interesting for demonstrating quantum mechanical coupling between quantum dots, a prerequisite for optical quantum computing [47]. However, random fluctuations in time of the local dielectric environment of the particle provide a further form of inhomogeneous spectral broadening, spectral diffusion. Single particle measurements can therefore, in principle, offer access to the energetic *limitations* of homogeneous broadening [48]. Time-dependent studies of the single particle emission open a window to some of the spectral broadening processes.

3. Experimental approach

There are two principal routes to study the emission of a single quantum dot: either the optical collection volume is minimized, e.g. through the use of suitable spatial apertures imposed either on the emitters [49, 50] or on the detector as in the case of near-field scanning microscopy; or by reducing the number of emitters present in the far-field optical collection volume. The latter is rather facile to achieve in solution-based quantum dots (but not so in vapour phase, epitaxially grown nanostructures) by simple serial dilution, keeping in mind the limitations with regards to stability of the ligands mentioned above. As the spatial resolution of an optical microscope is roughly limited to the dimensions of the wavelength of light involved, spacing particles further than ~500 nm apart should allow optical spectroscopy to differentiate between the electronic properties of individual particles. Ideally, the particles are fixed in space, as particles in dilute solutions would drift in and out of the microscope focus. Fortunately, most particles are readily dispersed in common polar (e.g. poly(vinyl-alcohol)) or non-polar (e.g. polystyrene, poly(methyl-methacrylate), Zeonex) matrices, which can be spin-coated on quartz glass substrates to yield highly uniform films of optical quality, tens to hundreds of nanometers thick. These films are then studied using either a confocal scanning or a wide-field imaging microscope [24]. The experiments reported in the following were all carried out using a wide-field setup, which offers the key advantage of being able to record multiple emitters (i.e. microscope fluorescence spots) simultaneously [51]. In brief, the microscope used employed a long working distance (8 mm) microscope objective lens with a (comparatively small) numerical aperture of 0.55. This lens allowed the collection of light emitted by a nanoparticle film mounted in a cold-finger helium cryostat under vacuum. The nanoparticles were excited by an argon ion laser (typically at 488 nm), focused to a spot of approximately 100 μm in diameter incident at a non-vertical angle on the substrate. The fluorescence, collected by the microscope objective lens, was spectrally dispersed in a monochromator and recorded by a CCD camera of appropriate sensitivity.

Figure 1 illustrates the key properties and strengths of the experimental setup. Figure 1a shows the CCD image of a substrate covered with a highly dilute nanocrystal–polystyrene dispersion. The entrance slit of the detector was closed in horizontal direction such that in the collection plane along the y-axis typically only

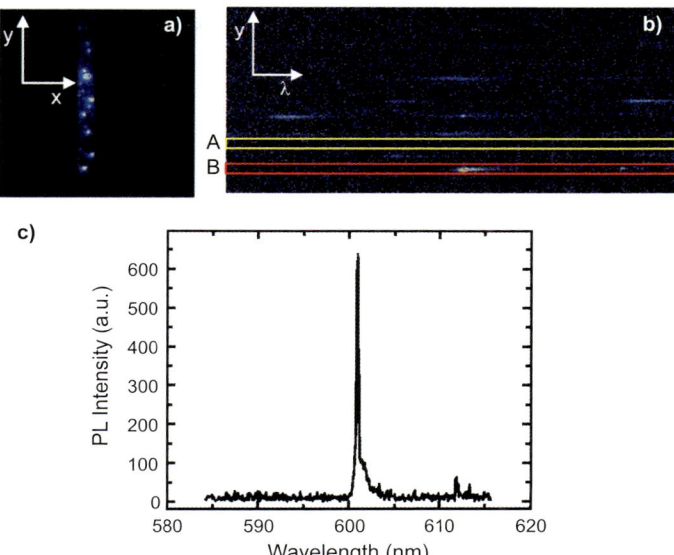

Fig. 1. Single particle detection and spectroscopy. **a** The spatial region of the microscope image projected onto the CCD camera is defined by a horizontal entrance slit so that only one single particle (seen as a bright spot) is visible in the horizontal direction. The image corresponds to a region of approximately 60 μm height on the substrate. The fluorescence light is subsequently passed through a monochromator and spectrally dispersed so that the *x*-component of the image now corresponds to wavelength λ **b**. The spatial information of the location of the particles is maintained in the *y*-direction. The regions A and B mark the background fluorescence and a typical single particle emission spectrum, respectively. **c** shows a typical single particle photoluminescence spectrum at 5 K. Adapted from [51]

one nanocrystal is detected at once, manifested by a fluorescence spot on the CCD camera. The typical spacing between single nanocrystals is 5–10 μm. In order to carry out spectroscopy on the single particle, the emitted light is dispersed spectrally in a grating spectrometer. Figure 1b shows exactly the same spatial image as recorded in (a), but now resolved by emission wavelength. Consequently, the *x*-axis of the emission now corresponds to the wavelength of emission, whereas the spatial information in *y*-direction is maintained. The emission intensity is encoded in the colour of the image. The image illustrates the role of the entrance slit in spectroscopy: if more than one particle is present in the *x*-direction of the image in (a), multiple fluorescence peaks or spectral broadening would be observed in (b). As a large area is excited by the impinging laser light, the imaging technique offers a facile way to correct for any residual background emanating from the matrix or the substrate. Choosing a region of interest (A) immediately besides the nanocrystal image (B) in Fig. 1b defines the fluorescence background, which can subsequently simply be subtracted from the emission spectrum. Figure 1c illustrates a typical example of a single particle fluorescence spectrum, recorded for a CdSe/CdS nanodot/nanorod at 5 K: a narrow transition is observed around 602 nm, which is somewhat broadened to longer wavelengths.

As one expects, this emission spectrum of the single particle at low temperatures is substantially narrower than the bulk emission. Figure 2a illustrates an ensemble

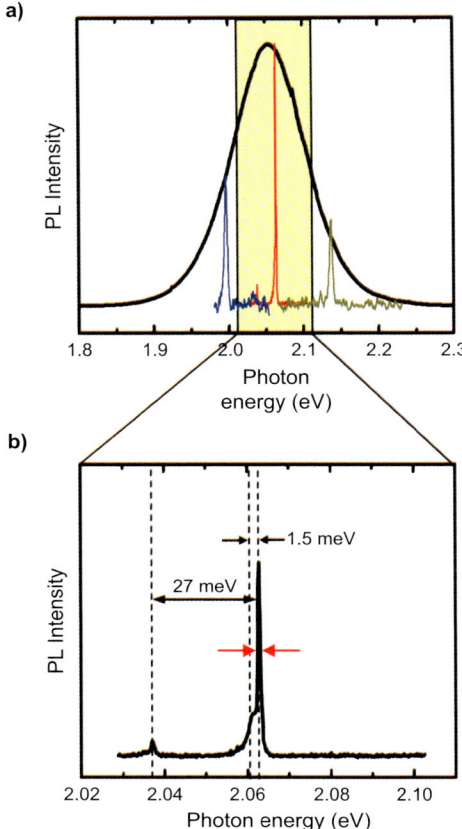

Fig. 2. Comparison of single particle line width to the inhomogeneous broadening of the ensemble. **a** Three characteristic single particle spectra, recorded at 5 K, superimposed on the ensemble fluorescence spectrum. **b** A typical low temperature fluorescence spectrum, showing a narrow zero-phonon line, an acoustic phonon side band, and an optical phonon progression. Adapted from [51]

emission spectrum at room temperature with three characteristic single particle spectra superimposed. The bulk emission spectrum constitutes the sum over many single particle spectra, recorded from different spots seen in the CCD image. The width of the ensemble spectrum is primarily defined by the energetic scatter from one particle to the next. Figure 2b displays a close-up of one of the emission spectra. The transition line width in this case is 800 µeV, over 130 times narrower than the ensemble spectrum. A hump is visible on the low energy side of the spectrum, characteristically offset by roughly 1–2 meV. This hump, which is well-known from spectral hole burning measurements [52], an alternative way of determining the elementary transition width of a particular energetic subset of the ensemble, corresponds to the acoustic phonon side band, or phonon wing. Interestingly, due to the finite size of the semiconductor nanocrystal, the acoustic phonon modes become quantized [53] rather than forming the continuum of states familiar from the

Debye approximation of heat capacities in solid state physics. The discrete acoustic phonons, a collective breathing motion of all of the atoms constituting the crystal, are typically observed around 1 meV and simply depend on the size of the crystal. Significantly offset from the narrow peak by approximately 27 meV is a further emissive feature, which corresponds to the optical phonon side band of the emission. This optical phonon is characteristic of the CdSe ionic bond, and immediately reveals that recombination, and thus electron–phonon coupling, occurs in the CdSe part of a CdSe/CdS core/shell nanocrystal. To a first approximation, the narrow transition can be referred to as a zero-phonon line, as it is purely electronic in nature. As will be discussed in the following, the line, which is still two orders of magnitude broader than spectra reported using the technique of resonant spectral hole burning [54], is significantly broadened by spectral diffusion. In contrast to spectral hole burning, however, single particle measurements immediately provide us with information relating both to the random scatter from particle to particle, as well as the intrinsic electronic properties and dynamics. Most importantly, spectroscopy on a single particle can enable us to relate far-field spectral information to the nanoscale physical properties of the particle such as its shape. As spectral hole burning probes a homogeneous subset of an inhomogeneous ensemble, shape effects are generally masked in amorphous systems through random orientations. It is also worth noting that fluorescence excitation spectroscopy is a further interesting technique to study the electronic structure of single nanocrystals [55]. Although all of the initial single molecule fluorescence experiments were carried out in excitation using a narrow-band, tuneable laser [7], nanocrystals are usually studied non-resonantly as the broad absorption spectrum of nanocrystals arising from a continuum of states makes it possible to observe the fluorescence from the relaxed state of lower energy. Remarkably, excitonic features can be discerned within the continuum of states of a single particle, providing insight into higher lying states [55].

4. Spectral dynamics of single nanocrystal emitters: signatures of particle shape

The following discussion will focus on a particularly interesting class of CdSe/CdS nanocrystals, in which a spherical CdSe core is capped on one end by an elongated CdS rod-like structure [38]. The transmission electron microscope image in Fig. 3 displays a typical structure of a single particle. The high degree of crystallinity of the nanoparticle is visible through the diffraction planes in the microscope image. One end of the particle is somewhat more bulky than the other and corresponds to the CdSe nucleus, with a slightly smaller band gap than the CdS structure growing out of one side on the $\{00\bar{1}\}$ facet of the CdS nanocrystal. The typical aspect ratio of these particles is 1:4, with lengths ranging between 15 and 20 nm. These nanocrystals of mixed dimensionality are particularly interesting for single particle experiments – and, as will be discussed later on, for FRET studies – as the absorption cross section of a particle increases with physical size. Because of the electronic structure of the particle – the CdS band gap is larger than the CdSe gap, forming a type I heterostructure – emission always occurs from the CdSe unit, as witnessed by the CdSe optical phonon seen in the single particle spectra. The

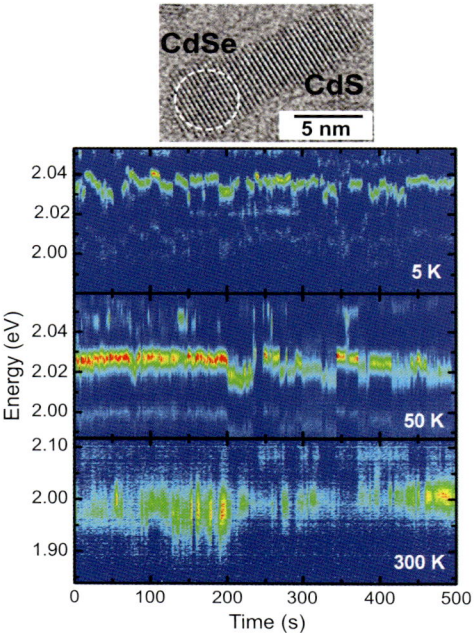

Fig. 3. Spectral diffusion in the fluorescence of single elongated CdSe/CdS nanocrystals at 5, 50 and 300 K. The photoluminescence spectra are plotted as a function of time in a two-dimensional representation with the intensity encoded in colour scale (blue: low intensity, red: high intensity). The temporal resolution of the data is 1 s. The transmission electron microscope image of a single nanocrystal is shown, with the CdSe core region indicated by a white circle. Adapted from [27]

absorption can therefore be enhanced without fundamentally changing the emission properties. Figure 3 illustrates three typical single particle fluorescence time traces, recorded in 1 s integration windows. The x-axis of the plots corresponds to time, whereas the y-axis indicates the energy of the emitted photon. Dark blue marks low emission intensity, and yellow-red indicates high intensity. The two-dimensional representation therefore displays the time evolution of the emission spectra. Three different temperatures are shown, which primarily control the spectral width and the temporal stability of the transition.

At first sight, the spectra appear to be rather similar for the different temperatures. However, as the temperature is raised from 50 to 300 K, the spectra broaden 30-fold and appear to exhibit more rapid fluctuations. The spectral time traces display both spectral jitter, which is the focus of the present discussion, and fluorescence intermittency, in which the emission suddenly vanishes and the intensity, marked by the colour code in the two-dimensional plot, returns to the background level. Whereas spectral transitions appear more continuous at room temperature, the low temperature traces suggest both continuous jitter and sudden, larger, spectral jumps [56]. Qualitative inspection of these time traces is limited, but as discussed in the following, there are a few facile analyses which can be applied to reveal a plethora of phenomena and correlations.

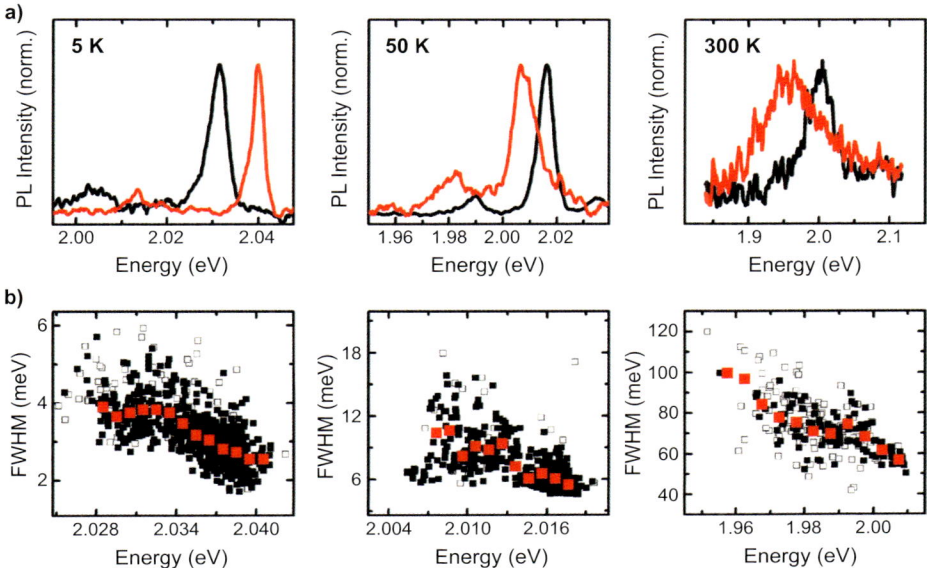

Fig. 4. Correlation between peak position and peak line width for three different particles recorded at 5, 50 and 300 K. **a** Two representative spectra taken from the traces in Fig. 3, illustrating a spectral broadening concomitant with the spectral red shift. **b** Correlation between spectral line width and peak position. Adapted from [27]

The first piece of information which can be derived from the spectral traces is the fluorescence peak energy and the spectral line width. Figure 4a displays two example spectra extracted at each temperature. As is particularly visible at room temperature, the lower energy spectra appear to correlate with a broader emission band. Figure 4b provides a more systematic analysis of the traces shown in Fig. 3 by plotting the spectral line width against the spectral peak position for each of the spectra comprising Fig. 3. The correlation in the data is unambiguous: as the emission shifts to the red, the spectrum broadens. The open symbols indicate the raw data, which clearly display significant scatter. The scatter can be reduced by limiting the consideration to data points (solid squares) for which only spectral shifts were considered which occurred on time scales longer than the measurement window of 1 s. The correlation can be further highlighted by displaying a binned average, as indicated by the solid squares in the figure.

Much of scientific methodology revolves around the identification of correlations, so a clear correlation between spectral line width and transition energy of the single particle must carry a physical meaning. It was soon realised in the spectroscopy of single nanocrystals [57–59], and indeed of single epitaxially grown quantum dots [60–62] and molecules in general [63], that the spectral properties of the individual emitter can be modified by fluctuations in the local dielectric environment. Nanocrystals carry their own local dielectric environment in the form of organic ligands and a large density of surface defects. Empedocles and Bawendi were able to relate the spectral shifts observed in single particle fluorescence as a function of time to electric field effects, by studying the influence of external electric fields on the emission properties at low temperatures [64]. Local charges on the surface of the nanocrystal result in an effective electric field [65, 66] which the exciton within the nanocrystal experiences. The charges therefore provide an additional potential which controls the electron-hole wave function overlap, and therefore the transition energy [62]. Core-shell nanocrystals are prototypical materials to study the quantum-confined Stark effect [67], in which the exciton gains a high degree of polarisability due to the ability to push one wave function into the core and the other into the shell under application of a field. To a first approximation, there are two principal routes by which local charges, accumulated on the surface of the nanocrystal, can influence the emission [64]. The appearance or disappearance of a surface charge will result in a sudden jump of the transition energy. On the other hand, once surface charge has been formed on the nanocrystal, it may redistribute with time, leading to a slight variation in distance to the exciton. Consequently, the exciton experiences a temporally varying electric field and thus displays a change in transition energy with time. If this change in transition energy occurs on time scales shorter than the measurement window (in the present discussion, typically 1 s), the spectral shift of the transition will manifest itself as a spectral broadening. In a perfectly symmetric particle, changes in the mean position of the surface charge should not influence the transition energy, but only changes in the surface charge density. Consequently, in a perfectly spherical nanocrystal there should not be a correlation between transition line width and peak energy, as remarked by Empedocles and Bawendi [64]. Most particles are, however, not perfectly spherical – least of all the CdSe/CdS heterostructure nanorods. Interestingly, similar spectral correlations were recently reported for much larger CdS nanorods which are outside of the regime of quantum confinement [68]. This observation suggests that emission in these large rods occurs from localised regions, which can then in turn become sensitive in their emission to surface charges.

Figure 5 illustrates a possible scenario for the physical origin of the correlation of line width with transition energy. Figure 5b summarises the electronic structure of the nanocrystals. The offset between conduction bands of the CdSe and the CdS is minimal so that the electron can delocalise from the CdSe core into the CdS shell. The CdSe valence band is, however, significantly lower in energy so that the hole of the exciton becomes localised in the CdSe – hence the CdSe optical phonon in the single particle emission. As a potential is applied to the particle ($F > 0$), the electron wave function can delocalise further into the CdS shell as the external potential screens the internal Coulombic attraction between electron and hole. This reduces the confinement energy of the electron–hole pair, leading to a red shift in the emission. Surface charges, as indicated in the cartoon in Fig. 5a, lead to the formation of an effective potential and thus to local electric fields. As the exciton is localised to one end of the elongated nanocrystal, in the CdSe core, the spatial position of the surface charge along the shell determines the transition energy. On the right hand side of Fig. 5a, the surface charge is situated far from the core. Consequently, the emission appears at the higher end of the spectrum. As the surface charge moves around the surface of the nanocrystal randomly in the course of the spectral measurement, different effective quantum-confined Stark shifts are probed. The envelope of these

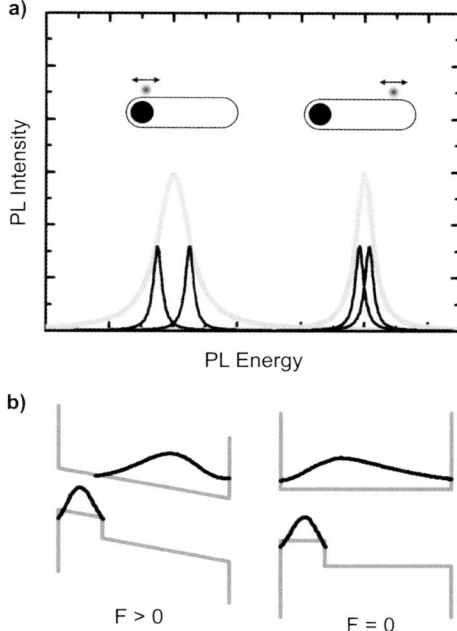

Fig. 5. Schematic representation of the influence of surface charge density on the emission from a single elongated nanocrystal. **a** Surface charge present at two different distances from the core of the nanocrystal, in which the hole of the exciton is localised, results in different magnitudes of a Stark shift in the emission. As the detection of the emission spectrum occurs over a finite time, spatial jitter in the charge density will result in spectral broadening. The closer the surface charge is located to the core of the nanocrystal, the stronger the red shift and the stronger the line broadening. **b** The electric field F induced by the surface charge also results in a perturbation of the electron–hole wave function overlap within the nanocrystal and thus of the radiative rate. A large local electric field leads to a displacement of the electron with respect to the hole and therefore reduces the wave function overlap and the radiative rate. Under the assumption of constant non-radiative decay, this change in wave function overlap results in a reduction in photoluminescence intensity. Adapted from [27]

different effective spectra seen within a particular temporal measurement window is indicated. This spectral envelop is significantly narrower if the surface charges are situated far from the CdSe core, assuming a random spatial fluctuation amplitude of the charges which is independent of the position on the nanoparticle surface. As the surface charges move closer to the core, the emission shifts to the red, but as the effective Stark shifts become larger, the overall envelope function broadens: the peak position correlates with the spectral width.

Although it was initially noted that such a correlation should not be observed in spherical nanocrystals [64], it was recently reported in room temperature spectroscopy [69]. However, as the transmission electron microscope images of [69] show, the term "spherical" is somewhat open to interpretation. From the cartoon in Fig. 5 it can be inferred that even the slightest breach in particle symmetry will lead to a correlation of line width with transition energy. The smaller the particle, the greater the sensitivity to a miniscule perturbation of shape. In addition, the internal field of the ionic lattice [70] of the nanocrystal should also play a role, although this aspect has thus far only received little attention. Spectral fluctuations can be equally influenced by fluctuations in the equilibrium coordinates of the atoms constituting the ionic crystal. This phenomenon is primarily electrostatic – or electrostrictive – in nature, and could also lead to a correlation of line width with energy, but only under the premise of a breach in spherical symmetry. Such a breach is conceivable through an atomic defect [71].

A simple electrostatic model can qualitatively explain the correlation observed [72]. Figure 6 shows a plot of line width–peak position correlations for three different temperatures, with the data accumulated over five particles each comprising a total of approximately 7000 spectra. The peak shift was determined by subtracting the peak position of the bluest spectrum recorded for the particle. The spectral range of 30 ± 10 meV over which peak shifts occur is virtually independent of temperature, although the average line width of the spectrum increases by almost two orders of magnitude with increasing temperature. For all temperatures, the spectral width

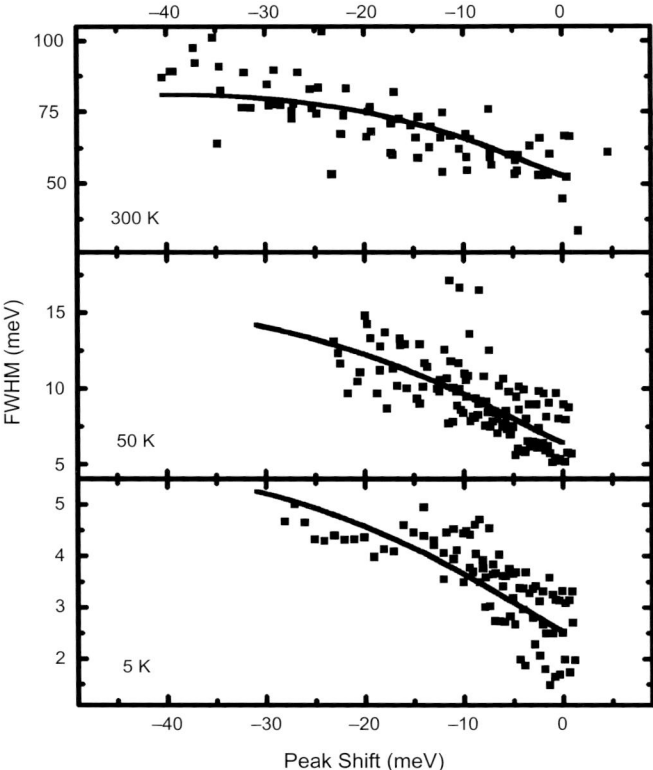

Fig. 6. Temperature dependence of line width–peak shift correlations. The data extracted for a total of five single nanocrystals are shown for each temperature. The solid line represents the quantum-confined Stark effect shift and resulting spectral broadening calculated for a single charge moving along the long axis of the nanocrystal. Adapted from [28]

varies by a factor of two during the spectral diffusion. The observation of correlated spectral diffusion at room temperature is somewhat surprising, yet fully consistent with the model proposed above. The spatial amplitude of surface charge oscillations, i.e. the average distance moved randomly by the charge distribution in a given time, should depend on temperature, but not the effect of the local field on the spectral position as induced by the quantum-confined Stark effect. The larger oscillations simply imply that the spectra broaden with increasing temperature. The maximum and minimum fields that surface charges can exert on the CdSe core, however, must be independent of temperature, and therefore the overall energy range over which drifts can occur during a measurement [28].

Although one can only speculate on the nature of the surface charges, the consistency of the phenomenological model can be verified by estimating the field effect of a single charge migrating along the long axis of the particle. Assuming a polarisability of the nanorods comparable to spherical NCs [64] and following the equations in [64, 72], a single charge must be separated from the core by about 3 nm to induce a Stark shift of 40 meV. This spatial separation is in good agreement with the diameter of the nanorods, as seen in the microscope image in Fig. 3. By moving the charge along the nanocrystal and considering a constant fluctuation in position one may estimate the dependence of line width on relative peak position, as shown by the solid lines in Fig. 6. As the quantum-confined Stark effect is minimal when the charge is situated at the far end of the nanorod (i.e. far from the core), yet a finite line width is measured, a line width offset has to be considered due to residual broadening at zero peak shift. Sources of this residual broadening may be charge fluctuations on time scales shorter than the measurement; or electronic dephasing induced by strong phonon scattering. The free fitting parameter in this simple model is the magnitude of spatial oscillations, which is determined as 0.2, 0.6 and 2.2 nm for 5, 50 and 300 K, respectively [28]. The 5 K value is of the order of the typical distance between ligands, respectively surface defects [73]. The simplest conceivable model of one-dimensional charge transfer along the nanorod thus provides both qualitative and quantitative agreement with the data. As it is the overall change in field which is of primary interest and not the actual charge distribution, the dynamics of the charge distribution can be well approximated by the dynamics of a single point charge. In addition, it is helpful to note that even if the spatial amplitude of charge oscillations were not constant along the rod, the experimental correlation would still provide a measure of charge *migration*, although this would complicate the microscopic model.

Evidently, the room temperature spectra are broader than the maximal achievable spectral shift due to the Stark effect of ~40 meV. Even if the oscillations of surface charge were so strong that the charge were completely annihilated and created during the detection window of 1 s, it is unlikely that local field-induced oscillations account for the observed spectral widths of 100 meV. This implies a second contribution to line broadening at room temperature, which is due to efficient exciton scattering on optical phonons, representing a very rapid dephasing process [74]. Whereas the low temperature single nanocrystal lines are inhomogeneously broadened by spectral diffusion and do not provide insight into dephasing processes, the observed spectral diffusion dynamics suggest that the opposite is true for room temperature measurements. To a first approximation, one may conclude that the narrowest spectra at room temperature (~50 meV width) are predominantly homogeneously broadened due to the phonon scattering-induced ultrafast dephasing [28].

Although the correlation between transition energy and fluorescence line width is clearly discernible in a wide range of single particle experiments, the data points in the correlation plots scatter widely. This scatter results from the fact that the fluorescence traces also contain temporal information, besides the spectral information relevant to the correlation. If the proposed model of a surface charge-induced change in electron–hole wave function overlap is correct, a spectral shift should also go hand in hand with a change in the radiative rate and thus a change in emission intensity. In addition, the optical phonon seen in the single particle spectra at 27 meV to the red of the zero-phonon line should be influenced by a change in local electric field, as it results from the vibration of a the polar CdSe bond. The coupling strength of electrons to such Fröhlich phonons changes with electric field so that a red shift in the emission energy should result in an increase in phonon coupling strength in the emission [45]. Such more detailed correlations are not immediately apparent from the raw data traces. However, there is a straightforward means of improving the data quality by performing a manipulation akin to boxcar averaging to remove the redundant temporal information from the fluorescence trace. To do this, the spectra in the trace are sorted with respect to the peak position, thereby removing the time dimension [27]. Figure 7 shows a plot of the normalised raw data, simply sorted by emission peak energy (which is marked on the x-axis). Without performing any further manipulation, two important features are immediately obvious from this trace: as the peak energy shifts to lower energy, the spectra become broader, giving the two-dimensional representation a funnel-like appearance. In addition, the

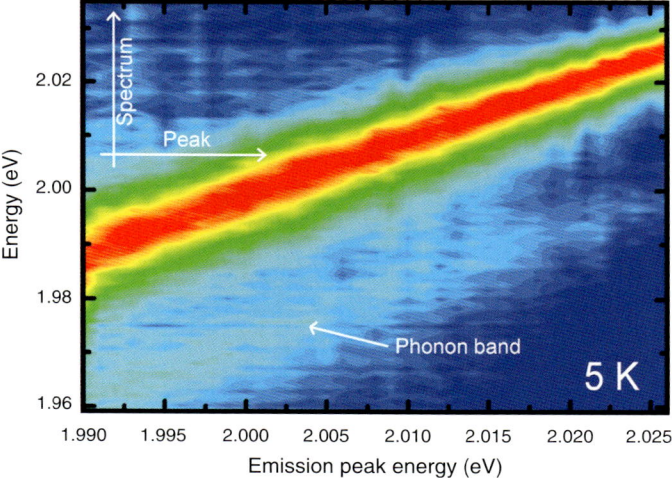

Fig. 7. Boxcar averaging of a single particle fluorescence trace to remove the temporal noise. The photoluminescence spectra of one single nanocrystal are shown. These were recorded during the spectral diffusion process and sorted by peak energy, plotted in the two-dimensional representation of spectrum versus peak energy, and smoothed in an energy window of 1 meV. As the peak energy shifts to the red, the optical phonon side band increases in intensity and the spectra broaden, giving rise to a funnel-like shape of the plot. Adapted from [27]

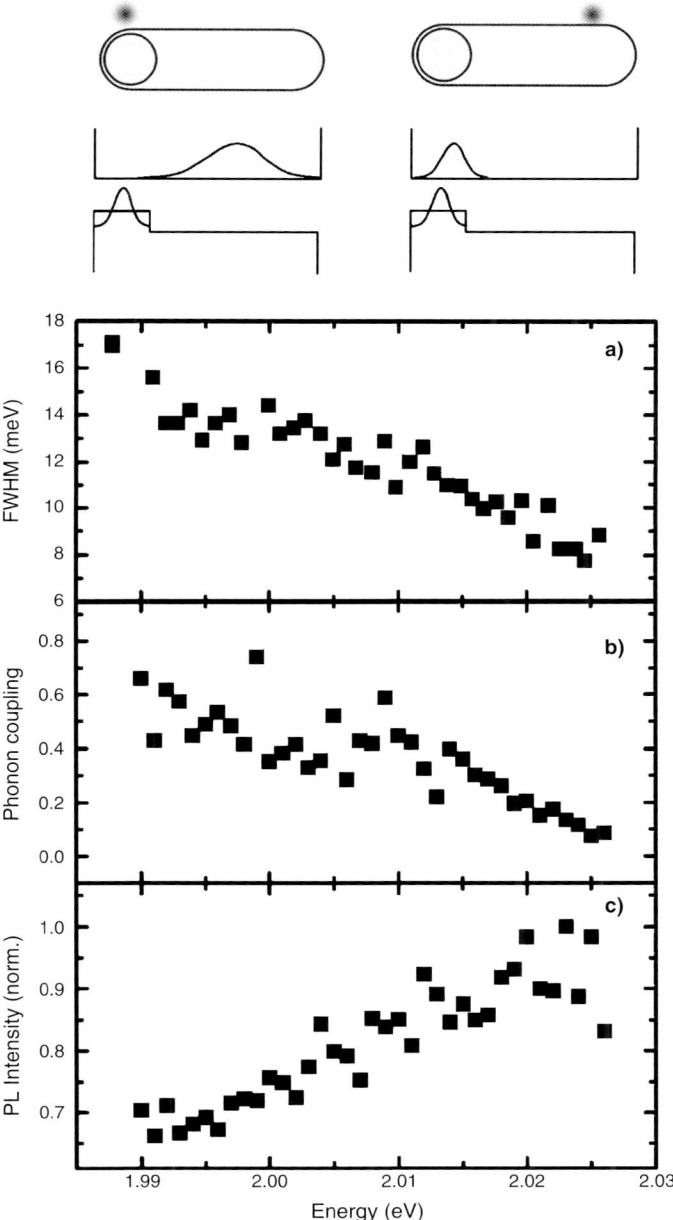

Fig. 8. Correlation between line width, phonon coupling strength and photoluminescence intensity for the sorted spectral trace measured at 5 K and shown in Fig. 7. Sorting of the spectra allows a statistical binning of energetically related spectra *prior* to fitting to the line shape and extracting the line width, peak position, phonon coupling and emission intensity. The scheme reiterates the mechanism responsible for the correlations, explained in more detail in Fig. 5. As the hole is confined to the CdSe core whereas the electron is free to penetrate the CdS shell, surface charges modify the overlap of the electron and hole wave functions, which in turn control line width, phonon coupling, peak energy and photoluminescence intensity. Adapted from [27]

phonon side band in the emission broadens and increases in intensity relative to the zero-phonon line as the emission shifts to the red.

The sorted spectra with the time dimension removed allow a much more facile and accurate extraction of the spectral parameters such as the line width, the phonon coupling strength (the Huang-Rhys factor), and the emission intensity. In addition, the sorted spectra can be smoothed further by binning spectra related in energy, i.e. by taking the spectral average over a sliding window of width of, e.g. 1 meV prior to determining the line width, peak position, etc. by fitting a Lorenzian curve to the spectrum. Figure 8 illustrates the data extracted from a sorted trace of one single particle. Under these conditions, the line width–peak position correlation forms an almost straight line with minimal scatter. As expected from the raw data shown in Fig. 7, the phonon coupling increases monotonously as the emission shifts to the red. Most importantly, the average single particle emission intensity decreases as the emission shifts to the red. All three of these observables are perfectly consistent with the simple surface charge model put forward above and reiterated in the cartoon in Fig. 8.

Charging in quantum dots is known to influence the transition energy. In large, vapour phase-grown quantum dots, charged exciton transitions become visible [75, 76]. In smaller colloids, charging can vastly promote Auger recombination, thereby leading to a quenching of the emission. This phenomenon has been studied intensively using time-resolved pump-probe spectroscopy [77]. Auger recombination has also been made responsible for the blinking of the fluorescence of single nanocrystals [56]. If a charge carrier is ejected from the nanocrystal following photoexcitation, the particle becomes charged, and consequently non-radiative Auger recombination competes with spontaneous emission [78]. If, however, the rate of spontaneous emission is accelerated by, for example, placing the nanocrystals on suitable plasmonic substrates which enhance electromagnetic coupling and therefore the radiative rate [79], emission from the charged exciton can become visible. It is important to distinguish between discrete charging events *within* the quantum dot and continuous redistribution of charges in the *vicinity* of the excitonic species. As demonstrated by Neuhauser et al., a blinking event in the nanocrystal emission correlates with a discrete spectral jump, whereas continuous spectral fluctuations occur at almost constant emission intensity [56]. The elongated nanocrystals allow a more detailed quantification of the role of surface charge fluctuations, i.e. overall changes in the charge of the nanocrystal. The net magnitude of surface charge need not change, but merely its position, in order to influence the emission intensity of the single particle.

The spectral meandering can be further analyzed by returning to the raw data traces. By making the differentiation proposed by Neuhauser et al. [56], we can identify the mean spectral shift between subsequent measurements $\Delta E = E_{\text{peak}, n+1} - E_{\text{peak}, n}$ and associate this with either a blinking event or a period of constant emission intensity. The spectral shift ΔE is plotted in a histogram. As there are long periods in the emission which are uninterrupted by an intermittency, the statistics are significantly better for the continuous spectral drift (the spectral jitter) than for the discrete spectral changes (the spectral jump). Figure 9 displays histograms for these two cases as derived from one single fluorescence trace of an individual particle (the 5 K trace shown in Fig. 3). As expected, the spectral jump

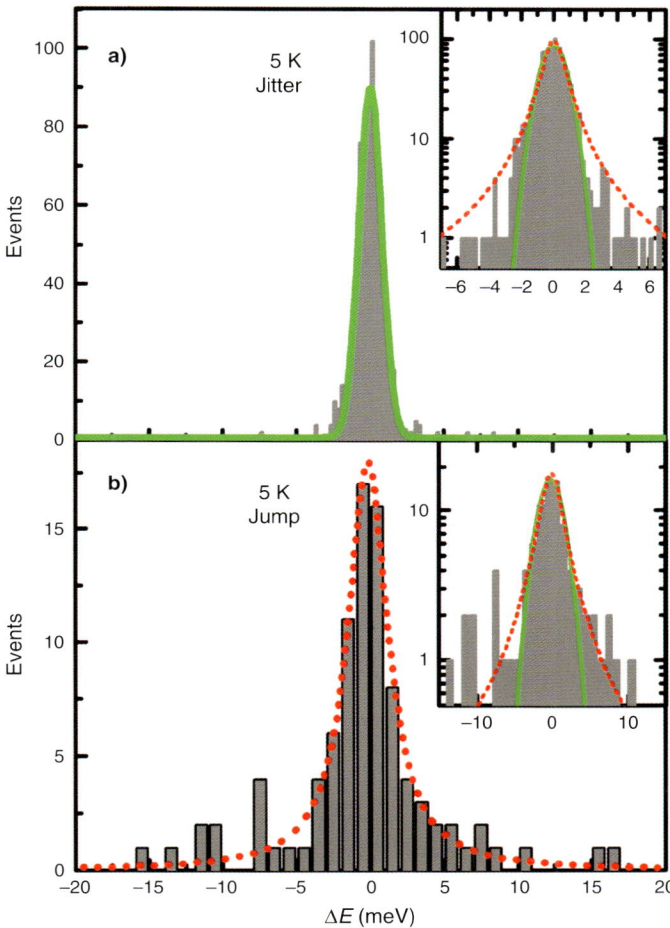

Fig. 9. Spectral diffusion statistics of the photoluminescence trace shown in Fig. 3a. The histograms illustrate the difference in energy $\Delta E = E_{\text{peak},n+1} - E_{\text{peak},n}$ between two subsequent spectral peaks recorded in the trace. **a** Overall trace consisting of 800 events. **b** Events interrupted by fluorescence intermittency (total of 90). The insets show the histograms on a logarithmic scale with a Gaussian (solid line) and a Lorenzian (dotted line) superimposed. Adapted from [27]

histogram is significantly broader than the spectral jitter histogram. A Gaussian function is superimposed on the spectral jitter histogram, which clearly provides a good match. In contrast, the spectral jump histogram appears to be more readily described by a Lorenzian function. The insets show the same histograms on a logarithmic scale, with both Gaussian and Lorenzian functions superimposed. These results illustrate that spectral diffusion in such nanoscale emitters follows a universal distribution of events. Both jump and jitter distributions increase slightly with increasing temperature (as the temperature is raised from 5 to 300 K), but depend strongly on the incident power [27, 45]. Whereas the jump distribution appears to follow a sublinear dependence on excitation density, the spectral jitter increases linearly [27]. The intensity dependence illustrates that spectral diffusion is a primarily optically driven process [63], resulting from the dissipation of excess photon energy, i.e. the energy difference between the absorbed and emitted photon.

The spectral jitter, which is resolvable on the time scale of 1 s of the experiment, is most likely related to the line width of the single particle spectrum and thus to random spatial fluctuations or oscillations of the surface charge density [64]. The overall histogram of the spectral jitter should therefore contain information on where the surface charge responsible for the quantum-confined Stark effect is located with respect to the exciton in the nanoparticle core. Assuming that the random spatial fluctuations of the surface charge in time, which are responsible for spectral broadening, are independent of the actual location of the charges on the nanocrystal surface, the width of the spectral jitter histogram itself should correlate with the overall spectral red shift. The underlying assumption appears reasonable as the fluctuations in surface charge are most likely due to a redistribution of the population of trap states [80, 81], which should exhibit comparable kinetics along the long axis of the nanoparticle. The way to test for evidence that spectral broadening originates from spectral jitter on time scales which cannot be resolved by the experiment is to study the spectral jitter distribution within different regions of the line width–peak position correlation. Figure 10a shows a typical correlation, which was arbitrarily dissected into three regions. Within these three spectral regions, the spectral jitter histogram is extracted, using the methodology discussed above. Figure 10b displays the spectral jitter histograms for the three regions, superimposed with Gaussian functions. It is clearly seen that as the emission shifts to higher energies, the individual spectral lines narrow, but so does also the actual spectral jitter distribution. The mean spectral shift from one measurement to the next is reduced as the emission moves to the blue. The higher energy emission corresponds to surface charges located at the greatest distance to the emitting core. As the surface charges move closer, the same random spatial fluctuation leads to a greater spectral shift ΔE from one measurement to the next. The closer the surface charges are to the emitting core, the greater the effect of a slight lateral redistribution of the charges on the transition energy will be. Consequently, the further away the charges are from the core, the more uniform the spectral jitter appears. Figure 10c shows the histogram of the most distant region labelled "3", plotted on a logarithmic scale. The distribution is accurately described by a Gaussian function over two orders of magnitude in event frequency, illustrating that the local fluctuations in electric field responsible for the emission dynamics are purely random in nature.

The elongated shape of the nanoparticle therefore allows a direct spatial tracking of surface charges, leading to correlations of peak position, line width, phonon coupling strength, emission intensity, and spectral jitter. An optical technique, which is, in principle, limited to a spatial resolution of a few hundred nanometers, can

Fig. 10. Dissection of the spectral diffusion of a single nanocrystal at 5 K into three regions, depending on the magnitude of the Stark effect. **a** The line width–peak energy correlation appears continuous and is arbitrarily divided into three equally large regions. **b** Spectral jitter ΔE histograms of the continuous spectral diffusion (blinking events and subsequent discrete spectral jumps were discarded) of the three regions indicated in panel **a**. **c** ΔE distribution of Sect. 3 shown on a logarithmic scale with a Gaussian superimposed, which clearly describes the histogram over two orders of magnitude. Adapted from [27]

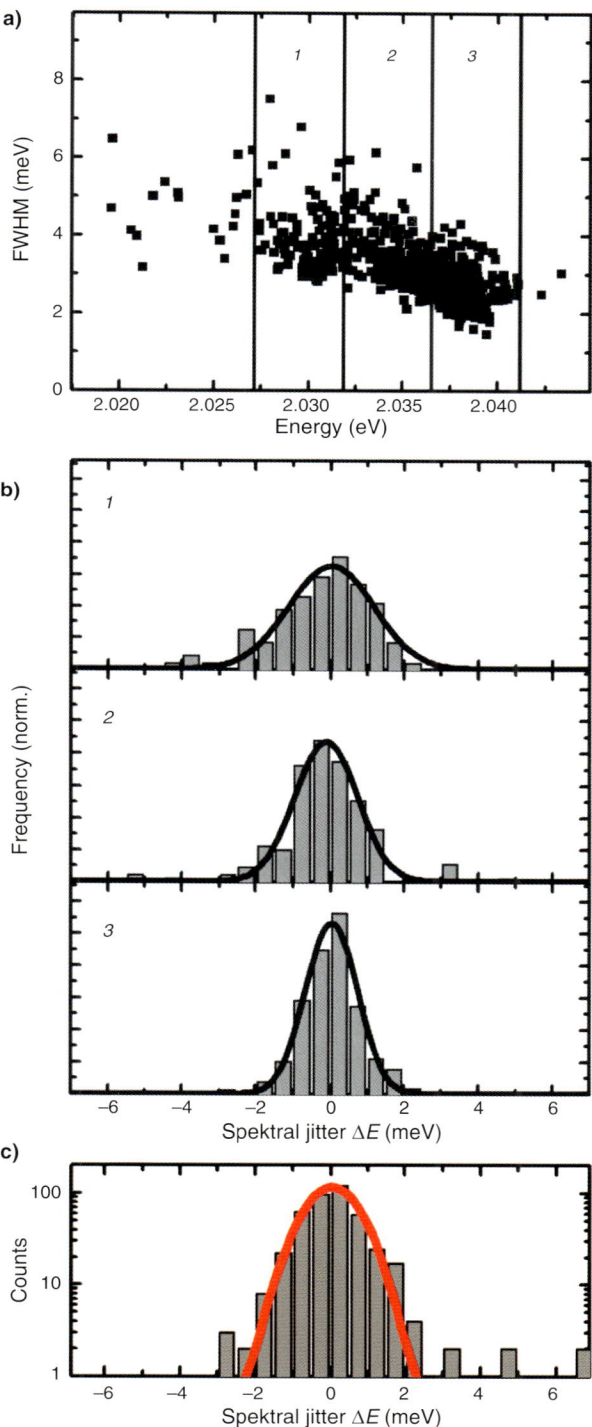

therefore be used to derive information on charging-related phenomena on length scales of a few nanometers. Random spectral diffusion need not necessarily be a detrimental thing. As discussed below, it can influence excitation energy transfer. The spectral jitter has also been used in quantum optical experiments to drive the excitonic transition in and out of resonance with a photonic mode, providing signatures of strong exciton–photon coupling in a single emitter [82].

An open question is how the clear spectral dynamics correlate with the fluorescence lifetime of single nanocrystals. It is conceivable that spatial redistribution of surface charge leads to a change in both radiative and non-radiative decay rates of the single nanocrystal [83]. Time-resolved fluorescence spectroscopy will provide ultimate insight into how the wave function overlap of the exciton is modified by external charges, and most importantly whether changes in surface charge density do modify non-radiative decay dynamics. Most of the present investigations do indeed suggest that blinking events correlate with a shortening of the fluorescence lifetime due to an increase in non-radiative decay [84–87], but to date it has not been possible to correlate this directly with the low-temperature fluorescence spectroscopy.

5. Universal spectral fluctuations in nanoscale systems

The spectral fluctuations observed in the single nanoparticle emission suggest some level of generality, so the question naturally arises whether similar fluctuations can be observed in different material systems. The fluctuations most likely originate from a spatial redistribution of charges, i.e. an optically driven electron transfer process. Dynamic electron transfer phenomena are well known from photochemical studies of molecules and surfaces [88], but are also observed in the noise of electrical transport measurements on small systems [89, 90]. Under certain conditions single molecules have been found to exhibit similar blinking dynamics to nanocrystals [91], suggesting some unifying behaviour. Interestingly, very similar spectral fluctuations are observed in the emission of a single chain of the conjugated polymer MEH-PPV [92, 93], a prototypical material system used in organic light-emitting diodes and photovoltaic devices. The excited state structure of such a macromolecule is very different to that of a quantum dot. The excitation in the polymer is highly anisotropic, the intramolecular bonds are covalent rather than ionic, and the dielectric constants are much lower so that the Coulombic interaction between electron and hole is stronger. Figure 11 displays the fluorescence jitter of such a single-polymer molecule, emitting in the spectral range around 530 nm at 5 K. Discrete spectral jumps are observed along with a pronounced spectral jitter of the primary transition line, which has a width of approximately 0.5 nm. The histogram of spectral diffusion, which does not differentiate between jump and jitter as in the case of Fig. 9, exhibits a pronounced Gaussian peak corresponding to the continuous spectral jitter. Side lobes are observed around ±20 meV, which result from the discrete spectral jumps. The histogram of spectral diffusion is evidently similar to that observed for the nanocrystals, but the typical jumps are approximately twice as large.

On the one hand, the spectral dynamics of single nanocrystals do not appear to depend strongly on the immediate dielectric environment [24, 69], and are merely controlled by the immediate density of surface charge on the nanoparticle. On the

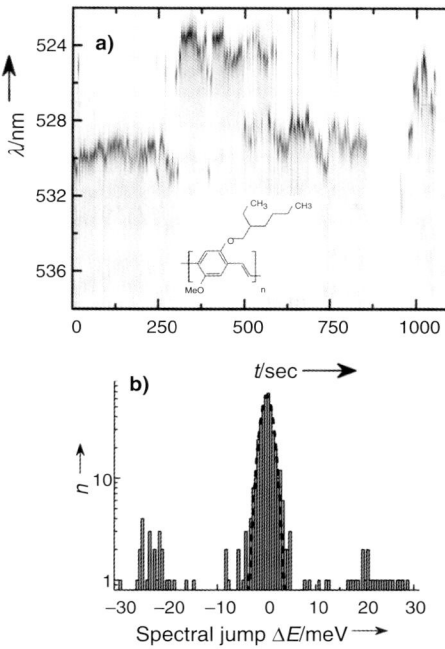

Fig. 11. Spectral diffusion in single chains of a conjugated polymer. **a** Single chain fluorescence of the polymer MEH-PPV as a function of time at 5 K, measured with a temporal resolution of 2 s. The emission spectrum is shown in a grey scale representation with darkening tones corresponding to increasing intensity. Switching between seemingly alike emissive states is observed along with a strong spectral jitter. By fitting a Lorenzian line to the individual spectra, both the spectral width and the spectral position can be extracted. **b** Histogram of the energy difference ΔE between two consecutive spectra, shown on a logarithmic scale. The distribution is accurately described by a Gaussian of width 2.4 meV. The large jumps in the spectral trace lead to a second group of events at ± 22 meV. Adapted from [93]

other hand, very similar temporal fluctuations are observed in rather different distinct nanoscale light sources, such as the conjugated polymer MEH-PPV. An interesting comparison can be drawn between the polymer and the nanocrystal. In effect, the polymer also possesses surface charge, harboured by the backbone substituents which interface the conjugated π-electron system with the outside world. One way to test this hypothesis is to carry out Stark spectroscopy of single chains by monitoring the molecular fluorescence under an electric field [94]. For the case of MEH-PPV, this experiment reveals that a linear Stark shift of the emission, indicative of the presence of a permanent polarisation, is only observed when the highly polarisable π-system is oriented orthogonal to the external electric field. The conclusion drawn from this result is that the polar alkoxy groups linking the side chains to the conjugated backbone induce a permanent polarisation within the molecule. Molecules without such polar substituents display a smaller Stark shift [94]. Interestingly, spontaneous fluctuations of the magnitude of the Stark shift, i.e. the magnitude of polarisation of the molecule, have also been observed for some material systems [95], indicating dynamic charge redistribution events. In general, the polarisation of a nanoscale entity is an extremely important parameter in understanding its interaction with individual charges and external electric fields. Conjugated polymers, for example, are used in light-emitting diodes, field-effect transistors, and photovoltaic devices. In a light-emitting diode, the question is how a charge carrier is actually injected from the electrode into the highest occupied or lowest unoccupied orbital of a molecule. Over two decades of research it has become clear that the injection efficiency is not merely defined by an offset of the energy of the molecular orbital with respect to the work function of the metal [96]. A dipole, oriented orthogonally to the polymer chain, induces an internal electric field which can either increase or reduce the externally applied field. Injection of charge carriers from the electrodes therefore only occurs into suitably oriented molecules. The effect is equally important in field-effect transistors. More so, such devices are based on a field-induced (electrostatic) polarisation of the medium. If the semiconducting layer is already polarised due to chemical substitutions or the formation of the trapped charges within the vicinity of the conjugated segment, an electric gate field will not be able to switch the conductivity sufficiently. Finally, molecular polarisations could be rather beneficial to charge separation, the elementary process in photovoltaic devices. Molecular semiconductors are characterised by vast exciton (electron–hole) binding energies, typically of order 0.5 eV [97]. Dipoles may promote carrier dissociation and then guide the optically generated charges on a suitable pathway towards the electrodes under the action of the built-in field.

These examples serve to illustrate the subtle complexity of describing the interaction of a nanoscale electronic system with the outside world, but also underline the occurrence of some universal signatures. Noise arising from random charge fluctuations is important in transport spectroscopy of semiconductor nanosystems [90] and can even induce decoherence in quantum information systems such as Josephson junction qubits [98]. The random spectral fluctuations, associated with charging phenomena, are therefore relevant to a wide range of physical systems, and even find analogues in biophysical relaxation dynamics [99].

6. Control of single particle emission by electric fields

The previous discussion illustrated that the remarkable optical dynamics of single quantum dots can be understood in terms of temporally varying electric fields. However, quantitative confirmation that it is indeed electric fields and the quantum-confined Stark effect which are responsible for the spectral dynamics, can only be derived from actually applying external electric fields to the particles. This approach was initially taken by Empedocles and Bawendi in 1998, who demonstrated that the spontaneous spectral shifts observed in the fluorescence can be reproduced entirely by external electric fields [64]. Similar effects have also been discussed for epitaxially grown quantum dots [100]. The authors defined two types of local charge effects: periodic fluctuations in surface charge; and polarisation of the particle through (semi-)permanent rearrangement, i.e. generation or annihilation, of surface charge [64]. The unique system of elongated semiconductor nanocrystals with broken dimensionality now allows us to address the question of the role of particle shape in the fundamental interaction with electric fields [101]. As will be shown later on, this interaction can then be exploited to design a new category of energy transfer-

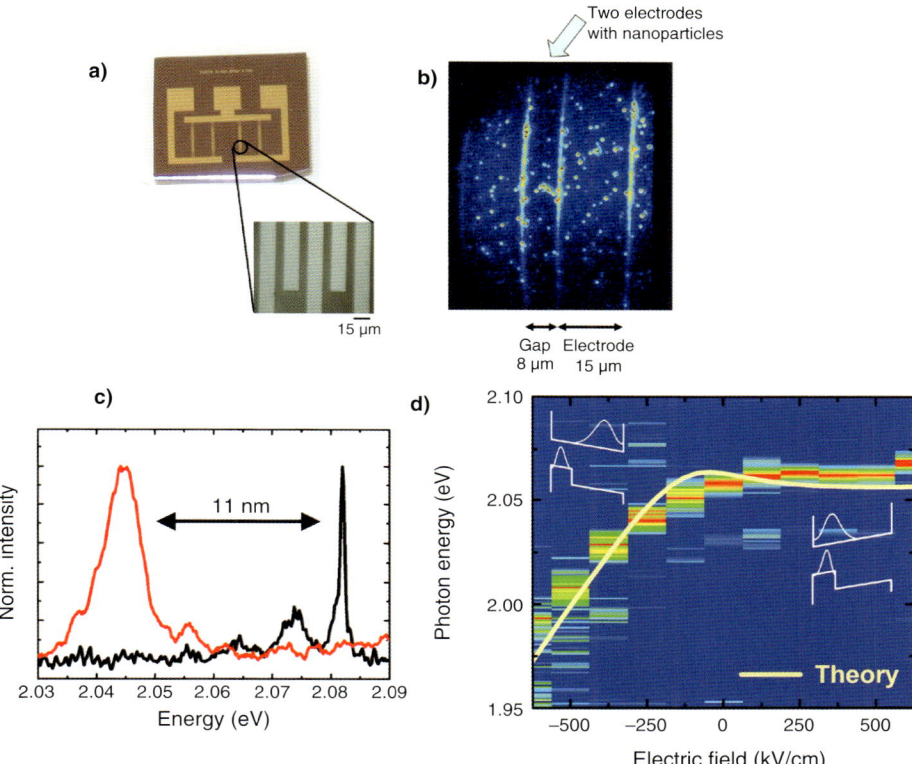

Fig. 12. The quantum-confined Stark effect in the emission of single elongated nanocrystals. **a** Photograph of the interdigitated finger electrode structure used to apply large lateral electric fields of up to 0.5 MV/cm. **b** Image of the electrode structure as seen beneath the fluorescence microscope. The spots correspond to emission from single nanocrystals. To study field effects, nanocrystals are chosen which are located in between two electrode fingers, which are clearly identified in the image through enhanced light scattering. **c** Quantum-confined Stark effect in the single particle emission at 5 K. The normalized emission spectra of a single nanocrystal in an external electric field show a large shift to the red by approximately 11 nm as the bias applied is changed from +200 to −500 V. **d** Normalised fluorescence spectra as a function of electric field applied, indicating a highly asymmetric electric field effect due to the asymmetry of the nanoparticle shape and composition. Negative fields can move electron and hole apart in the structure, as indicated in the inset showing the band structure and the calculated wave functions, whereas positive fields barely modify the confinement. The solid line shows the results of calculations using the effective mass approximation in a selfconsistent field method. Adapted from [101]

based optoelectronic devices [102]. In addition, it will be shown that Stark spectroscopy provides direct insight into the shape of the nanoparticle, information which is hard to access by other means.

Figure 12 summarizes the experimental approach to perform Stark spectroscopy on single elongated nanocrystals. The nanocrystals are spin-coated in their polymer solution onto an array of interdigitated finger electrodes, a photograph of which is shown in Fig. 12a. These fingers have a spacing of 8 μm and enable the application of uniform electric fields of over 0.5 MV/cm. It is not entirely trivial to switch from the conventional microscope samples to the finger electrodes due to the different surface wetting properties of the lithographically prepared electrodes, requiring an adjustment of nanoparticle concentration and spinning speeds. Figure 12b illustrates a microscope image of the nanoparticles deposited on top of the finger electrodes. Fluorescence spots are observed both between the electrodes and on top of the aluminium fingers. As one would expect from spin-coating a highly dilute, non-viscous solution, accumulation of particles occurs near the edges of the fingers. Stark spectra are only considered from particles situated at the centre between the two electrodes. Figure 12c displays a typical Stark shift spectrum, recorded under an applied bias of +200 V (black) and −500 V (red). Evidently, the electric field can strongly modulate the fluorescence spectrum, leading to a shift of the emission maximum by over 11 nm. As will be shown later on, this is a significant spectral shift, many times the transition line width, which can be exploited in applications. In contrast to spherical nanocrystals, which typically exhibit a small quadratic (isotropic) Stark shift superimposed on a strong linear Stark shift depending on the polarity of the surface charge [64], suitably oriented elongated nanocrystals generally display a highly anisotropic Stark effect, as shown in Fig. 12d. A spectral shift is only observed for negative electric fields, not for positive fields. This effect is a direct consequence of the spatial asymmetry of the nanoparticle, which outweighs any influence of polarising surface charges. The quantum-confined Stark effect leads to a red shift of the excitonic transition when the electron (or hole) wave function can penetrate the quantum-confining barrier layer. This is only the case for negative fields in the example given. There is a substantial energetic barrier for holes between the CdSe core and the CdS shell, whereas the electron is effectively isoenergetic in the two materials. Application of a negative field for a particle with the CdSe core pointing towards the cathode results in the electron within the exciton being pushed out of the CdSe core. This lowers the quantum mechanical confinement and results in a spectral red shift. Reversal of the field does not alter the transition energy significantly, as the electron and hole cannot be pushed arbitrarily close together due to an increase in the correlation energy with decreasing carrier spacing [101]. A relatively straightforward effective mass Hamiltonian, solved using an iterative selfconsistent field method and described in [101], reproduces the experimental observations accurately as shown by the solid line in Fig. 12d. Some approximations have to be made in this comparison, however, as the surrounding matrix and the substrate partially screen the electric field. The effective electric field experienced by the particles is therefore somewhat smaller than the mere ratio of the potential applied to the electrodes and the electrode separation. The sketches inset in Fig. 12d illustrate the calculated wave functions of electron and hole in the nanoparticle. A direct consequence of the separation of electron and hole through application of an electric field is the reduction in wave function overlap, which determines the radiative rate. Assuming that the non-radiative rate remains constant, reduction in the wave function overlap should therefore result in a decrease in emission intensity. Note that the reduction in emission intensity was also described above for the case of spontaneous rearrangements of surface charges, and could clearly be correlated with the change in emission wavelength as shown in Fig. 8 [27]. The externally induced spectral and intensity fluctuations are, however, significantly larger than the spontaneously occurring phenomena.

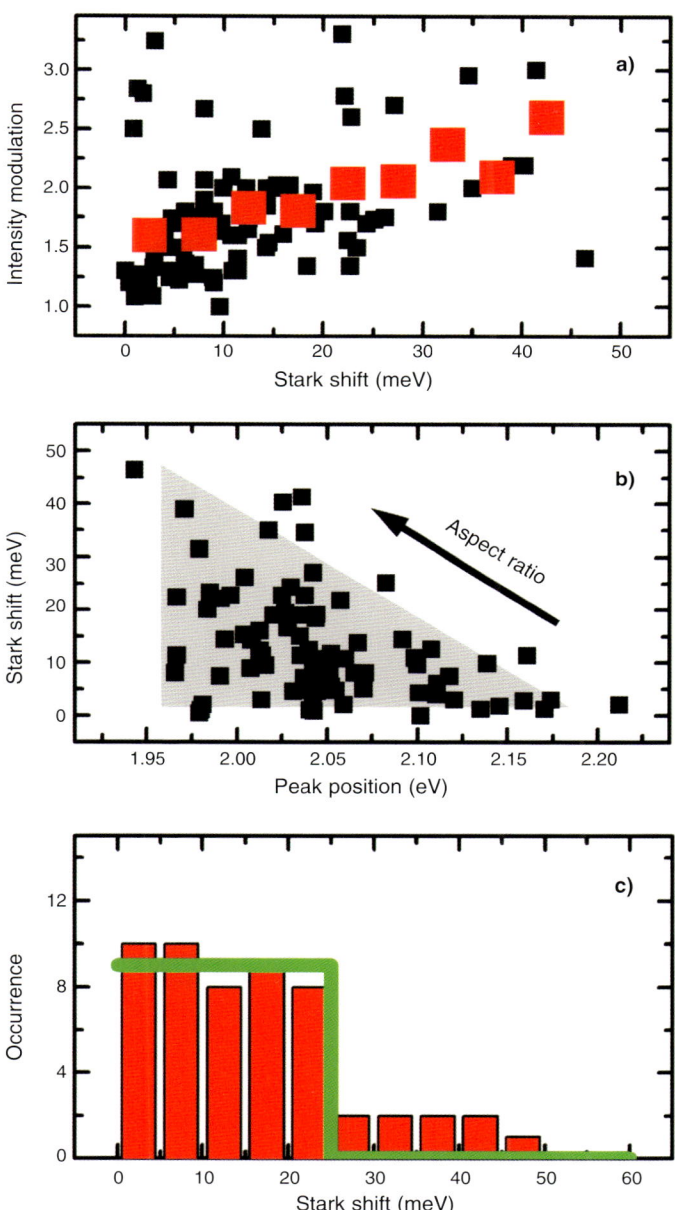

Fig. 13. Statistics of the quantum-confined Stark effect for 88 single nanoparticles. **a** Correlation between the maximum Stark shift and the overall intensity modulation (black squares). Large squares indicate 5 meV average bins. **b** Statistical distribution of the QCSE magnitude as a function of emission wavelength. The grey triangle highlights the increase of the QCSE with the decrease in emission energy. This decrease in energy corresponds to an increasing aspect ratio and thus an increasing electron penetration volume. The triangular scatter arises because of the random particle orientation. **c** Frequency of occurrence of a particular Stark shift illustrating an isotropic spatial distribution of the particles with respect to the field. The green line indicates the statistically expected distribution. Adapted from [101]

The average spectral shifts per unit field are several times greater for the elongated nanoparticles than for spherical particles [64]. As the elongated particles are larger, the overall effect of surface charge on the spectral characteristics is weaker than for spherical particles, in particular when compared to core-only spheres [64]. The influence of a large external electric field can therefore be described accurately without needing to include surface charge effects, which only serve to induce a particular electric field offset to the spectral shift. Nanoengineering of the electronic structure of the particle allows us to induce a new functionality, in this case an enhanced quantum-confined Stark effect.

The Stark effect is anisotropic in these anisotropic particles and can be used to extract information on the shape and shape distribution of the nanocrystals. Although transmission electron microscopy provides access to the shape distribution of the particles, it cannot yield information on the microscopic distribution in composition and in particular on three-dimensional shape. Stark spectroscopy probes the polarisability of the nanoparticle, which is a three-dimensional quantity directly related to the volume of the particle. Figure 13 provides detailed statistics of the modulation of single particle photoluminescence by an electric field of 350 kV/cm. The electric field leads both to a spectral shift (to the red) and to a modulation in intensity (a reduction), defined as I_{max}/I_{min}. For the 88 particles shown in the figure it is apparent that the intensity modulation correlates directly with the Stark shift. The further electron and hole can be moved apart within the nanocrystal, the larger the Stark shift, the greater the intensity modulation. The large squares in the figure indicate the average binned over a region of 5 meV to reduce the statistical scatter in the data points. The scatter in the Stark shift and in the intensity modulation can have two origins: either the particles are not aligned suitably with respect to the electric field; or the polarisable volume varies strongly from particle to particle. These two effects can be distinguished. The polarisable volume must correlate with the particle size, which is in turn defined by the aspect ratio. The larger the particle, the weaker the overall quantum confinement, the further in the red emission occurs. Figure 13b provides a plot of the observed Stark shift against the zero-field fluorescence peak position, which is expected to correlate directly with the size of the nanoparticle. It is seen that the larger the particle is (i.e. the further in the red the emission occurs), the greater the maximum observed Stark shift is. However, the particles are oriented randomly with respect to the electric field, so if a particle has its long axis orthogonal to the field only a very weak spectral modulation will be recorded [101]. Consequently, the scatter of points in the plot is described by the shape of a triangle. To address the question of orientation, we can now consider the most polarisable particles, i.e. the particles in Fig 13b with a transition energy smaller than 2.05 eV. Figure 13c shows a histogram of the frequency of a particular Stark shift for these largest particles. The histogram resembles a flat step, switching at a Stark shift of 25 meV. The histogram can be accurately reproduced by semiempirical calculations [101] by assuming a random distribution of the particles in the plane with respect to the electric field, as indicated by the green line.

Semiconductor nanocrystals are highly polarisable so that their emission characteristics can be controlled by external electric fields, making them suitable building blocks for optoelectronic devices. Engineering of the shape breaks the particle symmetry so that only a certain subset of nanoparticles respond to electric fields. On the one hand, one would expect the highly anisotropic Stark shifts to average out in the ensemble; no change of photoluminescence with electric field should be observed in a bulk film [64]. However, temporal modulation of the electric field allows one to pick out a particular shape subset of nanocrystals, namely just those nanocrystals which display the strongest electric field response. This enables fluorescence modulation and the Stark effect to be observed at room temperature in the ensemble [103]. In brief, a thin film consisting of a blend of nanocrystals in an inert polymer matrix is deposited between two vertical electrodes. Application of an electric field quenches the fluorescence of the ensemble, as is readily observed. Time-resolved (gated) spectroscopy allows one to study both the intensity and the spectral dynamics. If the electric field applied to the device is removed after a duration of one microsecond, much of the initially quenched fluorescence is recovered, leading to a pronounced fluorescence burst. The anisotropic shape of the nanocrystals therefore provides a pathway to storing excitons in semiconductor nanostructures, over time scales well beyond the typical durations of radiative decay. This is made possible by reversibly converting the direct, emissive exciton into an indirect, dark state [104]. Interestingly, the fact that the temporal gating technique accesses a particular subset of nanocrystals allows the observation of the quantum-confined Stark effect in the ensemble emission at room temperature [103]. This room-temperature demonstration of an electric field effect illustrates the power of low temperature single particle spectroscopy in identifying new application areas of nanoscale semiconductors.

7. Single nanocrystals as a probe of excitation energy transfer in disordered systems: the FRET gate

Much of the research interest in semiconductor nanocrystals has been driven by their characteristics as bleach-resistant nanoscale light beacons [105]. Single molecule fluorescence opened an entirely new dimension to the physical characterisation of biological systems [106] and has enabled experimenters to observe the hybridisation of single DNA strands, folding of proteins, and even the nature of motion of various molecular motors [107, 108]. Optical experiments in the limit of linear excitation are confined to length scales comparable to the wavelength of light, dimensions much larger than a typical displacement of a molecular entity. However, by studying the combined optical response of two fluorescent species, the spatial resolution of optical microscopy can be enhanced dramatically [109]. Labelling a DNA strand with two different fluorescent molecules, a donor and an acceptor, can provide information on hybridisation of the DNA. Provided an electronic resonance exists between the two labels, i.e. the donor can pass its excitation energy on to the acceptor given sufficient spectral overlap between the donor emission and the acceptor absorption, a small rearrangement in the spatial orientation or separation of the donor with respect to the acceptor will result in a substantial change in the acceptor emission intensity. Such experiments assume that donor and acceptor labels can be excited selectively, i.e. that the donor can be excited optically without photopumping the acceptor. On the other hand, the donor should absorb as much light as possible to maximize the fluorescence of the beacon while minimizing residual and undesirable background fluorescence. Semiconductor nanoparticles have very broad absorption spectra, with the absorption increasing continuously as the wavelength of light is reduced below the optical gap. In addition, nanocrystals are larger than most molecular dyes and can hence absorb more light energy per particle. Semiconductor nanocrystals are therefore interesting as labels for biophysical investigations [110–115]. As will be discussed in the following, the FRET properties of nanocrystals can even be used to learn more about the nanoparticle material characteristics and its interaction with the surrounding.

FRET can be likened to transmission and reception of radio waves on the nanoscale. Only when transmitter and receiver are in resonance, i.e. the transmitting and absorbing antennae and oscillating circuitry are tuned to have the same frequency, can energy and information be transferred. In addition, FRET is a near-field phenomenon. As light passes through a block of glass and is reflected at the boundary to air due to total internal reflection, only the travelling part of the light-field is considered. However, press a second block of glass against the first block of glass, and light transmission through the thin air gap will be observed. As in the analogy of quantum mechanical tunnelling, the solution to Maxwell's equation simply becomes real, i.e. exponentially decaying, in the region of total internal reflection. An index-matched material can enable the wave solution to become imaginary and therefore propagating again, in effect tunnelling the light across a non-propagating gap. Electrodynamic interactions over this gap are extremely sensitive to distance. Efficient coupling between transmitter and receiver can occur over the gap, i.e. for the real, evanescent wave solution of the wave equation.

Most considerations of FRET in the past have only taken the spectral overlap between donor emission and acceptor absorption in the ensemble into account. FRET is, however, a fundamentally microscopic process. A complete understanding of FRET therefore requires that the microscopic spectral overlap between the absorption of one single acceptor unit and the emission of one single donor unit are considered. As discussed above, going from the ensemble of nanocrystals to single particles and lowering the measurement temperature dramatically reduces the transition line width. If nanocrystals are used as donors in a FRET experiment, one may expect the strength of resonant incoherent dipole–dipole coupling to change with temperature. This change will depend on the properties of the acceptor transition. Typically, one may expect that the absorption line width of a single acceptor will also narrow with decreasing temperature. One single donor will therefore not be able to pass excitation energy to any arbitrary acceptor, but only to proximal acceptors with suitable transitions. The density of these acceptors will decrease as the temperature is lowered. On the other hand, as the temperature drops, FRET becomes much more selective and can therefore be used to probe a certain subset of the ensemble. Fortunately, the electronic transitions of nanocrystals are tuneable through the quantum-confined Stark effect. We therefore expect to be able to push an individual donor–acceptor pair in and out of resonance.

Figure 14 summarizes the concept of electrically controllable Förster-type energy transfer between a single semiconductor nanorod and a suitable dye acceptor [102]. As FRET becomes selective at low temperatures to a small subset of the acceptors,

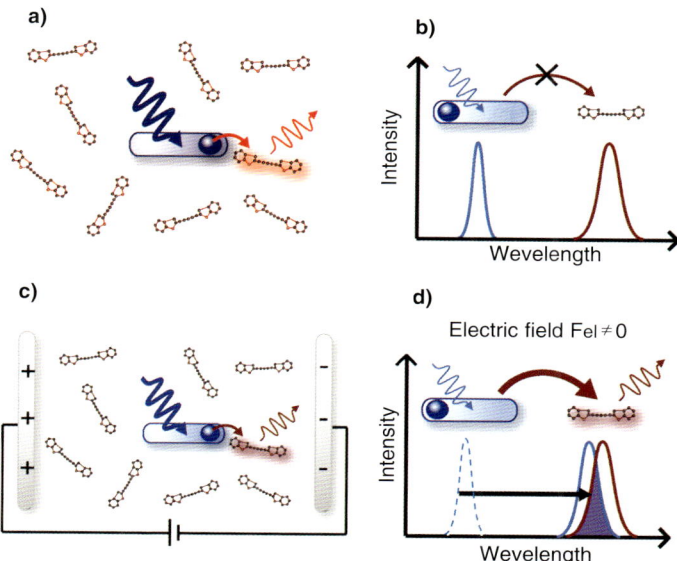

Fig. 14. Electrical control of energy transfer in a single FRET couple consisting of an absorbing dye molecule and an emitting nanocrystal. **a** The nanocrystals are dispersed in a dye film so that the nanocrystal concentration by far exceeds the concentration of dye molecules. Each nanocrystal should therefore, in principle, have an adjacent acceptor molecule. **b** At low temperatures, however, the electronic transitions of donor and acceptor are too narrow to enable a sufficient spectral overlap, required for efficient FRET. **c** Application of an electric field to the nanocrystal–dye mixture can facilitate FRET by shifting the emission of the nanocrystal into resonance with the absorption of an adjacent dye molecule **d**. Adapted from [102]

we can consider the nanocrystal donor in a mixture with dye molecules at higher concentration. The nanocrystal donor with the much larger absorption cross section (in the present case, for the choice of excitation wavelength of 458 nm and the cyanine dye derivative used as an acceptor, the absorption ratio is over 1:1000) acts as a form of excitonic optical nanoantenna, absorbing the incident laser radiation and passing it on selectively to a dye molecule in its vicinity [116]. The nanocrystal nanoantenna is therefore able to pick a single dye molecule out of a seemingly homogeneous film of dye molecules, effectively forming a subdiffraction optical probe addressed in the optical far-field. A typical nanocrystal will not display suitable spectral overlap with the dye emission as the transitions narrow at low temperature, as indicated schematically in Fig. 14b. Application of an electric field can remedy this deficiency, driving the nanocrystal donor into resonance with the dye acceptor (Fig. 14c, d) so that electrically controlled FRET occurs.

The experimental implementation of this FRET switch is demonstrated in Fig. 15. A periodic electric field is applied to a mixture of nanocrystals and dye molecules. Figure 15a shows the modulation of the fluorescence of a nanocrystal in a plot of fluorescence wavelength against time. Turning the electric field on and off leads to a red shift of the nanocrystal emission by approximately 10 nm. This spectral shift can drive the nanocrystal into resonance with the absorption of a nearby dye acceptor. Figure 15b displays the fluorescence of such a dye molecule as a function of time,

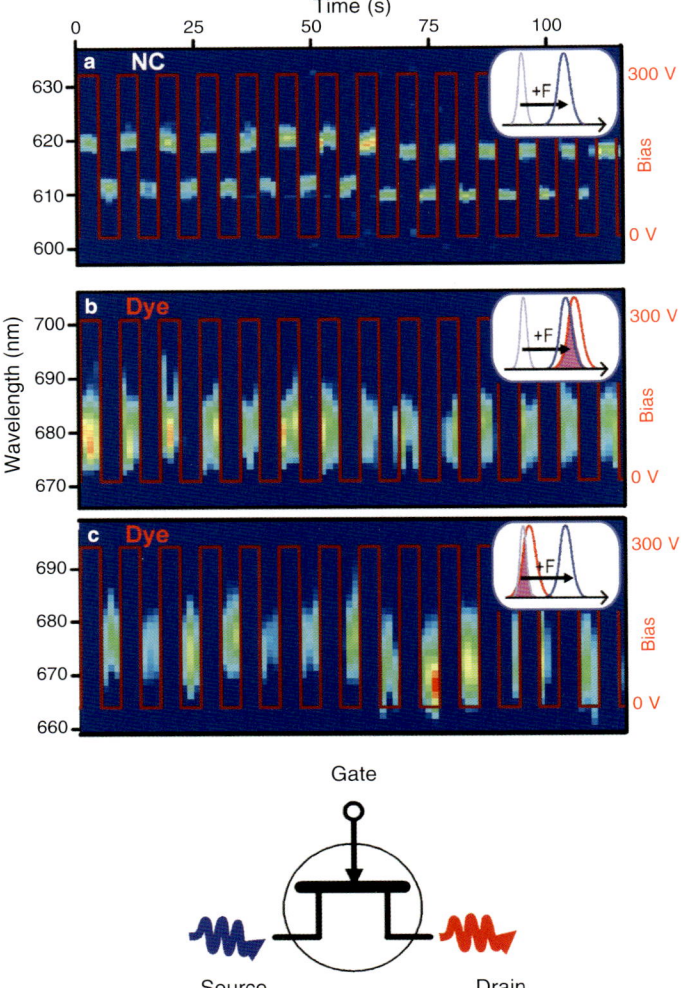

Fig. 15. Electrical control of energy transfer. **a** A periodically modulated electric field shifts the emission a typical nanocrystal by 10 nm. This shift in emission spectrum controls the FRET efficiency from nanocrystal to dye, depending on the zero-field spectral overlap of nanocrystal donor and dye acceptor. **b** If the adjacent dye molecule is not in resonance with the nanocrystal emission at zero-field, the applied electric field can switch FRET on, thereby making the dye fluorescence appear. Note that the dye spectrum is distinct from the nanocrystal spectrum: it is shifted to the red to approximately 680 nm and significantly broadened. **c** If donor and acceptor are in resonance at zero-field, application of an electric field can destroy the resonance, thereby turning the acceptor emission off. The overall phenomenon corresponds to a form of a field-effect switch, in which the electrical gate – the Stark effect – controls the excitation energy flow from source (nanocrystal) to drain (dye molecule). Adapted from [102]

recorded at a temperature of 50 K. Note that the dye spectrum is significantly broader than the nanocrystal spectrum and shifted to the red by over 60 nm. In contrast to the nanocrystal emission, the dye luminescence does not exhibit any spectral shift with applied bias. However, the dye emission is switched on and off depending on the magnitude of the electric field. Application of the field drives the nanocrystal into resonance with the dye so that excitation energy is transferred – the dye molecule lights up. As in the case of the nanocrystals (see Fig. 2a), every single dye molecule has a distinct transition energy as every molecule has a slightly different microscopic conformation which controls its electronic structure. A certain dye molecule may therefore turn out to be in resonance with the nanocrystal donor in the absence of an electric field. In this case, which is illustrated in Fig. 15c, the electric field drives the nanocrystal out of resonance with the dye acceptor; the dye fluorescence vanishes under application of a field.

This concept of voltage-switchable energy transfer can be thought of as a nanophotonic transistor in the form of a FRET gate, as illustrated in Fig. 15. A voltage gate, the quantum-confined Stark effect, controls the flow of excitation energy from the source (the nanocrystal absorber of external radiation) to the drain (the dye emitter of radiation to the outside world). It is now up to materials chemists to devise routes of linking multiple such logic elements to each other to construct versatile nanophotonic circuitry. This will be an interesting challenge, as it is already hard to link precisely one molecule to one single nanocrystal [117], which is physically much larger. As an aside, this FRET gate also constitutes an exquisite sensor of the local dielectric environment, which controls the electric field experienced by the nanocrystal. The FRET functionality in effect serves to amplify the quantum-confined Stark effect: the spectral shift of the nanocrystal by ~10 nm results in a shift of the emission of the hybrid couple from the nanocrystal to the dye molecule, i.e. by over 60 nm. The recent demonstration of the possibility to sense the photonic environment, i.e. the photonic mode density, with a single nanocrystal [118], provides an additional avenue of applications for tuneable FRET couples. It is also worth noting that electrical pumping of single nanocrystals was recently demonstrated [119, 120]. In combination with lateral electric fields which tune FRET, such a device could offer a voltage tuneable light-emitting diode for single photons on demand.

The FRET gate can also be used to gain insight into the disorder limitation of energy transfer, which is as important in biological light-harvesting systems as it is in synthetic energy conversion devices. The electric field can effectively be used to map out the inhomogeneous broadening of the local acceptor ensemble. Interestingly, spectral diffusion as discussed above can reduce the constraints on resonance matching of donor and acceptor. In addition, thermally activated spectral broadening (which may be due to increased electron–phonon coupling, accelerated spectral diffusion, or even accelerated electronic dephasing) can enhance microscopic dipole–dipole coupling [2]. It was found that the FRET gate experiments were ideally performed at temperatures around 50 K. Below this temperature, the electronic transitions particularly of the acceptor become too narrow so that spontaneous spectral diffusion leads to a random temporal modulation of the FRET efficiency [116]. For higher temperatures, on the other hand, the Stark shift becomes too small with respect to the transition line width, diminishing the FRET gate effect.

It is often assumed that spectral overlap in the ensemble is a sufficient criterion for FRET to occur. This is clearly not the case. Increased polydispersity of the particles could readily improve the ensemble spectral overlap. If anything, this would reduce the microscopic density of suitable donor–acceptor pairs, leading to an overall reduction in FRET efficiency. Biophysical nanocrystal beacon experiments therefore require a careful trade-off between single particle and single molecule line widths of donor and acceptor as well as their spectral separation to optimize the probability of creating a functional FRET couple.

Returning to the original discussion of this chapter, it is interesting to note the influence of spectral diffusion on FRET. Depending on the temperature and the line widths relative to the typical magnitudes of spectral diffusion, a random spectral jitter of either donor or acceptor may increase or decrease FRET. For a case where spectral overlap is generally poor, random spectral fluctuations may sporadically enhance FRET. In an ensemble, these fluctuations in FRET are clearly averaged out, but they directly determine the overall FRET efficiency. A microscopic control over spectral diffusion, which may be achievable through chemical modification of the surface groups, could therefore be of interest to tuning FRET efficiencies of nanocrystals to molecular materials, for example, in hybrid polymer–nanocrystal solar cells [121]. Finally, it is worth noting that electrical control of dipole–dipole coupling in molecular systems [122] and in epitaxially grown quantum dots [123, 124] has been discussed in the context of quantum computing applications. The synthetic versatility of semiconductor nanocrystals certainly merits taking a deeper look at the surprising range of functionality achievable with these systems in terms of applications which may not be immediately apparent – such as quantum computation [125].

8. Conclusions

Semiconductor nanocrystals constitute a fascinating example of nanoscale light sources. The control of physical shape and chemical composition availed by ever increasingly versatile synthetic procedures goes hand in hand with nanoscale optical characterisation techniques. Single particle fluorescence probes the immediate electrostatic environment of the particle, but in turn also provides direct information on the shape of the nanostructure. Studies of single particles directly reveal the intrinsic electronic properties, which are important for describing the electronic transitions with suitable theoretical models [101]. The single particle studies reveal that a significant contribution to spectral broadening of the ensemble arises from spectral diffusion of the individual excitonic transition. This spectral diffusion correlates directly with the shape of the nanocrystal and can be thought of as originating from the spatial redistribution of surface charge [27]. The elementary transition in the nanocrystal is highly polarisable, making electric field modulation possible on the single particle level. These quantum-confined Stark spectroscopic investigations also bear strong signatures of the nanoparticle geometry, promoting Stark spectroscopy to a versatile tool for studying nanoscale shape [101]. The electric modulation of single particle fluorescence constitutes an example of a nanoscale optoelectronic device, where electrical information is encoded onto an optical output. While usable single particle modulators are still a long way off, this remarkable ability to control the electronic properties of a nanoscale light source enables the construction of a unique functional device, the FRET gate, which switches the flow of excitation energy. This proof of principle, which is also important to understand the general disorder limitation of FRET and the way in which this limitation can be overcome by spontaneous spectral diffusion, points the way to a new class of hybrid nanoscale devices which will exploit the full potential of molecular and colloidal selfassembly. Having demonstrated such novel functionality, which is only made possible by bringing together material constituents with differing individual properties [102], materials chemists are now challenged to develop routes to link different nanoscale FRET gates together to construct logic circuitry. Such building blocks could find applications in both computation and communication, and are particularly interesting for the development of novel dielectric and biophysical nanoscale sensors.

Acknowledgements

The authors are indebted to Klaus Becker, Florian Schindler, Jochen Feldmann, Andrey Rogach and Dmitri Talapin for many stimulating discussions in the course of the research described in this chapter, and express their gratitude to Dmitri Talapin and Horst Weller for the kind provision of the excellent samples used. JML thanks Florian Schindler for illuminating the analogy of spectral jitter in conjugated polymers, shown in Fig. 11. JML thanks Klaus Becker for investigating applications of the quantum-confined Stark effect and realising the FRET gate, described in Figs. 14, 15.

References

[1] Scholes GD, Rumbles G (2006) Excitons in nanoscale systems. Nat Mater 5: 683–696
[2] Jang SJ, Newton MD, Silbey RJ (2004) Multichromophoric Forster resonance energy transfer. Phys Rev Lett 92: 218301
[3] Fleming GR, Scholes GD (2004) Physical chemistry – quantum mechanics for plants. Nature 431: 256–257
[4] Kühlbrandt W (1995) Photosynthesis – many wheels make light work. Nature 374: 497–498
[5] van Oijen AM, Ketelaars M, Kohler J, Aartsma TJ, Schmidt J (1999) Unraveling the electronic structure of individual photosynthetic pigment-protein complexes. Science 285: 400–402
[6] Hofmann C, Aartsma TJ, Michel H, Köhler J (2003) Direct observation of tiers in the energy landscape of a chromoprotein: a single-molecule study. Proc Natl Acad Sci USA 100: 15534–15538
[7] Moerner WE, Orrit M (1999) Illuminating single molecules in condensed matter. Science 283: 1670
[8] Müller JG, Lemmer U, Raschke G, Anni M, Scherf U, Lupton JM, Feldmann J (2003) Linewidth-limited energy transfer in single conjugated polymer molecules. Phys Rev Lett 91: 267403
[9] Brokmann X, Giacobino E, Dahan M, Hermier JP (2004) Highly efficient triggered emission of single photons by colloidal CdSe/ZnS nanocrystals. Appl Phys Lett 85: 712–714
[10] Lounis B, Bechtel HA, Gerion D, Alivisatos P, Moerner WE (2000) Photon antibunching in single CdSe/ZnS quantum dot fluorescence. Chem Phys Lett 329: 399–404
[11] Nirmal M, Dabbousi BO, Bawendi MG, Macklin JJ, Trautman JK, Harris TD, Brus LE (1996) Fluorescence intermittency in single cadmium selenide nanocrystals. Nature 383: 802–804
[12] Stefani FD, Knoll W, Kreiter M, Zhong X, Han MY (2005) Quantification of photoinduced and spontaneous quantum-dot luminescence blinking. Phys Rev B 72: 125304
[13] Stefani FD, Zhong XH, Knoll W, Han MY, Kreiter M (2005) Memory in quantum-dot photoluminescence blinking. J Phys 7: 197
[14] Chung I, Witkoskie JB, Cao JS, Bawendi MG (2006) Description of the fluorescence intensity time trace of collections of CdSe nanocrystal quantum dots based on single quantum dot fluorescence blinking statistics. Phys Rev E 73: 011106
[15] Verberk R, van Oijen AM, Orrit M (2002) Simple model for the power-law blinking of single semiconductor nanocrystals. Phys Rev B 66: 233202
[16] Shimizu KT, Neuhauser RG, Leatherdale CA, Empedocles SA, Woo WK, Bawendi MG (2001) Blinking statistics in single semiconductor nanocrystal quantum dots. Phys Rev B 63: 205316
[17] Pelton M, Grier DG, Guyot-Sionnest P (2004) Characterizing quantum-dot blinking using noise power spectra. Appl Phys Lett 85: 819–821
[18] Kuno M, Fromm DP, Hamann HF, Gallagher A, Nesbitt DJ (2000) Nonexponential "blinking" kinetics of single CdSe quantum dots: a universal power law behavior. J Chem Phys 112: 3117–3120

[19] Kuno M, Fromm DP, Hamann HF, Gallagher A, Nesbitt DJ (2001) "On"/"off" fluorescence intermittency of single semiconductor quantum dots. J Chem Phys 115: 1028–1040
[20] Tang J, Marcus RA (2005) Mechanisms of fluorescence blinking in semiconductor nanocrystal quantum dots. J Chem Phys 123: 054704
[21] Efros AL, Rosen M (1997) Random telegraph signal in the photoluminescence intensity of a single quantum dot. Phys Rev Lett 78: 1110–1113
[22] Pelton M, Smith G, Scherer NF, Marcus RA (2007) Evidence for a diffusion-controlled mechanism for fluorescence blinking of colloidal quantum dots. Proc Nat Acad Sci USA 104: 14249–14254
[23] Müller J, Lupton, JM, Rogach AL, Feldmann J, Talapin DV, Weller H (2004) Air induced fluorescence bursts from single semiconductor nanocrystals. Appl Phys Lett 85: 381
[24] Gomez DE, Califano M, Mulvaney P (2006) Optical properties of single semiconductor nanocrystals. Phys Chem Chem Phys 8: 4989–5011
[25] Basché T (1998) Fluorescence intensity fluctuations of single atoms, molecules and nanoparticles. J Lumin 76–77: 263–269
[26] Frantsuzov PA, Marcus RA (2005) Explanation of quantum dot blinking without the long-lived trap hypothesis. Phys Rev B 72: 155321
[27] Müller J, Lupton JM, Rogach A, Feldmann J, Talapin DV, Weller H (2005) Signatures of surface charge migration in the spectral diffusion of single elongated CdSe/CdS nanocrystals. Phys Rev B 72:205339
[28] Müller J, Lupton JM, Rogach AL, Feldmann J, Talapin DV, Weller H (2004) Monitoring surface charge movement in single elongated semiconductor nanocrystals. Phys Rev Lett 93: 167402
[29] Hohng S, Ha T (2004) Near-complete suppression of quantum dot blinking in ambient conditions. J Am Chem Soc 126: 1324–1325
[30] He H, Qian HF, Dong CQ, Wang KL, Ren JC (2006) Single nonblinking CdTe quantum dots synthesized in aqueous thiopropionic acid. Angew Chem Int Ed 45: 7588–7591
[31] Kiraz A, Ehrl M, Bräuchle C, Zumbusch A (2003) Low temperature single molecule spectroscopy using vibronic excitation and dispersed fluorescence detection. J Chem Phys 118: 10821–10824
[32] Krishnan R, Hahn MA, Yu ZH, Silcox J, Fauchet PM, Krauss TD (2004) Polarization surface-charge density of single semiconductor quantum rods. Phys Rev Lett 92: 216803
[33] Krauss TD, O'Brien S, Brus LE (2001) Charge and photoionization properties of single semiconductor nanocrystals. J Phys Chem B 105: 1725–1733
[34] Krauss TD, Brus LE (1999) Charge, polarizability, and photoionization of single semiconductor nanocrystals. Phys Rev Lett 83: 4840–4843
[35] Shim M, Guyot-Sionnest P (1999) Permanent dipole moment and charges in colloidal semiconductor quantum dots. J Chem Phys 111: 6955–6964
[36] Empedocles SA, Norris DJ, Bawendi MG (1996) Photoluminescence spectroscopy of single CdSe nanocrystallite quantum dots. Phys Rev Lett 77: 3873–3876
[37] Peng XG, Manna L, Yang WD, Wickham J, Scher E, Kadavanich A, Alivisatos AP (2000) Shape control of CdSe nanocrystals. Nature 404: 59–61
[38] Talapin DV, Koeppe R, Götzinger S, Kornowski A, Lupton JM, Rogach AL, Benson O, Feldmann J, Weller H (2003) Highly emissive colloidal CdSe/CdS heterostructures of mixed dimensionality. Nano Lett 3: 1677
[39] Alivisatos AP (1996) Semiconductor clusters, nanocrystals, and quantum Dots. Science 271: 933–937
[40] Koberling F, Mews A, Philipp G, Kolb U, Potapova I, Burghard M, Basché T (2002) Fluorescence spectroscopy and transmission electron microscopy of the same isolated semiconductor nanocrystals. Appl Phys Lett 81: 1116–1118
[41] Chung IH, Bawendi MG (2004) Relationship between single quantum-dot intermittency and fluorescence intensity decays from collections of dots. Phys Rev B 70: 165304
[42] Tang J, Marcus RA (2006) Determination of energetics and kinetics from single-particle intermittency and ensemble-averaged fluorescence intensity decay of quantum dots. J Chem Phys 125: 044703
[43] Margolin G, Barkai E (2004) Aging correlation functions for blinking nanocrystals, and other on–off stochastic processes. J Chem Phys 121: 1566–1577
[44] Tang J, Marcus RA (2005) Diffusion-controlled electron transfer processes and power-law statistics of fluorescence intermittency of nanoparticles. Phys Rev Lett 95: 107401
[45] Empedocles SA, Neuhauser R, Shimizu K, Bawendi MG (1999) Photoluminescence from single semiconductor nanostructures. Adv Mater 11: 1243–1256
[46] Gammon D, Snow ES, Shanabrook BV, Katzer DS, Park D (1996) Homogeneous linewidths in the optical spectrum of a single gallium arsenide quantum dot. Science 273: 87–90

[47] Li XQ, Wu YW, Steel D, Gammon D, Stievater TH, Katzer DS, Park D, Piermarocchi C, Sham LJ (2003) An all-optical quantum gate in a semiconductor quantum dot. Science 301: 809–811
[48] Woggon U (2007) Single semiconductor nanocrystals: physics and applications. J Appl Phys 101: 081737
[49] Leon R, Petroff PM, Leonard D, Fafard S (1995) Spatially-resolved visible luminescence of self-assembled semiconductor quantum dots. Science 267: 1966–1968
[50] Shields AJ, Stevenson RM, Thompson RM, Ward MB, Yuan Z, Kardynal BE, See P, Farrer I, Lobo C, Cooper K, Ritchie DA (2003) Self-assembled quantum dots as a source of single photons and photon pairs. Phys Stat Sol B 238: 353–359
[51] Müller J (2005) Elektrische Manipulation der Lichtemission von einzelnen CdSe/CdS Nanostäbchen. Ph.D. Thesis, University of Munich, http://edoc.ub.uni-muenchen.de/5129/
[52] Tavenner-Kruger S, Park YS, Lonergan M, Woggon U, Wang HL (2006) Zero-phonon linewidth in CdSe/ZnS core/shell nanorods. Nano Lett 6:2154–2157
[53] Le Thomas N, Allione M, Fedutik Y, Woggon U, Artemyev MV, Ustinovich EA (2006) Multiline spectra of single CdSe/ZnS core-shell nanorods. Appl Phys Lett 89: 263115
[54] Palinginis P, Tavenner S, Lonergan M, Wang HL (2003) Spectral hole burning and zero phonon linewidth in semiconductor nanocrystals. Phys Rev B 67: 201307
[55] Htoon H, Cox PJ, Klimov VI (2004) Structure of excited-state transitions of individual semiconductor nanocrystals probed by photoluminescence excitation spectroscopy. Phys Rev Lett 93: 187402
[56] Neuhauser RG, Shimizu KT, Woo WK, Empedocles SA, Bawendi MG (2000) Correlation between fluorescence intermittency and spectral diffusion in single semiconductor quantum dots. Phys Rev Lett 85: 3301–3304
[57] Empedocles SA, Bawendi MG (1999) Influence of spectral diffusion on the line shapes of single CdSe nanocrystallite quantum dots. J Phys Chem B 103: 1826–1830
[58] Koberling F, Mews A, Basché T (1999) Single-dot spectroscopy of CdS nanocrystals and CdS/HgS heterostructures. Phys Rev B 60: 1921–1927
[59] Tittel J, Gohde W, Koberling F, Basché T, Kornowski A, Weller H, Eychmuller A (1997) Fluorescence spectroscopy on single CdS nanocrystals. J Phys Chem B 101: 3013–3016
[60] Robinson HD, Goldberg BB (2000) Light-induced spectral diffusion in single self-assembled quantum dots. Phys Rev B 61: R5086–R5089
[61] Bayer M, Forchel A (2002) Temperature dependence of the exciton homogeneous linewidth in In0.60Ga0.40As/GaAs self-assembled quantum dots. Phys Rev B 65: 041308
[62] Kammerer C, Voisin C, Cassabois G, Delalande C, Roussignol P, Klopf F, Reithmaier JP, Forchel A, Gerard JM (2002) Line narrowing in single semiconductor quantum dots: toward the control of environment effects. Phys Rev B 66: 041306
[63] Moerner WE (1994) Examining nanoenvironments in solids on the scale of a single isolated impurity molecule. Science 265: 46–53
[64] Empedocles SA, Bawendi MG (1997) Quantum-confined stark effect in single CdSe nanocrystallite quantum dots. Science 278: 2114–2117
[65] Turck V, Rodt S, Stier O, Heitz R, Engelhardt R, Pohl UW, Bimberg D, Steingruber R (2000) Effect of random field fluctuations on excitonic transitions of individual CdSe quantum dots. Phys Rev B 61: 9944–9947
[66] Blome PG, Wenderoth M, Hubner M, Ulbrich RG, Porsche J, Scholz F (2000) Temperature-dependent linewidth of single InP/GaxIn1-xP quantum dots: interaction with surrounding charge configurations. Phys Rev B 61: 8382–8387
[67] Miller DA, Chemla DS, Damen TC, Gossard AC, Wiegmann W, Wood TH, Burrus CA (1984) Band-edge electroabsorption in quantum well structures: the quantum-confined stark effect. Phys Rev Lett 53: 2173–2176
[68] Kulik D, Htoon H, Shih CK, Li YD (2004) Photoluminescence properties of single CdS nanorods. J Appl Phys 95: 1056–1063
[69] Gomez DE, van Embden J, Mulvaney P (2006) Spectral diffusion of single semiconductor nanocrystals: the influence of the dielectric environment. Appl Phys Lett 88: 154106
[70] Klingshirn C (1998) Some selected aspects of the optical properties of II–VI semiconductor structures of reduced dimensionality. Ann Phys 23: 3–17
[71] Puzder A, Williamson AJ, Gygi F, Galli G (2004) Self-healing of CdSe nanocrystals: first-principles calculations. Phys Rev Lett 92: 217401
[72] Rothenberg E, Kazes M, Shaviv E, Banin U (2005) Electric field induced switching of the fluorescence of single semiconductor quantum rods. Nano Lett 5: 1581–1586
[73] Eychmüller A, Hässelbarth A, Katsikas L, Weller H (1991) Photochemistry of semiconductor colloids. 36. Fluorescence investigations on the nature of electron and hole traps in Q-sized colloidal CdS particles. Ber Bunsenges Phys Chem 95: 79–88

[74] Patton B, Langbein W, Woggon U (2003) Trion, biexciton, and exciton dynamics in single self-assembled CdSe quantum dots. Phys Rev B 68: 125316
[75] Warburton RJ, Schaflein C, Haft D, Bickel F, Lorke A, Karrai K, Garcia JM, Schoenfeld W, Petroff PM (2000) Optical emission from a charge-tunable quantum ring. Nature 405: 926–929
[76] Karrai K, Warburton RJ, Schulhauser C, Högele A, Urbaszek B, McGhee EJ, Govorov AO, Garcia JM, Gerardot BD, Petroff PM (2004) Hybridization of electronic states in quantum dots through photon emission. Nature 427: 135–138
[77] Klimov VI, Mikhailovsky AA, McBranch DW, Leatherdale CA, Bawendi MG (2000) Quantization of multiparticle Auger rates in semiconductor quantum dots. Science 287: 1011–1013
[78] Kraus RM, Lagoudakis PG, Müller J, Rogach AL, Lupton JM, Feldmann J, Talapin DV, Weller H (2005) Interplay between Auger and ionization processes in nanocrystal quantum dots. J Phys Chem B 109: 18214
[79] Shimizu KT, Woo WK, Fisher BR, Eisler HJ, Bawendi MG (2002) Surface-enhanced emission from single semiconductor nanocrystals. Phys Rev Lett 89: 117401
[80] Kuno M, Fromm DP, Johnson ST, Gallagher A, Nesbitt DJ (2003) Modeling distributed kinetics in isolated semiconductor quantum dots. Phys Rev B 67: 125304
[81] Zhang K, Chang HY, Fu AH, Alivisatos AP, Yang H (2006) Continuous distribution of emission states from single CdSe/ZnS quantum dots. Nano Lett 6: 843–847
[82] LeThomas N, Woggon U, Schops O, Artemyev MV, Kazes M, Banin U (2006) Cavity QED with semiconductor nanocrystals. Nano Lett 6: 557–561
[83] Biebricher A, Sauer M, Tinnefeld P (2006) Radiative and nonradiative rate fluctuations of single colloidal semiconductor nanocrystals. J Phys Chem B 110: 5174–5178
[84] Fisher BR, Eisler HJ, Stott NE, Bawendi MG (2004) Emission intensity dependence and single-exponential behavior in single colloidal quantum dot fluorescence lifetimes. J Phys Chem B 108: 143–148
[85] Brokmann X, Coolen L, Dahan M, Hermier JP (2004) Measurement of the radiative and nonradiative decay rates of single CdSe nanocrystals through a controlled modification of their spontaneous emission. Phys Rev Lett 93: 107403
[86] Labeau O, Tamarat P, Lounis B (2003) Temperature dependence of the luminescence lifetime of single CdSe/ZnS quantum dots. Phys Rev Lett 90: 257404
[87] Schlegel G, Bohnenberger J, Potapova I, Mews A (2002) Fluorescence decay time of single semiconductor nanocrystals. Phys Rev Lett 88: 137401
[88] Adams DM, Brus L, Chidsey CED, Creager S, Creutz C, Kagan CR, Kamat PV, Lieberman M, Lindsay S, Marcus RA, Metzger RM, Michel-Beyerle ME, Miller JR, Newton MD, Rolison DR, Sankey O, Schanze KS, Yardley J, Zhu XY (2003) Charge transfer on the nanoscale: current status. J Phys Chem B 107: 6668–6697
[89] Weissman MB (1988) 1/F noise and other slow, nonexponential kinetics in condensed matter. Rev Mod Phys 60: 537–571
[90] Jung SW, Fujisawa T, Hirayama Y, Jeong YH (2004) Background charge fluctuation in a GaAs quantum dot device. Appl Phys Lett 85: 768–770
[91] Hoogenboom JP, van Dijk EMHP, Hernando J, van Hulst NF, García-Parajo MF (2005) Power-law-distributed dark states are the main pathway for photobleaching of single organic molecules. Phys Rev Lett 95: 097401
[92] Schindler F, Lupton JM, Feldmann J, Scherf U (2004) A universal picture of chromophores in pi-conjugated polymers derived from single molecule spectroscopy. Proc Natl Acad Sci USA 101: 14695
[93] Schindler F, Lupton JM (2005) Single Chromophore Spectroscopy of MEH-PPV: homing-in on the elementary emissive species in conjugated polymers. Chem Phys Chem 6: 926
[94] Schindler F, Lupton JM, Müller J, Feldmann J, Scherf U (2006) How single conjugated polymer molecules respond to electric fields. Nat Mater 5: 141
[95] Schindler F, Lupton JM, Feldmann J (2006) Spontaneous switching of permanent dipoles in single conjugated polymer molecules. Chem Phys Lett: 428: 405–410
[96] Koch N (2007) Organic electronic devices and their functional interfaces. Chem Phys Chem 8: 1438–1455
[97] Arkhipov VI, Emelianova EV, Bässler H (1999) Hot exciton dissociation in a conjugated polymer. Phys Rev Lett 82: 1321–1324
[98] Falci G, D'Arrigo A, Mastellone A, Paladino E (2005) Initial decoherence in solid state qubits. Phys Rev Lett 94: 167002
[99] Yang H, Luo GB, Karnchanaphanurach P, Louie TM, Rech I, Cova S, Xun LY, Xie XS (2003) Protein conformational dynamics probed by single-molecule electron transfer. Science 302: 262–266

[100] Heller W, Bockelmann U, Abstreiter G (1998) Electric-field effects on excitons in quantum dots. Phys Rev B 57: 6270–6273
[101] Müller J, Lupton JM, Schindler F, Koeppe R, Lagoudakis PG, Rogach AL, Feldmann J, Talapin DV, Weller H (2005) Wavefunction engineering in elongated semiconductor nanocrystals with heterogeneous carrier confinement. Nano Lett 5: 2044–2050
[102] Becker K, Lupton JM, Müller J, Rogach AL, Talapin DV, Weller H, Feldmann J (2006) Electrical control of Förster energy transfer. Nat Mater 5: 777
[103] Kraus RM, Lagoudakis PG, Rogach AL, Talapin DV, Weller H, Lupton JM, Feldmann J (2007) Room-temperature exciton storage in elongated semiconductor nanocrystals. Phys Rev Lett 98: 017401
[104] Scholes GD, Jones M, Kumar S (2007) Energetics of photoinduced electron-transfer reactions decided by quantum confinement. J Phys Chem C 111: 13777–13785
[105] Michalet X, Pinaud FF, Bentolila LA, Tsay JM, Doose S, Li JJ, Sundaresan G, Wu AM, Gambhir SS, Weiss S (2005) Quantum dots for live cells, in vivo imaging, and diagnostics. Science 307: 538–544
[106] Barkai E, Jung YJ, Silbey R (2004) Theory of single-molecule spectroscopy: beyond the ensemble average. Ann Rev Phys Chem 55: 457–507
[107] Ha TJ, Ting AY, Liang J, Caldwell WB, Deniz AA, Chemla DS, Schultz PG, Weiss S (1999) Single-molecule fluorescence spectroscopy of enzyme conformational dynamics and cleavage mechanism. Proc Natl Acad Sci USA 96: 893–898.
[108] Mori T, Vale RD, Tomishige M (2007) How kinesin waits between steps. Nature 450: 750–755
[109] Deniz AA, Dahan M, Grunwell JR, Ha TJ, Faulhaber AE, Chemla DS, Weiss S, Schultz PG (1999) Single-pair fluorescence resonance energy transfer on freely diffusing molecules: observation of Förster distance dependence and subpopulations. Proc Natl Acad Sci USA 96: 3670–3675
[110] Clapp AR, Medintz IL, Mauro JM, Fisher BR, Bawendi MG, Mattoussi H (2004) Fluorescence resonance energy transfer between quantum dot donors and dye-labeled protein acceptors. J Am Chem Soc 126: 301–310
[111] Medintz IL, Uyeda HT, Goldman ER, Mattoussi H (2005) Quantum dot bioconjugates for imaging, labelling and sensing. Nat Mater 4: 435–446
[112] Willard DM, Carillo LL, Jung J, Van Orden A (2001) CdSe–ZnS Quantum dots as resonance energy transfer donors in a model protein–protein binding assay. Nano Lett 1: 469–474
[113] Kloepfer JA, Cohen N, Nadeau JL (2004) FRET between CdSe quantum dots in lipid vesicles and water- and lipid-soluble dyes. J Phys Chem B 108: 17042–17049
[114] Medintz IL, Konnert JH, Clapp AR, Stanish I, Twigg ME, Mattoussi H, Mauro JM, Deschamps JR (2004) A fluorescence resonance energy transfer-derived structure of a quantum dot-protein bioconjugate nanoassembly. Proc Natl Acad Sci USA 101: 9612–9617
[115] Zhang CY, Yeh HC, Kuroki MT, Wang TH (2005) Single-quantum-dot-based DNA nanosensor. Nat Mater 4: 826–831
[116] Soujon D, Becker K, Rogach AL, Feldmann J, Weller H, Talapin DV, Lupton JM (2007) Time resolved Förster energy transfer from individual semiconductor nanoantennae to single dye molecules. J Phys Chem C 111: 11511
[117] Potapova I, Mruk R, Hübner C, Zentel R, Basché T, Mews A (2005) CdSe/ZnS nanocrystals with dye-functionalized polymer ligands containing many anchor groups. Angew Chem Int Ed 44: 2437–2440
[118] Barth M, Schuster R, Gruber A, Cichos F (2006) Imaging single quantum dots in three-dimensional photonic crystals. Phys Rev Lett 96: 243902
[119] Gudiksen MS, Maher KN, Ouyang L, Park H (2005) Electroluminescence from a single-nanocrystal transistor. Nano Lett 5: 2257–2261
[120] Huang H, Dorn A, Bulovic V, Bawendi MG (2007) Electrically driven light emission from single colloidal quantum dots at room temperature. Appl Phys Lett 90: 023110
[121] Huynh WU, Dittmer JJ, Alivisatos AP (2002) Hybrid nanorod-polymer solar cells. Science 295: 2425–2427
[122] Hettich C, Schmitt C, Zitzmann J, Kuhn S, Gerhardt I, Sandoghdar V (2002) Nanometer resolution and coherent optical dipole coupling of two individual molecules. Science 298: 385–389.
[123] Ortner G, Bayer M, Lyanda-Geller Y, Reinecke TL, Kress A, Reithmaier JP, Forchel A (2005) Control of vertically coupled InGaAs/GaAs quantum dots with electric fields. Phys Rev Lett 94: 157401
[124] Krenner HJ, Sabathil M, Clark EC, Kress A, Schuh D, Bichler M, Abstreiter G, Finley JJ (2005) Direct observation of controlled coupling in an individual quantum dot molecule. Phys Rev Lett 94: 057402
[125] Fernee MJ, Rubinsztein-Dunlop H (2006) Quantum gate based on Stark tunable nanocrystal interactions with ultrahigh-Q/V field modes in fused silica microcavities. Phys Rev B 74: 115321

Applications of quantum dots in biomedicine

By

Angela O. Choi, Dusica Maysinger

Department of Pharmacology and Therapeutics, McGill University, Montreal, QC, Canada

1. Introduction

Research and development in nanotechnology has become an increasingly popular trend in the last 5 years as the demand and production of nanometer-sized materials continue to grow. Nanotechnology is an area of research encompassing multi-disciplinary studies (including chemistry, physics, engineering, and biotechnology), and has diverse applications in agriculture, automobile, clothing, defense and more recently, biology and biomedicine [1, 2]. Among many different nanotechnological products, quantum dots (QD) have gained a lot of popularity as imaging probes in biology due to their very special physico-chemical and optical properties [3, 4]. They are stable, highly fluorescent, tunable and can be functionalized via surface modifications. Despite the numerous ongoing studies on QD synthesis to improve their physical properties, the biological effects of QDs are poorly investigated. Thus far, it is known that QD biocompatibility is largely dependent on their size, surface charge, core and surface materials [2]. Currently, extensive studies on the interactions (or interference) of QDs with cellular processes are under investigation in many scientific centers.

The understanding of cellular processes and molecular mechanisms is essential for drug discovery, particularly for disease diagnosis and treatment; however current development in biomedicine is hindered by the lack of tools to visualize cellular events and signaling of individual molecules [5]. Integration of nanotechnology in biomedicine is thus timely and inevitable, as high resolution biomedical imaging, from microscopic to nanoscopic and from two-dimensional to spatio-temporal [6, 7], is rapidly progressing.

2. Quantum dots as imaging tools in biology and medicine

2.1 Advantages and limitations of quantum dots and fluorescent dyes. Among the current array of nanotechnology products, semiconductor nanocrystal quantum dots were first reported to be a very promising tool for cellular imaging by two groups of scientists (Alivisatos and Nie) in 1998 [8, 9]. The colloidal QD core typically ranges from 2 to 10 nm in diameter, and is typically composed of atoms from groups II–VI (e.g. CdTe, CdSe) and III–V (e.g. InP, InAs) of the periodic table. These QD cores are often capped with an additional layer or "shell" of inorganic material (e.g. ZnS) to enhance their quantum yield, resulting in enhanced signal-to-noise ratio (robust signal). Depending on the size and composition, QDs can emit at distinct and different wavelengths, all the way from UV through visible to near-infrared (NIR). Unlike traditional organic fluorophores, QDs absorb wavelengths from a broad spectrum and in turn, emit in symmetrical and narrow spectra. Taken altogether, QDs of different sizes can be excited simultaneously by a single wavelength and emit with distinctly different colors, allowing for concurrent labeling of multiple species [10, 11]. In addition to their novel and unique optical properties, QDs are also highly photostable due to their inorganic composition, rendering them less susceptible to photobleaching and providing them with significantly longer fluorescent lifetimes (10–40 ns) compared to organic fluorescent dyes, thereby permitting their use for long-term, repeated imaging [9, 12].

QDs are often synthesized in an organic environment, and in order for biological applications, QD surfaces must be modified with hydrophilic material (e.g. mercaptoproprionic acid, cysteamine) or micelle-forming polymeric materials to enhance their water solubility [13, 14]. To prevent aggregation of these nanoparticles, surface conjugations with synthetic polymers such as polyethylene glycol (PEG) are often advantageous, allowing QDs to remain as finely dispersed individual nanoparticles (Fig. 1).

The extent of cellular internalization and subcellular distribution of non-functionalized, hydrophilic QDs is largely dependent on nanoparticle size and surface charge. Studies from our group showed that different charges on CdTe QD surfaces can regulate the extent of nanoparticle uptake such that the more positively charged (cysteamine-capping) nanoparticles are taken up more readily [15]. QD internalization can also be enhanced by surface conjugation with phospholipids, synthetic polymers (i.e. PEG) [14] or other synthetic material like silica [16]. Lovrić et al. showed that non-functionalized, cationic CdTe QDs (cysteamine-capping) are internalized readily, within 1 h of incubation with cells, suggesting uptake mechanisms involving phagocytosis in microglia and macrophages in peripheral sites [17]. The larger, red-fluorescing QDs (∼5 nm in diameter) are retained in the cytoplasm,

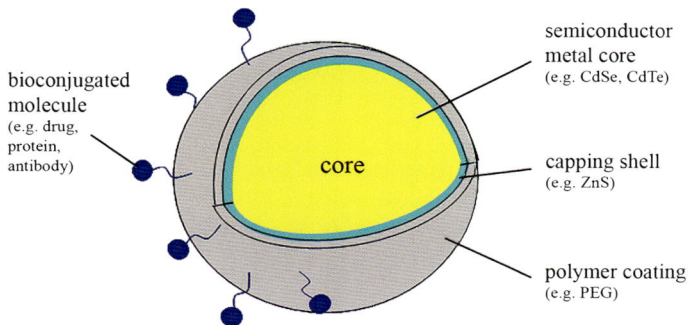

Fig. 1. Quantum dot "anatomy"

whereas the smaller, green-fluorescing QDs (∼2 nm in diameter) are localized in the nucleus. These findings were corroborated by the recent work by Volkov et al., showing that non-functionalized CdTe QDs exploit the cellular active transport machinery for delivering these QDs to specific intracellular destination [18]. Both of these studies point towards a critical role of the size, charge and surface properties of QDs which together with the cell-type specific properties will determine the fate of these nanocrystals.

2.2 Imaging of cellular and subcellular structures. Mammalian cells are typically 10 μm in diameter and contain a variety of subcellular machineries in the sub-micron range, which act to control cellular function and maintain homeostasis [19]. Pathak et al. showed that QDs bioconjugated with cell-type specific antibodies can be used to distinguish between neurons and glia in primary cultures without the use of secondary antibodies [20]. Antibodies against β-tubulin (ubiquitous cytoskeletal protein specific to neurons) and glial fibrillary acidic protein (GFAP, specific to glia) were conjugated to streptavidin-conjugated QDs to label primary cortical cultures. Compared to the blurry signals obtained from traditional fluorophore-tagged secondary antibodies, cells labeled with QDs were brighter and exhibited sharper and finer features. Although these are the pioneering studies exploiting QD-conjugates to explore neurons and glia, such approach has a number of limitations for in vivo studies in whole animals. For instance, the size of the QD-antibody structure may be too large to cross the blood brain barrier unless a targeting moiety with penetrating properties (e.g. TAT peptide, transferrin receptor) was added to facilitate the transport. Secondly, stability of the QD-antibody bond may not be adequate to preserve the integrity of the complex long enough for delivery to the destination (e.g. in deeper structures of the central nervous system).

Biological function cannot be determined by simply elucidating cellular and molecular structures without studying the spatio-temporal organization and distribution of intracellular molecules, and more importantly, tracking dynamic molecular interactions in real-time. Intracellular organelles are composed of and are regulated by nanometer-sized molecules such as proteins (1–20 nm) (Fig. 2). Current imaging techniques, including electron and confocal microscopy, have helped to elucidate the structure and the specific localization of these nanomolecules. Highly fluorescent and photostable QDs can allow live imaging of individual cellular

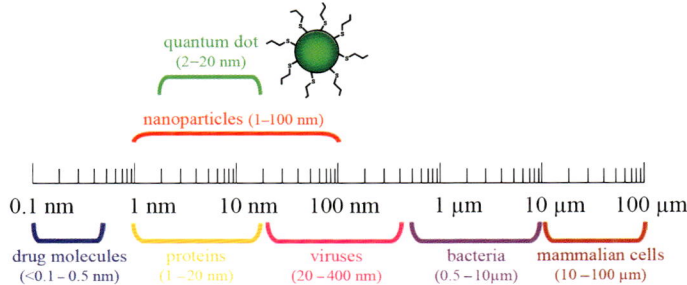

Fig. 2. QD sizes relative to drug molecules and mammalian cells

components with high resolution, selectivity, precision and bright fluorescence. Dahan et al. studied and compared the dynamic action of individual glycine receptors (GlyR) in rat spinal cord neuronal cultures, using an antibody against the GlyR α1 subunit, tagged either with QDs or a commonly used fluorescent dye (Cy3) [21]. In addition to the enhanced brightness in fluorescence of QD-GlyR (almost an order of magnitude higher than Cy3-GlyR), the authors were able to extend the live tracking of GlyR lateral dynamics in the neuronal membranes to 20 min using QD probes, compared to the much shorter 5 s fluorescence lifetime of the Cy3 probe. Diffusion coefficients of the QD-GlyRs localized within the synaptic cleft were also found to be larger compared with bead-GlyRs, suggesting that there is little or no interference of receptor dynamics by QDs compared to beads.

Imaging cell surface receptors dynamics is only one of the many aspects of signal transduction; trafficking and transport of ligands are also important for localizing the function of a specific molecule in real-time. Cui et al. conjugated QDs with nerve growth factor (NGF) and tracked the uptake and retrograde transport of NGF in rat dorsal root ganglia cultures (DRG) [22]. DRG cultures are often used as a model system in neuroscience to explore signal transduction pathways involved in nerve growth and survival [23, 24]. These primary cultures consist of mixed neurons and Schwann cells and provide a superior model over the immortalized cell line, as the mixed cultures conserve the interactions between the cell types, thereby better representing the actual environment cells are normally exposed to [25]. NGF-QDs were found to be taken up by the TrkA receptors and these receptors were transported along the axon by endosome-like vesicles ranging between 50 and 150 nm in diameter. The rate of uptake was comparable to studies using radiolabeled NGF (^{125}I-NGF), suggesting that QDs do not restrict or profoundly alter NGF structure, and more importantly, do not hinder NGF trafficking. In addition, the studies show that colocalization and activation with TrkA receptors, and phosphorylation of Erk1/2 were not abolished, indicating that the functionality of NGF was not impeded by the QDs [25].

As shown by these studies, target-conjugated QDs can be and has been used not only as a cellular marker, but as a molecular marker which can track the live, dynamic action of a molecule with bright fluorescence for a relatively long time without interfering with their endogenous function or motion. Similar approaches can be taken to explore other receptors, their distributions and functional changes under experimental conditions. Such studies are invaluable to gain insights into molecular mechanisms at the cellular level and how they are conducted under relatively controlled conditions. The limitation of such studies is that it does not provide functional connections and communications in situ, as it is in a living whole animal. In the next section, we will highlight several studies and discuss some of the advantages and limitations of whole animal studies with QDs.

2.3 Functional cell imaging in living animals in real-time.
Whole animal imaging was limited for a long time mainly because of the poor signal resolution, resulting from the photounstable dyes, despite the quality of the microscopes used. With the advances of improved contrast agents, new opportunities arose and provided

a better handle to explore normal and diseased tissues, as well as the entire body of experimental animals and humans.

One of the objectives in imaging normal and pathologic sites in the body is not only to detect the site, but also to provide means of detecting dynamic changes as a response of progressive tissue deterioration or gradual recovery from the injury. In this regard, our laboratory has recently devised a way of merging nanotechnology with transgenic technology and investigated the responsiveness of glial cells in living mice. QDs were administered directly into brain parenchyma [26]. The objective of this work was to establish a sensitive in vivo assay for the responsiveness of astrocytes to the

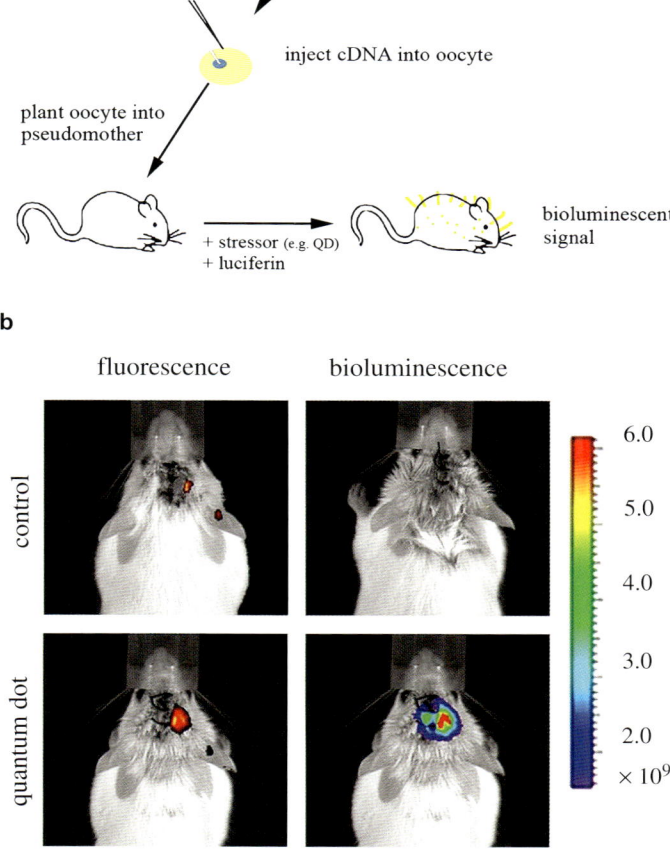

Fig. 3. In vivo neuroimaging of injected quantum dots in transgenic animals. Expression of luciferase (Luc) is driven by the glial fibrillary acidic protein (GFAP) promoter in the GFAP-Luc transgenic mice (promoter is activated in response to stress) (**a**). The substrate for luciferase (i.e. luciferin) is injected and bioluminescence is detected. **b** GFAP-Luc mice (Xenogen-Calipers LS, Alameda, CA) were imaged using the IVIS in vivo imaging system, 24 h after injection with 16 pM PEGylated CdSe/ZnS QDs (emission wavelength 705 nm) [26]

nanoparticle-induced brain injury. Astrocytes are glial cells which are activated around the site of injury or more widely in brain inflammation [27]. Currently, there are commercially available a number of transgenic mice expressing luciferase (Luc) under the control of different promoters, allowing for real-time imaging of specific tissues in the whole animal, depending on the specificity of the promoters.

The transgenic animals (GFAP-Luc; Xenogen-Caliper LS) used in our study express luciferase under the control of a promoter specifically expressed in astrocytes (i.e. GFAP). Once activated, the GFAP promoter induces luciferase expression and upon injection of the substrate (i.e. luciferin), a strong luminescent signal is generated and can be quantified. The illustration shows the principle of the luciferase expression and detection in GFAP-Luc mice (Fig. 3a) and provides an example of astrocyte activation surrounding the QD administration (Fig. 3b). It was noted that the rate of astrocyte activation is very different depending on the type of surface on the QDs [26]. This finding underlines the importance of thorough characterization of nanoparticles to be used since QDs with even comparable sizes and core materials but different surface properties can markedly change the kinetics and intensity of astrocyte activation.

Among the many QDs and other fluorescent or non-fluorescent nanoparticles, infrared-emitting QDs were used in our in vivo studies. Emission in the near-infrared is necessary to overcome autofluorescence, to penetrate the skull and to provide for long-term, repeated live imaging of the brain [26]. We recently reported that NIR QDs can be used for deep tissue imaging and can penetrate up to 6 mm of tissue [28]. As promising as these initial in vivo results with QDs may be, further research is necessary to characterize and optimize QD properties and to study how these properties are altered in an in vivo system, where these QDs end up, how long they stay at one site or whether they are eliminated and do not present any hazard to the normal functions of the surrounding tissues and of the whole organism.

3. Quantum dots as diagnostic tools

Investigation of nanoparticles for diagnostic purposes is currently the most advanced and well-studied in the field of oncology [29–34]. Cancer is presently the leading cause of death in North America and the number of new cases in the United States is expected to be over 1.5 million in 2008 (American Cancer Society Inc.). Current cancer therapies are lacking due to inadequate understanding of the multi-modality disease, particularly failing to detect tumor formation with early diagnosis and accurate prognosis, and in turn impeding the effectiveness of anticancer drugs.

Existing diagnostic approaches are mostly limited to the detection of relatively large, solid tumors, which often involve invasive techniques such as tissue biopsies [29]. In most cases, this detectable tumor is at a late stage, at which the cancer has metastasized to other tissues, resulting in a greater challenge for both tumor detection and proper therapeutics. In addition, tissue biopsies are difficult to obtain from deep tissues, bioanalytical assays from urine and blood samples are often not providing reliable results, and imaging with contrast agents are limiting as current dyes cannot distinguish between the highly invasive and benign types of tumor. More recently, high throughput genomic and proteomic analyses have revealed that many of these subtypes can be distinguished based on expression profile rather than presence of a single protein [30]. It would therefore be useful to have diagnostic tools which could allow for simultaneous detection of multiple proteins with sufficient sensitivity. QDs, among other nanotechnological products, could become versatile tools for screening cancer markers in biological fluids (urine, blood) and tissue biopsies, as well as high resolution contrast agents for medical imaging of metastatic tumors [31].

The rationale for using QDs as diagnostics in cancer are the following: (i) they are highly fluorescent and can be used for deep tissue imaging in vivo, (ii) they can serve as sensitive probes for multiple cell types because of their multiplexing abilities and wide range of tunable emissions [3], and (iii) utilization of functionalized QDs to target tumors in experimental animals shows promising results for future developments and eventual applications in humans.

Voura et al. reported the use of non-functionalized CdSe/ZnS QDs for multiphoton tracking of "metastasis" of different tumor cell populations in animals [31]. Different populations of murine lung melanoma (B16F10) were matured in vitro and transfected with QDs emitting in different wavelengths (510, 550, 570, 590, and 610 nm), after which these were injected into syngeneic mice. Tumor cell invasion into the lung was assessed using fluorescence emission-scanning microscopy 5 h after QD injection, and the individual tumor populations could be clearly identified. Another study by Stroh et al. also reported the use of non-functionalized CdSe/ZnS QDs in imaging murine mammary adenocarcinoma vasculature in vivo [32]. QDs with different colors were encapsulated into micelles and injected into GFP-transgenic animals xenografted with tumors and multiphoton microscopy was used to track the uptake of these nanoparticles into the tumor vasculature. The group reported that the QD-labeled vasculature could be clearly distinguished from perivascular cells in vivo, and labeling of the bone marrow with another type of QDs also showed recruitment of precursor cells to the vasculature. These studies not only again emphasize the promising implications of QD imaging in vivo, but also show the ready uptake of QDs by tumors via passive targeting mechanisms. It is well documented that macromolecules and nanoparticles can progressively accumulate in tumors due to the hypervasculature and enhanced permeability, a process known as enhanced permeability and retention (EPR) [33].

In addition to passive uptake of nanoparticles by the EPR effect, active targeting of tumors in vivo had also been reported using highly fluorescent QDs. Gao et al. reported in their study the use of QDs, conjugated with a tumor-targeting ligand, to actively localize at tumor sites in live animals [34]. CdTe/ZnS QDs were encapsulated in a polymer micelle conjugated with an antibody targeting the prostate-specific membrane antigen (PSMA), and injected systemically in mice xenografted with a prostate tumor. In vivo fluorescence of the brightly fluorescent QD-PSMA Ab probes was measured with a high signal-to-background ratio, and QD probes were found to localize specifically at the sites of tumor growth (i.e. prostates). Additionally, microbeads (0.5 µm in diameter) linked to different color QD probes were injected into three adjacent locations in the animal and imaged with multi-photon microscopy, suggesting the possibility of in vivo tracking of therapeutic action of drugs linked with QDs.

Despite the success of in vivo imaging studies with QDs, currently the application of QDs in diagnostic assays yielded more practical results. Immunoassays using bioconjugated QDs have been developed to assess variety of cellular states and functions, including protein–protein interaction [35], protein function [36], and more relevant to cancer, cell motility [37]. Pellegrino and colleagues reported in three studies the use of CdSe/ZnS QDs to track the motility of tumor cells in vitro, which in turn could be useful in determining the invasiveness of the cancer cells [37–39]. Based on the concept of EPR, cancer cells engulf the highly photostable QDs readily, and combined with real-time tracing of the fluorescent trail (or the disappearance of this trail), the metastatic potential of the tumors can be staged.

In summary, QDs together with transgenic animals as presented in this section could be used as versatile screening platforms for the assessments of effectiveness of chemotherapeutic, surgical, and radiation therapies to facilitate diagnosis and possibly treatment of solid and metastatic tumors. Combining genomic, proteomic and nanomedical tools for in vitro and in vivo imaging, will eventually contribute to future developments in achieving more personalized medicine in cancer and other diseases.

4. Quantum dots as nanotherapeutics

4.1 Quantum dots as drug delivery systems. Unfortunately, poor diagnosis and prognosis are only small parts of the overall "cancer problem," as this inevitably leads to inadequate development of treatments. Current chemotherapeutic agents are highly cytotoxic; however, most are lacking in specificity to cancerous cells and resulting in systemic toxicity and adverse side effects [40]. Nanoparticles, such as QDs and the often reported liposomes and polymeric micelles, may not only improve tumor targeting, but may also act as a new drug delivery tool, and even as direct therapeutics against tumor cells [41].

There is a growing trend for the development of multifunctional nanoparticles to image, diagnose and deliver treatment to cancer cells. The major struggle underlying the design of drug delivery systems is the same problem encountered in developing tools for imaging and diagnosis, and that is, target specificity. Encapsulating drugs in nanosized micelles was a big step forward in the research and development of drug delivery systems, as these nanoparticles can be easily surface-conjugated with ligands for targeted delivery, but more importantly, the release of drugs can be localized at the targeted region, reducing side effects due to non-specific drug action [42]. However, the current trend in the design of nano-delivery systems is focused on yet another level, which is the monitoring of drug action. QDs, among the array of nanomaterials, may be the optimal tool for all these purposes: imaging, diagnosis, drug delivery and tracking drug action [43].

Bagalkot et al. recently proposed in their study the design of a cancer imaging, therapy, and sensor system based on functionalized CdSe/ZnS QDs [44]. QDs were first surface conjugated with A10 RNA aptamers, which target the PSMA specific on prostate tumors, and subsequently, a fluorescent anticancer agent, doxorubicin (Dox), was intercalated with the aptamer to yield the QD-Apt(Dox) probe. Based on the concept of a bi-fluorescence resonance energy transfer (Bi-FRET), both QD and Dox fluorescence are quenched by their close proximity with each other in the intact

QD-Apt(Dox) probe. Upon the PSMA-mediated internalization of the nanosystem into the tumor cell, there is a release of the drug from the QD, thereby unquenching the fluorescence of both. The group reported that this nanosystem is indeed functional and specific to PSMA(+) cells in vitro, and showed that the fluorescence can be unmistakably distinguished. Another study by Derfus et al. described the use of QDs to deliver and monitor the delivery of siRNA, with the potential to knockout overexpressed oncogenes [45]. QDs, multiconjugated with a tumor-homing peptide (F3) and siRNA against an artificially transfected gene (enhanced green fluorescent protein, EGFP), were added to HeLa cells in vitro and the expression of EGFP was measured as the outcome of the knockdown. Fluorescence micrographs showed that cells containing the functionalized QDs also had no EGFP fluorescence, indicating the effectiveness of the system.

4.2 Photodynamic therapy using quantum dots. The photophysical properties, specifically the high photoluminescence and the energy-transfer potential, of QDs can be harnessed for therapeutic purposes, especially in the case of cancer, and this have been shown by a number of studies [46–48]. QDs are photosensitive energy donors, which can offer useful photodynamic therapy tools (PDT), at least for now in experimental animals. The principles of such a therapeutic approach has been proposed by several teams [47, 49, 50] and it is briefly summarized: in response to light and in the presence of oxygen, energy is released from QDs and transferred to cellular molecules, leading to the formation of reactive oxygen species (ROS) [51, 52]. Excessive production of ROS can induce cell apoptosis via oxidative stress-linked mechanisms, which when targeted to tumors, can lead to destruction of the specific tissues in a non-invasive manner [53]. In fact, a number of studies, including those from our laboratory, have shown evidence of the photosensitive and oxygen-sensitive nature of QDs, leading to the degradation of the QD core and subsequent release of free metal ions, and ultimately inducing cell death via apoptotic ROS signaling [15, 51, 54–56] (Fig. 4). Despite the lack of concrete evidence of the actual effectiveness of QDs as a PDT agent in anticancer therapy, one can envision the potential application of QDs conjugated with a targeting molecule (i.e. against an oncogene such as epithelial growth factor) in targeting and imaging the tumor sites.

One potential problem in this regard is that QDs may induce cytotoxicity and damage the surrounding and distant tissue at the initial photoactivation site. To avoid such undesirable effects, thorough biodistribution analyses and pharmacokinetic studies are required for every new biotechnological product to be used in nano-oncology, including QDs.

5. Biodistribution of quantum dots

A major concern regarding the safe use of QDs is their accumulation in the body and the poor understanding of the pharmacokinetics of nanoparticles after different routes of administration. The first in vivo imaging study of QDs was reported by Ballou et al., and they showed the distribution of intravenously (tail vein) injected non-functionalized QDs in mice and found that QD fluorescence can be measure in vivo for at least 4 months [57]. Live animal imaging with fluorescence microscopy shows that QDs distribute to different sites in the body immediately after injection, and the circulation lifetime of QDs was monitored and found to vary greatly (12–70 min) depending on the length of the polymer (i.e. PEG) conjugated on QD surfaces. Circulation lifetimes, in turn, determine the rates of QD deposition in the liver, spleen, lymph nodes and bone marrow. Accumulation of QDs in the liver and spleen was detected by necropsy and electron microscopy as early as 24 h after injection. After 1 month, QD fluorescence was mostly found in the lymph nodes, bone marrow and intestinal contents, with residual fluorescence in the liver and spleen, suggesting eventual excretion of these nanoparticles with time [57].

In contrast to these earlier findings, Fischer and Chan reported sequestration of non-functionalized QDs in rats after intravenous injection (jugular vein cannula) [58]. The group used a quantitative method (atomic emission spectroscopy) to assess cadmium content (correlated to QD concentration) in different organs and found that the liver alone takes up the majority of the injected QDs within 90 min (ranging from 40 to 90% depending on QD-surface conjugates) despite comparable fluorescence measured in the liver and the spleen. Daily analysis of the fecal and urinal materials for up to 10 days after injection did not yield detectable QD content, and additional experiments using transmission electron microscopy and digestion-ultracentrifugation show that intact QDs were taken up and retained by Kupffer cells after long-term circulation in the body, suggesting that these nanocrystals are poorly metabolized, retained in the reticuloendothelial system, and likely re-distributed in the body. Recent studies by Soo Choi et al. provided data on the renal clearance of intravenously injected QDs in rats [59]. The major finding from these studies shows that several requirements must be fulfilled before renal filtration and urinary elimination of these inorganic, metal-containing nanoparticles can be achieved. For instance, a final hydrodynamic diameter greater than 5.5 nm hinders renal excretion, whereas nanoparticles smaller than 5.5 nm are effectively excreted in urine. In addition, QD surfaces with zwitterionic charge are superior over positively and negatively charged surfaces, as QD interaction with plasma proteins is improved.

In summary, these studies highlight the notion that total body clearance of nanoparticles is not trivial and point towards the need to analyze biological fluids, including urine and bile, as a part of human risk assessment after environmental exposure or intended nanoparticle use for diagnostic (imaging) purposes. Analyses of these biological materials together with other routine clinical biochemical tests will help to estimate the total amount of retained nanoparticles (if the exposure dose is unknown), thereby indicating which ones are hazardous and which are harmless.

6. Nanoparticle-induced cytotoxicity

6.1 Experimental approaches to assess cytotoxicity: advantages and limitations. Conflicting results on the biodistribution and clearance of QDs are owed to the variety of methods and assays available for assessing cytotoxicity. Despite the variety in methods, one should keep in mind that most, if not all, of these assays simply evaluate the functions and structural integrity of different subcellular organelles

Table 1. Some biochemical methods for the assessment of nanoparticle-induced cell toxicity

Tools	Subcellular target	Outcome measures	References
Annexin V[a]	Plasma membrane lipid (phosphatidylserine, PS)	Extracellular PS due to "flipping" of membrane	[73]
Lactate dehydrogenase release	Plasma membrane	Membrane integrity	[74]
Propidium iodide[a] exclusion	Plasma membrane	Membrane permeability and integrity	[51, 75]
Trypan blue exclusion	Plasma membrane	Membrane permeability and integrity	[15, 76]
Alamar blue[a]	Cytosol (dehydrogenase)	Metabolic activity	[77]
JC-1[a] aggregation	Mitochondrion (membrane potential)	Membrane depolarization	[15]
MitoTracker®[a]	Mitochondrion	Morphological structure	[17, 51]
MTT (tetrazolium) reduction	Mitochondrion (dehydrogenase)	Metabolic activity	[15, 78]
LysoTracker®[a]	Lysosome	Morphological structure	[2]
DRAQ5[a]	Nucleus (DNA)	Morphological structure	[63, 79]
Dihydroethidium[a] oxidation	Nucleus (DNA)	Oxidative stress: detection of superoxide	[51, 80]
Hoechst[a]	Nucleus (DNA)	Morphological structure	[54]

[a] Commercially available fluorescent dyes

(examples of some of these methods used by our laboratory to evaluate nanotoxicity are selected and compiled in Table 1). If used individually, these techniques are restricting and results may be misleading. For example, the MTT assay is often used to assess cell viability, but an increase in formazan conversion (usually associated with improved viability) simply represents the increased activity of mitochondrial enzymes, which could very well be an initial cell defence response, as the cell struggles to boost its survival chances against the stressor [51]. It is important, therefore, to use a number of other approaches in concert with in vivo pharmacokinetic studies to evaluate the safety or the extent of toxicity of nanoparticles. A brief overview of the most commonly used techniques in assessing nanoparticle toxicity and examples of studies employing them is provided in a recent review [60].

6.2 Molecular mechanisms in quantum dot-induced cytotoxicity.

In response to the demand in establishing screening procedures for nanotoxicity, the focus is gradually moving towards the detection of early molecular changes induced by QDs, and development of pharmacological interventions to reverse or prevent the changes leading to cell death. Elucidating the mechanisms underlying QD-induced cytotoxicity is therefore an important first step as nanotoxicity is becoming a prominent concern in the scientific community, especially with the growing number of studies highlighting the toxicity of nanoproducts [60–62]. However, one must emphasize that tremendous efforts are being made to minimize and eventually eliminate current concerns regarding biohazard of some nanomaterials, especially in limiting the production of QDs with undesirable surface properties, core composition and poor stability in complex biological environments [2].

Table 2. Selected examples of biologically important reactive species

	Free radicals	Non-radicals
Reactive oxygen species (ROS)	Superoxide, $\cdot O_2^-$ Hydroxyl, $\cdot OH$ Peroxyl, $\cdot RO_2$ Hydroperoxyl, $\cdot HO_2^-$	Hydrogen peroxide, H_2O_2 Hydrochlorous acid, HOCl
Reactive nitrogen species (RNS)	Nitric oxide, $\cdot NO$ Nitrogen dioxide, $\cdot NO_2^-$	Peroxynitrite, $OONO^-$ Nitrous oxide, HNO_2

The cytotoxic potential of semiconductor QDs is of no surprise as their cores are composed of known toxic metals such as cadmium, tellurium and mercury [55]. However, only in the past few years had we begun to understand the mechanisms underlying the toxicity of these QDs. Studies in our laboratory first showed that QDs with different core composition, core size, surface coating, and surface charge induce different levels of toxicity [15, 17, 51, 55]. CdTe QDs enter cells readily, localize in different subcellular organelles [17], and in turn, cause lipid peroxidation [15, 54], mitochondrial damage [15], nuclear damage, epigenetic and genetic changes, even if they do not enter the organelles in detectable quantities [63]. This triggering of cytotoxic events stems, in part, from the initial degradation of QDs upon exposure to light and oxygen (photosensitization), leading to the release of free metal ions (i.e. Cd^{++}) and the formation of excessive reactive species, including reactive nitrogen species (RNS) and ROS, both intracellularly and extracellularly [55, 64] (Table 2 provides a brief list of some examples of RNS and ROS that are important for stress-activated cellular signaling). Extracellular ROS can damage the cell membrane and induce plasma membrane lipid peroxidation, leading to the production of more cell-damaging molecules such as aldehydes like ONE (4-oxo-2-nonenal), which would trigger the p53-dependent apoptotic signaling cascade [15]. Extracellular ROS can also trigger other pro-apoptotic events like the activation of cell surface Fas death receptors, which leads to subsequent activation of caspases, eventually leading to mitochondria-dependent apoptosis [65]. ROS can passively cross the plasma membrane and can lead to organelle damage. Due to the lack of choice and specificity of markers available currently, one of the urgent needs in biological sciences and nanomedicine is to develop suitable probes to detect specific types of reactive oxygen and nitrogen species.

Small, green-emitting QDs with a diameter ≤ 5 nm can enter the nucleus via the nuclear pore and induce damage including DNA nicking [66]. More recently, we suggested that cells exposed to small amounts of QDs for a prolonged period, undergo epigenetic changes which will modify gene expression [63]. The epigenome regulates the expression of genes via DNA methylation and posttranslational modifications of histones, which can have lasting effects on the organism and its offspring [67]. The epigenetic changes observed by our group further indicated that non-functionalized CdTe QDs induced upregulation of pro-apoptotic genes and a downregulation of antiapoptotic genes, thereby shifting the cellular homeostasis to be more cell death-favourable [63]. Simultaneous damage at other subcellular organelles is also occurring, most notably at the mitochondrial and lysosomal levels [15, 55]. Mitochondrial and lysosomal enlargement was observed early following CdTe QD treatment suggesting likely functional impairment in these organelles. We observed that mitochondrial function was indeed compromised in the presence of CdTe QDs as shown by the decrease in mitochondrial membrane potential [15]. This depolarization of the membrane leads to increased permeability across the mitochondrial membrane, and the subsequent release of apoptotic factors such as cytochrome c, triggering caspase-dependent apoptosis [68] (Fig. 4). However, caspase-independent cell death (e.g. necrosis) and several other modes of cell death can also be detected in cells exposed for a prolonged time to poorly protected QDs, particularly in those cells which have been predisposed to trophic factor deprivation.

6.3 Ways to overcome quantum dot-induced cytotoxicity. With the above-mentioned mechanisms underlying QD-toxicity, the outlook on developing

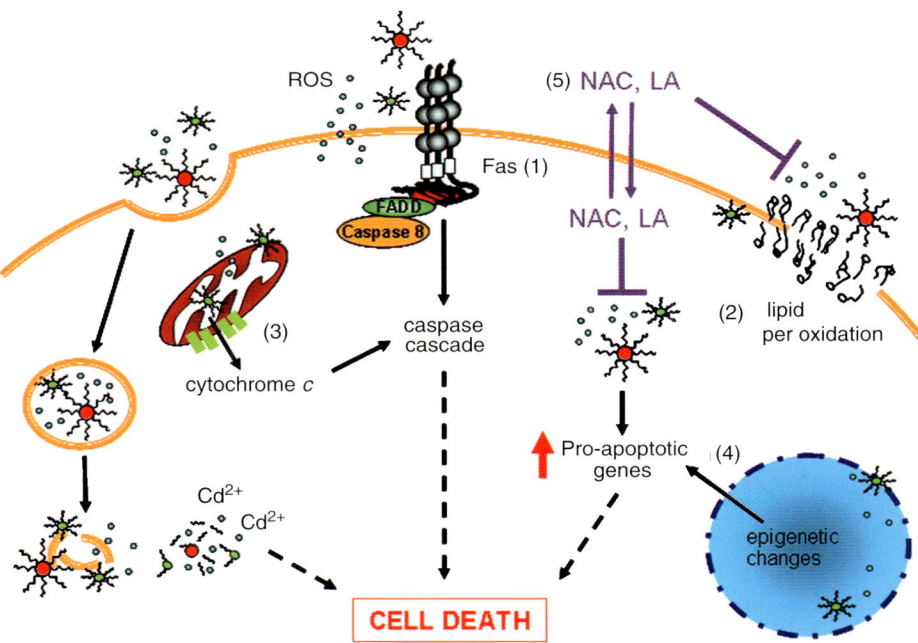

Fig. 4. Non-functionalized CdTe QDs interact and can interfere with cellular functions. (**1**) CdTe QD can upregulate the Fas death receptor, leading to the recruitment of the Fas-associated death domain (FADD) and initiating the caspase cascade [15]. (**2**) QD-induced production of reactive oxygen species (ROS) can induce lipid peroxidation of the plasma membrane and those of subcellular organelles. Internalization of CdTe QDs can be via endocytosis or active transport as well, and may result in the retention of the QDs in individual organelles. (**3**) CdTe QDs can impair mitochondrial function and enhance mitochondrial membrane permeability, thereby facilitating the release of pro-apoptotic factors such as cytochrome c [51]. (**4**) Nuclear damage by QDs is seen as chromatin condensation and epigenetic changes which favor the expression of pro-apoptotic genes (e.g. p53 and Bax) [63]. (**5**) Cell preconditioning with antioxidants such as N-acetylcysteine (NAC) and α-lipoic acid (LA) can protect cells from CdTe QD-induced cell death

preventive measures against QD-induced cytotoxicity is quite positive. QD-toxicity is dependent on QD stability, size, surface materials and charge among many factors. Not all QDs are toxic. In fact, most current studies now show that QDs with zwitterionic surfaces [59] or coated with synthetic polymers (i.e. PEG) are mostly inert and do not induce toxic response in most cell types under investigated conditions [16, 69, 70].

QD surface modifications can affect their cellular internalization, which in turn determines the extent of toxicity such that positively charged QDs can cross the plasma membrane very readily and induced more toxicity than negatively charged QDs. Choi et al. demonstrated that by conjugating or capping the surface of CdTe QDs with a small drug molecule, *N*-acetylcysteine, the overall charge of the QDs can be changed, in turn affecting their internalization and improving QD biocompatibility [15]. In addition, NAC can act as an antioxidant against the ROS produced by QDs. It is well documented that NAC acts with different modes of action as an antioxidant: (i) directly scavenge ROS with its cysteine moiety, (ii) regenerate endogenous antioxidants as a precursor to glutathione, (iii) regulate gene transcription to maintain cellular homeostasis, and (iv) promote cell survival by inhibiting JNK and p38 pathways [71]. Cell preconditioning with clinically relevant antioxidants such as NAC and LA can also prevent and reduce the cell damage induced by QDs [15, 51].

7. Current status and prospective

Most of the studies conducted with nanoparticles in cell cultures and animals so far, were carried out for relatively "short" time (up to several days) except some distribution studies which covered much longer time periods (several months [58, 72]). As with many pharmacological agents, it could also be the case with certain types of QDs that very small concentrations of QDs, undetectable by common chemical or imaging techniques, could lead to small changes critical to cellular function (e.g. epigenetic modifications). Advances in chemical and bioengineering approaches providing fine-tuning of the QD surfaces and other properties should allow for more positive rather than deleterious long-term effects at cellular level. Epigenetics is an evolving area of research and "nano-epigenetics" is in its infancy. Among the first studies addressing the epigenetic changes by QDs is one by Choi et al. [63]. These authors pointed out possible long-term consequences of cellular exposure to small QD concentrations and showed that histone acetylation is altered. This finding complements the more recent study pointing towards the active transport of non-functionalized nanoparticles and subsequent interaction with histone proteins [18]. Since histones play an important regulatory role in the normal cell cycle and tumor growth, consequences of epigenetic changes induced by QDs (and possibly other nanomaterials) and interactions between nanoparticles and histones remain to be explored in more detail. Our laboratory has initiated a number of studies, in both cell cultures and live animals, with the aim to provide an additional screening platform, including "nano-epigenetics," to complement common toxicological assay systems in defining hazardous versus well-tolerated nanomaterials.

Acknowledgments

The authors acknowledge the financial supports of the Natural Sciences and Engineering Research Council of Canada (NSERC), the Canadian Institutes of Health Research (CIHR) and the Alzheimer's Association (USA).

References

[1] Leary SP, Liu CY, Apuzzo ML (2006) Toward the emergence of nanoneurosurgery: part II – nanomedicine: diagnostics and imaging at the nanoscale level. Neurosurgery 58: 805–823; discussion 805–823
[2] Maysinger D (2008) Nanoparticles and cells: good companions and doomed partnerships. Org Biomol Chem 5: 2335–2342
[3] Alivisatos AP, Gu W, Larabell C (2005) Quantum dots as cellular probes. Annu Rev Biomed Eng 7: 55–76
[4] Zhang L, Gu FX, Chan JM, Wang AZ, Langer RS, Farokhzad OC (2007) Nanoparticles in medicine: therapeutic applications and developments. Clin Pharmacol Ther 83: 761–769
[5] Zhang J, Campbell RE, Ting AY, Tsien RY (2002) Creating new fluorescent probes for cell biology. Nature Rev 3: 906–918
[6] Lang P, Yeow K, Nichols A, Scheer A (2006) Cellular imaging in drug discovery. Nat Rev Drug Discov 5: 343–356
[7] Whitesides GM (2003) The 'right' size in nanobiotechnology. Nat Biotechnol 21: 1161–1165
[8] Bruchez M Jr, Moronne M, Gin P, Weiss S, Alivisatos AP (1998) Semiconductor nanocrystals as fluorescent biological labels. Science 281: 2013–2016
[9] Chan WC, Nie S (1998) Quantum dot bioconjugates for ultrasensitive nonisotopic detection. Science 281: 2016–2018
[10] Alivisatos P (2004) The use of nanocrystals in biological detection. Nat Biotechnol 22: 47–52
[11] Chan WC, Maxwell DJ, Gao X, Bailey RE, Han M, Nie S (2002) Luminescent quantum dots for multiplexed biological detection and imaging. Curr Opin Biotechnol 13: 40–46
[12] Jaiswal JK, Mattoussi H, Mauro JM, Simon SM (2003) Long-term multiple color imaging of live cells using quantum dot bioconjugates. Nat Biotechnol 21: 47–51
[13] Larson DR, Zipfel WR, Williams RM et al (2003) Water-soluble quantum dots for multiphoton fluorescence imaging in vivo. Science 300: 1434–1436
[14] Dubertret B, Skourides P, Norris DJ, Noireaux V, Brivanlou AH, Libchaber A (2002) In vivo imaging of quantum dots encapsulated in phospholipid micelles. Science 298: 1759–1762
[15] Choi AO, Cho SJ, Desbarats J, Lovric J, Maysinger D (2007) Quantum dot-induced cell death involves Fas upregulation and lipid peroxidation in human neuroblastoma cells. Nanobiotechnol 5: 1–13
[16] Zhang T, Stilwell JL, Gerion D et al (2006) Cellular effect of high doses of silica-coated quantum dot profiled with high throughput gene expression analysis and high content cellomics measurements. Nano Lett 6: 800–808
[17] Lovric J, Bazzi HS, Cuie Y, Fortin GR, Winnik FM, Maysinger D (2005) Differences in subcellular distribution and toxicity of green and red emitting CdTe quantum dots. J Mol Med 83: 377–385
[18] Nabiev I, Mitchell S, Davies A et al (2007) Nonfunctionalized nanocrystals can exploit a cell's active transport machinery delivering them to specific nuclear and cytoplasmic compartments. Nano Lett 7: 3452–3461
[19] Alberts B (2002) Molecular biology of the cell. Garland Science, New York, pp. xxxiv, 1463
[20] Pathak S, Cao E, Davidson MC, Jin S, Silva GA (2006) Quantum dot applications to neuroscience: new tools for probing neurons and glia. J Neurosci 26: 1893–1895
[21] Dahan M, Levi S, Luccardini C, Rostaing P, Riveau B, Triller A (2003) Diffusion dynamics of glycine receptors revealed by single-quantum dot tracking. Science 302: 442–445
[22] Cui B, Wu C, Chen L et al (2007) One at a time, live tracking of NGF axonal transport using quantum dots. Proc Natl Acad Sci USA 104: 13666–13671
[23] Bradbury EJ, McMahon SB, Ramer MS (2000) Keeping in touch: sensory neurone regeneration in the CNS. Trends Pharmacol Sci 21: 389–394
[24] Guertin AD, Zhang DP, Mak KS, Alberta JA, Kim HA (2005) Microanatomy of axon/glial signaling during Wallerian degeneration. J Neurosci 25: 3478–3487
[25] Heron PM, Sutton BM, Curinga GM, Smith GM, Snow DM (2007) Localized gene expression of axon guidance molecules in neuronal co-cultures. J Neurosci Meth 159: 203–214

[26] Maysinger D, Behrendt M, Lalancette-Hebert M, Kriz J (2007) Real-time imaging of astrocyte response to quantum dots: in vivo screening model system for biocompatibility of nanoparticles. Nano Lett 7: 2513–2520
[27] Williams K, Alvarez X, Lackner AA (2001) Central nervous system perivascular cells are immunoregulatory cells that connect the CNS with the peripheral immune system. Glia 36: 156–164
[28] Sandros MG, Behrendt M, Maysinger D, Tabrizian M (2007) InGaP@ZnS-enriched chitosan nanoparticles: a versatile fluorescent probe for deep-tissue imaging. Adv Funct Mater 17: 3724–3730
[29] Henry NL, Hayes DF (2006) Uses and abuses of tumor markers in the diagnosis, monitoring, and treatment of primary and metastatic breast cancer. Oncologist 11: 541–552
[30] Heath JR, Davis ME (2008) Nanotechnology and cancer. Annu Rev Med 59: 251–265
[31] Voura EB, Jaiswal JK, Mattoussi H, Simon SM (2004) Tracking metastatic tumor cell extravasation with quantum dot nanocrystals and fluorescence emission-scanning microscopy. Nat Med 10: 993–998
[32] Stroh M, Zimmer JP, Duda DG et al (2005) Quantum dots spectrally distinguish multiple species within the tumor milieu in vivo. Nat Med 11: 678–682
[33] Matsumura Y, Maeda H (1986) A new concept for macromolecular therapeutics in cancer chemotherapy: mechanism of tumoritropic accumulation of proteins and the antitumor agent smancs. Cancer Res 46: 6387–6392
[34] Gao X, Cui Y, Levenson RM, Chung LW, Nie S (2004) In vivo cancer targeting and imaging with semiconductor quantum dots. Nat Biotechnol 22: 969–976
[35] Anikeeva N, Lebedeva T, Clapp AR et al (2006) Quantum dot/peptide-MHC biosensors reveal strong CD8-dependent cooperation between self and viral antigens that augment the T cell response. Proc Natl Acad Sci USA 103: 16846–16851
[36] Medintz IL, Clapp AR, Brunel FM et al (2006) Proteolytic activity monitored by fluorescence resonance energy transfer through quantum-dot-peptide conjugates. Nat Mater 5: 581–589
[37] Pellegrino T, Parak WJ, Boudreau R et al (2003) Quantum dot-based cell motility assay. Differentiation 71: 542–548
[38] Gu W, Pellegrino T, Parak WJ et al (2005) Quantum-dot-based cell motility assay. Sci STKE 2005: l5
[39] Gu W, Pellegrino T, Parak WJ et al (2007) Measuring cell motility using quantum dot probes. Methods Mol Biol 374: 125–131
[40] Nie S, Xing Y, Kim GJ, Simons JW (2007) Nanotechnology applications in cancer. Annu Rev Biomed Eng 9: 257–288
[41] Vicent MJ, Duncan R (2006) Polymer conjugates: nanosized medicines for treating cancer. Trends Biotechnol 24: 39–47
[42] Duncan R (2006) Polymer conjugates as anticancer nanomedicines. Nat Rev Cancer 6: 688–701
[43] Portney NG, Ozkan M (2006) Nano-oncology: drug delivery, imaging, and sensing. Anal Bioanal Chem 384: 620–630
[44] Bagalkot V, Zhang L, Levy-Nissenbaum E et al (2007) Quantum dot-aptamer conjugates for synchronous cancer imaging, therapy, and sensing of drug delivery based on bi-fluorescence resonance energy transfer. Nano Lett 7: 3065–3070
[45] Derfus AM, Chen AA, Min DH, Ruoslahti E, Bhatia SN (2007) Targeted quantum dot conjugates for siRNA delivery. Bioconjugate Chem 18: 1391–1396
[46] Bakalova R, Ohba H, Zhelev Z, Ishikawa M, Baba Y (2004) Quantum dots as photosensitizers? Nat Biotechnol 22: 1360–1361
[47] Samia AC, Dayal S, Burda C (2006) Quantum dot-based energy transfer: perspectives and potential for applications in photodynamic therapy. Photochem Photobiol 82: 617–625
[48] Clarke SJ, Hollmann CA, Zhang Z et al (2006) Photophysics of dopamine-modified quantum dots and effects on biological systems. Nat Mater 5: 409–417
[49] Hsieh JM, Ho ML, Wu PW, Chou PT, Tsai TT, Chi Y (2006) Iridium-complex modified CdSe/ZnS quantum dots; a conceptual design for bi-functionality toward imaging and photosensitization. Chem Commun 6: 615–617
[50] Dayal S, Lou Y, Samia AC, Berlin JC, Kenney ME, Burda C (2006) Observation of non-Forster-type energy-transfer behavior in quantum dot-phthalocyanine conjugates. J Am Chem Soc 128: 13974–13975
[51] Lovric J, Cho SJ, Winnik FM, Maysinger D (2005) Unmodified cadmium telluride quantum dots induce reactive oxygen species formation leading to multiple organelle damage and cell death. Chem Biol 12: 1227–1234

[52] Ipe BI, Lehnig M, Niemeyer CM (2005) On the generation of free radical species from quantum dots. Small 1: 706–709
[53] Samia AC, Chen X, Burda C (2003) Semiconductor quantum dots for photodynamic therapy. J Am Chem Soc 125: 15736–15737
[54] Funnell WR, Maysinger D (2006) Three-dimensional reconstruction of cell nuclei, internalized quantum dots and sites of lipid peroxidation. J Nanobiotechnol 4: 1–19
[55] Cho SJ, Maysinger D, Jain M, Roder B, Hackbarth S, Winnik FM (2007) Long-term exposure to CdTe quantum dots causes functional impairments in live cells. Langmuir 23: 1974–1980
[56] Maysinger D, Lovric J, Eisenberg A, Savic R (2007) Fate of micelles and quantum dots in cells. Eur J Pharm Biopharm 65: 270–281
[57] Ballou B, Lagerholm BC, Ernst LA, Bruchez MP, Waggoner AS (2004) Noninvasive imaging of quantum dots in mice. Bioconjugate Chem 15: 79–86
[58] Fischer HC, Liu L, Pang KS, Chan WC (2006) Pharmacokinetics of nanoscale quantum dots: in vivo distribution, sequestration, and clearance in the rat. Adv Funct Mater 16: 1299–1305
[59] Soo Choi H, Liu W, Misra P et al (2007) Renal clearance of quantum dots. Nat Biotechnol 25: 1165–1170
[60] Lewinski N, Colvin V, Drezek R (2007) Cytotoxicity of nanoparticles. Small 4: 26–49
[61] Nel A, Xia T, Madler L, Li N (2006) Toxic potential of materials at the nanolevel. Science 311: 622–627
[62] Hardman R (2006) A toxicologic review of quantum dots: toxicity depends on physicochemical and environmental factors. Environ Health Perspect 114: 165–172
[63] Choi AO, Brown SE, Szyf M, Maysinger D (2008) Quantum dot-induced epigenetic and genotoxic changes in human breast cancer cells. J Mol Med 86: 291–302
[64] Evans JL, Goldfine ID, Maddux BA, Grodsky GM (2002) Oxidative stress and stress-activated signaling pathways: a unifying hypothesis of type 2 diabetes. Endocrine Rev 23: 599–622
[65] Nagata S, Golstein P (1995) The Fas death factor. Science 267: 1449–1456
[66] Green M, Howman E (2005) Semiconductor quantum dots and free radical induced DNA nicking. Chem Commun 1: 121–123
[67] Callinan PA, Feinberg AP (2006) The emerging science of epigenomics. Human molecular genetics 15 Spec No 1: R95–R101
[68] Sen T, Sen N, Tripathi G, Chatterjee U, Chakrabarti S (2006) Lipid peroxidation associated cardiolipin loss and membrane depolarization in rat brain mitochondria. Neurochem Int 49: 20–27
[69] Ryman-Rasmussen JP, Riviere JE, Monteiro-Riviere NA (2007) Surface coatings determine cytotoxicity and irritation potential of quantum dot nanoparticles in epidermal keratinocytes. J Invest Dermatol 127: 143–153
[70] Susumu K, Uyeda HT, Medintz IL, Pons T, Delehanty JB, Mattoussi H (2007) Enhancing the stability and biological functionalities of quantum dots via compact multifunctional ligands. J Am Chem Soc 129: 13987–13996
[71] Zafarullah M, Li WQ, Sylvester J, Ahmad M (2003) Molecular mechanisms of N-acetylcysteine actions. Cell Mol Life Sci 60: 6–20
[72] Ballou B, Ernst LA, Andreko S et al (2007) Sentinel lymph node imaging using quantum dots in mouse tumor models. Bioconjugate Chem 18: 389–396
[73] van Engeland M, Nieland LJ, Ramaekers FC, Schutte B, Reutelingsperger CP (1998) Annexin V-affinity assay: a review on an apoptosis detection system based on phosphatidylserine exposure. Cytometry 31: 1–9
[74] Boyles S, Lewis GP, Westcott B (1970) Intracellular enzymes in local lymph after chemical injury. Br J Pharmacol 38: 441P–442P
[75] Darzynkiewicz Z, Bruno S, Del Bino G et al (1992) Features of apoptotic cells measured by flow cytometry. Cytometry 13: 795–808
[76] O'Brien R, Gottlieb-Rosenkrantz P (1970) An automatic method for viability assay of cultured cells. J Histochem Cytochem 18: 581–589
[77] O'Brien J, Wilson I, Orton T, Pognan F (2000) Investigation of the Alamar Blue (resazurin) fluorescent dye for the assessment of mammalian cell cytotoxicity. Eur J Biochem 267: 5421–5426
[78] Mosmann T (1983) Rapid colorimetric assay for cellular growth and survival: application to proliferation and cytotoxicity assays. J Immunol Methods 65: 55–63
[79] Smith PJ, Blunt N, Wiltshire M et al (2000) Characteristics of a novel deep red/infrared fluorescent cell-permeant DNA probe, DRAQ5, in intact human cells analyzed by flow cytometry, confocal and multiphoton microscopy. Cytometry 40: 280–291
[80] Budd SL, Castilho RF, Nicholls DG (1997) Mitochondrial membrane potential and hydroethidine-monitored superoxide generation in cultured cerebellar granule cells. FEBS Lett 415: 21–24

Subject index

A

ab initio calculations 218
Absorption 37
Acceptor 292, 301, 337, 339
Acoustic phonons 219, 233, 317
AlN 54
All-optical temperature sensor 269
Alloys 51, 88, 89
Amorphous (glassy) 133
Antibodies 351
Amplifier media for telecommunications 90
Anti-Stokes photoluminescence (ASPL) 245, 257
Aqueous synthesis 73
Arrested precipitation 43
Artificial solids 119
Atom transfer radical polymerisation (ATRP) 174, 182
Auger recombination 258, 261, 326
2,2′-azobisisobutyronitrile (AIBN) 177

B

Band-edge emission 86
Band gap 37, 38
bi-excitons 289, 291
Bilayers (or cationic–anionic pairs) 198, 201
Binary nanocrystal superlattices (BNSLs) 152, 153
Binary systems 47
Biodistribution of nanocrystals 358
Bio-imaging 90, 270, 301, 351

Bio-labelling 90
Biomedical applications 203, 349
Birefringent 146
Blinking of the fluorescence 39, 326
Blending of nanocrystals and polymers 184
Block-copolymer micelles 178
Bohr exciton radii 35, 54
Born–Oppenheimer approximation 233
Bowing parameter 51
Branched nanocrystals 20
"Bright" exciton state 281, 291
Broadening of the size distribution 9, 81

C

Cadmium oxide 45, 47
Capping layer 283
Capping ligands 4, 40, 76, 140
Carbon nanotubes 208
$Cd_xHg_{1-x}Te$ 88
CdS 43, 47, 49, 59, 74, 75
CdS/HgS quantum dot quantum well (QDQW) 102
CdSe 20, 43, 45, 47, 128, 175, 180, 187, 221, 262, 279, 281, 284, 296
CdSe nanocrystal superlattices 136
CdSe/CdS 46, 47, 58, 317
CdSe/CdS heterostructure nanorods 320
CdSe/ZnS 42, 57, 267
CdSe/ZnSe 59

368 Subject index

CdSe/ZnTe 42, 61, 62
CdTe 20, 43, 47, 78, 176, 186, 190, 200, 202, 211, 261, 262, 279, 283, 300, 350
CdTe nanowires 208, 305
CdTe/CdSe 42, 61, 62
CdTe/ZnS 60
Cell imaging 352
Cell surface receptors 352
Charge–dipole interactions 122, 123
Charge–induced dipole interactions 122, 123
Charge transport 165
Coarse-grain model 127
Colloidal crystals 138
Common anion alloys 51
Common cation alloys 51
Conduction band 35
Control of single particle emission by electric fields 332
Controllable oversaturation 138, 139
Core 36, 46
Core/shell nanocrystals of II–VI semiconductors 57
Core/shell nanocrystals of III–V semiconductors 61
Core/shell nanocrystals of IV–VI semiconductors 61
Core shell shell (CSS) 101
Core/shell structures 36, 41, 62
Coulomb interaction 221, 224
Coulomb potential 121, 158
Critical size 24
Crystalline solids 133
Curie Gibbs Wulff theorem 19
Cysteamine 79

D

Dangling bonds 26, 42, 77, 313
"Dark" exciton state 266, 281, 291
Debye–Sherrer approximation 141
Decay kinetics of the luminescence 277
Decoupling spacer layers 115
Dendrites 130

Diagnostic tools 354
Dielectric continuum (DC) model 230
Diffusion 7
Diffusion limited energy migration 295
Dilute magnetic semiconductors (DMS) 51
Dimethylcadmium 4, 45, 47
Dipole–dipole interactions 122–124, 131, 298
Dodecylamine (DDA) 49
Donor 292, 301, 337
Doped nanocrystals 51
Double quantum dot quantum well (double-QDQW) systems 101
Drug delivery systems 356
Dye 339

E

Effective mass approximation (EMA) 38, 83, 218, 221, 280
Electron–hole exchange interaction 225
Electron–phonon coupling 317
Electron–phonon interaction 264
Electron–phonon scattering 217
Electrostatic interactions 121
Electrostatic repulsion 171
Emission of single nanocrystals 39, 311
Energy transfer 292, 295, 337
Enhanced permeability and retention (EPR) 355
Epigenetics 362
Epitaxial-type shell growth 45
Exciton 35
Exciton Bohr radius 221
Exciton dynamics 277, 278
Exciton life time 305
Exciton states 224

F

Förster distance 303
Förster energy transfer 303, 338–340

Förster equation 293
Förster theory 294
Facets 125
Face centered cubic (*fcc*) lattices 136, 137, 152, 153
Fe$_2$O$_3$ 20
Fermi's Golden Rule 220, 278
Fluorescence blinking 312
Fluorescence line-narrowing spectroscopy 105
Fluorescence (or Förster) resonance energy transfer (FRET) 301, 311, 338
Fluorescence quantum yield 42, 57, 59
Focusing of size distribution 9, 81
Formation of rings 129
Four-wave mixing 220
Franck–Condon progression 237
Fröhlich mechanism 237, 239, 268
Fröhlich phonons 324
FRET gate 341
FRET switch 339

G

GaAs 61
GaP 54
Gibbs–Thompson equation 9, 44
Glassy films 147
Grafting-from strategy 180
Grafting polymers on nanocrystals 177
Grafting-to strategy 178
Growth mechanism of nanocrystals 5, 7

H

Hard spheres 121, 134, 137, 142, 152
Heating-up method 44, 47
Heisenberg uncertainty principle 288, 313
Heterodimers 23
Heterogeneous nucleation 24
Hexadecylamine (HDA) 40, 47, 50
Hexagonally close-packed (*hcp*) lattices 134, 136, 137, 152

HgTe 51, 88, 202
Histone proteins 362
HOMO (highest occupied molecular orbital) 36
Hot carriers 266, 279, 286
Hot carrier relaxation 247
Hot-injection method 44, 47, 55
HRTEM 105
Huang-Rhys parameter (HPR) 219, 238, 240, 326
Hybrid materials 23

I

InAs 54, 61
Indirect-gap materials 226
Infrared-emitting nanocrystals 55, 60, 90
Inhomogeneous broadening 261, 313
Inkjet printing 190
InN 54
Inokuti–Hirayama model 296
InP 40, 54, 61, 261, 292
Inverse micelles 43
Inversed quantum dot quantum well (QDQW) systems 101
Isotropic fluid 145

J

Josephson junction 332

L

Langmuir Blodget (LB) film 297
Langmuir–Blodgett (LB) technique 127
Laser-assisted catalytic growth (LCG) 209
Lattice adapting spacer layers 110
Laser cooling 271
Lattice mismatch 45
Laurylmethacrylate 175
Layer-by-layer (LbL) assembly 185, 197
Layer-by-layer deposition technique 271

LbL assembly on curved substrates 187
LbL assembly on planar substrates 186
Lennard-Jones interaction 6
Ligands 35, 313
Ligand exchange 40, 81, 180
Local density of states 284
Longitudinal optical (LO) phonons 265, 268
Luminescence 277
Luminescence decay 284
Luminescence life times 289
Luminescence linewidths 38
Luminescence temperature anti-quenching 283
LUMO (lowest unoccupied molecular orbital) 36

M

Magic size clusters 11, 13, 57
Maltese cross 147, 148
Marangoni effect 130
MEH-PPV 330, 331
2-Mercaptoethanol 76
Mercaptopropionic acid (MPA) 79
Mercury chalcogenides 43
Metal nanoparticles 305
Metamaterials 164
Micelles 180
Microspheres 189
2D monolayers 128
Monomers 4, 7, 9, 19, 44
Monte Carlo calculations 131
Multi-phonon Raman scattering 246
Multiple exciton generation (MEG) 279, 289, 291

N

NaHTe 78
Nanocrystal–polymer composites 171, 173
Nanocrystal superlattices 136
Nanoparticle-induced cytotoxicity 359
Nanorods 19, 26, 144, 323
Nanotubes 208
Nanowires 126, 208
Nematic liquid crystals 145
Nonresonant Stokes shift 259
Non-solvent 138, 139
Nucleation 5, 17, 43
Nucleation-doping strategy 52

O

Octadecylamine (ODA) 50
Octa-twin model 21
Oleic acid (OA) 49
Olive oil 49
One-pot synthesis 58
Onsager theory 145
Optically detected magnetic resonance spectroscopy (ODMR) 106
Optical phonons 219, 230, 234
Optical spectroscopy 278
Organometallic precursors 45, 53
Oscillator strength 39
Ostwald ripening 10, 25, 44, 53, 88

P

Particle-in-a-box model 35, 83, 104
PbS 55, 261, 289, 301
PbSe 55, 289
PbSe/PbS 61
PbTe 55
Phonons 227
Phonon-assisted ASPL 258
Phonon bottleneck 217, 248, 279, 285
Phosphonic acids 21, 45
Photo-assisted etching 61
Photo-bleaching 57
Photodynamic therapy tools 358
Photoluminescence 38, 74, 176
Photonic crystals 285
Photo-stability 42
Plasmon–exciton interaction 270
Plasmonics 303
Polaron 248

Polaron states 237
Poly(acrylic acid) (PAA) 201
Poly(allylamine hydrochloride) (PAH) 201
Poly(diallyldimethylammonium chloride) (PDDA) 186, 201
Polyethylene glycol (PEG) 172, 179, 180, 350
Polyethyleneimine (PEI) 201
Polylaurylmethacrylate (PLMA) 176
Polymer matrices 174
Polymerization 172, 173, 176
Polymorphism 21
Polyphosphates 73, 102
Poly(styrene sulfonate) (PSS) 201
Potential barrier 17
Powder XRD 83, 87
Pt 20

Q

Quantum confinement effect 35, 37, 39, 73, 217
Quantum-confined Stark effect (QCSE) 303, 320, 323, 328, 332, 336, 341
Quantum dots 35
Quantum dot quantum well (QDQW) 101
Quantum rods 303
Quenching of the emission 283

R

Rabi splitting 242
Radiative decay 279
Radiative decay rate 284
Raman scattering 219, 228, 229, 236, 268
Raman spectra 224, 267
Reaction controlled growth 7
Reactive oxygen species (ROS) 358, 361
Resonant Raman scattering (RRS) 246

S

SAXS 139
Schlenk-line 4
Schlieren structures 147
Seeded growth 26, 27, 64
Selected area electron diffraction (SAED) 159
Self-assembly 119, 137, 145
Semiconductor clusters 73
II–VI Semiconductor nanocrystals 47
III–V Semiconductor nanocrystals 53
IV–VI Semiconductor nanocrystals 54
Semiconductor nanocrystal–polymer composites 173
Shape control 18
Shell 36, 42, 46
SILAR (successive ion layer adsorption and reaction) 47, 61, 106, 113
Simple hexagonal packing (shp) 137
Single particle spectroscopy 313
Singlet state 226
Single-walled carbon nanotube (SWCNT) 212
Size distribution 9, 10, 44, 81
Size focussion regime 9
Size selective precipitation 11, 13, 75
Sizing curve 82
Smectic phase 145, 147, 148
Spectral diffusion 314, 342
Spectral fluctuations 330
Spectral hole burning 220, 317
Spectral jitter 327, 328
Spectral line width 319
Stark effect 287
Stark shift 336
Stark spectroscopy 333, 336
Steric repulsion 171
Stokes shift 38, 219, 257
Stokes-shifted photoluminescence (SSPL) 257
Strong confinement regime 224
3D supercrystals 138
Superlattices 119, 127, 139, 152
Supersaturation 27, 44
Supersaturated solution 44

Surface energy 5
Surface optical (SO) phonon modes 268
Surface passivation 40, 57
Surface tension 5, 9, 19
Surfactants 4, 15, 18, 35, 43
Synthesis in aqueous media 73
Synthesis in organic solvents 35
Synthetic opals 135

T

TEM 82, 156
TEM images 20
Temperature dependence of ASPL 262
Ternary systems 51
Tetrapods 3, 151
Thermally populated defect states 264
1-Thioglycerol (TG) 76, 81, 88
Thioglycolic acid (TGA) 76, 79, 186
Thiols 73, 77, 79, 186, 200
Tight-binding methods 224
Time-correlated single photon counting 278
Time-resolved pump-probe spectroscopy 326
TOPSe 50, 51
Transient absorption 286, 291
Transient hole burning 105
Transient photobleaching 104
Trap emission 86
Trap states 42, 45
tri-excitons 289
tri-n-butylphosphine (TBP) 4
tri-n-octylphosphine (TOP) 4, 47, 50
tri-n-octylphosphine oxide (TOPO) 4, 40, 47
Triplet state 226
Twinning 21
Two-photon excitation 261

Type I band alignment 42, 57, 317
Type II band alignment 42, 61

U

Up-conversion 257

V

Valence band 35
van der Waals attraction 121
Végard's law 51

W

Wurtzite (W) – zink blende (ZB) polytypism 40
Whispering gallery mode (WGM) microcavities 270
White light emission 116
Wide-angle electron diffraction 142
Wide-angle XRD 141
Wide-field imaging microscope 314
Wurtzite (W) structure 21, 40, 124

X

X-ray diffraction 140

Z

Zero acoustic phonon line (ZAPL) 240, 244
Zero growth rate 9
Zero optical phonon line (ZPL) 240, 243, 244
Zinc blende (ZB) structure 18, 21, 40
ZnO 20, 22
ZnS 20, 22, 49
ZnSe 20, 50, 59, 85
ZnTe 50
ZnTe/CdSe 63
ZnTe/CdTe 63